BIBLIOGRAPHIC RECORD TARGET

ACQ7072
UL FMT B RT a BL m T/C DT 07/18/88 R/DT 07/18/88 CC STAT mm E/L 1
010: : |a 14006490//r852
035/1: : |a (RLIN)MIUG86-B27879
035/2: : |a (CaOTULAS)160536225
040: : |a DLC/ICU |c ICU |d CMfNASA |d MiU
050/1:1 : |a QA343 |b .G7
100:1 : |a Greenhill, G. |q (George), |c Sir, |d 1847-1927.
245:04: |a The applications of elliptic functions, |c by Alfred George Greenhill.
260: : |a London, |a New York, |b Macmillan and co., |c 1892.
300/1: : |a xi, 357 p. |b diagrs. |c 23 cm.
504/1: : |a Includes bibliographical references and index.
650/1: 0: |a Functions, Elliptic
998: : |c DPJ |s 9120

THE APPLICATIONS OF ELLIPTIC FUNCTIONS.

Alexander Ziwet

THE APPLICATIONS

OF

ELLIPTIC FUNCTIONS

BY

ALFRED GEORGE GREENHILL

M.A., F.R.S., PROFESSOR OF MATHEMATICS IN THE ARTILLERY COLLEGE, WOOLWICH

London

MACMILLAN AND CO.

AND NEW YORK

1892

CONTENTS.

CHAPTER IX.

CHAPTER X.

INTRODUCTION.

"L'ÉTUDE approfondie de la nature est la source la plus féconde des découvertes mathématiques.

Non seulement cette étude, en offrant aux recherches un but déterminé, a l'avantage d'exclure les questions vagues et les calculs sans issue; elle est encore un moyen assuré de former l'Analyse elle-même, et d'en découvrir les éléments qu'il nous importe le plus de connaître et que cette science doit toujours conserver.

Ces éléments fondamentaux sont ceux qui se reproduisent dans tous les effets naturels." (Fourier.)

These words of Fourier are taken as the text of the present treatise, which is addressed principally to the student of Applied Mathematics, who will in general acquire his mathematical equipment as he wants it for the solution of some definite actual problem; and it is in the interest of such students that the following Applications of Elliptic Functions have been brought together, to enable them to see how the purely analytical formulas may be considered to arise in the discussion of definite physical questions.

The Theory of Elliptic Functions, as developed by Abel and Jacobi, beginning about 1826, although now nearly seventy years old, has scarcely yet made its way into the ordinary curriculum of mathematical study in this country; and is still considered too advanced to be introduced to the student in elementary text-books.

In consequence of this omission, many of the most interesting problems in Dynamics are left unfinished, because the complete solution requires the use of the Elliptic Functions; these could not be introduced without a long digression, unless a considerable knowledge is presupposed of a course of Pure Mathematics in this subject.

But by developing the Analysis as it is required for some particular problem in hand, the student of Applied Mathematics will obtain a working knowledge of the subject of Elliptic Functions, such as he would probably never acquire from a study of a treatise like Jacobi's *Fundamenta Nova*, where the formulas are established and the subject is developed in strictly logical order as a branch of Pure Mathematical Analysis, without any digression on the application of the formulas, or on the manner in which they originate independently, as the expression of some physical law.

In introducing these applications we are following, to some extent, the plan of Durège's excellent treatise on Elliptic Functions (Leipsic, Teubner); and also of Halphen's *Traité des fonctions elliptiques et de leurs applications* (Paris, 1886-1891).

But while volume I. of Halphen's treatise is devoted entirely to the establishment of the formulas and analytical properties of the functions, and the applications are not discussed till volume II.; in the following pages it is proposed to develop the formulas immediately from some definite physical or geometrical problem; and the reader who wishes to follow up the purely analytical development of the subject is referred to such treatises as Abel's *Œuvres*, Jacobi's *Fundamenta Nova*, already mentioned, or the Treatises on Elliptic Functions of Cayley, Enneper, Königsberger, H. Weber, etc.

The following works also may be mentioned as having been consulted in the preparation of this work:—

Legendre: *Theorie des fonctions elliptiques;* 1825.

Thomæ: *Abriss einer Theorie der complexen Functionen und der Thetafunctionen einer Veränderlichen;* 1873.

Schwarz: *Formeln und Lehrsätze zum Gebrauche der elliptischen Functionen.*

Klein (Morrice): *Lectures on the Icosahedron;* 1888.

Klein und Fricke; *Vorlesungen über die Theorie der elliptischen Modalfunctionen;* 1890.

Despeyrous et Darboux: *Cours de mécanique;* 1886.

R. A. Roberts: *Integral Calculus;* 1887.

Bjerknes: *Niels Hendrik Abel; tableau de sa vie et de son action scientifique;* 1885.

We shall begin by the discussion of the Problem of the Simple Circular Pendulum, as the problem best calculated to define the Elliptic Functions, and to give the student an idea of their nature and importance.

Previously to the introduction of the Elliptic Functions, the Circular Pendulum could only be treated by means of the circular functions, by considering the oscillations as indefinitely small, and by assimilating its motion to that of Huygens' Cycloidal Pendulum, of 1673.

But now the employment of the Elliptic Functions renders the ordinary discussion of the Cycloidal Pendulum antiquated and of mere historical interest, and banishes from our treatises such expressions as "an integral which cannot be found," or "reducible to a matter of quadrature" in describing an elliptic integral, expressions which aroused the indignation of Clifford (*Mathematical Papers*, p. 562).

According to the new regulations for the Mathematical Tripos at Cambridge, to come into force in the examination in May 1893, the schedule II. of Part I. includes "Elementary Elliptic Functions, excluding the Theta Functions and the theory of Transformation"; so it is to be hoped that this reintroduction of Elliptic Functions into the ordinary mathematical curriculum will cause the subject to receive more general attention and study. These Applications have been put together with the idea of covering this ground by exhibiting their practical importance in Applied Mathematics, and of securing the interest of the student, so that he may if he wishes follow with interest the analytical treatises already mentioned.

We begin with Abel's idea of the inversion of Legendre's elliptic integral of the first kind, and employ Jacobi's notation, with Gudermann's abbreviation, for a considerable extent at the outset.

The more modern notation of Weierstrass is introduced subsequently, and used in conjunction with the preceding notation, and not to its exclusion; as it will be found that sometimes one notation and sometimes the other is the more suitable for the problem in hand.

At the same time explanation is given of the methods by which a change from the one to the other notation can be speedily carried out.

It has been considered sufficient in many places, for instance in the reduction of the Integrals in Chapter II., to write down the results without introducing the intermediate analysis; as the trained mathematical student to whom this book is addressed will have no difficulty in supplying the connecting steps, and this work will at the same time provide instructive exercises in the subject; and further, in the interest of such students, many important problems have been introduced in the text, forming immediate applications of theorems already developed previously.

I have to thank Mr. A. G. Hadcock for his assistance in preparing the diagrams, and in drawing them carefully to scale.

ERRATA.

Page 6. Line 9 from bottom, read Huygens.

42. Line 6, read $\sin^{-1}\sqrt{\frac{\alpha - x}{x - \gamma}}$.

48. Line 5 from bottom, read $-4n^2(9e^2+4n^2)^2$.

64. Line 19, read *Fonctions elliptiques*.

99. The diagram must be replaced by the one given below. The Nodoid in fig. 12, p. 99, was described by a point which was not a focus of the rolling hyperbola.

107. Line 2 from bottom, delete minus sign before radical.

138. Equation (7), read $(c_2^2 - c_1^2)/D$.

158. Line 12, read $36K(x, y)$.

205. Line 6 from bottom, read $\wp(u-v) - \wp(u+v)$.

213. Line 7 from bottom, read $G + Lx' - X(yz' - y'z) = 0$ with the corresponding subsequent corrections.

227. Line 7, read $P\sqrt{X_1} + Q\sqrt{X_2} = 0$.

282. Line 5 from top, for rectangle read ribbon.

328. Line 12 from bottom, read *Proc. L. M. S.*, IX.

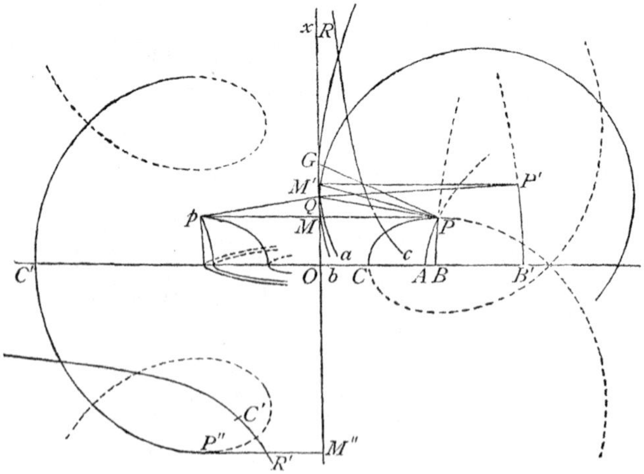

ABBREVIATIONS.

Q. J. M.,	Quarterly Journal of Mathematics.
Proc. L. M. S.,	Proceedings of the London Mathematical Society.
Proc. C. P. S.,	Proceedings of the Cambridge Philosophical Society.
Am. J. M.,	American Journal of Mathematics.
F. E.,	Fonctions elliptiques (Legendre and Halphen).
Math. Ann.,	Mathematische Annalen.
Phil. Mag.,	Philosophical Magazine.
Phil. Trans.	Philosophical Transactions of the Royal Society of London.
Berlin Sitz.,	Sitzungsberichte der Berliner Akademie.

CHAPTER I.

THE ELLIPTIC FUNCTIONS.

1. *The Pendulum; introducing Elliptic Functions into Dynamics.*

When a pendulum OP swings through a finite angle about a horizontal axis O, the determination of the motion introduces the *Elliptic Functions* in such an elementary and straightforward manner, that we may take the elliptic functions as defined by pendulum motion, and begin the investigation of their use and theory by their application to this problem.

Denote by W the weight in lb. of the pendulum, and let $OG=h$ (feet), where G is the centre of gravity; let Wk^2 denote the moment of inertia of the pendulum about the horizontal axis through G, so that $W(h^2+k^2)$ is the moment of inertia about the parallel axis through O (fig. 1).

Then if OG makes with the vertical OA an angle θ *radians* at the time t seconds, reckoned from an instant at which the pendulum was vertical; and if we employ the absolute unit of force, the *poundal,* and denote by g (32 *celoes,* roughly) the acceleration of gravity, the equation of motion obtained by taking moments about O is

$$W(h^2+k^2)\frac{d^2\theta}{dt^2}=-Wgh\sin\theta,$$

since the impressed force of gravity is Wg poundals, acting vertically through G; so that

$$\left(h+\frac{k^2}{h}\right)\frac{d^2\theta}{dt^2}=-g\sin\theta;$$

or, on putting $$h+k^2/h=l,$$

$$l\frac{d^2\theta}{dt^2}=-g\sin\theta.\qquad\qquad(1)$$

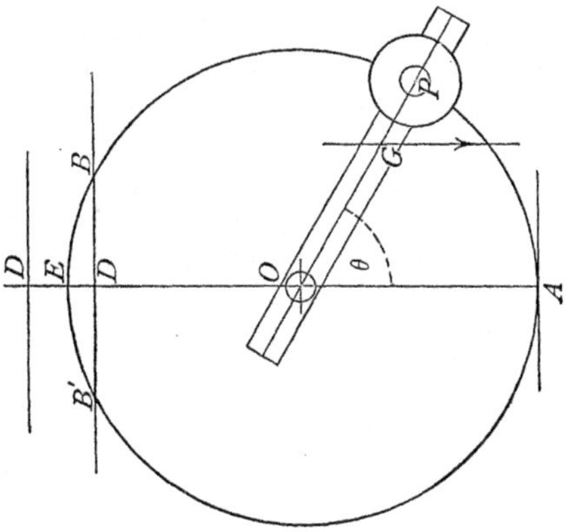

Fig. 1.

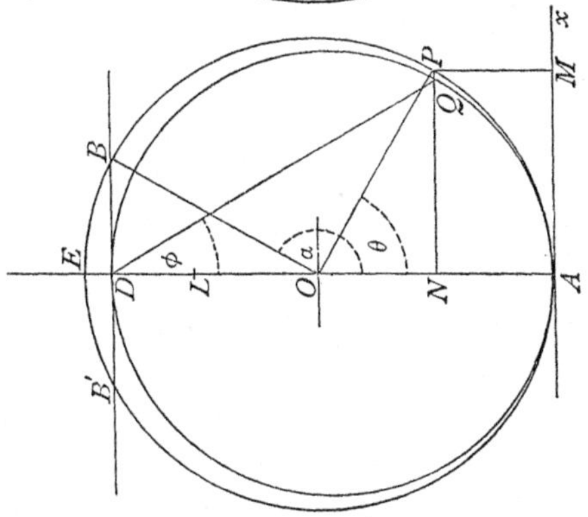

Fig. 2

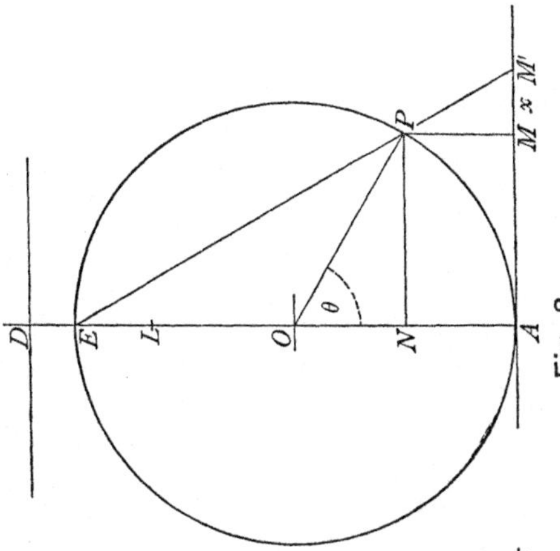

Fig. 3.

If the gravitation unit of force, the *force* of a pound, is employed, then the equation of motion is written

$$\frac{W}{g}(h^2+k^2)\frac{d^2\theta}{dt^2}=-Wh\sin\theta,$$

reducing to (1) as before.

2. Producing OG to P, so that $OP=l$, $GP=k^2/h$, the point P is called the *centre of oscillation* (or of *percussion*); and l is called the *length of the simple equivalent pendulum*, because the point P oscillates on the circle AP in exactly the same manner as a small plummet suspended by a fine thread from O (fig. 2); as is seen immediately by resolving tangentially along the arc $AP=s=l\theta$; when the equation of motion of the plummet is

$$\frac{d^2s}{dt^2}=-g\sin\theta=-g\sin\frac{s}{l},$$

or

$$l(d^2\theta/dt^2)=-g\sin\theta;\dots\dots(1)$$

and integrating,

$$\tfrac{1}{2}l(d\theta/dt)^2=C-g\operatorname{vers}\theta.\dots\dots(2)$$

These theorems are explained in treatises on Analytical Mechanics, such as Routh's *Rigid Dynamics*, or Bartholomew Price's *Infinitesimal Calculus*, vol. IV., and might have been assumed here; but now we proceed further, to the complete integration of equation (2).

3. First suppose the pendulum to oscillate, the angle of oscillation $BOA+AOB'$ being denoted by 2α (fig. 2); the angle of oscillation is purposely made large, as in early clocks, in the Navez Ballistic Pendulum, in a swing, or as in ringing a church bell, so as to emphasize the difference from small oscillations, the only case usually considered in the text-books; in fig. 2 the angle of oscillation is made 300°.

Then $d\theta/dt=0$ when $\theta=\alpha$, so that in equation (2)

$$C=g\operatorname{vers}\alpha;$$

and now denoting g/l by n^2, so that n is what Sir W. Thomson calls the *speed* (angular) of the pendulum,

$$\tfrac{1}{2}(d\theta/dt)^2=n^2(\operatorname{vers}\alpha-\operatorname{vers}\theta)$$

$$=2n^2(\sin^2\tfrac{1}{2}\alpha-\sin^2\tfrac{1}{2}\theta),\dots\dots(3)$$

since $\operatorname{vers}\theta=2\sin^2\tfrac{1}{2}\theta$;

$$d\theta/dt=2n\sqrt{(\sin^2\tfrac{1}{2}\alpha-\sin^2\tfrac{1}{2}\theta)},$$

and

$$nt=\int_0^\theta\frac{d\tfrac{1}{2}\theta}{\sqrt{(\sin^2\tfrac{1}{2}\alpha-\sin^2\tfrac{1}{2}\theta)}}\dots\dots(4)$$

and (4) is called by Legendre an *elliptic integral of the first kind;* it is not expressible by any of the algebraical, circular, or hyperbolic functions of elementary mathematics.

4. To reduce this elliptic integral to the standard form considered by Legendre, we put

$$\sin\tfrac{1}{2}\theta = \sin\tfrac{1}{2}\alpha \sin\phi,$$

equivalent geometrically to denoting the angle ADQ by ϕ (fig. 2), where AQD is the circle on AD as diameter, touching BB' in D, and cutting the horizontal line PN in Q.

For, in the circle AP,

$$AN = l \operatorname{vers}\theta = 2l\sin^2\tfrac{1}{2}\theta;$$

and, in the circle AQ,

$$AN = \tfrac{1}{2}AD \operatorname{vers} 2\phi = AD\sin^2\phi$$
$$= l \operatorname{vers}\alpha \sin^2\phi = 2l\sin^2\tfrac{1}{2}\alpha\sin^2\phi.$$

Now $\qquad \sin^2\tfrac{1}{2}\alpha - \sin^2\tfrac{1}{2}\theta = \sin^2\tfrac{1}{2}\alpha\cos^2\phi,$

and $\qquad \tfrac{1}{2}\theta = \sin^{-1}(\sin\tfrac{1}{2}\alpha\sin\phi),$

so that $\qquad d\tfrac{1}{2}\theta = \dfrac{\sin\tfrac{1}{2}\alpha\cos\phi\, d\phi}{\sqrt{(1-\sin^2\tfrac{1}{2}\alpha\sin^2\phi)}},$

and therefore $\qquad nt = \displaystyle\int_0^\phi \frac{d\phi}{\sqrt{(1-\sin^2\tfrac{1}{2}\alpha\sin^2\phi)}},$

which is now an elliptic integral of the first kind, in the *standard form* employed by Legendre.

(*Fonctions Elliptiques*, t. I., chap VI.)

5. In Legendre's notation, $\sin\tfrac{1}{2}\alpha$ is replaced by κ; the quantity $\sqrt{(1-\kappa^2\sin^2\phi)}$ is denoted by $\Delta\phi$ or $\Delta(\phi, \kappa)$; and the integral $\int_0 d\phi/\Delta\phi$ or $\int_0 (1-\kappa^2\sin^2\phi)^{-\frac{1}{2}}d\phi$ is denoted by $F\phi$ or $F(\phi, \kappa)$, and called the *elliptic integral of the first kind*, ϕ being called the *amplitude* and κ the *modulus*.

Thus, in the pendulum motion,

$$nt = F\phi, \text{ or } F(\phi, \sin\tfrac{1}{2}\alpha).$$

Legendre employs c instead of κ, and puts $\kappa = \sin\theta$ (a different θ to what we have just employed) and calls θ the *modular angle;* and he has tabulated the numerical values of $F(\phi, \kappa)$ for every degree of ϕ and θ. (*Fonctions Elliptiques*, t. II. Table IX.)

Legendre spent a long life in investigating the properties of the function $F\phi$, the elliptic integral of the first kind; but the subject was revolutionised by the single remark of Abel (in

1823), that $F\phi$ is of the nature of an *inverse* function; and that if we put $u = F\phi$, then we should study the properties of ϕ, the amplitude, as a function of u, and not of u as a function of ϕ, as carried out by Legendre in his *Fonctions Elliptiques*.

6. Jacobi proposed the notation $\phi = \text{am}\, u$, or $\text{am}(u, \kappa)$ when the modulus κ is required to be put in evidence; and now, considered as functions of u, we have Jacobi's notation

$$\cos\phi = \cos \text{am}\, u, \sin\phi = \sin \text{am}\, u, \Delta\phi = \Delta\, \text{am}\, u,$$

the three *elliptic functions* of u; and in Jacobi's *Fundamenta Nova* (1829) the properties of these functions,

$$\cos \text{am}\, u,\ \sin \text{am}\, u,\ \Delta\, \text{am}\, u,$$

are developed, the elegance of Jacobi's notation tending greatly to the popularity of this treatise.

7. *Definition of the Elliptic Functions.*

Jacobi's notation is rather lengthy, so that nowadays, in accordance with Gudermann's suggestion (*Theorie der Modular Functionen*, Crelle, t. 18), $\cos \text{am}\, u$ is abbreviated to $\text{cn}\, u$, $\sin \text{am}\, u$ to $\text{sn}\, u$, and $\Delta\, \text{am}\, u$ to $\text{dn}\, u$; and

$$\text{cn}\, u,\quad \text{sn}\, u,\quad \text{dn}\, u$$

are the three elliptic functions (pronounced, according to Halphen, with separate letters, as c, n, u; s, n, u; d, n, u); and they are defined by

$$\text{cn}\, u = \cos\phi,\ \text{sn}\, u = \sin\phi,\ \text{dn}\, u = \Delta\phi = \sqrt{(1-\kappa^2\sin^2\phi)};$$

where ϕ is a function of u, denoted by $\text{am}\, u$, and defined by the relation

$$u = \int_0^\phi (1-\kappa^2\sin^2\phi)^{-\frac{1}{2}} d\phi,$$

so that
$$u = \int_0^{\text{am}\, u} \sqrt{(1-\kappa^2\sin^2\phi)^{-\frac{1}{2}}} d\phi;$$

and
$$\frac{d\, \text{am}\, u}{du} = \frac{d\phi}{du} = \sqrt{(1-\kappa^2\sin^2\phi)} = \text{dn}\, u.$$

Thence
$$\frac{d\, \text{cn}\, u}{du} = \frac{d\cos\phi}{du} = -\sin\phi\frac{d\phi}{du} = -\text{sn}\, u\, \text{dn}\, u;$$

and similarly

$$\frac{d\, \text{sn}\, u}{du} = \frac{d\sin\phi}{du} = \cos\phi\frac{d\phi}{du} = \text{cn}\, u\, \text{dn}\, u;$$

and
$$\frac{d\, \text{dn}\, u}{du} = \frac{d\Delta\phi}{du} = -\frac{\kappa^2\sin\phi\cos\phi}{\Delta\phi}\frac{d\phi}{du} = -\kappa^2 \text{sn}\, u\, \text{cn}\, u$$

8. Returning now with these definitions and this notation to the motion of the pendulum, we have, on comparison, $u=nt$, while $\kappa=\sin\frac{1}{2}\alpha$, so that the modular angle is $\frac{1}{2}\alpha$;

and $\quad \kappa=AD/AB=AB/AE,\ \kappa^2=AD/AE$ (fig. 2);

also $\quad \phi=\text{am}\,u,\ \cos\phi=\text{cn}\,u,\ \sin\phi=\text{sn}\,u,\ d\phi/dt=n\,\text{dn}\,u$;

$$d\theta/dt=2n\kappa\,\text{cn}\,u=2n\kappa\,\text{cn}\,nt,$$

$$\sin\tfrac{1}{2}\theta=\kappa\,\text{sn}\,u=\kappa\,\text{sn}\,nt,$$

$$\cos\tfrac{1}{2}\theta=\text{dn}\,u=\text{dn}\,nt;$$

$$AP=AE\sin\tfrac{1}{2}\theta=AB\,\text{sn}\,nt,\ PE=AE\cos\tfrac{1}{2}\theta=AE\,\text{dn}\,nt;$$

$$AN=AD\,\text{sn}^2nt,\ ND=AD\,\text{cn}^2nt,\ NE=AE\,\text{dn}^2nt;$$

$$NQ=\surd(AN\,.\,ND)=AD\,\text{sn}\,nt\,\text{cn}\,nt,\ NP=AB\,\text{sn}\,nt\,\text{dn}\,nt;$$

giving these quantities as elliptic functions of u or nt.

9. We notice that $\kappa=0$ for infinitely small oscillations of the pendulum, the only case usually treated in the text-books; and now $\phi=u=nt$, so that

$$\text{cn}\,u=\cos u,\ \text{sn}\,u=\sin u,\ \text{while}\ \text{dn}\,u=1;$$

and the elliptic functions have degenerated into the ordinary *circular* functions of Trigonometry.

But in finite oscillations of the pendulum, where κ is not zero, these new functions are required, which are called the *elliptic functions;* and their geometrical definition is exhibited in fig. 2, in a manner similar to that employed in Trigonometry for the *circular functions.*

The name *elliptic function* is somewhat of a misnomer; but arose from the functions having been first approached by mathematicians in their attempt at the rectification of the ellipse (§ 77).

For finite oscillations the circular functions are applicable only to *cycloidal* oscillations, as discovered by Huyghens, 1673, whence the motion on the arc of a cycloid is generally investigated at length in elementary treatises; but this discussion may be considered as of mere antiquarian interest, now that we are proceeding to discuss the finite oscillations of the pendulum by the aid of the elliptic functions.

We may however make here a slight digression on cycloidal oscillations, treated in the manner we have employed for circular oscillations.

10. *Cycloidal Oscillations.*

In the cycloid, fig. 4, the angle ADQ or $\phi = nt$ (not am nt, as in the circular pendulum) for all finite oscillations; for as P oscillates on the arc BAB' of the inverted cycloid described by the rolling of the circle AE, Q follows P at the same level on the circle AD with constant velocity.

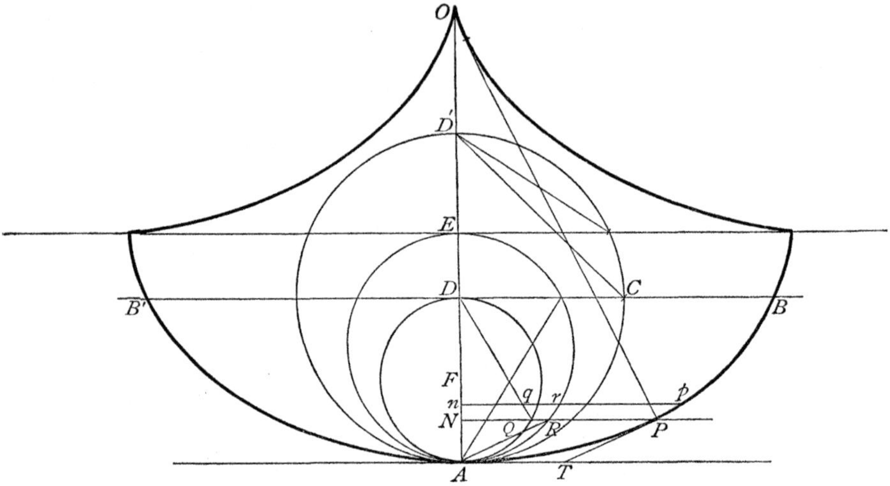

Fig. 4.

For if PQN meets the circle on AE as diameter in R, then, from a well-known property of the cycloid, the tangent TP is equal and parallel to AR, and half the arc AP; and if n, p, q, r denote simultaneous consecutive positions of N, P, Q, R,

$$\frac{\text{the velocity of } Q}{\text{the velocity of } P} = \text{lt}\frac{Qq}{Pp} = \text{lt}\frac{Qq}{Nn}\text{lt}\frac{Nn}{Pp}$$

$$= \operatorname{cosec} qQP \sin pPQ = \operatorname{cosec} AFQ \sin AER$$

$$= \frac{\frac{1}{2}AD}{NQ}\frac{AR}{AE} = \frac{\frac{1}{2}AD}{AE}\sqrt{\frac{AN.\,AE}{AN.\,ND}} = \frac{\frac{1}{2}AD}{\sqrt{(AE.ND)}}.$$

Now the velocity of $P = \sqrt{(2g\,.\,ND)}$

and therefore the velocity of $Q = \frac{1}{2}AD\sqrt{(2g/AE)}$

$$= AD\sqrt{(g/l)} = n\,.\,AD, \text{ a constant,}$$

if $AE = \frac{1}{2}l$; and therefore the angular velocity of Q about D is n, and the angle $ADQ = \phi = nt$.

Therefore the oscillations are *isochronous*, since the *period* $2\pi/n = 2\pi\sqrt{(l/g)}$ is independent of the amplitude of oscillation.

But in the circular pendulum the period increases with the amplitude or angle of oscillation; because in the circle AP (fig. 2) the versed sine AN varies as the square of the chord AP, while in the cycloid AP (fig. 4) the versed sine AN varies as the square of the arc AP.

The time from P to A on the cycloid is equal to the c.m. (circular measure) of the angle ADQ divided by n or $\sqrt{(g/l)}$; and generally the time over any finite arc Pp of the cycloid will be equal to the c.m. of the corresponding angle QDq divided by n, supposing the body to start from the level of D.

This will be true even when the point D is above E, as at D', so that the body enters the cycloid with given velocity; as for instance in the case of a railway train entering with given velocity V a cycloidal tunnel BAB' under a river.

Making $DD'=\frac{1}{2}V^2/g$, the *impetus* of the velocity V, then the time occupied by the train in the tunnel from B to B' is twice the c.m. of $AD'C$ divided by n.

Also if the length of the tunnel is $2s$, then $s=\sqrt{(2lh)}$, if AD, the depth or versed sine of the tunnel, is h; so that the time occupied is

$$\frac{2}{n}\tan^{-1}\frac{DC}{DD'}=2\sqrt{\frac{l}{g}}\tan^{-1}\sqrt{\frac{AD}{DD'}}=\frac{2s}{\sqrt{(2gh)}}\tan^{-1}\sqrt{\left(\frac{h}{\frac{1}{2}V^2/g}\right)}.$$

11. *The Period of the Pendulum, and of the Elliptic Functions.*

The *period* of the pendulum is the name now given to the time of a *double* swing, according to the report of a Committee at the *Conference of Electricians in Paris*, 1889; thus, if the swing is small, the period is $2\pi\sqrt{(l/g)}$ seconds.

But if the angle of vibration 2α is finite, the period is increased; denoting the period by T, and therefore the quarter-period, or time of motion of P from A to B (fig. 2) by $\frac{1}{4}T$, then as t increases from 0 to $\frac{1}{4}T$, θ increases from 0 to α, and ϕ from 0 to $\frac{1}{2}\pi$, so that nt or u increases from 0 to K, where (§ 4)

$$K=\int_0^{\frac{1}{2}\pi}(1-\kappa^2\sin^2\phi)^{-\frac{1}{2}}d\phi;$$

and K (or $F^1\kappa$ in Legendre's notation, and called by him the *complete* elliptic integral of the first kind) is now called the real *quarter period* of the elliptic functions, to the modulus κ.

Now, expanding by the Binomial Theorem,

$$(1-\kappa^2\sin^2\phi)^{-\frac{1}{2}}=1+\sum_{n=1}^{n=\infty}\frac{1\,.\,3\,.\,5\,\ldots\,(2n-1)}{2\,.\,4\,.\,6\,\ldots\quad 2n}\kappa^{2n}(\sin\phi)^{2n},$$

and, by Wallis's Theorem,

$$\int_0^{\frac{1}{2}\pi}(\sin\phi)^{2n}d\phi=\frac{1\,.\,3\,.\,5\,\ldots\,(2n-1)}{2\,.\,4\,.\,6\,\ldots\quad 2n}\tfrac{1}{2}\pi;$$

so that $$K=\tfrac{1}{2}\pi\left[1+\sum\left\{\frac{1\,.\,3\,.\,5\,\ldots\,(2n-1)}{2\,.\,4\,.\,6\,\ldots\quad 2n}\right\}^2\kappa^{2n}\right].$$

Thus the period of a pendulum of length l, oscillating through an angle $2a$, is

$$T=\frac{4K}{n}=2\pi\sqrt{\frac{l}{g}}\left\{1+\left(\frac{1}{2}\right)^2(\sin\tfrac{1}{2}a)^2+\left(\frac{1\,.\,3}{2\,.\,4}\right)^2(\sin\tfrac{1}{2}a)^4\right.$$
$$\left.+\left(\frac{1\,.\,3\,.\,5}{2\,.\,4\,.\,6}\right)^2(\sin\tfrac{1}{2}a)^6+\ldots\right\}.$$

As a first approximation therefore in the correction for amplitude of swing, the period must be increased by the fraction $\frac{1}{4}(\sin\frac{1}{2}a)^2$ of itself, or by $100(\frac{1}{4}$ chord of $a)^2$ per cent.

Thus a pendulum, which beats seconds when swinging through an angle of 6°, will lose 11 to 12 seconds a day if made to swing through 8°, and 26 seconds a day if made to swing through 10°. (Simpson's *Fluxions*, § 464.)

The value of K or $F^1\kappa$ has been tabulated by Legendre for every degree and tenth of a degree in the modular angle (*Fonctions Elliptiques*, t. II., Table I.).

We denote the modular angle by $\frac{1}{2}a$, and put $\kappa=\sin\frac{1}{2}a$; while $\cos\frac{1}{2}a$ is denoted by κ' and called the *complementary modulus*, so that

$$\kappa^2+\kappa'^2=1;$$

and then $F^1\kappa'$ is denoted by K', and called the *complementary quarter period*.

The following table (from Bertrand's *Calcul Intégral*, p. 714), gives the logarithms of the quarter periods K and K', corresponding to every half degree in $\frac{1}{2}a$, the quarter angle of swing; and then

$$2\kappa\kappa'=\sin a,\ \kappa=\sin\tfrac{1}{2}a,\ \kappa'=\cos\tfrac{1}{2}a,$$

and $\frac{1}{2}a$ is the modular angle.

The modular angle in the Table is given from 0 to 45°; to determine K for a modular angle greater than 45°, we look out the value of K' corresponding to the complementary modular angle.

TABLE I.

$\frac{1}{2}\alpha$	log K.	log K'	$\frac{1}{2}(\pi-\alpha)$	$\frac{1}{2}\alpha$	log K.	log K'	$\frac{1}{2}(\pi-\alpha)$	$\frac{1}{2}\alpha$	log K	log K'	$\frac{1}{2}(\pi-\alpha)$
0·0	0·1961199	Infinite.	90·0	15·5	0·2041274	0·4373581	74·5	31·0	0·2289427	0·3283840	59·0
0·5	1281	0·7873031	89·5	16·0	6567	·4326595	74·0	31·5	·2300476	·3257624	58·5
1·0	1530	·7351923	89·0	16·5	·2052034	·4280740	73·5	32·0	·2311728	·3231815	58·0
1·5	1943	·7015560	88·5	17·0	7675	·4235961	73·0	32·5	·2323184	·3206403	57·5
2·0	2522	·6760272	88·0	17·5	·2063492	·4192208	72·5	33·0	·2334847	·3181381	57·0
2·5	3266	·6551599	87·5	18·0	9484	·4149432	72·0	33·5	·2346716	·3156741	56·5
3·0	4176	·6373550	87·0	18·5	·2075652	·4107592	71·5	34·0	·2358795	·3132474	56·0
3·5	5252	·6217319	86·5	19·0	·2081997	·4066647	71·0	34·5	·2371084	·3108575	55·5
4·0	6493	·6077507	86·0	19·5	8519	·4026560	70·5	35·0	·2383586	·3085037	55·0
4·5	7900	·5950549	85·5	20·0	·2095220	·3987297	70·0	35·5	·2396301	·3061851	54·5
5·0	9474	·5833963	85·0	20·5	·2102099	·3948825	69·5	36·0	·2409233	·3039013	54·0
5·5	·1971213	·5725943	84·5	21·0	9158	·3911115	69·0	36·5	·2422382	·3016515	53·5
6·0	3119	·5625136	84·0	21·5	·2116398	·3874139	68·5	37·0	·2435751	·2994353	53·0
6·5	5191	·5530498	83·5	22·0	·2123818	·3837869	68·0	37·5	·2449341	·2972520	52·5
7·0	7430	·5441205	83·0	22·5	·2131421	·3802283	67·5	38·0	·2463154	·2951012	52·0
7·5	9836	·5356595	82·5	23·0	9206	·3767357	67·0	38·5	·2477193	·2929822	51·5
8·0	·1982409	·5276129	82·0	23·5	·2147175	·3733069	66·5	39·0	·2491460	·2908945	51·0
8·5	5149	·5199360	81·5	24·0	·2155329	·3699400	66·0	39·5	·2505956	·2888377	50·5
9·0	8057	·5125914	81·0	24·5	·2163668	·3666329	65·5	40·0	·2520684	·2868114	50·0
9·5	·1991134	·5055474	80·5	25·0	·2172194	·3633838	65·0	40·5	·2535647	·2848150	49·5
10·0	4378	·4987770	80·0	25·5	·2180907	·3601912	64·5	41·0	·2550846	·2828480	49·0
10·5	7791	·4922569	79·5	26·0	9808	·3570533	64·0	41·5	·2566285	·2809102	48·5
11·0	·2001373	·4859667	79·0	26·5	·2198899	·3539686	63·5	42·0	·2581965	·2790011	48·0
11·5	5124	·4798888	78·5	27·0	·2208181	·3509356	63·0	42·5	·2597889	·2771202	47·5
12·0	9044	·4740077	78·0	27·5	·2217654	·3479531	62·5	43·0	·2614061	·2752673	47·0
12·5	·2013135	·4683095	77·5	28·0	·2227319	·3450196	62·0	43·5	·2630482	·2734418	46·5
13·0	7396	·4627819	77·0	28·5	·2237179	·3421340	61·5	44·0	·2647155	·2716436	46·0
13·5	·2021828	·4574142	76·5	29·0	·2247233	·3392950	61·0	44·5	·2664085	·2698722	45·5
14·0	6431	·4521964	76·0	29·5	·2257484	·3365016	60·5	45·0	·2681272	·2681272	45·0
14·5	·2031206	·4471196	75·5	30·0	·2267933	·3337526	60·0	45·5	·2698722	·2664085	44·5
15·0	6154	·4421760	75·0	30·5	·2278580	·3310471	59·5	46·0	·2716436	·2647155	44·0

TABLE II.

ϕ	u	ϕ	u	ϕ	u	ϕ	u	ϕ	u	ϕ	u
0·0	0·00000	15·5	0·27218	30·5	0·54496	45·5	0·83611	60·5	1·15348	75·5	1·49984
0·5	·00873	16·0	·28107	31·0	·55432	46·0	·84623	61·0	·16457	76·0	·51183
1·0	·01745	16·5	·28997	31·5	·56370	46·5	·85638	61·5	·17569	76·5	·52383
1·5	·02618	17·0	·29889	32·0	·57310	47·0	·86656	62·0	·18682	77·0	·53586
2·0	·03491	17.5	·30781	32·5	·58253	47·5	·87677	62·5	·19804	77·5	·54792
2·5	·04634	18·0	·31675	33·0	·59197	48·0	·88701	63·0	·20926	78·0	·55999
3·0	·05237	18·5	·32570	33·5	·60144	48·5	·89729	63·5	·22051	78·5	·57208
3·5	·06111	19·0	·33466	34·0	·61093	49·0	·90759	64·0	·23180	79·0	·58419
4·0	·06984	19·5	·34363	34·5	·62044	49·5	·91792	64·5	·24312	79·5	·59633
4·5	·07858	20·0	·35262	35·0	·62998	50·0	·92829	65·0	·25447	80·0	·60848
5·0	·08732	20·5	·36162	35·5	·63954	50·5	·93869	65·5	·26585	80·5	·62064
5·5	·09607	21·0	·37063	36·0	·64912	51·0	·94912	66·0	·27727	81·0	·63283
6·0	·10482	21·5	·37966	36·5	·65873	51·5	·95958	66·5	·28872	81·5	·64503
6·5	·11357	22·0	·38871	37·0	·66836	52·0	·97007	67·0	·30020	82·0	·65725
7·0	·12233	22·5	·39776	37·5	·67801	52·5	·98060	67·5	·31171	82·5	·66948
7·5	·13109	23·0	·40683	38·0	·68769	53·0	·99115	68·0	·32325	83·0	·68172
8·0	·13985	23·5	·41592	38·5	·69740	53·5	1·00175	68·5	·33482	83·5	·69398
8·5	·14863	24·0	·42503	39·0	·70713	54·0	·01237	69·0	·34642	84·0	·70625
9·0	·15740	24·5	·43415	39·5	·71689	54·5	·02302	69·5	·35806	84·5	·71853
9·5	·16619	25·0	·44328	40·0	·72667	55·0	·03371	70·0	·36972	85·0	·73082
10·0	·17498	25·5	·45244	40·5	·73648	55·5	·04443	70·5	·38141	85·5	·74312
10·5	·18377	26·0	·46161	41·0	·74632	56·0	·05519	71·0	·39313	86·0	·75542
11·0	·19258	26·5	·47079	41·5	·75618	56·5	·06598	71·5	·40488	86·5	·76774
11·5	·20139	27·0	·48000	42·0	·76608	57·0	·07680	72·0	·41666	87·0	·78006
12·0	·21021	27·5	·48922	42·5	·77600	57·5	·08765	72·5	·42846	87·5	·79239
12·5	·21903	28·0	·49846	43·0	·78594	58·0	·09854	73·0	·44030	88·0	·80472
13·0	·22787	28·5	·50772	43·5	·79592	58·5	·10946	73·5	·45215	88·5	·81705
13·5	·23671	29·0	·51700	44·0	·80592	59·0	·12042	74·0	·46404	89·0	·82939
14·0	·24556	29·5	·52630	44·5	·81596	59·5	·13141	74·5	·47595	89·5	·84173
14·5	·25443	30·0	·53562	45·0	·82602	60·0	·14243	75·0	·48788	90·0	1·85407
15·0	·26330										

12. We notice that when the modular angle is 15°, then

$$\log K'/K = \cdot 2385606 = \tfrac{1}{2} \log 3, \text{ so that } K'/K = \sqrt{3}:$$

this will be proved subsequently; but it shows here that the period of a pendulum oscillating through 300° is $\sqrt{3}$ times the period when the pendulum oscillates through 60°.

Again we shall prove subsequently that,

if $$K'/K = \sqrt{7}, \text{ then } 2\kappa\kappa' = \tfrac{1}{8};$$

so that equal parallel horizontal chords, BB' the higher, and bb' the lower, each of length one-eighth the diameter, cut off arcs of the circle below them, which would be swung through by the pendulum in times which are in the ratio of $\sqrt{7}$ to 1.

Many other similar numerical examples can be constructed when the Theory of the *Complex Multiplication* of Elliptic Functions is studied.

13. When $\alpha = \frac{1}{2}\pi$, the pendulum drops from a horizontal position and swings through two right angles, as in the Navez Electro-Ballistic Pendulum; and now $K = K'$, and the modular angle is $\frac{1}{4}\pi$.

Table II. from Legendre's *Fonctions Elliptiques*, t. II., gives to five decimals the value of $u = F\phi$ for every half degree in the value of ϕ, when the modular angle is 45°; and thence by means of the preceding formulas which determine the motion of the pendulum by elliptic functions, the pendulum can be graduated so as to measure small intervals of time $\Delta t = \Delta u/n$, as required for electro-ballistic experiments.

Then from Table II., when $K = K'$, and $\kappa = \kappa' = \frac{1}{2}\sqrt{2}$,

$$\text{cn } u = \cos \phi, \text{ sn } u = \sin \phi, \text{ dn } u = \sqrt{(1 - \tfrac{1}{2} \sin^2\phi)}.$$

14. Generally in the pendulum, $K = \frac{1}{4}nT$, so that the period

$$T = 4K/n = 4K\sqrt{(l/g)}.$$

When $\kappa = 0$, $K = \frac{1}{2}\pi$, and the period is $2\pi\sqrt{(l/g)}$, as proved otherwise in the ordinary elementary treatises, for small oscillations of the pendulum.

But in the finite oscillations of the pendulum, with

$$u = nt = 4Kt/T,$$

then (§ 8) $$d\theta/dt = 2n\kappa \text{ cn } 4Kt/T,$$

$$\sin\tfrac{1}{2}\theta = \kappa \text{ sn } 4Kt/T,$$

$$\cos\tfrac{1}{2}\theta = \text{dn } 4Kt/T, \text{ etc.}$$

Putting $t = 0$, $u = 0$, we find

$$\text{cn } 0 = 1, \text{ sn } 0 = 0, \text{ dn } 0 = 1;$$

and putting $\qquad t=\frac{1}{4}T,\ u=K,\ \phi=\frac{1}{2}\pi,$

when the pendulum has swung to OB,

$$\text{cn } K=\cos\tfrac{1}{2}\pi=0,\ \text{sn } K=1,\ \text{dn } K=\kappa' = \sqrt{1-k^2}$$

while putting $\qquad t=\frac{1}{2}T,\ u=2K,$

when the pendulum is swinging backwards through the vertical OA, $\qquad \text{cn } 2K=-1,\ \text{sn } 2K=0,\ \text{dn } 2K=1$;

analogous to the values of $\cos\theta$ and $\sin\theta$, for $\theta=0,\ \frac{1}{2}\pi,\ \pi$; so that $2K$ is the *half period* of the elliptic functions, corresponding to the *half period* π of the circular functions.

Since $\displaystyle\int_0^{\pi\pm\phi} d\phi/\Delta\phi=\int_0^{\pi} d\phi/\Delta\phi\pm\int_0^{\phi} d\phi/\Delta\phi=2K\pm u$, if $\phi=\text{am } u$,

therefore $\qquad \text{am}(2K\pm u)=\ \pi\pm\phi=\ \pi\pm\text{am } u$;

and generally $\text{am}(2mK\pm u)=m\pi\pm\phi=m\pi\pm\text{am } u$;

so that $\qquad \text{cn}(2mK\pm u)=\cos(m\pi\pm\text{am } u)=\ (-1)^m\text{cn } u,$

$\qquad \text{sn}(2mK\pm u)=\sin(m\pi\pm\text{am } u)=\pm(-1)^m\text{sn } u,$

while $\qquad \text{dn}(2mK\pm u)=\text{dn } u$;

analogous to $\qquad \cos(m\pi\pm\theta)=\ (-1)^m\cos\theta,$

$\qquad \sin(m\pi\pm\theta)=\pm(-1)^m\sin\theta$;

and representing the motion, m half periods, past or future.

15. *The degenerate Circular and Hyperbolic Functions.*

As α increases from 0 to π, κ increases from 0 to 1, and K from $\frac{1}{2}\pi$ to infinity; the pendulum has now, with $\kappa=1$, just sufficient velocity to carry it to the highest position, and this will take an infinite time.

For with $\alpha=\pi$, equation (3), page 3, becomes

$$\tfrac{1}{2}(d\theta/dt)^2=n^2(1+\cos\theta)=2n^2\cos^2\tfrac{1}{2}\theta;$$

$$nt=\int_0^{\theta}\sec\tfrac{1}{2}\theta\, d\tfrac{1}{2}\theta$$

$$=\log\tan\tfrac{1}{4}(\pi+\theta)=\log(\sec\tfrac{1}{2}\theta+\tan\tfrac{1}{2}\theta),$$

which is infinite when $\theta=\pi$.

In small oscillations the period is $2\pi/n$, and the motion of M, the projection of P on the horizontal axis Ax, is then a *Simple Harmonic Motion* (S.H.M.) given by the differential equation

$$\frac{d^2x}{dt}+n^2x=0,$$

the solution of which is

$$x=A\cos nt, \text{ or } B\sin nt, \text{ or } A\cos nt+B\sin nt, \text{ or } a\cos(nt+\epsilon);$$

so that n is the constant angular velocity round D of the point Q on the infinitesimal circle AQD, as in the cycloid.

In Kepler's Problem in Astronomy, n represents what is called the *mean motion* of a planet or satellite, and nt or $nt+\epsilon$ the *mean anomaly*; a satellite of Jupiter, when observed in the plane of its orbit, supposed circular, will appear to move with a S. H. M.

But with $\kappa=1$, putting $\frac{1}{2}\theta=\phi=$ angle AEP (fig. 3)

$$nt=\int_0 \sec\phi d\phi=\log(\sec\phi+\tan\phi),$$

so that

$$\sec\phi+\tan\phi=e^{nt},$$
$$\sec\phi-\tan\phi=e^{-nt},$$
$$\sec\phi=\tfrac{1}{2}(e^{nt}+e^{-nt})=\cosh nt,$$
$$\tan\phi=\tfrac{1}{2}(e^{nt}-e^{-nt})=\sinh nt,$$
$$\sin\phi=\tanh nt,\ \cos\phi=\operatorname{sech} nt,$$
$$\tan\tfrac{1}{2}\phi=\tanh\tfrac{1}{2}nt, \text{ and so on.}$$

Also $d\theta/dt=2n\cos\frac{1}{2}\theta=2n\operatorname{sech} nt$;

so that if the angular velocity of the pendulum in the lowest position OA is $2n$, the pendulum will just reach the highest position OE; but the time occupied in reaching it will be infinite, since $\theta=\pi$, $\phi=\frac{1}{2}\pi$ makes nt and therefore t infinite.

The velocity of P in any position is

$$l(d\theta/dt)=2nl\cos\tfrac{1}{2}\theta=n\,.\,EP,$$

and therefore varies as EP.

If EP in fig. 3 is produced to meet Ax in M', then

$$AM'=AE\tan\tfrac{1}{2}\theta=2l\sinh nt,\ EM'=EA\sec\tfrac{1}{2}\theta=2l\cosh nt;$$

so that, if AM' or EM' is denoted by x,

$$\frac{d^2x}{dt^2}-n^2x=0,$$

the general solution of which differential equation is

$$x=A\cosh nt+B\sinh nt.$$

16. When the pendulum just reaches the highest position OE, $\kappa=1$; and u (or nt) and ϕ, the c.m. of the angle AEP, are connected by the relations

$$u=\int_0 \sec\phi\, d\phi=\log(\sec\phi+\tan\phi)$$
$$=\cosh^{-1}\sec\phi=\sinh^{-1}\tan\phi=\tanh^{-1}\sin\phi=2\tanh^{-1}\tan\tfrac{1}{2}\phi.$$

Conversely

$$\phi=\cos^{-1}\operatorname{sech} u=\sin^{-1}\tanh u=\tan^{-1}\sinh u=2\tan^{-1}\tanh\tfrac{1}{2}u;$$

and then ϕ is called by Professor Cayley the *Gudermannian* of u, and denoted by $\operatorname{gd} u$; so that if $\phi=\operatorname{gd} u$, then

$$u=\operatorname{gd}^{-1}\phi=\log(\sec\phi+\tan\phi)=\cosh^{-1}\sec\phi, \text{ etc.}$$

Hoüel proposes for ϕ the name of *hyperbolic amplitude* of u, with the notation $\phi = \text{amh}\, u$, instead of $\text{gd}\, u$; so that

$$u = \int_0^{\text{amh}\, u} \sec \phi d\phi;$$

or $\phi = \text{amh}\, u = \int_0^u \text{sech}\, u du = \cos^{-1} \text{sech}\, u = \sin^{-1} \tanh u$, etc;

analogous in the general case of the elliptic functions, for any modulus κ, to (§ 7)

$$F^{-1}u = \text{am}\, u = \int_0 \quad \text{dn}\, u du = \quad \cos^{-1} \text{cn}\, u = \quad \sin^{-1} \text{sn}\, u, \text{ etc.}$$

As degenerate forms, when $\kappa = 1$,

$$\text{cn}\, u = \text{sech}\, u,\ \text{sn}\, u = \tanh u,\ \text{dn}\, u = \text{sech}\, u;$$

while, with $\kappa = 0$,

$$\text{cn}\, u = \cos u,\ \text{sn}\, u = \sin u,\ \text{dn}\, u = 1.$$

Thus, when $\kappa = 1$, the elliptic functions degenerate into the hyperbolic functions; and, when $\kappa = 0$, into the circular functions; but with any other value of the modulus κ, the elliptic functions must be considered as new functions, of a higher order of complexity than the circular or hyperbolic functions.

The following Table, from Legendre, *F. E.*, t. II., Table IV., gives the values of

$$u = \log(\sec \phi + \tan \phi) = \log \tan(\tfrac{1}{4}\pi + \tfrac{1}{2}\phi)$$

for every degree of ϕ radians; whence the numerical values of the hyperbolic functions of u can be determined, by aid of a table of circular functions, and by the relations

$$\cosh u = \sec \phi,\ \sinh u = \tan \phi,\ \tanh u = \sin \phi, \ldots.$$

For values of u greater than about 4 the Table fails; but then it is sufficient, to two decimals, to take

$$\cosh u = \qquad \sinh u = \tfrac{1}{2}e^u;$$

$$\log_{10} \cosh u = \log_{10} \sinh u = Mu - \log 2;$$

or, to a closer approximation,

$$\log_{10} \cosh u = Mu - \log 2 + Me^{-2u}, \ldots,$$

$$\log_{10} \sinh u = Mu - \log 2 - Me^{-2u}, \ldots,$$

$$\log_{10} \tanh u = \qquad -2Me^{-2u} \ldots,$$

M denoting the modulus $\log_{10} e$.

(*Proposed Tables of Hyperbolic Functions*, Report to the British Association, 1888, by Prof. Alfred Lodge.)

TABLE III.

	ϕ	u		ϕ	u		ϕ	u
0°	0·00000	0·00000	30°	0·52360	0·54931	60°	1·04720	1·31696
1	0·01745	0·01745	31	0·54105	0·56956	61	1·06465	1·35240
2	0·03491	0·03491	32	0·55851	0·59003	62	1·08210	1·38899
3	0·05236	0·05238	33	0·57596	0·61073	63	1·09956	1·42679
4	0·06981	0·06987	34	0·59341	0·63166	64	1·11701	1·46591
5	0·08727	0·08738	35	0·61087	0·65284	65	1·13446	1·50645
6	0·10472	0·10491	36	0·62832	0·67428	66	1·15192	1·54855
7	0·12217	0·12248	37	0·64577	0·69599	67	1·16937	1·59232
8	0·13963	0·14008	38	0·66323	0·71799	68	1·18682	1·63794
9	0·15708	0·15773	39	0·68068	0·74029	69	1·20428	1·68557
10	0·17453	0·17543	40	0·69813	0·76291	70	1·22173	1·73542
11	0·19199	0·19318	41	0·71558	0·78586	71	1·23918	1·78771
12	0·20944	0·21099	42	0·73304	0·80917	72	1·25664	1·84273
13	0·22689	0·22886	43	0·75049	0·83284	73	1·27409	1·90079
14	0·24435	0·24681	44	0·76794	0·85690	74	1·29154	1·96226
15	0·26180	0·26484	45	0·78540	0·88137	75	1·30900	2·02759
16	0·27925	0·28295	46	0·80285	0·90628	76	1·32645	2·09732
17	0·29671	0·30116	47	0·82030	0·93163	77	1·34390	2·17212
18	0·31416	0·31946	48	0·83776	0·95747	78	1·36136	2·25280
19	0·33161	0·33786	49	0·85521	0·98381	79	1·37881	2·34040
20	0·34907	0·35638	50	0·87266	1·01068	80	1·39626	2·43625
21	0·36652	0·37501	51	0·89012	1·03812	81	1·41372	2·54209
22	0·38397	0·39377	52	0·90757	1·06616	82	1·43117	2·66031
23	0·40143	0·41266	53	0·92502	1·09483	83	1·44862	2·79422
24	0·41888	0·43169	54	0·94248	1·12418	84	1·46608	2·94870
25	0·43633	0·45088	55	0·95993	1·15423	85	1·48353	3·13130
26	0·45379	0·47021	56	0·97738	1·18505	86	1·50098	3·35467
27	0·47124	0·48972	57	0·99484	1·21667	87	1·51844	3·64253
28	0·48869	0·50939	58	1·01229	1·24916	88	1·53589	4·04813
29	0·50615	0·52925	59	1·02974	1·28257	89	1·55334	4·74135
30	0·52360	0·54931	60	1·74720	1·31696	90	1·57080	infinite.

Considered as a function of the latitude ϕ, u was called the *meridional part* by Edward Wright, 1599, who first employed it for the accurate construction of the parallels of latitude on the Mercator Chart, by making the ratio of the distance from the equator of the parallel of latitude ϕ to the distance between the meridians whose difference of longitude is ϕ equal to the ratio of u/ϕ (§ 98).

17. Returning to the general elliptic functions, we notice that

$$\mathrm{cn}^2u+\ \mathrm{sn}^2u=1,$$
$$\mathrm{dn}^2u+\kappa^2\mathrm{sn}^2u=1,$$
$$\mathrm{dn}^2u-\kappa^2\mathrm{cn}^2u=\kappa'^2;$$

or, in a tabular form,

	cn	sn	dn
$\mathrm{cn}\,u=$	$\mathrm{cn}\,u$	$\surd(1-\mathrm{sn}^2u)$	$\surd(\mathrm{dn}^2u-\kappa'^2)/\kappa$
$\mathrm{sn}\,u=$	$\surd(1-\mathrm{cn}^2u)$	$\mathrm{sn}\,u$	$\surd(1-\mathrm{dn}^2u)/\kappa$
$\mathrm{dn}\,u=$	$\surd(\kappa'^2+\kappa^2\mathrm{cn}^2u)$	$\surd(1-\kappa^2\mathrm{sn}^2u)$	$\mathrm{dn}\,u$

whence any one of the three elliptic functions cn, sn, dn, can be expressed in terms of any other; the three functions are thus not absolutely necessary, but all three are retained and utilized for simplicity of expression, as sometimes one and sometimes another is most appropriate for the particular problem in hand; in the same way, of the circular functions

$$\cos\theta,\ \sin\theta,\ \tan\theta,\ \cot\theta,\ \sec\theta,\ \mathrm{cec}\,\theta,\ \mathrm{vers}\,\theta,$$

one would be sufficient, but all are useful; and so also with the hyperbolic functions $\cosh u$, $\sinh u$, $\tanh u$, ...

For the reciprocals and quotients of the elliptic functions cn, sn, dn, a convenient notation has been invented by Dr. Glaisher, according to which $1/\mathrm{cn}\,u$ is represented by $\mathrm{nc}\,u$, $1/\mathrm{sn}\,u$ by $\mathrm{ns}\,u$, $1/\mathrm{dn}\,u$ by $\mathrm{nd}\,u$, $\mathrm{cn}\,u/\mathrm{dn}\,u$ by $\mathrm{cd}\,u$, and so on.

In this manner $\mathrm{sn}\,u/\mathrm{cn}\,u$ would be denoted by $\mathrm{sc}\,u$; but it is more commonly denoted by $\mathrm{tanam}\,u$, abbreviated to $\mathrm{tn}\,u$; while $\mathrm{cn}\,u/\mathrm{sn}\,u$ or $\mathrm{cs}\,u$ would be denoted by $\mathrm{cotam}\,u$, or $\mathrm{ctn}\,u$.

According to Clifford (*Dynamic*, p. 89) we might abbreviate the designation of the hyperbolic cosine, sine, and tangent to hc, hs, and ht; or we may write them ch, sh, th; with cn, sn, tn for the elliptic functions; and merely c, s, t for the circular functions.

18. *Pendulum performing complete revolutions.*

Secondly, suppose the pendulum performs complete revolutions (fig. 3).

We have seen previously (§ 15) that if the pendulum has an angular velocity $2n=2\sqrt{(g/l)}$ in the lowest position, it will just reach the highest position; and therefore if this angular velocity is increased, the pendulum will perform complete revolutions.

The integration of equation (1) in the form

$$\tfrac{1}{2}l^2(d\theta/dt)^2=C-gl \text{ vers } \theta$$

or $$\tfrac{1}{2}v^2/g+AN=AD, \text{ a constant, denoted by } 2R,$$

shows that the velocity of P is that which would be acquired in falling freely from the level of a certain horizontal line BDB', which now does not cut the circle, as in fig. 2 when the pendulum oscillated, but lies entirely above the circle, as in fig. 3, at a height $2R$ above the lowest point A; and the *impetus* of the velocity of P is the depth of P below BB'.

Denoting the angle AEP by ϕ, so that $\phi=\tfrac{1}{2}\theta$, then

$$2l^2(d\phi/dt)^2=g(2R-l \text{ vers } 2\phi)=2g(R-l\sin^2\phi),$$

or $$\left(\frac{d\phi}{dt}\right)^2=\frac{gR}{l^2}\left(1-\frac{l}{R}\sin^2\phi\right)=\frac{n^2}{\kappa^2}(1-\kappa^2\sin^2\phi),$$

on putting $\kappa^2=l/R=AE/AD$; and $n^2=g/l$, as before;

so that $$nt/\kappa=\int_0(1-\kappa^2\sin^2\phi)^{-\frac{1}{2}}d\phi=F(\phi,\kappa),$$

in Legendre's notation; and inverting the function according to Abel's suggestion, with Jacobi's notation,

$$\tfrac{1}{2}\theta=\phi=\text{am}(nt/\kappa,\kappa);$$

and now, with Gudermann's abbreviated notation,

$$\cos\tfrac{1}{2}\theta=\text{cn}\, nt/\kappa,$$
$$\sin\tfrac{1}{2}\theta=\text{sn}\, nt/\kappa,$$
$$\frac{d\theta}{dt}=2\frac{n}{\kappa}\text{dn}\, nt/\kappa,$$
$$AN=l \text{ vers } \theta=2l\sin^2\phi=AE\,\text{sn}^2 nt/\kappa,$$
$$NE=AE\,\text{cn}^2 nt/\kappa,\quad ND=AD\,\text{dn}^2 nt/\kappa,$$
$$AP=AE\,\text{sn}\, nt/\kappa,\quad PE=AE\,\text{cn}\, nt/\kappa,$$
$$NP=2l\sin\tfrac{1}{2}\theta\cos\tfrac{1}{2}\theta\quad=AE\,\text{sn}\, nt/\kappa\,\text{cn}\, nt/\kappa.$$

19. The time of moving from A to E is obtained by putting $\phi=\frac{1}{2}\pi$, and is therefore $K\kappa/n$; and therefore the *period*, or time of a complete revolution, is $2K\kappa/n$ (not $4K\kappa/n$).

With the series for K as given in § 11, and with $\kappa^2=l/R$, the period of the pendulum for a complete revolution is

$$\pi\sqrt{\left(\frac{l^2}{gR}\right)}\left\{1+\left(\frac{1}{2}\right)^2\frac{l}{R}+\left(\frac{1\,.\,3}{2\,.\,4}\right)^2\frac{l^2}{R^2}+\left(\frac{1\,.\,3\,.\,5}{2\,.\,4\,.\,6}\right)^2\frac{l^3}{R^3}+\ldots\right\}.$$

The analogous expression for the period when the pendulum oscillates, rising on each side to a height $2R$, less than $2l$, is, as in § 11,

$$2\pi\sqrt{\left(\frac{l}{g}\right)}\left\{1+\left(\frac{1}{2}\right)^2\frac{R}{l}+\left(\frac{1\,.\,3}{2\,.\,4}\right)^2\frac{R^2}{l^2}+\left(\frac{1\,.\,3\,.\,5}{2\,.\,4\,.\,6}\right)^2\frac{R^3}{l^3}+\ldots\right\}.$$

Putting $\kappa=1$, and $R=l$, makes K infinite, and brings us back again to the separating case between oscillations and complete revolutions of the pendulum; and we thus regain for this case the original expressions involving hyperbolic functions, previously investigated in § 15.

But as κ now diminishes again from 1 to 0, the pendulum revolves faster and faster, until finally, when $\kappa=0$, we must suppose the pendulum to revolve with infinite angular velocity, the fluctuations of which for different positions of P are insensible; and the period is now zero.

20. We notice that, in the circle AQ (fig. 2) the point Q moves according to the law

$$\phi=\text{am}\, nt,$$

so that Q moves round in a circle, centre C, in fig. 2 like the point P making complete revolutions in fig. 3.

But now, in the motion of Q, gravity must be supposed diluted from g to $\kappa^4 g$; for if R or $\kappa^2 l$ denotes the radius of the circle AQ, g' the diluted value of gravity, and $n'=\sqrt{(g'/R)}$ the speed of the pendulum CQ, then we must have

$$\phi=\text{am}\, nt=\text{am}\, n't/\kappa,$$

so that

$$n'=\kappa n,$$

$$g'/R=\kappa^2 g/l,$$

$$g'/g=\kappa^2 R/l=\kappa^4.$$

We may dilute gravity in the circle AQ by inclining the plane of the circle to the vertical at an appropriate angle.

21. Another way of diluting gravity would be to replace the circle AQ by a fine tube in the form of a uniform helix with horizontal axis through its centre C perpendicular to the plane of the circle AQ, and to suppose the particle Q to move in this helix under gravity.

Then we shall find that if the length of one complete turn of this helical tube is equal to the circumference of the circle AP, the particle Q moving with velocity due to the level of E will follow the motion of the particle P moving on the circle AP with velocity due to the level of D, so that PQ will always be horizontal, if once it is horizontal, and P, Q will always be at the same level during the motion.

For in this case the mechanical similitude is secured by increasing the square of the velocity of Q in the ratio of 1 to $1/\kappa^4$, instead of diluting gravity to $\kappa^4 g$.

We may secure the same effect by supposing Q to be a point on a pendulum CQ', of length greater than CQ; or else of length CQ, but of which the axis C is cut into a smooth screw of appropriate pitch; or else engaging with teethed wheels, so as to increase the angular inertia about C.

22. If we produce CQ to any fixed distance $CQ'=l'$, then Q' will also perform complete revolutions like a pendulum of length l', with gravity changed in a certain fixed ratio depending on l'; and we can keep gravity unchanged by choosing l' so that

$$n'^2 = g/l' = \kappa^2 n^2 = \kappa^2 g/l,$$

or

$$l' = l/\kappa^2 = l \operatorname{cosec}^2 \tfrac{1}{2}\alpha\,;$$

and now Q' revolves with velocity due to a level at a height $2l/\kappa^4 = 2l \operatorname{cosec}^4 \frac{1}{2}\alpha$ above its lowest position; so that the period of revolution of a simple pendulum of length $l \operatorname{cosec}^2 \frac{1}{2}\alpha$, when the velocity is due to the level of a line at a height $2l \operatorname{cosec}^4 \frac{1}{2}\alpha$ above its lowest point is equal to the time of oscillation of a simple pendulum of length l through an angle 2α from rest to rest.

These problems on the pendulum have been developed here at some length, in accordance with the idea of this Treatise, that it is simple pendulum motion which affords the best concrete illustration of the Elliptic Functions.

Similar principles are involved in the following three theorems, which the student can prove as an exercise in the manner employed for the cycloid in § 10.

1. If two vertical circles, of diameters AD and AE, touch at their lowest points A, the time of oscillation from rest to rest of a particle in the circle AE with velocity due to the level of D will be to the time of revolution of a particle in the circle AD with velocity due to the level of E in the ratio of AE to AD (fig. 2).

2. Two particles move, under gravity, in vertical circles. The one oscillates; the other performs complete revolutions. Prove that if the height to which the velocity of the first is due bears to the diameter of the first circle the same ratio as the diameter of the second circle bears to the height to which the velocity in it is due (the heights being measured from the lowest points of the circles) the ratio of the squares of the times in corresponding small arcs—and therefore the squares of the whole times of oscillation and revolution—will be that compounded of either of the before-mentioned equal ratios and the ratio of the diameters of the circles.

3. Two equal smooth circles are fixed so as to touch the same horizontal plane, their planes being at different inclinations; two small heavy beads are projected at the same instant along these circles from their lowest points, the velocity of each bead being that due to the height of the highest point of the other circle above the horizontal plane, show that during the motion the two beads will always be at equal heights above the horizontal plane.

23. We have compared the motion of the pendulum in fig. 1 with that of the simple equivalent pendulum composed of the particle P moving on a smooth circle, or at the end of a fine thread or wire OP; oscillating from B to B' in fig. 2, and performing complete revolutions in fig. 3, the velocity of P at any point being that acquired in falling from the level of D.

Taking as coordinate axes the horizontal and vertical axes Ax and Ay through A, and referring the motion of P to the coordinates x and y, then since P describes the circle AP of radius l,

$$x^2 = 2ly - y^2.$$

Denoting by $v = ds/dt$ the velocity of P, then by the principle of energy

$$\tfrac{1}{2}v^2/g = 2R - y,$$

$2R$ denoting the height of D above A.

But since
$$\frac{dx}{dy}=\frac{l-y}{\sqrt{(2ly-y^2)}},$$
$$\frac{ds^2}{dy^2}=1+\frac{dx^2}{dy^2}=\frac{l^2}{2ly-y^2};$$
while
$$\tfrac{1}{2}(ds/dt)^2=g(2R-y);$$
so that
$$\tfrac{1}{2}l^2(dy/dt)^2=g(2R-y)(2ly-y^2),$$
$$\frac{dt}{dy}=\frac{l}{\sqrt{(2g)}}\frac{1}{\sqrt{\{(2R-y)(2ly-y^2)\}}},$$
$$t=\frac{l}{\sqrt{(2g)}}\int_0^y\frac{dy}{\sqrt{\{(2R-y)(2ly-y^2)\}}},$$
called an *elliptic integral* in y, and of the *first kind.*

24. Firstly, if the pendulum oscillates, R is less than l, and y oscillates between 0 and $2R$; and the integral is reduced to Legendre's canonical form by putting $y=2R\sin^2\phi$; when
$$nt=\int_0(1-\kappa^2\sin^2\phi)^{-\frac{1}{2}}d\phi=F(\phi,\kappa),$$
where
$$\kappa^2=R/l,\ n^2=g/l;$$
and therefore with Jacobi's and Gudermann's notation,
$$\phi=\text{am}(nt,\kappa)$$
and
$$y=2R\,\text{sn}^2nt=2l\kappa^2\,\text{sn}^2nt,\ x=2l\kappa\,\text{sn}\,nt\,\text{dn}\,nt;$$
or
$$AN=AD\,\text{sn}^2nt,\ ND=AD\,\text{cn}^2nt,\ NE=AE\,\text{dn}^2nt,$$
as before, in § 8.

25. When $\kappa=0$, the oscillations are indefinitely small; and now
$$y=2R\sin^2nt,$$
where R is a very small quantity;
and
$$nt=\int_0\frac{\frac{1}{2}dy}{\sqrt{\{y(2R-y)\}}}=\sin^{-1}\sqrt{\frac{y}{2R}},$$
an ordinary circular integral.

It was Abel who pointed out (about 1823) that in looking only at the *Elliptic Integrals,* mathematicians had been taking the same difficult point of view as if they had begun to deduce the theorems of elementary Trigonometry from an examination of the properties of the *inverse* circular functions, as deduced from the *circular integrals.*

(*Niels-Henrik Abel. Tableau de sa vie et de son action scientifique.* Par C. A. Bjerknes. 1885.)

26. Secondly, if the pendulum performs complete revolutions, as in fig. 3, R is greater than l, and y oscillates in value between 0 and $2l$; we now reduce the elliptic integral in § 23 to Legendre's standard form by putting $y=2l\sin^2\phi$,

when $$nt/\kappa=\int_0(1-\kappa^2\sin^2\phi)^{-\frac{1}{2}}d\phi=F(\phi,\kappa)$$

where $$\kappa^2=l/R,$$

the reciprocal of its former expression; and now

$$\phi=\text{am}(nt/\kappa,\kappa),\ y=2l\,\text{sn}^2 nt/\kappa,\ x=2l\,\text{sn}\,nt/\kappa\,\text{cn}\,nt/\kappa;$$

or $$AN=AE\,\text{sn}^2 nt/\kappa,\ NE=AE\,\text{cn}^2 nt/\kappa,\ ND=AD\,\text{dn}^2 nt/\kappa,$$

as proved before, in § 18.

27. In the separating case between oscillations and complete revolutions, $R=l$, and now $\kappa=1$;

and $$y=2l\sin^2\phi=l\,\text{vers}\,2\phi=l\,\text{vers}\,\theta;$$

also (§ 23) $$nt=\int_0\sec\phi d\phi=\log(\sec\phi+\tan\phi)$$

$$=\cosh^{-1}\sec\phi=\sinh^{-1}\tan\phi=\tanh^{-1}\sin\phi=2\tanh^{-1}\tan\tfrac{1}{2}\phi;$$

so that $$\phi=\text{gd}\,nt,\ \text{or amh}\,nt,$$

and $$\sec\phi=\cosh nt,\ \tan\phi=\sinh nt,\ \sin\phi=\tanh nt,$$

$$y=2l\tanh^2 nt,\ x=2l\,\text{sech}\,nt\tanh nt,$$

as before, in § 15.

28. *Landen's Point.*

With centre E in fig. 2 and radius EB describe a circle cutting the vertical AE in L; then L is an important point in the theory of pendulum motion and elliptic functions, called *Landen's point.*

Since $$EB^2=ED\,.\,EA=EC^2-CA^2,$$

therefore the circle, centre E and radius EB, will cut the circle AQD, centre C, at right angles; and

$$LQ^2=LC^2+CQ^2+2LC\,.\,CN=2LC\,.\,EN=2l(1-\kappa')^2EN;$$

since $$LC^2+CQ^2=LC^2+EC^2-EL^2=2LC\,.\,EC,$$

and $$EL=EB=2l\kappa',\ EC=l(1+\kappa'^2),\ LC=l(1-\kappa')^2.$$

Now, by § 20, the velocity of Q

$$=\sqrt{(2g'\,.\,EN)}=\sqrt{(2g\kappa^4\,.\,EN)}=n\kappa^2\sqrt{(2l\,.\,EN)}$$

$$=n\,.\,LQ(1+\kappa').$$

Similarly in fig. 3, where P makes complete revolutions, the velocity of $P=n\,.\,LP(1+\kappa')/\kappa$, where the Landen point L is obtained by drawing a circle with centre D, cutting the circle AE orthogonally, and the vertical AD in L.

We shall prove subsequently that any straight line through L divides the circle APE in fig. 3 (or the circle AQD in fig. 2) into two parts, each described in half the period.

29. *Change from one modulus to its reciprocal.*

It is important for the simplicity and for convenience of tabulation of the elliptic functions that the modulus κ should not exceed unity; but the preceding reductions of the motion of the pendulum to elliptic functions, in the two cases in which the pendulum oscillates and performs complete revolutions, show us how to make the elliptic functions to a modulus κ, which is greater than unity, depend on the elliptic functions to the reciprocal modulus $1/\kappa$, which is less than unity.

For, on comparing the two expressions for y, according as the pendulum oscillates or performs complete revolutions,

$$y = 2R\,\mathrm{sn}^2(nt, \kappa), \text{ or } 2l\,\mathrm{sn}^2(\kappa nt, 1/\kappa),$$

where $$\kappa^2 = R/l;$$

so that $$\kappa^2\mathrm{sn}^2(nt, \kappa) = \mathrm{sn}^2(\kappa nt, 1/\kappa);$$

or, putting $$nt = u,$$

$$\kappa\,\mathrm{sn}(u, \kappa) = \mathrm{sn}(\kappa u, 1/\kappa),$$

so that $$\mathrm{dn}(u, \kappa) = \mathrm{cn}(\kappa u, 1/\kappa),$$

$$\mathrm{cn}(u, \kappa) = \mathrm{dn}(\kappa u, 1/\kappa).$$

Independently, if we suppose $\phi = \mathrm{am}(u, \kappa)$, and if we put

$$\kappa \sin\phi = \sin\psi,$$

then $$\kappa\cos\phi\, d\phi = \cos\psi\, d\psi,$$

and $$\cos\phi = \sqrt{(1-\kappa^{-2}\sin^2\psi)} = \Delta(\psi, 1/\kappa),$$

$$\cos\psi = \sqrt{(1-\kappa^2\sin^2\phi)} = \Delta(\phi, \kappa);$$

so that $$u = \int_0 (1-\kappa^2\sin^2\phi)^{-\frac{1}{2}} d\phi = \int \sec\psi\, d\phi;$$

$$\kappa u = \int \sec\phi\, d\psi = \int (1-\kappa^{-2}\sin^2\psi)^{-\frac{1}{2}} d\psi,$$

or $$\psi = \mathrm{am}(\kappa u, 1/\kappa);$$

and since $$\kappa\sin\phi = \sin\psi, \text{ etc.},$$

therefore $$\kappa\,\mathrm{sn}(u, \kappa) = \mathrm{sn}(\kappa u, 1/\kappa), \text{ etc.}$$

When $u = K$, $\phi = \frac{1}{2}\pi$, and $\psi = \sin^{-1}\kappa$; so that, if κ is less than unity, $$K\kappa = \int_0^{\sin^{-1}\kappa} (1-\kappa^{-2}\sin^2\psi)^{-\frac{1}{2}} d\psi.$$

30. *Rectilinear Oscillations expressed by Elliptic Functions.*

In simple pendulum motion, referred to horizontal and vertical axes Ax, Ay, drawn through the lowest point A, we have shown in §§ 24, 26, that

$$y = 2l\kappa^2 \text{sn}^2 nt,\ x = 2l\kappa \text{ sn } nt \text{ dn } nt;$$

or

$$y = 2l \text{sn}^2 nt/\kappa,\ x = 2l \text{ sn } nt/\kappa \text{ cn } nt/\kappa;$$

according as the pendulum oscillates or performs complete revolutions.

Treating the vertical motions separately, and differentiating according to the rules established in § 7, we find, on taking

$$y = 2l\kappa^2 \text{sn}^2 nt,$$

$$dy/dt = 4ln\kappa^2 \text{sn } nt \text{ cn } nt \text{ dn } nt$$

$$d^2y/dt^2 = 4ln^2\kappa^2(\text{cn}^2 nt \text{ dn}^2 nt - \text{sn}^2 nt \text{ dn}^2 nt - \kappa^2 \text{sn}^2 nt \text{ cn}^2 nt)$$

$$= 4ln^2\kappa^2\left\{\left(1 - \frac{y}{2l\kappa^2}\right)\left(1 - \frac{y}{2l}\right) - \frac{y}{2l\kappa^2}\left(1 - \frac{y}{2l}\right) - \frac{y}{2l}\left(1 - \frac{y}{2l\kappa^2}\right)\right\}$$

$$= 4ln^2\kappa^2\left(1 - \frac{y}{l} - \frac{y}{l\kappa^2} + \frac{3y^2}{4l^2\kappa^2}\right), \text{ by § 17.}$$

Taking $y = 2l \text{ sn}^2 nt/\kappa$, we find in a similar manner

$$\frac{d^2y}{dt^2} = \frac{4ln^2}{\kappa^2}\left(1 - \frac{y}{l} - \frac{\kappa^2 y}{l} + \frac{3\kappa^2 y^2}{4l^2}\right);$$

both immediately obtainable from the equation of § 23,

$$\tfrac{1}{2}l^2(dy/dt)^2 = g(2R - y)(2ly - y^2)$$

whence

$$l^2(d^2y/dt^2) = 4g(Rl - Ry - ly + \tfrac{3}{4}y^2).$$

We shall find similar expressions for d^2y/dt^2 when y varies as $\text{cn}^2 nt$ or $\text{dn}^2 nt$, all of the form

$$d^2y/dt^2 = A + By + Cy^2.$$

Let us determine then, as exercises in the differentiation of the elliptic functions, the acceleration d^2x/dt^2, and thence the force at a distance x, which will make a body oscillate in a straight line according to one of the laws

$$x = a \text{ cn } nt,\ \text{sn } nt,\ \text{dn } nt,\ \text{tn } nt,\ \text{nc } nt,\ \text{ns } nt, \ldots.$$

Taking

$$x = a \text{ cn } nt,$$

$$dx/dt = -na \text{ sn } nt \text{ dn } nt$$

$$d^2x/dt^2 = -n^2a(\text{cn } nt \text{ dn}^2 nt - \kappa^2 \text{sn}^2 nt \text{ cn } nt)$$

$$= -n^2x\left(\kappa'^2 - \kappa^2 + 2\kappa^2\frac{x^2}{a^2}\right),$$

so that $$\frac{d^2x}{dt^2}+n^2x=2n^2\kappa^2x\left(1-\frac{x^2}{a^2}\right);$$
reducing to zero when $\kappa=0$.

It is often simpler to find dx/dt, and then to express $\frac{1}{2}(dx/dt)^2$ as a function of x; and then a differentiation with respect to t will give d^2x/dt^2 immediately as a function of x.

Thus, if $$x=a \operatorname{sn} nt,$$
$$dx/dt=na \operatorname{cn} nt \operatorname{dn} nt$$
$$\tfrac{1}{2}\left(\frac{dx}{dt}\right)^2=\tfrac{1}{2}n^2a^2\left(1-\frac{x^2}{a^2}\right)\left(1-\frac{\kappa^2x^2}{a^2}\right),$$
so that $$\frac{d^2x}{dt^2}=-n^2(1+\kappa^2)x+\frac{2n^2\kappa^2x^3}{a^2},$$
$$\frac{d^2x}{dt^2}+n^2x=-n^2\kappa^2\left(x-\frac{2x^3}{a^2}\right),$$
reducing to zero, when $\kappa=0$.

Similarly, if $$x=a \operatorname{dn} nt,$$
$$\frac{d^2x}{dt^2}=n^2(1+\kappa'^2)x-\frac{2n^2x^3}{a^2}.$$

Generally, when x varies also as $\operatorname{tn} nt$, $\operatorname{nc} nt$, ..., we shall find a relation of the form
$$d^2x/dt^2=\mu x+2\nu x^3,$$
which, when multiplied by dx/dt and integrated, gives
$$\tfrac{1}{2}(dx/dt)^2=C+\tfrac{1}{2}\mu x^2+\tfrac{1}{2}\nu x^4$$
or $$dx/dt=\sqrt{(2C+\mu x^2+\nu x^4)},$$
$$t=\int(2C+\mu x^2+\nu x^4)^{-\frac{1}{2}}dx,$$
an *elliptic integral*, of which the different expressions are given in Chapter II.

31. *A Special Minimum Surface.*

Another interesting exercise in the differentiation of elliptic functions is to verify that the surface discovered by Schwarz (*Gesammelte Mathematische Abhandlungen*, vol. I., p. 77),
$$\operatorname{cn} x+\operatorname{cn} y+\operatorname{cn} z+\operatorname{cn} x \operatorname{cn} y \operatorname{cn} z=0,$$
with the modulus $\kappa=\frac{1}{2}$, is a *minimum surface*, having zero curvature at every point, and therefore satisfying the condition
$$(1+q^2)r-2pqs+(1+p^2)t=0,$$
p, q, r, s, t having their usual meaning as partial differential coefficients of z with respect to x and y.

Schwarz shows that this condition is equivalent to

$$\frac{1}{\rho_1}+\frac{1}{\rho_2}=\frac{(1+q^2)r-2pqs+(1+p^2)t}{1+p^2+q^2}=\sqrt{(1+p^2+q^2)}\left(\frac{\partial X}{\partial x}+\frac{\partial Y}{\partial y}\right)=0,$$

ρ_1, ρ_2 denoting the principal radii of curvature of the surface (C. Smith, *Solid Geometry*, § 255), where

$$X=\frac{p}{\sqrt{(p^2+q^2+1)}},\quad Y=\frac{q}{\sqrt{(p^2+q^2+1)}}.$$

Let us write c_1, s_1, d_1, for $\operatorname{cn} x$, $\operatorname{sn} x$, $\operatorname{dn} x$; and c_2, s_2, d_2, c_3, s_3, d_3 for the same functions of y and z.

Then $$c_1+c_2+c_3+c_1c_2c_3=0;$$
and differentiating with respect to x,

$$-s_1d_1-s_3d_3p-s_1d_1c_2c_3-c_1c_2s_3d_3p=0,$$

or $$p=-\frac{s_1d_1(1+c_2c_3)}{s_3d_3(1+c_1c_2)}.$$

But $$c_3=-\frac{c_1+c_2}{1+c_1c_2},$$

$$s_3^2=1-c_3^2=\frac{(1+c_1c_2)^2-(c_1+c_2)^2}{(1+c_1c_2)^2}=\frac{s_1^2s_2^2}{(1+c_1c_2)^2};$$

so that $$s_3(1+c_1c_2)=s_1s_2, \text{ etc.};$$

$$p=-\frac{s_2s_3d_1}{s_1s_2d_3}=-\frac{s_3d_1}{s_1d_3}=-\frac{d_1/s_1}{d_3/s_3}=-\frac{s_3/d_3}{s_1/d_1}.$$

By symmetry, $$q=-\frac{d_2/s_2}{d_3/s_3};$$

so that we may write

$$X=\frac{-d_1/s_1}{\sqrt{\{(d_1/s_1)^2+(d_2/s_2)^2+(d_3/s_3)^2\}}},$$

$$Y=\frac{-d_2/s_2}{\sqrt{\{(d_1/s_1)^2+(d_2/s_2)^2+(d_3/s_3)^2\}}}.$$

Now $$\frac{d}{dx}\left(\frac{d_1}{s_1}\right)=\frac{-\kappa^2s_1^2c_1-c_1d_1^2}{s_1^2}=-\frac{c_1}{s_1^2},\quad \frac{d}{dx}\left(\frac{d_3}{s_3}\right)=-\frac{c_3p}{s_3^2};$$

so that $$\frac{\partial X}{\partial x}=\left\{\frac{c_1}{s_1^2}\left(\frac{d_1^2}{s_1^2}+\frac{d_2^2}{s_2^2}+\frac{d_3^2}{s_3^2}\right)-\frac{d_1}{s_1}\left(\frac{d_1}{s_1}\frac{c_1}{s_1^2}+\frac{d_3}{s_3}\frac{c_3}{s_3^2}p\right)\right\}\div D^{\frac{3}{2}},$$

where $$D=(d_1/s_1)^2+(d_2/s_2)^2+(d_3/s_3)^2;$$

or $$\frac{\partial X}{\partial x}=\left(\frac{c_1d_2^2}{s_1^2s_2^2}+\frac{c_1d_3^2+c_3d_1^2}{s_1^2s_3^2}\right)\div D^{\frac{3}{2}}.$$

By symmetry

$$\frac{\partial Y}{\partial y}=\left(\frac{c_2d_1^2}{s_1^2s_2^2}+\frac{c_2d_3^2+c_3d_2^2}{s_2^2s_3^2}\right)\div D^{\frac{3}{2}};$$

so that $\dfrac{\partial X}{\partial x}+\dfrac{\partial Y}{\partial y}=0$, provided that

$$\frac{c_2d_3^2+c_3d_2^2}{s_2^2s_3^2}+\frac{c_3d_1^2+c_1d_3^2}{s_3^2s_1^2}+\frac{c_1d_2^2+c_2d_1^2}{s_1^2s_2^2}=0,$$

or $$c_1(s_2^2d_3^2+s_3^2d_2^2)+\ldots=0;$$

or, since $$s_1^2=1-c_1^2,\ d_1^2=\tfrac{1}{4}(3+c_1^2),$$

$$c_1\{(1-c_2)^2(3+c_3^2)+(1-c_3^2)(3+c_2^2)\}+\ldots=0,$$

or $$(c_1+c_2+c_3+c_1c_2c_3)(3-c_2c_3-c_3c_1-c_1c_2)=0,$$

and this is true, in consequence of the original relation

$$c_1+c_2+c_3+c_1c_2c_3=0.$$

The other relation $3-c_2c_3-c_3c_1-c_1c_2=0$ represents isolated conjugate points, where

$$c_1=c_2=c_3=1.$$

Another minimum surface is

$$\text{tn}\,y\,\text{tn}\,z+\text{tn}\,z\,\text{tn}\,x+\text{tn}\,x\,\text{tn}\,y+3=0,$$

with $$\kappa=\tfrac{2}{3}\sqrt{2},\ \kappa'=\tfrac{1}{3}.$$

32. *Elliptic Function Solution of Euler's Equations of Motion.*

Before leaving the mechanical interpretation of elliptic functions, we may just mention here an important application, the application to the solution of *Euler's equations of motion*, for a body under no forces, moving about its centre of gravity, or about any fixed point.

Euler's equations for p, q, r, the component angular velocities about the principal axes, are (Routh, *Rigid Dynamics*)

$$Adp/dt=(B-C)qr,$$
$$Bdq/dt=(C-A)rp,$$
$$Cdr/dt=(A-B)pq;$$

where A, B, C denote the moments of inertia about the principal axes; and two first integrals of these equations are

$$Ap^2+Bq^2+Cr^2=T,\text{ a constant};$$
$$A^2p^2+B^2q^2+C^2r^2=G^2,\text{ a constant},$$

obtained by multiplying Euler's equations respectively by (i.) p, q, r, and adding, (ii.) by Ap, Bq, Cr, and adding; and then integrating.

Comparing these equations with the equations of § 7,

$$\text{cn}'u=-\text{sn}\,u\,\text{dn}\,u,\ \text{sn}'u=\text{cn}\,u\,\text{dn}\,u,\ \text{dn}'u=-\kappa^2\text{sn}\,u\,\text{cn}\,u,$$

where accents denote differentiation with respect to u, we notice that if $A>B>C$, and the polhode includes the axis C, so that $AT>BT>G^2>CT$, we may put $u=nt$, and

$$p=P\,\mathrm{cn}\,u,\ q=-Q\,\mathrm{sn}\,u,\ r=R\,\mathrm{dn}\,u;$$

and then, on substituting in Euler's equations of motion,

$$\frac{B-C}{A}=\frac{nP}{QR},\ \frac{A-C}{B}=\frac{nQ}{RP},\ \frac{A-B}{C}=\frac{\kappa^2 nR}{PQ}.$$

Putting $t=0$, and therefore $p=P$, $q=0$, $r=R$; then

$$AP^2+CR^2=T,\ A^2P^2+C^2R^2=G^2,$$

so that
$$P^2=\frac{G^2-CT}{A(A-C)},\ R^2=\frac{AT-G^2}{C(A-C)};$$

and then
$$Q^2=P^2\frac{A}{B}\frac{A-C}{B-C}=\frac{G^2-CT}{B(B-C)};$$

while
$$n^2=R^2\frac{(A-C)(B-C)}{AB}=\frac{(AT-G^2)(B-C)}{ABC},$$

and
$$\kappa^2=\frac{P^2}{R^2}\frac{A}{C}\frac{A-B}{B-C}=\frac{G^2-CT}{AT-G^2}\frac{A-B}{B-C}.$$

If the polhode encloses the axis of greatest moment A, so that $AT>G^2>BT>CT$, we must put

$$p=P\,\mathrm{dn}\,u,\ q=-Q\,\mathrm{sn}\,u,\ r=R\,\mathrm{cn}\,u;$$

and then determine P, Q, R, n, κ as before; when

$$n^2=\frac{(G^2-CT)(A-B)}{ABC},\ \kappa^2=\frac{AT-G^2}{G^2-CT}\frac{B-C}{A-B}.$$

In the separating case, when $G^2=BT$, then $\kappa=1$, and

$$p=P\,\mathrm{sech}\,nt,\ q=-Q\tanh nt,\ r=R\,\mathrm{sech}\,nt;$$

so that, when $t=0$,

$$p^2=\frac{G^2}{AB}\frac{B-C}{A-C},\ q=0,\ r^2=\frac{G^2}{BC}\frac{A-B}{A-C};$$

and initially or finally, when $t=\mp\infty$,

$$p=0,\ q=\pm G/B,\ r=0;$$

and the body is spinning about its mean axis B.

But when the body is spinning about the axis of greatest or least moment, $G^2=AT=A^2p^2$, or $G^2=CT=C^2r^2$, and $\kappa=0$; and the period of a small oscillation is $2\pi/n$, where

$$n^2=\frac{(A-B)(A-C)}{ABC}T=\frac{(A-B)(A-C)}{BC}p^2,$$

or
$$n^2=\frac{(A-C)(B-C)}{ABC}T=\frac{(A-C)(B-C)}{AB}r^2.$$

We shall return subsequently to these equations in Chap. III.

CHAPTER II.

THE ELLIPTIC INTEGRALS (OF THE FIRST KIND).

33. In Chapter I. we have immediately made use of Abel's valuable idea of the *Inversion of the Elliptic Integral,* which is the foundation of the modern theory of the *Elliptic Functions;* and we have considered the functions which are inverse to the elliptic integral, and treated them as the direct fundamental functions of our Theory.

Previously to Abel's discovery (1823) it was the elliptic integral which was studied, as in the writings of Euler and Legendre; and, in fact, in a physical and dynamical problem it is the elliptic integral which arises in the course of the work; for instance in the form of the Equation of Energy,

$$\tfrac{1}{2}(dx/dt)^2 = X, \text{ so that } \sqrt{2}\, t = \int dx/\sqrt{X};$$

and now, when X is a cubic or quartic function of x, so that d^2x/dt^2 is a quadratic or cubic, as in § 30, the integral is called an *elliptic integral of the first kind;* and we have to follow Abel and determine the *elliptic function* which expresses x as a function of t.

To accomplish this, it will be useful to employ the notation of the inverse functions, given by Clifford (*Proc. London Math. Society,* vol. vii., p. 29; *Mathematical Papers,* p. 207) analogous to those used in Trigonometry for the inverse circular functions; and to make a collection of all the important cases that can occur.

34. *The Circular and Hyperbolic Integrals.*

Starting with the circular functions, $\sin x$, $\cos x$, $\tan x$, $\cot x$, ..., we have, in the ordinary notation,

$$\int_0^x \frac{dx}{\sqrt{(1-x^2)}} = \sin^{-1}x = \cos^{-1}\sqrt{(1-x^2)},$$

$$\int_x^1 \frac{dx}{\sqrt{(1-x^2)}} = \cos^{-1}x = \sin^{-1}\sqrt{(1-x^2)},$$

$$\int_0^x \frac{dx}{1+x^2} = \tan^{-1}x = \cot^{-1}\frac{1}{x},$$

$$\int_x^\infty \frac{dx}{x^2+1} = \cot^{-1}x = \tan^{-1}\frac{1}{x}, \text{ etc.}$$

We can employ a similar notation with the hyperbolic functions, $\cosh x$, $\sinh x$, $\tanh x$, $\coth x$, ..., and write

$$\int_1^x \frac{dx}{\sqrt{(x^2-1)}} = \cosh^{-1}x = \sinh^{-1}\sqrt{(x^2-1)} = \log\{x+\sqrt{(x^2-1)}\},$$

$$\int_0^x \frac{dx}{\sqrt{(1+x^2)}} = \sinh^{-1}x = \cosh^{-1}\sqrt{(1+x^2)} = \log\{\sqrt{(1+x^2)}+x\},$$

$$\int_0^x \frac{dx}{1-x^2} = \tanh^{-1}x = \tfrac{1}{2}\log\frac{1+x}{1-x}\ (x<1),$$

$$\int_x^\infty \frac{dx}{x^2-1} = \coth^{-1}x = \tfrac{1}{2}\log\frac{x+1}{x-1}\ (x>1);\ \text{etc.};$$

and the analogy with the circular functions is now complete, and the results can be more easily remembered and written down, than when the logarithmic function alone is employed.

To avoid complications due to the *multiplicity of the values* of these and subsequent integrals, in consequence of the variable x assuming complex values and performing circuits of contours round the *poles* of the integral, we suppose for the present that x is real, and increases or diminishes continually, so as to assume all real values once only between the limits of integration; also that the positive sign is taken with the radical under the sign of integration; we thus obtain what is called the *principal value* of the integral or inverse function.

35. *The Elliptic Integrals.*

With the elliptic functions, $\operatorname{sn} u$, $\operatorname{cn} u$, $\operatorname{dn} u$, we have (§ 7)

$$\frac{d\operatorname{sn} u}{du} = \operatorname{cn} u \operatorname{dn} u,\ \frac{d\operatorname{cn} u}{du} = -\operatorname{sn} u \operatorname{dn} u,\ \frac{d\operatorname{dn} u}{du} = -\kappa^2\operatorname{sn} u \operatorname{cn} u;$$

and $\quad \text{cn}^2 u = 1 - \text{sn}^2 u, \ \text{dn}^2 u = 1 - \kappa^2 \text{sn}^2 u$;

so that, if $x = \text{sn}\, u$, then $\text{cn}\, u = \surd(1-x^2)$, $\text{dn}\, u = \surd(1-\kappa^2 x^2)$;

$$\frac{dx}{du} = \surd(1-x^2 \,.\, 1-\kappa^2 x^2),$$

and $$\int_0^x \frac{dx}{\surd(1-x^2 \,.\, 1-\kappa^2 x^2)} = u = \text{sn}^{-1} x, \text{ or } \text{sn}^{-1}(x, \kappa), \quad\text{........(1)}$$

when the modulus κ is required to be put in evidence.

Putting $x=1$ makes the integral equal to K, the quarter period corresponding to the modulus κ (§ 11).

Similarly, with

$$x = \text{cn}\, u, \text{ then } \text{sn}\, u = \surd(1-x^2),\ \text{dn}\, u = \surd(\kappa'^2 + \kappa^2 x^2),$$

$$\frac{dx}{du} = -\text{sn}\, u\, \text{dn}\, u = -\surd(1-x^2 \,.\, \kappa'^2 + \kappa^2 x^2),$$

and $$\int_x^1 \frac{dx}{\surd(1-x^2 \,.\, \kappa'^2 + \kappa^2 x^2)} = u = \text{cn}^{-1} x, \text{ or } \text{cn}^{-1}(x, \kappa), \quad\text{.......(2)}$$

so that the integral is K when the lower limit is 0.

Again, with

$$x = \text{dn}\, u, \text{ then } \kappa\, \text{sn}\, u = \surd(1-x^2),\ \kappa\, \text{cn}\, u = \surd(x^2 - \kappa'^2);$$

and $$\frac{dx}{du} = -\kappa^2 \text{sn}\, u\, \text{cn}\, u = -\surd(1-x^2 \,.\, x^2 - \kappa'^2),$$

$$\int_x^1 \frac{dx}{\surd(1-x^2 \,.\, x^2-\kappa'^2)} = u = \text{dn}^{-1} x, \text{ or } \text{dn}^{-1}(x, \kappa). \quad\text{........(3)}$$

We may also put $x = \text{tn}\, u$, using Gudermann's abbreviation of $\text{tn}\, u$ for $\tan \text{am}\, u$; and now

$$\text{cn}\, u = \frac{1}{\surd(1+x^2)},\quad \text{dn}\, u = \frac{\surd(1+\kappa'^2 x^2)}{\surd(1+x^2)};$$

$$\frac{dx}{du} = \frac{\text{dn}\, u}{\text{cn}^2 u} = \surd(1+x^2 \,.\, 1+\kappa'^2 x^2),$$

$$\int_0 \frac{dx}{\surd(1+x^2 \,.\, 1+\kappa'^2 x^2)} = u = \text{tn}^{-1} x, \text{ or } \text{tn}^{-1}(x, \kappa) \quad\text{........(4)}$$

and the integral is K when the upper limit is ∞.

Putting $x = \sin\phi$, $\cos\phi$, $\Delta\phi$, or $\tan\phi$ in (1), (2), (3), or (4), reduces the integral to

$$\int_0 (1-\kappa^2 \sin^2\phi)^{-\frac{1}{2}} d\phi = u = \text{F}(\phi, \kappa)$$

$$= \text{am}^{-1}(\phi, \kappa) = \text{sn}^{-1}(\sin\phi, \kappa) = \text{cn}^{-1}(\cos\phi, \kappa) = \text{dn}^{-1}(\Delta\phi, \kappa);$$

so that

$\phi = \text{am}\, u$, and $\cos\phi = \text{cn}\, u$, $\sin\phi = \text{sn}\, u$, $\Delta\phi = \text{dn}\, u$, $\tan\phi = \text{tn}\, u$.

36. Thus, with $a > b > x$,

$$\int_0^x \frac{dx}{\sqrt{(a^2-x^2 \, . \, b^2-x^2)}} = \frac{1}{a} \operatorname{sn}^{-1}\left(\frac{x}{b}, \frac{b}{a}\right), \ldots\ldots\ldots\ldots\ldots (5)$$

indicating that we must put $x = b\sin\phi$; and then the integral is reduced to

$$\frac{1}{a}\int_0^\phi \left(1 - \frac{b^2}{a^2}\sin^2\phi\right)^{-\frac{1}{2}} d\phi = \frac{1}{a}\operatorname{sn}^{-1}\left(\sin\phi, \frac{b}{a}\right) = \frac{1}{a}\operatorname{sn}^{-1}\left(\frac{x}{b}, \frac{b}{a}\right).$$

Similarly, with $\infty > x > a$,

$$\int_x^\infty \frac{dx}{\sqrt{(x^2-a^2 \, . \, x^2-b^2)}} = \frac{1}{a}\operatorname{sn}^{-1}\left(\frac{a}{x}, \frac{b}{a}\right), \ldots\ldots\ldots\ldots\ldots (6)$$

indicating the substitution $x = a \operatorname{cosec}\phi$ (or $a \operatorname{cec}\phi$, as Dr. Glaisher writes it).

Thus, for instance, with $\infty > x > 1/\kappa$,

$$\int_x^\infty \frac{dx}{\sqrt{(1-x^2 \, . \, 1-\kappa^2 x^2)}} = \operatorname{sn}^{-1}\left(\frac{1}{\kappa x}, \kappa\right).$$

Again,

$$\int_x^b \frac{dx}{\sqrt{(a^2+x^2 \, . \, b^2-x^2)}} = \frac{1}{\sqrt{(a^2+b^2)}} \operatorname{cn}^{-1}\left\{\frac{x}{b}, \frac{b}{\sqrt{(a^2+b^2)}}\right\}; \ldots\ldots (7)$$

$$\int_b^x \frac{dx}{\sqrt{(a^2+x^2 \, . \, x^2-b^2)}} = \frac{1}{\sqrt{(a^2+b^2)}} \operatorname{cn}^{-1}\left\{\frac{b}{x}, \frac{a}{\sqrt{(a^2+b^2)}}\right\}; \ldots\ldots (8)$$

$$\int_x^a \frac{dx}{\sqrt{(a^2-x^2 \, . \, x^2-b^2)}} = \frac{1}{a}\operatorname{dn}^{-1}\left\{\frac{x}{a}, \sqrt{\left(1-\frac{b^2}{a^2}\right)}\right\}; \ldots\ldots (9)$$

$$\int_0^x \frac{dx}{\sqrt{(x^2+a^2 \, . \, x^2+b^2)}} = \frac{1}{a}\operatorname{tn}^{-1}\left\{\frac{x}{b}, \sqrt{\left(1-\frac{b^2}{a^2}\right)}\right\} \ldots (10)$$

37. As numerical examples,

$$\int_x^1 \frac{dx}{\sqrt{(1-x^4)}} = \tfrac{1}{2}\sqrt{2}\operatorname{cn}^{-1}(x, \tfrac{1}{2}\sqrt{2}),$$

the integration required in the rectification of the *lemniscate* $r^2 = a^2\cos 2\theta$; so that $r = a\operatorname{cn}(\sqrt{2}\,s/a, \tfrac{1}{2}\sqrt{2})$.

$$\int_1^x \frac{dx}{\sqrt{(x^4-1)}} = \tfrac{1}{2}\sqrt{2}\operatorname{cn}^{-1}\left(\frac{1}{x}, \tfrac{1}{2}\sqrt{2}\right) = \tfrac{1}{2}\sqrt{2}\operatorname{nc}^{-1}(x, \tfrac{1}{2}\sqrt{2}),$$

with Dr. Glaisher's notation (§ 17) of $\operatorname{nc} u$ for $1/\operatorname{cn} u$.

Consider also the vibrations given by the dynamical equation $d^2x/dt^2 = -2n^2x(c^2-x^2)$,
as in § 30; so that $x=0$ gives the point of stable equilibrium, and $x=\pm c$ gives the points of unstable equilibrium.

Integrating, supposing the motion to start from rest where $x=b$,
$$\tfrac{1}{2}(dx/dt)^2 = C - n^2c^2x^2 + \tfrac{1}{2}n^2x^4$$
$$= \tfrac{1}{2}n^2(b^2-x^2)(2c^2-b^2-x^2).$$

(i.) When $b^2 < c^2$, the motion is at the outset towards the origin, and $dx/dt = -n\surd(a^2-x^2 \,.\, b^2-x^2)$,
writing a^2 for $2c^2-b^2$; so that
$$nt = \int_x^b \frac{dx}{\surd(a^2-x^2 \,.\, b^2-x^2)} = \int_0^b \frac{dx}{\surd X} - \int_0^x \frac{dx}{\surd X}$$
$$= \frac{1}{a}\left(K - \text{sn}^{-1}\frac{x}{b}\right), \text{ with modulus } \frac{b}{a}, \text{ by (5);}$$
or
$$x = b\,\text{sn}(K-ant).$$

(ii.) When $b^2=c^2$, $dx/dt = \pm n(b^2-x^2)$;
and, by § 34, the ultimate state of motion is given by
$$x = b\tanh bnt, \text{ or } b\coth bnt,$$
according as the motion falls away from the position of unstable equilibrium, towards or away from the origin.

(iii.) When $c^2 < b^2 < 2c^2$,
$$dx/dt = +n\surd(x^2-a^2 \,.\, x^2-b^2),$$
$$nt = \int_b^x \frac{dx}{\surd(x^2-a^2 \,.\, x^2-b^2)} = \int_b^\infty \frac{dx}{\surd X} - \int_x^\infty \frac{dx}{\surd X}$$
$$= \frac{1}{b}\left(K - \text{sn}^{-1}\frac{b}{x}\right), \text{ mod. } \frac{a}{b}, \text{ by (6);}$$
or
$$x = b/\text{sn}(K-bnt) = b\,\text{ns}(K-bnt).$$

(iv.) When $b^2 = 2c^2$,
$$nt = \int_b^x \frac{dx}{x\surd(x^2-b^2)} = \frac{1}{b}\sec^{-1}\frac{x}{b},$$
or
$$x = b\sec bnt.$$

(v.) When $b^2 > 2c^2$, we must write a^2 for b^2-2c^2; and now
$$dx/dt = +n\surd(a^2+x^2 \,.\, x^2-b^2),$$
$$nt = \int_b^x \frac{dx}{\surd(a^2+x^2 \,.\, x^2-b^2)}$$
$$= \frac{1}{\surd(a^2+b^2)}\text{cn}^{-1}\left\{\frac{b}{x}, \frac{a}{\surd(a^2+b^2)}\right\},$$
or
$$x = b/\text{cn}\surd(a^2+b^2)nt = b\,\text{nc}\surd(a^2+b^2)nt.$$

38. So far the function X has been treated as an even quartic function of x, or as a quadratic function of x^2, resolved into two real factors; but according to Prof. Felix Klein there are certain advantages in considering the integrals obtained by writing $x^2=z$, in (1), (2), (3); and then, writing k for κ^2,

$$\int_0 \frac{dz}{\surd(z\,.\,1-z\,.\,1-kz)}=2\,\mathrm{sn}^{-1}\surd z,$$

$$\text{or } 2\,\mathrm{cn}^{-1}\surd(1-z), \text{ or } 2\,\mathrm{dn}^{-1}\surd(1-kz)\ldots\ldots\ldots\ldots(11)$$

Conversely, by writing for z the values x^2, $1-x^2$, $1-kx^2$, we reproduce the integrals (1), (2), (3) from (11), by the simplest *quadric transformations;* and it will not cause confusion if we sometimes call k the *modulus*.

For these and various other reasons, Prof. Klein suggests (*Math. Ann.* XIV., p. 116) that we should consider (11) as a more canonical form of the elliptic integral than (1), the form with which Legendre and Jacobi have worked.

39. Now, with $X=x-\alpha\,.\,x-\beta\,.\,x-\gamma$, and $\alpha>\beta>\gamma$, we have, if $\infty>x>\alpha$,

$$\int_x^\infty \frac{dx}{\surd X}=\frac{2}{\surd(\alpha-\gamma)}\,\mathrm{sn}^{-1}\sqrt{\frac{\alpha-\gamma}{x-\gamma}}$$

$$=\frac{2}{\surd(\alpha-\gamma)}\,\mathrm{cn}^{-1}\sqrt{\frac{x-\alpha}{x-\gamma}}=\frac{2}{\surd(\alpha-\gamma)}\,\mathrm{dn}^{-1}\sqrt{\frac{x-\beta}{x-\gamma}},\ (12)$$

with $\kappa^2=k=(\beta-\gamma)/(\alpha-\gamma)$;

indicating that we must put

$$x-\gamma=(\alpha-\gamma)\mathrm{cec}^2\phi,\ x-\alpha=(\alpha-\gamma)\cot^2\phi,$$

and then $$x-\beta=(\beta-\gamma)\Delta^2\phi\,\mathrm{cec}^2\phi,$$

to reduce the integral to Legendre's canonical form

$$F\phi=\int_0(1-k\sin^2\phi)^{-\frac{1}{2}}d\phi.$$

Similarly, by putting $x-\alpha=(\alpha-\beta)\tan^2\phi$, $x-\beta=(\alpha-\beta)\sec^2\phi$,

$$\int_\alpha^x \frac{Mdx}{\surd X}=\mathrm{sn}^{-1}\sqrt{\frac{x-\alpha}{x-\beta}}$$

$$=\mathrm{cn}^{-1}\sqrt{\frac{\alpha-\beta}{x-\beta}}=\mathrm{dn}^{-1}\sqrt{\frac{\alpha-\beta\,.\,x-\gamma}{\alpha-\gamma\,.\,x-\beta}},\ldots\ldots(13)$$

where M is used throughout to denote $\frac{1}{2}\surd(\alpha-\gamma)$.

Thus, with $\infty > x > 1/k$, integral (11) becomes

$$\int_x^\infty \frac{dx}{\sqrt{(x.1-x.1-kx)}} = 2\,\mathrm{sn}^{-1}\sqrt{\frac{1}{kx}}$$

$$= 2\,\mathrm{cn}^{-1}\sqrt{\frac{kx-1}{kx}} = 2\,\mathrm{dn}^{-1}\sqrt{\frac{x-1}{x}};$$

$$\int_{1/k}^x \frac{dx}{\sqrt{(x.1-x.1-kx)}} = 2\,\mathrm{sn}^{-1}\sqrt{\frac{kx-1}{k.x-1}}$$

$$= 2\,\mathrm{cn}^{-1}\sqrt{\frac{1-k}{k.x-1}} = 2\,\mathrm{dn}^{-1}\sqrt{\frac{1-k.x}{x-1}}.$$

40. When $\alpha > x > \beta$, X is negative, and

$$\int_x^\alpha \frac{M dx}{\sqrt{(-X)}} = \mathrm{sn}^{-1}\sqrt{\frac{\alpha-x}{\alpha-\beta}}$$

$$= \mathrm{cn}^{-1}\sqrt{\frac{x-\beta}{\alpha-\beta}} = \mathrm{dn}^{-1}\sqrt{\frac{x-\gamma}{\alpha-\gamma}}, \text{ .... (14)}$$

$$\int_\beta^x \frac{M dx}{\sqrt{(-X)}} = \mathrm{sn}^{-1}\sqrt{\frac{\alpha-\gamma.x-\beta}{\alpha-\beta.x-\gamma}}$$

$$= \mathrm{cn}^{-1}\sqrt{\frac{\beta-\gamma.\alpha-x}{\alpha-\beta.x-\gamma}} = \mathrm{dn}^{-1}\sqrt{\frac{\beta-\gamma}{x-\gamma}}, \text{ .... (15)};$$

and now the modulus κ' is given by $\kappa'^2 = k' = (\alpha-\beta)/(\alpha-\gamma)$, and the modulus is therefore complementary to the modulus in (12) and (13); and the form of the result in these and other subsequent integrals indicates the substitution required to reduce the integral to Legendre's standard form (§ 4); while the results can be verified by differentiation.

Thus, with $1/k > x > 1$, integral (11) is imaginary and may be written

$$\int_x^{1/k} \frac{dx}{\sqrt{(x.1-x.1-kx)}} = 2i\,\mathrm{sn}^{-1}\sqrt{\frac{1-kx}{1-k}}$$

$$= 2i\,\mathrm{cn}^{-1}\sqrt{\frac{k.x-1}{1-k}} = 2i\,\mathrm{dn}^{-1}\sqrt{(kx)}, \text{ mod. } k';$$

$$\int_1^x \frac{dx}{\sqrt{(x.1-x.1-kx)}} = 2i\,\mathrm{sn}^{-1}\sqrt{\frac{x-1}{1-k.x}}$$

$$= 2i\,\mathrm{cn}^{-1}\sqrt{\frac{1-kx}{1-k.x}} = 2i\,\mathrm{dn}^{-1}\sqrt{\frac{1}{x}}, \text{ mod. } k';$$

i denoting $\sqrt{(-1)}$.

41. When $\beta > x > \gamma$, X is again positive, and

$$\int_x^\beta \frac{M dx}{\sqrt{X}} = \text{sn}^{-1}\sqrt{\frac{\alpha-\gamma \,.\, \beta-x}{\beta-\gamma \,.\, \alpha-x}}$$

$$= \text{cn}^{-1}\sqrt{\frac{\alpha-\beta \,.\, x-\gamma}{\beta-\gamma \,.\, \alpha-x}} = \text{dn}^{-1}\sqrt{\frac{\alpha-\beta}{\alpha-x}}, \quad \text{..............(16)}$$

$$\int_\gamma^x \frac{M dx}{\sqrt{X}} = \text{sn}^{-1}\sqrt{\frac{x-\gamma}{\beta-\gamma}}$$

$$= \text{cn}^{-1}\sqrt{\frac{\beta-x}{\beta-\gamma}} = \text{dn}^{-1}\sqrt{\frac{\alpha-x}{\alpha-\gamma}}; \quad \text{..................(17)}$$

with $k=(\beta-\gamma)/(\alpha-\gamma)$, as in (12) and (13).

Thus

$$\int_x^1 \frac{dx}{\sqrt{(x \,.\, 1-x \,.\, 1-kx)}} = 2\,\text{sn}^{-1}\sqrt{\frac{1-x}{1-kx}}$$

$$= 2\,\text{cn}^{-1}\sqrt{\frac{1-k \,.\, x}{1-kx}} = 2\,\text{dn}^{-1}\sqrt{\frac{1-k}{1-kx}};$$

while the result is as in (11) when the lower limit is 0.

42. When $\gamma > x > -\infty$, X is negative, and

$$\int_x^\gamma \frac{M dx}{\sqrt{(-X)}} = \text{sn}^{-1}\sqrt{\frac{\gamma-x}{\beta-x}}$$

$$= \text{cn}^{-1}\sqrt{\frac{\beta-\gamma}{\beta-x}} = \text{dn}^{-1}\sqrt{\frac{\beta-\gamma \,.\, \alpha-x}{\alpha-\gamma \,.\, \beta-x}}, \quad \text{............(18)}$$

$$\int_{-\infty}^x \frac{M dx}{\sqrt{(-X)}} = \text{sn}^{-1}\sqrt{\frac{\alpha-\gamma}{\alpha-x}}$$

$$= \text{cn}^{-1}\sqrt{\frac{\gamma-x}{\alpha-x}} = \text{dn}^{-1}\sqrt{\frac{\beta-x}{\alpha-x}}; \quad \text{...................(19)}$$

with modulus $k'=(\alpha-\beta)/(\alpha-\gamma)$, as in (14) and (15).

Thus, with $0 > x > -\infty$, integral (11) becomes

$$\int_x^0 \frac{dx}{\sqrt{(x \,.\, 1-x \,.\, 1-kx)}} = 2i\,\text{sn}^{-1}\sqrt{\frac{-x}{1-x}}$$

$$= 2i\,\text{cn}^{-1}\sqrt{\frac{1}{1-x}} = 2i\,\text{dn}^{-1}\sqrt{\frac{1-kx}{1-x}}, \text{mod.}\, k';$$

$$\int_{-}^x \frac{dx}{\sqrt{(x \,.\, 1-x \,.\, 1-kx)}} = 2i\,\text{sn}^{-1}\sqrt{\frac{1}{1-kx}}$$

$$= 2i\,\text{cn}^{-1}\sqrt{\frac{-kx}{1-kx}} = 2i\,\text{dn}^{-1}\sqrt{\frac{k \,.\, 1-x}{1-kx}}, \text{mod.}\, k'.$$

43. We notice that the substitution

$$\frac{\alpha-\gamma}{x-\gamma}=\frac{y-\gamma}{\beta-\gamma}, \text{ or } \frac{x-\alpha}{x-\gamma}=\frac{\beta-y}{\beta-\gamma}, \text{ or } \frac{x-\beta}{x-\gamma}=\frac{\alpha-y}{\alpha-\gamma},$$

makes

$$\int_x^\infty \frac{dx}{\sqrt{(x-\alpha\,.\,x-\beta\,.\,x-\gamma)}}=\int_\gamma^y \frac{dy}{\sqrt{(y-\alpha\,.\,y-\beta\,.\,y-\gamma)}},$$

or changes (12) into (17), or (13) into (16).

Thus

$$\int_\alpha^\infty \frac{dx}{\sqrt{(x-\alpha\,.\,x-\beta\,.\,x-\gamma)}}=\int_\gamma^\beta \frac{dy}{\sqrt{(y-\alpha\,.\,y-\beta\,.\,y-\gamma)}}=\frac{2K}{\sqrt{(\alpha-\gamma)}}, \quad \text{...(20)}$$

where $\kappa^2=k=(\beta-\gamma)/(\alpha-\gamma)$.

Again the substitution

$$\frac{\alpha-x}{\alpha-\beta}=\frac{\alpha-\gamma}{\alpha-y}, \text{ or } \frac{x-\beta}{\alpha-\beta}=\frac{\gamma-y}{\alpha-y}, \text{ or } \frac{x-\gamma}{\alpha-\gamma}=\frac{\beta-y}{\alpha-y},$$

changes (14) into (19), or (15) into (18); and shows that

$$\int_\beta^\alpha \frac{dx}{\sqrt{(\alpha-x\,.\,x-\beta\,.\,x-\gamma)}}=\int_{-\infty}^\gamma \frac{dy}{\sqrt{(\alpha-y\,.\,\beta-y\,.\,\gamma-y)}}=\frac{2K'}{\sqrt{(\alpha-\gamma)}}, \quad \text{...(21)}$$

where $k'=\kappa'^2=(\alpha-\beta)/(\alpha-\gamma)$.

The substitution which changes any one integral into another is obvious by inspection of the preceding results.

44. Thus the integral $\int dx/\sqrt{X}$ can be written down, expressed by inverse elliptic functions, when X is a cubic form in x, resolved into its three real linear factors.

For example, with $a^2>b^2>c^2$,

$$\int_\lambda^\infty \frac{d\lambda}{\sqrt{(a^2+\lambda\,.\,b^2+\lambda\,.\,c^2+\lambda)}}=\frac{2}{\sqrt{(a^2-c^2)}}\,\text{cn}^{-1}\left(\sqrt{\frac{c^2+\lambda}{a^2+\lambda}},\ \sqrt{\frac{a^2-b^2}{a^2-c^2}}\right),$$

an integral occurring in the mathematical theories of Electricity, Magnetism, and Hydrodynamics, in connexion with ellipsoids.

As another example, the student may prove that

$$\int \frac{dS}{(x/a)^2+(y/b)^2+(z/c)^2}=\frac{4\pi abc}{\sqrt{(a^2-c^2)}}\,\text{cn}^{-1}\left(\frac{c}{a},\ \sqrt{\frac{a^2-b^2}{a^2-c^2}}\right),$$

when the integration is extended over the surface S of the sphere $$x^2+y^2+z^2=r^2$$

(W. Burnside, *Math. Tripos*, 1881).

45. When two of the roots, β and γ suppose, of the cubic $X=0$ are complex, we combine $(x-\beta)(x-\gamma)$ into the real quadratic $(x-m)^2+n^2$, suppose; so that $X=x-a\,.\,(x-m)^2+n^2$.

Now we substitute

$$y=\frac{X}{(x-a)^2}=\frac{(x-m)^2+n^2}{x-a},$$

a quadric substitution, the *graph* of which is a hyperbola, and find the *turning values* of y, say y_1 and y_3, the values of y which make the quadratic in x,

$$(x-m)^2+n^2-y(x-a)=0$$

have equal roots; so that y_1 and y_3 are the roots of

$$(\tfrac{1}{2}y+m)^2-(ay+m^2+n^2)=0, \text{ or } \tfrac{1}{4}y^2+(m-a)y-n^2=0.$$

Then $$y-y_1=\frac{(x-x_1)^2}{x-a},\quad y-y_3=\frac{(x-x_3)^2}{x-a},$$

and $$\frac{dy}{dx}=\frac{(x-x_1)(x-x_3)}{(x-a)^2};$$

x_1 and x_3 denoting the values of x corresponding to y_1 and y_3, and therefore denoting the roots of the quadratic equation

$$x^2-2ax+2am-m^2-n^2=0;$$

so that $$x_1=m+\tfrac{1}{2}y_1,\ x_3=m+\tfrac{1}{2}y_3.$$

Then $$\int^{\infty}\frac{dx}{\sqrt{X}}=\int\frac{dx}{(x-a)\sqrt{y}}=\int\frac{(x-a)dy}{(x-x_1)(x-x_3)\sqrt{y}}$$

$$=\int^{\infty}\frac{dy}{\sqrt{(y\,.\,y-y_1\,.\,y-y_3)}}$$

$$=\frac{2}{\sqrt{(y_1-y_3)}}\text{cn}^{-1}\left(\sqrt{\frac{y-y_1}{y-y_3}},\ \sqrt{\frac{-y_3}{y_1-y_3}}\right)$$

$$=\frac{\sqrt{2}}{\sqrt{(x_1-x_3)}}\text{cn}^{-1}\frac{x-x_1}{x-x_3},\quad\text{.....................(22)}$$

by (12), with $\quad k'=y_1/(y_1-y_3),\ k=-y_3/(y_1-y_3)$, since y_1 is positive and y_3 negative, or $y_1>y>0>y_3$.

Again, with the same substitution,

$$\int_{-\infty}\frac{dx}{\sqrt{\{a-x\,.\,(x-m)^2+n^2\}}}=\int_{-\infty}\frac{dy}{\sqrt{(-y\,.\,y_1-y\,.\,y_3-y)}}$$

$$=\frac{2}{\sqrt{(y_1-y_3)}}\text{cn}^{-1}\sqrt{\frac{y_3-y}{y_1-y}},$$

$$=\frac{\sqrt{2}}{\sqrt{(x_1-x_3)}}\text{cn}^{-1}\ \frac{x_3-x}{x_1-x},\quad\text{......(23)}$$

by (19), to a modulus k' the complementary modulus of (22), namely $$k'=y_1/(y_1-y_3).$$

46. We denote $(a-m)^2+n^2$ by H^2, and then

$$x_1=a+H,\quad x_3=a-H;$$

and by means of the same substitution as in § 45,

$$\int_a \frac{dx}{\sqrt{\{x-a\,.\,(x-m)^2+n^2\}}}=\frac{\sqrt{2}}{\sqrt{(x_1-x_3)}}\operatorname{cn}^{-1}\frac{x_1-x}{x-x_3}$$

$$=\frac{1}{\sqrt{H}}\operatorname{cn}^{-1}\left\{\frac{H-(x-a)}{H+(x-a)},\ \kappa\right\},$$

$$\kappa^2=\tfrac{1}{2}-\tfrac{1}{2}(a-m)/H,\ \ldots\ldots\ldots\ldots(24);$$

$$\int^a \frac{dx}{\sqrt{\{a-x\,.\,(x-m)^2+n^2\}}}=\frac{1}{\sqrt{H}}\operatorname{cn}^{-1}\left\{\frac{H-(a-x)}{H+(a-x)},\ \kappa'\right\},$$

$$\kappa'^2=\tfrac{1}{2}+\tfrac{1}{2}(a-m)/H,\ \ldots\ldots\ldots\ldots(25);$$

indicating that the substitutions $x-a$ or $a-x=H\left(\begin{smallmatrix}\tan\\ \cot\end{smallmatrix}\tfrac{1}{2}\phi\right)^2$ reduce the integrals to Legendre's standard form; also that

$$2\kappa\kappa'=n/H.$$

Thus, as numerical examples,

$$\int_x^\infty \frac{dx}{\sqrt{(x^3-1)}}=\frac{1}{\sqrt[4]{3}}\operatorname{cn}^{-1}\left(\frac{x-1-\sqrt{3}}{x-1+\sqrt{3}},\ \kappa\right),$$

$$\int_1^x \frac{dx}{\sqrt{(x^3-1)}}=\frac{1}{\sqrt[4]{3}}\operatorname{cn}^{-1}\left(\frac{\sqrt{3}+1-x}{\sqrt{3}-1+x},\ \kappa\right);$$

$$\int_x^1 \frac{dx}{\sqrt{(1-x^3)}}=\frac{1}{\sqrt[4]{3}}\operatorname{cn}^{-1}\left(\frac{\sqrt{3}-1+x}{\sqrt{3}+1-x},\ \kappa'\right),$$

$$\int_{-\infty}^x \frac{dx}{\sqrt{(1-x^3)}}=\frac{1}{\sqrt[4]{3}}\operatorname{cn}^{-1}\left(\frac{1-x-\sqrt{3}}{1-x+\sqrt{3}},\ \kappa'\right);$$

with $\quad 2\kappa\kappa'=\tfrac{1}{2}=\sin 30^\circ,\ \kappa=\sin 15^\circ,\ \kappa'=\sin 75^\circ.$

47. We notice that $\phi=\tfrac{1}{2}\pi$ when $x=a\pm H$; so that

$$\int_{a+H}^\infty \frac{dx}{\sqrt{\{x-a\,.\,(x-m)^2+n^2\}}}$$

$$=\int_a^{a+H} \frac{dx}{\sqrt{\{x-a\,.\,(x-m)^2+n^2\}}}=\frac{K}{\sqrt{H}};\ \ldots\ldots\ldots\ldots(26)$$

$$\int_{a-H}^a \frac{dx}{\sqrt{\{a-x\,.\,(x-m)^2+n^2\}}}$$

$$=\int_{-\infty}^{a+H} \frac{dx}{\sqrt{\{a-x\,.\,(x-m)^2+n^2\}}}=\frac{K'}{\sqrt{H}}\ \ldots\ldots\ldots\ldots(27)$$

Thus, $\displaystyle\int_{\sqrt{3}+1}^\infty \frac{dx}{\sqrt{(x^3-1)}}=\int_1^{\sqrt{3}+1}\frac{dx}{\sqrt{(x^3-1)}}=\frac{F(\sin 15^\circ)}{\sqrt[4]{3}};$

$$\int_{-\sqrt{3}+1}^{1}\frac{dx}{\sqrt{(1-x^3)}}=\int_{-\infty}^{-\sqrt{3}+1}\frac{dx}{\sqrt{(1-x^3)}}=\frac{F(\sin 75^\circ)}{\sqrt[4]{3}}.$$

But, by the *Cubic* substitution $x=(4-z^3)/3z^2$,

then $$1-x^3=\frac{(z^3-1)(z^3+8)^2}{27z^6},\quad \frac{dx}{dz}=-\frac{z^3+8}{3z^2};$$

and $$\int_{-\infty}^{x}\frac{dx}{\sqrt{(1-x^3)}}=\sqrt{3}\int_{z}^{\infty}\frac{dz}{\sqrt{(z^3-1)}};$$

so that $$\int_{-\infty}^{1}\frac{dx}{\sqrt{(1-x^3)}}=\sqrt{3}\int_{1}^{\infty}\frac{dz}{\sqrt{(z^3-1)}},$$

or $$F(\sin 75^\circ)=\sqrt{3}F(\sin 15^\circ),$$

that is, $K'/K=\sqrt{3}$, if $\kappa=\sin 15^\circ$, as stated in § 12.

48. *Degenerate Elliptic Integrals.*

When the middle root β of the cubic $X=0$ approaches to coincidence with either of the extreme roots, α or γ, or when the pair of imaginary roots become equal, the elliptic integrals degenerate into circular or hyperbolic integrals.

We notice, from § 16, that when $k=0$, $\mathrm{sn}^{-1}x$ becomes $\sin^{-1}x$, $\mathrm{cn}^{-1}x$ becomes $\cos^{-1}x$, etc.; and that, when $k=1$, $\mathrm{sn}^{-1}x$ becomes $\tanh^{-1}x$, $\mathrm{cn}^{-1}x$ or $\mathrm{dn}^{-1}x$ becomes $\mathrm{sech}^{-1}x$, and $\mathrm{tn}^{-1}x$ becomes $\sinh^{-1}x$.

Thus, when $k=1$, the integral (11)

$$\int_0\frac{dx}{\sqrt{(x\,.\,1-x\,.\,1-kx)}}=\int\frac{dx}{(1-x)\sqrt{x}}$$
$$=2\tanh^{-1}\sqrt{x}=2\,\mathrm{sech}^{-1}\sqrt{(1-x)}$$
$$=2\cosh^{-1}\sqrt{\frac{1}{1-x}}=2\sinh^{-1}\sqrt{\frac{x}{1-x}}=\sinh^{-1}\frac{2\sqrt{x}}{1-x}.$$

This supposes that $x<1$; but with $\infty>x>1$,

$$\int^{\infty}\frac{dx}{(x-1)\sqrt{x}}=2\coth^{-1}\sqrt{x}=2\,\mathrm{cosech}^{-1}\sqrt{(x-1)}$$
$$=2\sinh^{-1}\sqrt{\frac{1}{x-1}}=2\cosh^{-1}\sqrt{\frac{x}{x-1}}=\sinh^{-1}\frac{2\sqrt{x}}{x-1}.$$

But when $k=0$, the integral (11) becomes

$$\int_0^x\frac{dx}{\sqrt{(x\,.\,1-x)}}=2\sin^{-1}\sqrt{x}$$
$$=2\cos^{-1}\sqrt{(1-x)}=\sin^{-1}2\sqrt{(x\,.\,1-x)};$$
$$\int_x^1\frac{dx}{\sqrt{(x\,.\,1-x)}}=2\cos^{-1}\sqrt{x}$$
$$=2\sin^{-1}\sqrt{(1-x)}=\pi-\sin^{-1}2\sqrt{(x\,.\,1-x)}.$$

49. Making $\beta=\gamma$, or α, in the integrals (12) to (19), and still denoting $\frac{1}{2}\sqrt{(\alpha-\gamma)}$ by M, then

(i.) with $\infty > x > \alpha$,

$$\int_x^\infty \frac{Mdx}{(x-\gamma)\sqrt{(x-\alpha)}} = \cos^{-1}\sqrt{\frac{x-\alpha}{x-\gamma}} = \sin^{-1}\sqrt{\frac{\alpha-\gamma}{x-\gamma}}$$

$$= \tfrac{1}{2}\sin^{-1}\frac{\sqrt{(\alpha-\gamma\,.\,x-\alpha)}}{\frac{1}{2}(x-\gamma)}, \text{ etc.};$$

$$\int_\alpha^x \frac{Mdx}{(x-\gamma)\sqrt{(\alpha-x)}} = \sin^{-1}\sqrt{\frac{x-\alpha}{x-\gamma}} = \cos^{-1}\sqrt{\frac{\alpha-\gamma}{x-\gamma}};$$

$$\int_x^\infty \frac{Mdx}{(x-\alpha)\sqrt{(x-\gamma)}} = \tanh^{-1}\sqrt{\frac{\alpha-\gamma}{x-\gamma}} = \sinh^{-1}\sqrt{\frac{\alpha-\gamma}{x-\alpha}}$$

$$= \cosh^{-1}\sqrt{\frac{x-\gamma}{x-\alpha}} = \tfrac{1}{2}\sinh^{-1}\frac{\sqrt{(\alpha-\gamma.x-\gamma)}}{\frac{1}{2}(x-\alpha)},$$

this integral being infinite when $x=\alpha$.

(ii.) With $\alpha > x > \gamma$,

$$\int_\gamma^x \frac{Mdx}{(\alpha-x)\sqrt{(x-\gamma)}} = \sinh^{-1}\sqrt{\frac{x-\gamma}{\alpha-x}} = \cosh^{-1}\sqrt{\frac{\alpha-\gamma}{\alpha-x}},$$

which is infinite when $x=\alpha$;

$$\int_x^\alpha \frac{Mdx}{(x-\gamma)\sqrt{(\alpha-x)}} = \sinh^{-1}\sqrt{\frac{\alpha-x}{x-\gamma}} = \cosh^{-1}\sqrt{\frac{\alpha-\gamma}{x-\gamma}},$$

which is infinite when $x=\gamma$.

(iii.) With $\gamma > x > -\infty$,

$$\int_x^\gamma \frac{Mdx}{(\alpha-x)\sqrt{(\gamma-x)}} = \sin^{-1}\sqrt{\frac{\gamma-x}{\alpha-x}} = \cos^{-1}\sqrt{\frac{\alpha-\gamma}{\alpha-x}};$$

$$\int_{-\infty}^x \frac{Mdx}{(\alpha-x)\sqrt{(\gamma-x)}} = \cos^{-1}\sqrt{\frac{\gamma-x}{\alpha-x}} = \sin^{-1}\sqrt{\frac{\alpha-\gamma}{\alpha-x}};$$

$$\int_{-\infty}^x \frac{Mdx}{(\gamma-x)\sqrt{(\alpha-x)}} = \cosh^{-1}\sqrt{\frac{\alpha-x}{\gamma-x}} = \sinh^{-1}\sqrt{\frac{\alpha-\gamma}{\gamma-x}},$$

this last integral being infinite when $x=\gamma$.

The limits have been chosen so as to exclude these infinite values.

50. *Weierstrass's Elliptic Functions defined.*

When the general cubic expression X is given, not resolved into factors, then Weierstrass's notation becomes useful, and may be defined here.

Weierstrass writes $s+f$ for x, and chooses f so as to make s^2 disappear in the new value of X, which he denotes by $\frac{1}{4}S$; and thus

$$S=4s^3-g_2s-g_3,$$

where g_2 and g_3 are called the *invariants*; so that the integral

$$\int^\infty \frac{dx}{2\sqrt{X}}=\int\frac{ds}{\sqrt{S}}=\int^\infty\frac{ds}{\sqrt{(4s^3-g_2s-g_3)}}=u, \text{ suppose};$$

and now, inverting the function in Abel's manner, s is an elliptic function of u, denoted by $\wp u$ in Weierstrass's notation, so that

$$\int_s^\infty \frac{ds}{\sqrt{(4s^3-g_2s-g_3)}}=\wp^{-1}s, \text{ or } \wp^{-1}(s;\ g_2, g_3), \text{ ....(A)}$$

when the invariants g_2 and g_3 are to be put in evidence.

51. In Weierstrass's notation we are independent of the particular resolution of S into factors; but by what precedes in equation (12), if, when S is resolved into real factors,

$$S=4(s-e_1)(s-e_2)(s-e_3), \text{ with } e_1>e_2>e_3,$$

then, with $\infty>s>e_1$,

$$u=\int_s^\infty\frac{ds}{\sqrt{(4.s-e_1.s-e_2.s-e_3)}}=\frac{1}{\sqrt{(e_1-e_3)}}\text{sn}^{-1}\sqrt{\frac{e_1-e_3}{s-e_3}}$$

$$=\frac{1}{\sqrt{(e_1-e_3)}}\text{cn}^{-1}\sqrt{\frac{s-e_1}{s-e_3}}=\frac{1}{\sqrt{(e_1-e_3)}}\text{dn}^{-1}\sqrt{\frac{s-e_2}{s-e_3}},$$

by (12); so that

$$\text{sn}^2\sqrt{(e_1-e_3)}u=\frac{e_1-e_3}{\wp u-e_3},\quad \text{cn}^2\sqrt{(e_1-e_3)}u=\frac{\wp u-e_1}{\wp u-e_3},$$

$$\text{dn}^2\sqrt{(e_1-e_3)}u=\frac{\wp u-e_2}{\wp u-e_3}. \text{ ....(B)}$$

The value of u for $s=e_1$ is denoted by ω_1, and called the *real half period*; and by (20) we notice that

$$\omega_1=\int_{e_1}^\infty\frac{ds}{\sqrt{S}}=\int_{e_3}^{e_2}\frac{ds}{\sqrt{S}}=\frac{K}{\sqrt{(e_1-e_3)}}; \text{ ....(28)}$$

and by (13) and (B), $\displaystyle\int_{e_1}^s\frac{ds}{\sqrt{S}}=\wp^{-1}\left(\frac{e_1-e_2.e_1-e_3}{s-e_1}+e_1\right).$ ....(29)

With $e_2>s>e_3$, $\sqrt{S}$ is again real, and by (16), (17), and (B),

$$\int_s^{e_2}\frac{ds}{\sqrt{S}}=\wp^{-1}\left(\frac{e_1-e_2.e_2-e_3}{e_2-s}+e_2\right), \text{ ....(30)}$$

$$\int_e^s\frac{ds}{\sqrt{S}}=\wp^{-1}\left(\frac{e_1-e_3.e_2-e_3}{s-e_3}+e_3\right). \text{ ....(31)}$$

52. For values of s between e_1 and e_2, or between e_3 and $-\infty$, $\surd S$ is imaginary; however, the value of $\int ds/\surd S$ between the limits e_3 and $-\infty$ is denoted by ω_3, and called the *imaginary half period;* so that, by (21),

$$\omega_3 = \int_{e_2}^{e_1} \frac{ds}{\surd S} = \int_{-\infty}^{e_3} \frac{ds}{\surd S} = \frac{iK'}{\surd(e_1 - e_3)}, \quad \text{.............(32)}$$

and, from (12) and (14),

$$\kappa^2 = (e_2 - e_3)/(e_1 - e_3), \quad \kappa'^2 = (e_1 - e_2)/(e_1 - e_3).$$

Also, from (14) and (15), with $e_1 > s > e_2$,

$$\int_s^{e_1} \frac{ds}{\surd S} = i\,\wp^{-1}\left(\frac{e_1 - e_2 \,.\, e_1 - e_3}{e_1 - s} - e_1;\ g_2,\ -g_3\right), \quad \text{.........(33)}$$

$$\int_{e_2}^{s} \frac{ds}{\surd S} = i\,\wp^{-1}\left(\frac{e_1 - e_2 \,.\, e_2 - e_3}{s - e_2} - e_2;\ g_2,\ -g_3\right); \quad \text{.......(34)}$$

and, from (18) and (19), with $e_3 > s > -\infty$,

$$\int_s^{e_3} \frac{ds}{\surd S} = i\,\wp^{-1}\left(\frac{e_1 - e_3 \,.\, e_2 - e_3}{e_3 - s} + e_3;\ g_2,\ -g_3\right), \quad \text{.........(35)}$$

$$\int_{-\infty}^{s} \frac{ds}{\surd S} = i\,\wp^{-1}(-s;\ g_2,\ -g_3). \quad \text{........................(36)}$$

53. The quantity $g_2^3 - 27g_3^2$ is called the *discriminant*, and is denoted by Δ; it is called the discriminant, because the roots of $S = 0$ are all three real, or one real and two imaginary, according as Δ is positive or negative; and $\Delta = 0$, when two roots are equal.

Since $\quad S = 4s^3 - g_2 s - g_3 = 4(s - e_1)(s - e_2)(s - e_3),$

therefore $\quad e_1 + e_2 + e_3 = 0,$

and $\quad g_2 = -4(e_2 e_3 + e_3 e_1 + e_1 e_2) = 2(e_1^2 + e_2^2 + e_3^2), \quad g_3 = 4e_1 e_2 e_3,$

$$\Delta = 16(e_2 - e_3)^2(e_3 - e_1)^2(e_1 - e_2)^2.$$

Therefore

$$\kappa^2\kappa'^2 = (e_1 - e_2)(e_2 - e_3)/(e_1 - e_3)^2, \quad 1 - \kappa^2\kappa'^2 = \tfrac{3}{4} g_2/(e_1 - e_3)^2,$$

and

$$\frac{4}{27}\frac{(1 - \kappa^2\kappa'^2)^3}{\kappa^4\kappa'^4} = \frac{g_2^3}{\Delta}.$$

This quantity g_2^3/Δ is called by Klein the *absolute invariant*, and denoted by J; and then, with k for κ^2,

$$J = \frac{4}{27}\frac{(1 - k + k^2)^3}{k^2(1 - k)^2}, \quad J - 1 = \frac{(1 + k)^2(2 - k)^2(1 - 2k)^2}{27k^2(1 - k)^2}. \quad \text{....(C)}$$

54. For the present we reserve the difficulties of interpretation of the multiple values of the integral $u=\int ds/\sqrt{S}$, due to s being allowed to assume complex values, and to perform circuits round the *poles*, *branch points*, or *critical points*, so called, of the integral, given by the roots of $S=0$.

We suppose the variable s to pass once through all real values from ∞ to $-\infty$; and now

(i.) $\infty > s > e,$

$$u = \int_s^\infty ds/\sqrt{S} = \wp^{-1}(s\,;\ g_2, g_3),$$

or

$$u = \omega_1 - \int_{e_1}^s ds/\sqrt{S} = \omega_1 - \wp^{-1}\left(\frac{e_1-e_2 \,.\, e_1-e_3}{s-e_1} + e_1\right);\ \ldots\ldots\ldots(37)$$

which, employing the direct functions, expresses the relation

$$\wp(\omega_1 - u) - e_1 = \frac{e_1-e_2 \,.\, e_1-e_3}{\wp u - e_1} \ldots\ldots\ldots\ldots\ldots\ldots(38)$$

(ii.) $e_1 > s > e_2,$

$$u = \omega_1 + \int_s^{e_1} ds/\sqrt{S}$$

$$= \omega_1 + i\,\wp^{-1}\left(\frac{e_1-e_2 \,.\, e_1-e_3}{e_1-s} - e_1\,;\ g_2,\ -g_3\right);\ \ldots\ldots\ldots\ldots(39)$$

or

$$u = \omega_1 + \omega_3 - \int_{e_2}^s ds/\sqrt{S}$$

$$= \omega_1 + \omega_3 - i\,\wp^{-1}\left(\frac{e_1-e_2 \,.\, e_2-e_3}{s-e_2} - e_2\,;\ g_2,\ -g_3\right). \ \ldots\ldots\ldots(40)$$

(iii.) $e_2 > s > e_3,$

$$u = \omega_1 + \omega_3 + \int_s^{e_2} ds/\sqrt{S}$$

$$= \omega_1 + \omega_3 + \wp^{-1}\left(\frac{e_1-e_2 \,.\, e_2-e_3}{e_2-s} + e_2\,;\ g_2, g_3\right);\ \ldots\ldots\ldots\ldots(41)$$

or

$$u = 2\omega_1 + \omega_3 - \int_{e_3}^s ds/\sqrt{S}$$

$$= 2\omega_1 + \omega_3 - \wp^{-1}\left(\frac{e_1-e_3 \,.\, e_2-e_3}{s-e_3} + e_3\right)\ldots\ldots\ldots\ldots\ldots\ldots\ldots(42)$$

(iv.) $e_3 > s > -\infty\,;$

$$u = 2\omega_1 + \omega_3 + \int_s^{e_3} ds/\sqrt{S}$$

$$= 2\omega_1 + \omega_3 + i\,\wp^{-1}\left(\frac{e_1-e_3 \,.\, e_2-e_3}{e_3-s} + e_3\,;\ g_2,\ -g_3\right);\ \ldots\ldots(43)$$

or $$u = 2\omega_1 + 2\omega_3 - \int_{-\infty}^{s} ds/\sqrt{S}$$

$$= 2\omega_1 + 2\omega_3 - i\,\wp^{-1}(-s\,;\ g_2,\ -g_3). \quad\text{.....................(44)}$$

Thus $$\int_{-\infty}^{\infty} ds/\sqrt{S} = 2\omega_1 + 2\omega_3, \quad\text{.....................(45)}$$

and $2\omega_1$ is called the *real period*, and $2\omega_3$ the *imaginary period* of Weierstrass's elliptic function $\wp u$.

With Argand's geometrical representation of a complex quantity, such as $x+iy$, the complex quantity

$$u = t\omega_1 + t'\omega_3 \ (0 < t < 1,\ 0 < t' < 1)$$

represents all points lying inside a rectangle, called the *period parallelogram*.

As s or $\wp u$ diminishes continually from ∞ to $-\infty$, the argument u describes the contour of this rectangle; and for

$$u = \text{(i.)}\ t\omega_1 \quad (0<t<1),\ \text{(ii.)}\ \omega_1 + t'\omega_3\ (0<t'<1),$$
$$\text{(iii.)}\ t\omega_1 + \omega_3\ (1>t>0),\ \text{(iv.)}\quad t'\omega_3\ (1>t'>0),$$

the values of s or $\wp u$ are real, and lie in the intervals

(i.) $\infty > s > e_1$, (ii.) $e_1 > s > e_2$, (iii.) $e_2 > s > e_3$, (iv.) $e_3 > s > -\infty$;

while the corresponding values of $\wp' u$ are taken as

(i.) negative, (ii.) positive imaginary,
(iii.) positive, (iv.) negative imaginary.

For any point u inside the rectangle $\wp u$ assumes a complex value. (Schwarz, *Elliptische Functionen*, p. 74.)

55. In the same way, with the integral (11), denoting its value between the limits ∞ and z by u,

(i.) $\infty > z > 1/k$ (§ 39),

$$u = 2\,\text{sn}^{-1}\sqrt{\frac{1}{kz}} = 2K - 2\,\text{sn}^{-1}\sqrt{\frac{kz-1}{k\,.\,z-1}}. \quad\text{..................(46)}$$

(ii.) $1/k > z > 1$ (§ 40),

$$u = 2K + \qquad 2i\,\text{sn}^{-1}\left(\sqrt{\frac{1-kz}{1-k}},\ \kappa'\right)$$

$$= 2K + 2iK' - 2i\,\text{sn}^{-1}\left(\sqrt{\frac{z-1}{1-k\,.\,z}},\ \kappa'\right). \quad\text{................(47)}$$

(iii.) $1 > z > 0$ (§ 41),

$$u = 2K + 2iK' + 2\,\text{sn}^{-1}\sqrt{\frac{1-z}{1-kz}}$$

$$= 4K + 2iK' - 2\,\text{sn}^{-1}\sqrt{z}. \quad\text{..................................(48)}$$

(iv.) $0 > z > -\infty$ (§ 42),

$$u = 4K + 2iK + 2i\,\mathrm{cn}^{-1}\left(\sqrt{\frac{1}{1-z}},\ \kappa'\right)$$

$$= 4K + 4iK' - 2i\,\mathrm{sn}^{-1}\left(\sqrt{\frac{1}{1-kz}},\ \kappa'\right). \quad\text{.................(49)}$$

Therefore $\displaystyle\int_{-\infty}^{\infty}\frac{dz}{\sqrt{(z\,.\,1-z\,.\,1-kz)}} = 4K + 4iK'$, ...(50)

and $4K$ and $4iK'$ are called the real and imaginary periods of the corresponding elliptic function, in this case $\mathrm{sn}^2\tfrac{1}{2}u$.

56. But if we take Legendre's and Jacobi's fundamental integral $\int dx/\sqrt{X}$, where $X = 1 - x^2\,.\,1-\kappa^2 x^2$, and denote $\int^{\infty} dx/\sqrt{X}$ by u, then, by the preceding article, with x^2 for z,

(i.) $\infty > x > 1/\kappa$,

$$u = \mathrm{sn}^{-1}\frac{1}{\kappa x} = K - \mathrm{sn}^{-1}\sqrt{\frac{\kappa^2 x^2 - 1}{\kappa^2\,.\,x^2 - 1}}. \quad\text{.....................(51)}$$

(ii.) $1/\kappa > x > 1$,

$$u = K + \qquad i\,\mathrm{sn}^{-1}\left(\sqrt{\frac{1-\kappa^2 x^2}{\kappa'^2}},\ \kappa'\right)$$

$$= K + iK' - i\,\mathrm{sn}^{-1}\left(\sqrt{\frac{x^2-1}{\kappa'^2 x^2}},\ \kappa'\right). \quad\text{.....................(52)}$$

(iii.) $1 > x > -1$,

$$u = \ K + iK' + \mathrm{sn}^{-1}\sqrt{\frac{1-x^2}{1-\kappa^2 x^2}}$$

$$= 2K + iK' + \mathrm{sn}^{-1}x$$

$$= 3K + iK' - \mathrm{sn}^{-1}\sqrt{\frac{1-x^2}{1-\kappa^2 x^2}}. \quad\text{.........................(53)}$$

(iv.) $-1 > x > -1/\kappa$,

$$u = 3K + \ iK' + i\,\mathrm{sn}^{-1}\left(\sqrt{\frac{x^2-1}{\kappa'^2 x^2}},\ \kappa'\right)$$

$$= 3K + 2iK' - i\,\mathrm{sn}^{-1}\left(\sqrt{\frac{1-\kappa^2 x^2}{\kappa'^2}},\ \kappa'\right). \quad\text{.................(54)}$$

(v.) $-1/\kappa > x > -\infty$,

$$u = 3K + 2iK' + \mathrm{sn}^{-1}\sqrt{\frac{\kappa^2 x^2 - 1}{\kappa^2\,.\,x^2 - 1}}$$

$$= 4K + 2iK' - \mathrm{sn}^{-1}\frac{1}{\kappa x}. \quad\text{................................(55)}$$

Therefore $\displaystyle\int_{-\infty}^{\infty}(1-x^2\,.\,1-\kappa^2 x^2)^{-\frac{1}{2}}dx = 4K + 2iK'$; ....(56)

and $4K$ and $2iK'$ are called the periods of the elliptic function $\mathrm{sn}\,u$.

57. If, with $1 > x > -1$, and $X = 1 - x^2 . 1 - \kappa^2 x^2$, we denote the integral $\int_0^x dx/\surd X$ by u; then $\int_0^1 dx/\surd X = K$ (§ 11); and (§ 41)

$$K - u = \int_x^1 dx/\surd X = \text{sn}^{-1}\sqrt{\frac{1-x^2}{1-\kappa^2 x^2}};$$

or, employing the direct functions,

$$\text{sn}(K-u) = \sqrt{\frac{1-x^2}{1-\kappa^2 x^2}} = \frac{\text{cn}\,u}{\text{dn}\,u}, \quad \text{or} \quad \text{cd}\,u; \text{.........}(57)$$

and then (§ 17)

$$\text{cn}(K-u) = \sqrt{\frac{\kappa'^2 x^2}{1-\kappa^2 x^2}} = \frac{\kappa' \text{sn}\,u}{\text{dn}\,u}, \text{ or } \kappa' \text{sd}\,u; \text{.........}(58)$$

$$\text{dn}(K-u) = \sqrt{\frac{\kappa'^2}{1-\kappa^2 x^2}} = \frac{\kappa'}{\text{dn}\,u}, \quad \text{or } \kappa' \text{nd}\,u; \text{.........}(59)$$

relations analogous to equation (38); or to the relations

$$\sin(\tfrac{1}{2}\pi - \theta) = \cos\theta, \ \cos(\tfrac{1}{2}\pi - \theta) = \sin\theta,$$

of the circular functions of Trigonometry.

58. When the discriminant Δ of S is negative, and two of the roots of the equation $S = 0$ are imaginary, we take e_2 as the real root, and combine the product $s - e_1 . s - e_3$ into $(s-m)^2 + n^2$, as in § 45; and since

$$S = 4s^3 - g_2 s - g_3 = 4(s - e_2)\{(s-m)^2 + n^2\},$$

therefore $m = -\frac{1}{2}e_2$, $g_2 = 3e_2{}^2 - 4n^2$, $g_3 = e_2{}^3 + 4n^2 e_2$;

while $$H^2 = (e_2 - m)^2 + n^2 = \tfrac{9}{4}e_2{}^2 + n^2,$$

$$4\kappa^2\kappa'^2 = n^2/H^2 = 4n^2/(9e_2{}^2 + 4n^2),$$

$$1 - 16\kappa^2\kappa'^2 = 3g_2/(9e_2{}^2 + 4n^2),$$

$$\Delta = g_2{}^3 - 27g_3{}^2 = -4n^2/(9e^2 + 4n^2)^2,$$

so that $$J = \frac{g_2{}^3}{\Delta} = -\frac{(1 - 16\kappa^2\kappa'^2)^3}{108\kappa^2\kappa'^2} = -\frac{(1 - 16k + 16k^2)^3}{108k(1-k)}.$$

$$J - 1 = -\frac{(1-2k)^2(1 + 32k - 32k^2)^2}{108k(1-k)} \text{............}(D)$$

59. Now, as in § 45, by means of the quadric substitution,

$$\sigma - \epsilon_2 = \frac{\frac{1}{4}S}{(s-e_2)^2} = \frac{(s + \frac{1}{2}e_2)^2 + n^2}{s - e_2}, \text{..............}(60)$$

we find $$\frac{d\sigma}{ds}=\frac{(s-e_2)^2-H^2}{(s-e_2)^2}=\frac{(s-s_1)(s-s_3)}{(s-e_2)^2}, \text{ suppose };$$

while $$\sigma-\epsilon_1=\frac{(s-s_1)^2}{s-e_2},\ \sigma-\epsilon_3=\frac{(s-s_3)^2}{s-e_2},$$

provided $$s_1=e_2+H=\tfrac{1}{2}(\epsilon_1-\epsilon_2-e_2),$$
$$s_3=e_2-H=\tfrac{1}{2}(\epsilon_3-\epsilon_2-e_2).$$

Thence $s_1+s_3=2e_2=\frac{1}{2}(\epsilon_1+\epsilon_3)-\epsilon_2-e_2=-\frac{3}{2}\epsilon_2-e_2$;
or $e_2=-\frac{1}{2}\epsilon_2$; on the supposition that $\epsilon_1+\epsilon_2+\epsilon_3=0$;
and $$\epsilon_1=e_2+2H,\ \epsilon_2=-2e_2,\ \epsilon_3=e_2-2H.$$

Then $$\int_s^\infty\frac{ds}{\sqrt{S}}=\int\frac{(s-e_2)d\sigma}{2(s-s_1)(s-s_2)\sqrt{(\sigma-\epsilon_2)}}$$
$$=\int\frac{d\sigma}{2\sqrt{(\sigma-\epsilon_1\,.\,\sigma-\epsilon_2\,.\,\sigma-\epsilon_3)}}=\int_\sigma^\infty\frac{d\sigma}{\sqrt{\Sigma}}, \quad\text{...(61)}$$

where $\Sigma=4(\sigma-\epsilon_1)(\sigma-\epsilon_2)(\sigma-\epsilon_3)=4\sigma^3-\gamma_2\sigma-\gamma_3$,
suppose; and the discriminant Δ' of Σ is now positive.

60. Now, $\gamma_2=-4(\epsilon_2\epsilon_3+\epsilon_3\epsilon_1+\epsilon_1\epsilon_2)=12e_2^2+16H^2$,
$$\gamma_3=4\epsilon_1\epsilon_2\epsilon_3=32e_2H^2-8e_2^3,$$
$$\Delta'=\gamma_2^3-27\gamma_3^2=256H^2(4H^2-9e_2^2)^2.$$

Also with $$\lambda^2=\frac{\epsilon_2-\epsilon_3}{\epsilon_1-\epsilon_3}=\frac{2H-3e_2}{4H},\ \lambda'^2=\frac{\epsilon_1-\epsilon_2}{\epsilon_1-\epsilon_3}=\frac{2H+3e_2}{4H},$$

$$4\lambda^2\lambda'^2=\frac{4H^2-9e_2^2}{4H^2}=\frac{n^2}{H^2},\quad 1-\lambda^2\lambda'^2=\frac{3\gamma_2}{64H^2}.$$

Denoting by J' the *absolute invariant* of Σ, then (§ 53)
$$J'=\frac{\gamma_2^3}{\Delta'}=\frac{4}{27}\frac{(1-\lambda^2\lambda'^2)^3}{\lambda^4\lambda'^4}.$$

If we put $4\lambda^2\lambda'^2=1/\tau'$, then
$$J'=\frac{(4\tau'-1)^3}{27\tau'},\ J'-1=\frac{(\tau'-1)(8\tau'+1)^2}{27\tau'};$$

while, with $4\kappa^2\kappa'^2=\tau$ in (D),
$$J=\frac{(4\tau-1)^3}{27\tau},\quad J-1=\frac{(\tau-1)(8\tau+1)^2}{27\tau}. \quad\text{............(E)}$$

Now, if $2\kappa\kappa'=2\lambda\lambda'$, then $\tau\tau'=1$, the relation which holds in the transformation from a negative discriminant in S to a positive discriminant in Σ.

If we equate the values of J in (C) and (E), we find
$$\tau=-\frac{(1-k)^2}{4k},\ -\frac{k^2}{4(1-k)},\ \frac{1}{4k(1-k)}.$$

61. When Δ is negative, and when we know the real factor $s-e_2$ of S; so that, with $\frac{1}{4}e_2^2+n^2=\frac{1}{4}g_3/e_2$,

$$S=4(s-e_2)\{(s+\tfrac{1}{2}e_2)^2+n^2\};$$

then, with $H^2=\frac{1}{4}(9e_2^2+4n^2)$, and expressed as in § 46,

$$u=\int_s^\infty \frac{ds}{\surd S}=\frac{1}{2\surd H}\operatorname{cn}^{-1}\frac{s-e_2-H}{s-e_2+H}, \quad\text{.............(62)}$$

with $2\kappa\kappa'=n/H$; so that

$$\operatorname{cn}(2u\surd H)=\frac{\wp u-e_2-H}{\wp u-e_2+H}, \text{ or } \wp u=\frac{e_2+H-(e_2-H)\operatorname{cn}(2u\surd H)}{1-\operatorname{cn}(2u\surd H)}; \quad (63)$$

by means of which we change from Weierstrass's notation to Jacobi's and *vice versa*, when Δ is negative.

Thus, for example, if $g_2=0$, then $e_2=(\frac{1}{4}g_3)^{\frac{1}{3}}$, $n^2=\frac{3}{4}e_2^2$, $H^2=3e_2^2$; and, as in § 46,

$$\int_s^\infty \frac{ds}{\surd(4s^3-g_3)}=\wp^{-1}(s;\ 0,\ g_3)$$
$$=\frac{1}{2\sqrt[4]{3}(\frac{1}{4}g_3)^{\frac{1}{6}}}\operatorname{cn}^{-1}\left\{\frac{s-(\surd 3+1)(\frac{1}{4}g_3)^{\frac{1}{3}}}{s+(\surd 3-1)(\frac{1}{4}g_3)^{\frac{1}{3}}},\ \sin 15^\circ\right\},$$

$$\int_s^\infty \frac{ds}{\surd(4s^3+g_3)}=\wp^{-1}(s;\ 0,\ -g_3)$$
$$=\frac{1}{2\sqrt[4]{3}(\frac{1}{4}g_3)^{\frac{1}{6}}}\operatorname{cn}^{-1}\left\{\frac{s-(\surd 3-1)(\frac{1}{4}g_3)^{\frac{1}{3}}}{s+(\surd 3+1)(\frac{1}{4}g_3)^{\frac{1}{3}}},\ \sin 75^\circ\right\}.$$

62. Supposing s to range from ∞ to $-\infty$ in the integral $u=\int_s^\infty ds/\surd S$, when Δ is negative, then

(i.) $\infty > s > e_2$,

$$u=\wp^{-1}(s;\ g_2,\ g_3)$$
$$=\omega_2-\wp^{-1}\left(\frac{H^2}{s-e_2}+e_2\right), \quad\text{.........................(64)}$$

where ω_2 denotes $\int_{e_2}^\infty ds/\surd S$, the *real half period* of $\wp u$.

(ii.) $e_2 > s > -\infty$,

$$u=\omega_2+i\wp^{-1}\left(\frac{H^2}{e_2-s}-e_2;\ g_2,\ -g_3\right)$$
$$=\omega_2+\omega_2'-i\wp^{-1}(e_2-s;\ g_2,\ -g_3), \quad\text{.............(65)}$$

where ω_2' denotes $\int_{-\infty}^{e_2} ds/\surd S$, a pure imaginary quantity, called the *imaginary half period* of $\wp u$; and the period parallelogram (§ 55) is now bounded by ω_2 and ω_2', as adjacent sides.

Also (§ 47), $\omega_2=K/\surd H$, $\omega_2'=iK'/\surd H$. ....(66)

63. Treating in the same way the integral (2),

$$u=\int^{\infty}\frac{dx}{\sqrt{(1-x^2\,.\,\kappa'^2+\kappa^2x^2)}},$$

by replacing z by $1-x^2$ in §§ 38, 55;

(i.) $\infty > x > 1$,

$$u=i\,\mathrm{cn}^{-1}\left(\sqrt{\frac{\kappa^2\,.\,x^2-1}{\kappa'^2x^2+\kappa^2}},\ \kappa'\right)$$
$$=iK'-i\,\mathrm{cn}^{-1}(1/x,\ \kappa').\qquad(67)$$

(ii.) $1 > x > -1$,

$$u=iK'+\mathrm{cn}^{-1}x$$
$$=iK'+2K-\mathrm{cn}^{-1}(-x).\qquad(68)$$

(iii.) $-1 > \infty > -\infty$,

$$u=\ iK'+2K+i\,\mathrm{cn}^{-1}(-1/x,\ \kappa')$$
$$=2iK'+2K-i\,\mathrm{cn}^{-1}\left(\sqrt{\frac{\kappa'^2\,.\,x^2-1}{\kappa'^2x^2+\kappa^2}},\ \kappa'\right).\qquad(69)$$

64. By the substitution $x^2=1/y$, the integral

$$\int_0\frac{dx}{\sqrt{(A+Bx^2+Cx^4+Dx^6)}}=\int^{\infty}\frac{dy}{2\sqrt{(Ay^3+By^2+Cy+D)}}$$
$$=\frac{1}{\sqrt{A}}\int^{\infty}\frac{ds}{\sqrt{S}},\qquad(70)$$

on putting $y=s-\frac{1}{3}B/A$; which can be expressed by Weierstrass notation, or by the notation of Jacobi, when the factors of the denominator are known, as in equations (12) to (19);

$$\int\frac{E+Fx}{\sqrt{(A+Bx^2+Cx^4+Dx^6)}}dx$$

can thus be reduced to elliptic integrals, of the form considered in §§ 39-61, the first term by the substitution $x^2=1/y$, and the second term by the substitution $x^2=z$.

Thus $$\int^{a}\frac{a^3dr}{\sqrt{(a^6-r^6)}}=\frac{a}{2\sqrt[4]{3}}\,\mathrm{cn}^{-1}\left\{\frac{(\sqrt{3}+1)r^2-a^2}{(\sqrt{3}-1)r^2+a^2},\ \sin 15^\circ\right\},$$

the integration required in the rectification of $r^3=a^3\cos 3\theta$.

But by substituting $r^2/a^2=1/y$, we find

$$s=\int_0\frac{a^3dr}{\sqrt{(a^6-r^6)}}=\int^{\infty}\frac{ady}{\sqrt{(4y^3-4)}}=a\wp^{-1}(y;\ 0,\ 4);$$

so that $$\frac{a^2}{r^2}=\wp\left(\frac{s}{a};\ 0,\ 4\right).$$

65. Write X for $x^2-a^2 . x^2-b^2 . x^2-c^2$, where $a^2>b^2>c^2$; and write M for $b\sqrt{(a^2-c^2)}$; then we find, on substituting y for $1/x^2$, and taking α, β, γ for $1/c^2, 1/b^2, 1/a^2$;

(i.) $\infty > x^2 > a^2$, comparing with equation (18),

$$\int_a \frac{Mdx}{\sqrt{X}} = \text{sn}^{-1}\sqrt{\frac{a^{-2}-x^{-2}}{b^{-2}-x^{-2}}} = \text{sn}^{-1}\sqrt{\frac{b^2 . x^2-a^2}{a^2 . x^2-b^2}}$$

$$= \text{cn}^{-1}\sqrt{\frac{a^2-b^2 . x^2}{a^2 . x^2-b^2}} = \text{dn}^{-1}\sqrt{\frac{a^2-b^2 . x^2-c^2}{a^2-c^2 . x^2-b^2}}, \quad \text{...(71)}$$

to modulus $\sqrt{\dfrac{a^2 . b^2-c^2}{b^2 . a^2-c^2}}$.

(ii.) $a^2 > x^2 > b^2$, comparing with (17) and (16),

$$\int_x^a \frac{Mdx}{\sqrt{(-X)}} = \text{sn}^{-1}\sqrt{\frac{b^2 . a^2-x^2}{a^2-b^2 . x^2}}$$

$$= \text{cn}^{-1}\sqrt{\frac{a^2 . x^2-b^2}{a^2-b^2 . x^2}} = \text{dn}^{-1}\sqrt{\frac{a^2 . x^2-c^2}{a^2-c^2 . x^2}}, \quad \text{.........(72)}$$

$$\int_b^x \frac{Mdx}{\sqrt{(-X)}} = \text{sn}^{-1}\sqrt{\frac{a^2-c^2 . x^2-b^2}{a^2-b^2 . x^2-c^2}}$$

$$= \text{cn}^{-1}\sqrt{\frac{b^2-c^2 . a^2-x^2}{a^2-b^2 . x^2-c^2}} = \text{dn}^{-1}\sqrt{\frac{b^2-c^2 . x^2}{b^2 . x^2-c^2}}, \quad \text{....(73)}$$

to modulus $\sqrt{\dfrac{c^2 . a^2-b^2}{b^2 . a^2-c^2}}$.

(iii.) $b^2 > x^2 > c^2$, on comparison with (15) and (14),

$$\int_x^b \frac{Mdx}{\sqrt{(X)}} = \text{sn}^{-1}\sqrt{\frac{a^2-c^2 . b^2-x^2}{b^2-c^2 . a^2-x^2}}$$

$$= \text{cn}^{-1}\sqrt{\frac{a^2-b^2 . x^2-c^2}{b^2-c^2 . a^2-x^2}} = \text{dn}^{-1}\sqrt{\frac{a^2-b^2 . x^2}{b^2 . a^2-x^2}}, \quad \text{...(74)}$$

$$\int_c^x \frac{Mdx}{\sqrt{(X)}} = \text{sn}^{-1}\sqrt{\frac{b^2 . x^2-c^2}{b^2-c^2 . x^2}}$$

$$= \text{cn}^{-1}\sqrt{\frac{c^2 . b^2-x^2}{b^2-c^2 . x^2}} = \text{dn}^{-1}\sqrt{\frac{c^2 . a^2-x^2}{a^2-c^2 . x^2}}. \quad \text{.........(75)}$$

to modulus $\sqrt{\dfrac{a^2 . b^2-c^2}{b^2 . a^2-c^2}}$.

(iv.) $c^2 > x^2 > 0$, on comparison with (13) and (12),

$$\int_x^c \frac{Mdx}{\sqrt{(-X)}} = \text{sn}^{-1}\sqrt{\frac{b^2 . c^2-x^2}{c^2 . b^2-x^2}}$$

$$= \text{cn}^{-1}\sqrt{\frac{b^2-c^2 . x^2}{c^2 . b^2-x^2}} = \text{dn}^{-1}\sqrt{\frac{b^2-c^2 . a^2-x^2}{a^2-c^2 . b^2-x^2}}. \quad \text{...(76)}$$

$$\int_0^x \frac{Mdx}{\sqrt{(-X)}} = \text{sn}^{-1}\sqrt{\frac{a^2-c^2\,.\,x^2}{c^2\,.\,a^2-x^2}}$$

$$= \text{cn}^{-1}\sqrt{\frac{a^2\,.\,c^2-x^2}{c^2\,.\,a^2-x^2}} = \text{dn}^{-1}\sqrt{\frac{a^2\,.\,b^2-x^2}{b^2\,.\,a^2-x^2}} \quad \text{.........(77)}$$

to modulus $\sqrt{\dfrac{c^2\,.\,a^2-b^2}{b^2\,.\,a^2-c^2}}$.

66. When X is a quartic function of x, and we know a factor, $x-a$, of X, then the substitution $x-a=1/y$ reduces

$$\int dx/\sqrt{X} \text{ to the form } M\int dy/\sqrt{Y},$$

where Y is a cubic function of y; and this form can be treated by the preceding rules.

But, independently, if we can resolve X into four real linear factors, $\qquad x-\alpha,\ x-\beta,\ x-\gamma,\ x-\delta,$

so that $\qquad X = x-\alpha\,.\,x-\beta\,.\,x-\gamma\,.\,x-\delta,$

and we suppose that $\alpha>\beta>\gamma>\delta$; then with

(i.) $\infty > x > \alpha$,

$$\int_\alpha^x \frac{dx}{\sqrt{(x-\alpha\,.\,x-\beta\,.\,x-\gamma\,.\,x-\delta)}}$$

$$= \frac{2}{\sqrt{(\alpha-\gamma\,.\,\beta-\delta)}}\,\text{sn}^{-1}\sqrt{\frac{\beta-\delta\,.\,x-\alpha}{\alpha-\delta\,.\,x-\beta}}$$

$$= \frac{2}{\sqrt{(\alpha-\gamma\,.\,\beta-\delta)}}\,\text{cn}^{-1}\sqrt{\frac{\alpha-\beta\,.\,x-\delta}{\alpha-\delta\,.\,x-\beta}}$$

$$= \frac{2}{\sqrt{(\alpha-\gamma\,.\,\beta-\delta)}}\,\text{dn}^{-1}\sqrt{\frac{\alpha-\beta\,.\,x-\gamma}{\alpha-\gamma\,.\,x-\beta}}, \quad \text{...(78)}$$

indicating that we must put

$$\sin^2\phi = \frac{\beta-\delta\,.\,x-\alpha}{\alpha-\delta\,.\,x-\beta},\quad \cos^2\phi = \frac{\alpha-\beta\,.\,x-\delta}{\alpha-\delta\,.\,x-\beta},\quad \Delta^2\phi = \frac{\alpha-\beta\,.\,x-\gamma}{\alpha-\gamma\,.\,x-\beta},$$

to reduce the integral to the standard form (§ 4)

$$\frac{2}{\sqrt{(\alpha-\gamma\,.\,\beta-\delta)}}\int \frac{d\phi}{\sqrt{(1-k\sin^2\phi)}};$$

and then $\qquad \kappa^2 = k = \dfrac{\beta-\gamma\,.\,\alpha-\delta}{\alpha-\gamma\,.\,\beta-\delta},$

the *anharmonic ratio* of the four points A, B, C, D, the *poles* of the integral (§ 54), given by $x=\alpha, \beta\ \gamma, \delta$.

The verification by differentiation is a useful exercise for the student.

(ii.) With $\alpha > x > \beta$, we change the sign of X to make the integral real; and now, writing M for $\frac{1}{2}\surd(\alpha-\gamma.\beta-\delta)$ throughout,

$$\int_x^\alpha \frac{Mdx}{\surd(-X)}$$

$$=\text{sn}^{-1}\sqrt{\frac{\beta-\delta.\alpha-x}{\alpha-\beta.x-\delta}}=\text{cn}^{-1}\sqrt{\frac{\alpha-\delta.x-\beta}{\alpha-\beta.x-\delta}}=\text{dn}^{-1}\sqrt{\frac{\alpha-\delta.x-\gamma}{\alpha-\gamma.x-\delta}},\dots(79)$$

$$\int_\beta^x \frac{Mdx}{\surd(-X)}$$

$$=\text{sn}^{-1}\sqrt{\frac{\alpha-\gamma.x-\beta}{\alpha-\beta.x-\gamma}}=\text{cn}^{-1}\sqrt{\frac{\beta-\gamma.\alpha-x}{\alpha-\beta.x-\gamma}}=\text{dn}^{-1}\sqrt{\frac{\beta-\gamma.x-\delta}{\beta-\delta.x-\gamma}},\dots(80)$$

but now the modulus κ' is the complementary modulus to κ, so that

$$\kappa'^2=k'=\frac{\alpha-\beta.\gamma-\delta}{\alpha-\gamma.\beta-\delta};$$

the different forms of the result indicate the appropriate substitution required for reducing the integral to the Legendrian form.

(iii.) With $\beta > x > \gamma$, X is again positive, and

$$\int_x^\beta \frac{Mdx}{\surd X}$$

$$=\text{sn}^{-1}\sqrt{\frac{\alpha-\gamma.\beta-x}{\beta-\gamma.\alpha-x}}=\text{cn}^{-1}\sqrt{\frac{\alpha-\beta.x-\gamma}{\beta-\gamma.\alpha-x}}=\text{dn}^{-1}\sqrt{\frac{\alpha-\beta.x-\delta}{\beta-\delta.\alpha-x}},\dots(81)$$

$$\int_\gamma^x \frac{Mdx}{\surd X}$$

$$=\text{sn}^{-1}\sqrt{\frac{\beta-\delta.x-\gamma}{\beta-\gamma.x-\delta}}=\text{cn}^{-1}\sqrt{\frac{\gamma-\delta.\beta-x}{\beta-\gamma.x-\delta}}=\text{dn}^{-1}\sqrt{\frac{\gamma-\delta.\alpha-x}{\alpha-\gamma.x-\delta}},\dots(82)$$

with the same modulus κ as in (78).

(iv.) With $\gamma > x > \delta$, X is negative, and

$$\int_x^\gamma \frac{Mdx}{\surd(-X)}$$

$$=\text{sn}^{-1}\sqrt{\frac{\beta-\delta.\gamma-x}{\gamma-\delta.\beta-x}}=\text{cn}^{-1}\sqrt{\frac{\beta-\gamma.x-\delta}{\gamma-\delta.\beta-x}}=\text{dn}^{-1}\sqrt{\frac{\beta-\gamma.\alpha-x}{\alpha-\gamma.\beta-x}},\dots(83)$$

$$\int_\delta^x \frac{dx}{\surd(-X)}$$

$$=\text{sn}^{-1}\sqrt{\frac{\alpha-\gamma.x-\delta}{\gamma-\delta.\alpha-x}}=\text{cn}^{-1}\sqrt{\frac{\alpha-\delta.\gamma-x}{\gamma-\delta.\alpha-x}}=\text{dn}^{-1}\sqrt{\frac{\alpha-\delta.\beta-x}{\gamma-\delta.\alpha-x}},\dots(84)$$

with the modulus of (79) and (80).

(v.) With $\delta > x > -\infty$, X is positive, and

$$\int_x^\delta \frac{Mdx}{\sqrt{X}}$$

$$=\mathrm{sn}^{-1}\sqrt{\frac{\alpha-\gamma\,.\,\delta-x}{\alpha-\delta\,.\,\gamma-x}}=\mathrm{cn}^{-1}\sqrt{\frac{\gamma-\delta\,.\,\alpha-x}{\alpha-\delta\,.\,\gamma-x}}=\mathrm{dn}^{-1}\sqrt{\frac{\gamma-\delta\,.\,\beta-x}{\beta-\delta\,.\,\gamma-x}},\ldots.(85)$$

with the original modulus of (78), (81), and (82).

67. *Landen's Transformation.*

When Legendre's and Jacobi's standard integral (1) is treated as a particular case of these integrals (81) and (82), we write $\alpha=1/\lambda$, $\beta=1$, $\gamma=-1$, $\delta=-1/\lambda$, so that $M=\frac{1}{2}(1+\lambda)/\lambda$; and now, with y for variable,

$$\int_y^1 \frac{\frac{1}{2}(1+\lambda)dy}{\sqrt{(1-y^2\,.\,1-\lambda^2y^2)}}$$

$$=\mathrm{sn}^{-1}\sqrt{\frac{1+\lambda\,.\,1-y}{2\,.\,1-\lambda y}}=\mathrm{cn}^{-1}\sqrt{\frac{1-\lambda\,.\,1+y}{2\,.\,1-\lambda y}}=\mathrm{dn}^{-1}\sqrt{\frac{1-\lambda\,.\,1+\lambda y}{1+\lambda\,.\,1-\lambda y}},..(86)$$

$$\int_{-1}^y \frac{\frac{1}{2}(1+\lambda)dy}{\sqrt{(1-y^2\,.\,1-\lambda^2y^2)}}$$

$$=\mathrm{sn}^{-1}\sqrt{\frac{1+\lambda\,.\,1+y}{2\,.\,1+\lambda y}}=\mathrm{cn}^{-1}\sqrt{\frac{1-\lambda\,.\,1-y}{2\,.\,1+\lambda y}}=\mathrm{dn}^{-1}\sqrt{\frac{1-\lambda\,.\,1-\lambda y}{1+\lambda\,.\,1+\lambda y}},..(87)$$

where the modulus κ is now given by $\kappa^2=4\lambda/(1+\lambda)^2$, so that

$$\kappa=2\sqrt{\lambda}/(1+\lambda),\ \kappa'=(1-\lambda)/(1+\lambda),\ \text{or}\ (1+\kappa')(1+\lambda)=2;$$

and we are thus introduced to *Landen's transformation*, to be discussed hereafter.

Changing, in § 41, x into y^2, and k into λ^2, we find

$$\int_y^1 \frac{dy}{\sqrt{(1-y^2\,.\,1-\lambda^2y^2)}}$$

$$=\mathrm{sn}^{-1}\sqrt{\frac{1-y^2}{1-\lambda^2y^2}}=\mathrm{cn}^{-1}\sqrt{\frac{1-\lambda^2\,.\,y^2}{1-\lambda^2y^2}}=\mathrm{dn}^{-1}\sqrt{\frac{1-\lambda^2}{1-\lambda^2y^2}},\ldots\ldots(88)$$

with modulus λ; indicating, on comparison with (86), results such as

$$\tfrac{1}{2}(1+\lambda)\,\mathrm{sn}^{-1}\left(\sqrt{\frac{1-y^2}{1-\lambda^2y^2}},\ \lambda\right)=\mathrm{sn}^{-1}\left(\sqrt{\frac{1+\lambda\,.\,1-y}{2\,.\,1-\lambda y}},\ \frac{2\sqrt{\lambda}}{1+\lambda}\right),$$

$$\tfrac{1}{2}(1+\lambda)\,\mathrm{cn}^{-1}\left(\sqrt{\frac{1-\lambda^2\,.\,y^2}{1-\lambda^2y^2}},\ \lambda\right)=\mathrm{cn}^{-1}\left(\sqrt{\frac{1-\lambda\,.\,1+y}{2\,.\,1-\lambda y}},\ \frac{2\sqrt{\lambda}}{1+\lambda}\right),$$

$$\tfrac{1}{2}(1+\lambda)\,\mathrm{dn}^{-1}\left(\sqrt{\frac{1-\lambda^2}{1-\lambda^2y^2}},\ \lambda\right)=\mathrm{dn}^{-1}\left(\sqrt{\frac{1-\lambda\,.\,1+\lambda y}{1+\lambda\,.\,1-\lambda y}},\ \frac{2\sqrt{\lambda}}{1+\lambda}\right),(89)$$

which can be translated into the various forms of Landen's *quadric transformation.*

Denoting integrals (86) and (88) by u and v, then

$$u=\tfrac{1}{2}(1+\lambda)v,\ v=(1+\kappa')u;$$

$$\mathrm{sn}^2(u,\kappa)=\frac{1+\lambda\,.\,1-y}{2\,.\,1-\lambda y},$$

$$\mathrm{cn}^2(u,\kappa)=\frac{1-\lambda\,.\,1+y}{2\,.\,1-\lambda y},\ \mathrm{dn}^2(u,\kappa)=\frac{1-\lambda\,.\,1+\lambda y}{1+\lambda\,.\,1-\lambda y},\ \text{.........}(90)$$

$$\mathrm{sn}^2(v,\lambda)=\frac{1-y^2}{1-\lambda^2y^2},$$

$$\mathrm{cn}^2(v,\lambda)=\frac{1-\lambda^2\,.\,y^2}{1-\lambda^2y^2},\quad \mathrm{dn}^2(v,\lambda)=\frac{1-\lambda^2}{1-\lambda^2y^2},\ \text{..............}(91)$$

whence $$\mathrm{sn}(v,\lambda)=\frac{(1+\kappa')\mathrm{sn}(u,\kappa)\mathrm{cn}(u,\kappa)}{\mathrm{dn}(u,\kappa)},\ \text{etc............}(92)$$

We can easily prove, or verify by differentiation, that

$$\int_0^y\frac{\frac{1}{2}(1+\lambda)dy}{\sqrt{(1-y^2\,.\,1-\lambda^2y^2)}}$$

$$=\mathrm{sn}^{-1}\{\tfrac{1}{2}\sqrt{(1+y\,.\,1+\lambda y)}-\tfrac{1}{2}\sqrt{(1-y\,.\,1-\lambda y)}\}$$

$$=\mathrm{cn}^{-1}\{\tfrac{1}{2}\sqrt{(1+y\,.\,1-\lambda y)}+\tfrac{1}{2}\sqrt{(1-y\,.\,1+\lambda y)}\}$$

$$=\mathrm{dn}^{-1}\frac{\sqrt{(1-\lambda^2y^2)}+\lambda\sqrt{(1-y^2)}}{1+\lambda}=\mathrm{dn}^{-1}\frac{1-\lambda}{\sqrt{(1-\lambda^2y^2)}-\lambda\sqrt{(1-y^2)}},\ (93)$$

to the same modulus $\kappa=2\sqrt{\lambda}/(1+\lambda)$; so that, denoting this integral by u, and denoting $\mathrm{sn}(u,\kappa)$ by x, then

$$x=\tfrac{1}{2}\sqrt{(1+y\,.\,1+\lambda y)}-\tfrac{1}{2}\sqrt{(1-y\,.\,1-\lambda y)},$$

$$\sqrt{(1-x^2)}=\tfrac{1}{2}\sqrt{(1+y\,.\,1-\lambda y)}+\tfrac{1}{2}\sqrt{(1-y\,.\,1+\lambda y)},$$

$$\sqrt{(1-\kappa^2x^2)}=\frac{\sqrt{(1-\lambda^2y^2)}+\lambda\sqrt{(1-y^2)}}{1+\lambda}=\frac{1-\lambda}{\sqrt{(1-\lambda^2y^2)}-\lambda\sqrt{(1-y^2)}},\ \text{..}(94)$$

or $$\mathrm{dn}(u,\kappa)=\frac{\mathrm{dn}(v,\lambda)+\lambda\,\mathrm{cn}(v,\lambda)}{1+\lambda},\ \mathrm{nd}(u,\kappa)=\frac{\mathrm{dn}(v,\lambda)-\lambda\,\mathrm{cn}(v,\lambda)}{1-\lambda},\ (95)$$

since $y=\mathrm{sn}(v,\lambda)$, where $v=(1+\kappa')u$;

and thence

$$\mathrm{dn}(v,\lambda)=\tfrac{1}{2}(1+\lambda)\mathrm{dn}(u,\kappa)+\tfrac{1}{2}(1-\lambda)\mathrm{nd}(u,\kappa),\ \text{..............}(96)$$

$$\lambda\,\mathrm{cn}(v,\lambda)=\tfrac{1}{2}(1+\lambda)\mathrm{dn}(u,\kappa)-\tfrac{1}{2}(1-\lambda)\mathrm{nd}(u,\kappa);\ \text{..............}(97)$$

(Cayley, *Elliptic Functions*, p. 183).

The relation (92) between x and y, namely,

$$y=\frac{(1+\kappa')x\sqrt{(1-x^2)}}{\sqrt{(1-\kappa^2x^2)}},\ \text{..................}(92)^*$$

thus leads to the differential relation

$$\frac{\frac{1}{2}(1+\lambda)dy}{\sqrt{(1-y^2\,.\,1-\lambda^2y^2)}}=\frac{dx}{\sqrt{(1-x^2\,.\,1-\kappa^2x^2)}}.\ \text{.........}(98)$$

68. The six anharmonic ratios of $\alpha, \beta, \gamma, \delta$, arising by permutation or substitution, give rise to six values of the modulus k, given by

$$k, \frac{1}{k}, 1-k, \frac{1}{1-k}, 1-\frac{1}{k}, \frac{k}{k-1}, \text{..................}(99)$$

or $\sin^2\theta$, $\text{cec}^2\theta$, $\cos^2\theta$, $\sec^2\theta$, $-\cot^2\theta$, $-\tan^2\theta$, if $k=\sin^2\theta$;
or $\tanh^2 u$, $\coth^2 u$, $\text{sech}^2 u$, $\cosh^2 u$, $-\text{cech}^2 u$, $-\sinh^2 u$, if $k=\tanh^2 u$.

We may notice that the expression for J in (D) of § 53 is unaltered if for k we substitute any of these other five values; and, on comparison with Weierstrass's notation,

$$J=g_2{}^3/\Delta, \quad J-1=27g_3{}^2/\Delta,$$

so that we may put

$$g_2=\frac{1-k+k^2}{12}, \; g_3=\frac{(1+k)(1-2k)(2-k)}{432}, \; \Delta=\frac{k^2(1-k)^2}{256}; \text{...}(100)$$

and then $e_1=\frac{1}{12}(2-k)$, $e_2=\frac{1}{12}(-1+2k)$, $e_3=\frac{1}{12}(-1-k)$;
so that $k=(e_2-e_3)/(e_1-e_3)$, as in § 51.

69. *Degenerate Forms of the Elliptic Integral.*

When two of the roots $\alpha, \beta, \gamma, \delta$ become equal, the corresponding integrals degenerate into circular and hyperbolic integrals, which can easily be written down, on noticing as before (§ 48) that (i.) when $k=0$, $\text{sn}^{-1}x$ becomes $\sin^{-1}x$, $\text{cn}^{-1}x$ becomes $\cos^{-1}x$, etc; (ii.) when $k=1$, $\text{sn}^{-1}x$ becomes $\tanh^{-1}x$, $\text{cn}^{-1}x$ or $\text{dn}^{-1}x$ becomes $\text{sech}^{-1}x$, and $\text{tn}^{-1}x$ becomes $\sinh^{-1}x$.

When two of them are equal, we may replace the four quantities $\alpha, \beta, \gamma, \delta$ by the three distinct quantities a, b, c, suppose, where $a>b>c$; and now the degenerate elliptic integrals fall into three classes, I., II., III.

I. Writing M for $\frac{1}{2}\sqrt{(a-b\,.\,a-c)}$; then

(i.) $\infty>x>a$,

$$\int_x^\infty \frac{M dx}{(x-a)\sqrt{(x-b\,.\,x-c)}}=\sinh^{-1}\sqrt{\frac{a-b\,.\,x-c}{b-c\,.\,x-a}}=\cosh^{-1}\sqrt{\frac{a-c\,.\,x-b}{b-c\,.\,x-a}}.$$

(ii.) $a>x>b$,

$$\int_b^x \frac{M dx}{(a-x)\sqrt{(x-b\,.\,x-c)}}=\cosh^{-1}\sqrt{\frac{a-b\,.\,x-c}{b-c\,.\,a-x}}=\sinh^{-1}\sqrt{\frac{a-c\,.\,x-b}{b-c\,.\,a-x}}.$$

(iii.) $b>x>c$,

$$\int_x^b \frac{M dx}{(a-x)\sqrt{(b-x\,.\,x-c)}}=\cos^{-1}\sqrt{\frac{a-b\,.\,x-c}{b-c\,.\,a-x}}=\sin^{-1}\sqrt{\frac{a-c\,.\,b-x}{b-c\,.\,a-x}}.$$

$$\int_c^x \frac{Mdx}{(a-x)\surd(b-x.x-c)} = \sin^{-1}\sqrt{\frac{a-b.x-c}{b-c.a-x}} = \cos^{-1}\sqrt{\frac{a-c.b-x}{b-c.a-x}}.$$

(iv.) $c > x > -\infty$,

$$\int_x^c \frac{Mdx}{(a-x)\surd(b-x.c-x)} = \sinh^{-1}\sqrt{\frac{a-b.c-x}{b-c.a-x}} = \cosh^{-1}\sqrt{\frac{a-c.b-x}{b-c.a-x}}.$$

II. Writing M for $\frac{1}{2}\surd(a-b.b-c)$; then

(i.) $\infty > x > a$,

$$\int_a^x \frac{Mdx}{(x-b)\surd(x-a.x-c)} = \sin^{-1}\sqrt{\frac{b-c.x-a}{a-c.x-b}} = \cos^{-1}\sqrt{\frac{a-b.x-c}{a-c.x-b}}.$$

(ii.) $a > x > b$,

$$\int_x^a \frac{Mdx}{(x-b)\surd(a-x.x-c)} = \sinh^{-1}\sqrt{\frac{b-c.a-x}{a-c.x-b}} = \cosh^{-1}\sqrt{\frac{a-b.x-c}{a-c.x-b}}.$$

(iii.) $b > x > c$,

$$\int_c^x \frac{Mdx}{(b-x)\surd(a-x.x-c)} = \cosh^{-1}\sqrt{\frac{b-c.a-x}{a-c.b-x}} = \sinh^{-1}\sqrt{\frac{a-b.x-c}{a-c.b-x}}.$$

(iv.) $c > x > -\infty$,

$$\int_x^c \frac{Mdx}{(b-x)\surd(a-x.c-x)} = \cos^{-1}\sqrt{\frac{b-c.a-x}{a-c.b-x}} = \sin^{-1}\sqrt{\frac{a-b.c-x}{a-c.b-x}}.$$

III. Writing M for $\frac{1}{2}\surd(a-c.b-c)$; then

(i.) $\infty > x > a$,

$$\int_a^x \frac{Mdx}{(x-c)\surd(x-a.x-b)} = \cosh^{-1}\sqrt{\frac{a-c.x-b}{a-b.x-c}} = \sinh^{-1}\sqrt{\frac{b-c.x-a}{a-b.x-c}}.$$

(ii.) $a > x > b$,

$$\int_x^a \frac{Mdx}{(x-c)\surd(a-x.x-b)} = \cos^{-1}\sqrt{\frac{a-c.x-b}{a-b.x-c}} = \sin^{-1}\sqrt{\frac{b-c.a-x}{a-b.x-c}};$$

$$\int_b^x \frac{Mdx}{(x-c)\surd(a-x.x-b)} = \sin^{-1}\sqrt{\frac{a-c.x-b}{a-b.x-c}} = \cos^{-1}\sqrt{\frac{b-c.a-x}{a-b.x-c}}.$$

(iii.) $b > x > c$,

$$\int_x^b \frac{Mdx}{(x-c)\surd(a-x.b-x)} = \sinh^{-1}\sqrt{\frac{a-c.b-x}{a-b.x-c}} = \cosh^{-1}\sqrt{\frac{b-c.a-x}{a-b.x-c}}.$$

(iv.) $c > x > -\infty$,

$$\int \frac{Mdx}{(c-x)\surd(a-x.b-x)} = \cosh^{-1}\sqrt{\frac{a-c.b-x}{a-b.c-x}} = \sinh^{-1}\sqrt{\frac{b-c.a-x}{a-b.c-x}}.$$

70. When all four roots of the quartic $X=0$ are imaginary, so that

$$(x-\alpha)(x-\beta)=(x-m)^2+n^2,\ (x-\gamma)(x-\delta)=(x-p)^2+q^2,$$

$$\int dx/\surd X=\int\{(x-m)^2+n^2\,.\,(x-p)^2+q^2\}^{-\frac{1}{2}}dx$$

is reduced by the substitution

$$y=\frac{(x-m)^2+n^2}{(x-p)^2+q^2}. \qquad (101)$$

Let us suppose that X is resolved into two quadratic factors, so that X is of the form

$$X=(ax^2+2bx+c)(Ax^2+2Bx+C),$$

where, by supposition, $ac-b^2$ and $AC-B^2$ are negative, so that the roots of $X=0$ are all imaginary.

Let
$$y=\frac{ax^2+2bx+c}{Ax^2+2Bx+C}=\frac{N}{D}, \text{ suppose}, \qquad (101)^*$$

then the maximum and minimum of y, the *turning points* of y, being denoted by y_1 and y_2,

$$y_1-y=(Ay_1-a)(x_1-x)^2/D,\ y-y_2=(a-Ay_2)(x-x_2)^2/D, \qquad (102)$$

x_1 and x_2 denoting the values of x corresponding to y_1 and y_2 of y; and now

$$\frac{dy}{dx}=\frac{2(Ab-aB)(x_1-x)(x-x_2)}{(Ax^2+2Bx+C)^2}. \qquad (103)$$

For x is given in terms of y by the solution of

$$(Ay-a)x^2+2(By-b)x+Cy-c=0, \qquad (104)$$

and this equation has equal roots at the turning points of y, which are therefore given by the quadratic equation

$$(Ay-a)(Cy-c)-(By-b)^2=0,$$

or
$$(AC-B^2)y^2-(Ac+aC-2Bb)y+ac-b^2=0, \qquad (105)$$

and then

$$x=-\frac{By-b}{Ay-a}, \text{ or } y=\frac{ax+b}{Ax+B}=\frac{bx+c}{Bx+C}.$$

Now
$$\int\frac{dx}{\surd X}=\int\frac{dx}{\surd(ND)}=\int\frac{dx}{D\surd y}$$

$$=\int\frac{Ddy}{2(Ab-aB)(x_1-x)(x-x_2)\surd y}$$

$$=\frac{\surd(Ay_1-a\,.\,a-Ay_2)}{2(Ab-aB)}\int\frac{dy}{\surd(y\cdot y_1-y\cdot y-y_2)};$$

and
$$(Ay_1-a)(a-Ay_2)=-A^2y_1y_2+Aa(y_1+y_2)-a^2$$

$$=\frac{(Ab-aB)^2}{AC-B^2},$$

so that $\displaystyle\int\frac{dx}{\surd X}=\frac{1}{\surd(AC-B^2)}\int\frac{dy}{\surd(4y\,.\,y_1-y\,.\,y-y_2)}$, ...(106)

which, by (15), gives $\displaystyle\int^{x_1}\frac{\surd(AC-B^2)dx}{\surd X}=$

$$\frac{1}{\surd y_1}\operatorname{sn}^{-1}\sqrt{\frac{y_1-y}{y_1-y_2}}=\frac{1}{\surd y_1}\operatorname{cn}^{-1}\sqrt{\frac{y-y_2}{y_1-y_2}}=\frac{1}{\surd y_1}\operatorname{dn}^{-1}\sqrt{\frac{y}{y_1}},\quad(107)$$

with $\kappa^2=1-y_2/y_1,\ \kappa'^2=y_2/y_1$;

the last expression, by the inverse dn function, being the simplest, as expressing a function of an argument oscillating between two positive limits, y_1 and y_2.

71. For example, if

$$X=x^4+2a^2x^2\cos 2\alpha+a^4$$
$$=(x^2+2ax\sin\alpha+a^2)(x^2-2ax\sin\alpha+a^2),$$

and if $y=(x^2+2ax\sin\alpha+a^2)/(x^2-2ax\sin\alpha+a^2)$,

then $x_1=a,\ y_1=\tan^2(\frac{1}{4}\pi+\frac{1}{2}\alpha)$; $x_2=-a,\ y_2=\tan^2(\frac{1}{4}\pi-\frac{1}{2}\alpha)$;

so that $\kappa'=\tan^2(\frac{1}{4}\pi-\frac{1}{2}\alpha)=(1-\sin\alpha)/(1+\sin\alpha)$;

and $\displaystyle\int^a\frac{dx}{\surd(x^4+2a^2x^2\cos 2\alpha+a^4)}$

$$=\frac{1}{a^2(1+\sin\alpha)}\operatorname{dn}^{-1}\sqrt{\frac{1-\sin\alpha\,.\,x^2+2ax\sin\alpha+a^2}{1+\sin\alpha\,.\,x^2-2ax\sin\alpha+a^2}}.\ \text{......}(108)$$

But, by substituting $\displaystyle\frac{x^2}{a^2}=\frac{1+z}{1-z}$,

$$\int^\infty\frac{dx}{\surd(x^4+2a^2x^2\cos 2\alpha+a^4)}=\frac{1}{2a}\int^1\frac{dz}{\surd(1-z^2\,.\,\cos^2\alpha+z^2\sin^2\alpha)}$$

$$=\frac{1}{2a}\operatorname{cn}^{-1}(z;\ \sin\alpha)=\frac{1}{2a}\operatorname{cn}^{-1}\frac{x^2-a^2}{x^2+a^2},\ \text{......}(109)$$

by (2), a reduction of the elliptic integral to a different modulus, the modular angle being now α; affording another illustration of Landen's transformation of § 67.

Thus, with $\alpha=\frac{1}{4}\pi$, equation (108) gives

$$\int^1\frac{dx}{\surd(1+x^4)}=(2-\surd 2)\operatorname{dn}^{-1}\left\{(\surd 2-1)\sqrt{\frac{1+\surd 2\,x+x^2}{1-\surd 2\,x+x^2}}\right\}:$$

where $\kappa'=(\surd 2-1)^2$ (when $K'/K=\frac{1}{2}$); and by (109),

$$\int^\infty\frac{dx}{\surd(x^4+1)}=\tfrac{1}{2}\operatorname{cn}^{-1}\left(\frac{x^2-1}{x^2+1},\ \tfrac{1}{2}\surd 2\right),$$

$$\int^1\frac{dx}{\surd(1+x^4)}=\tfrac{1}{2}\operatorname{cn}^{-1}\frac{\surd 2x}{\surd(1+x^4)},\quad\int_0\frac{dx}{\surd(1+x^4)}=\tfrac{1}{2}\operatorname{cn}^{-1}\frac{1-x^2}{1+x^2},\ \text{etc.}$$

For other numerical examples, the student may take

$X=x^4+2x^2+2,\ x^4+3x^2+3,\ x^4+x^2+1,\ x^4+2x^2+3$, etc.

72. When two roots only of the quartic $X=0$ are imaginary, we may still make use of the substitution (§ 70)

$$y=N/D, \text{ where } X=ND;$$

but now take $ac-b^2$ negative, and $AC-B^2$ positive.

Proceeding as before we find that the maximum y_1 is positive, but the minimum y_3 is negative; and y oscillates between 0 and y_1 for real values of $\sqrt{X}$; and

$$\int\frac{dx}{\sqrt{X}}=\frac{1}{\sqrt{(AC-B^2)}}\int\frac{dy}{\sqrt{(4y\,.\,y_1-y\,.\,y-y_3)}},$$

so that, by (14),

$$\int\frac{\sqrt{(AC-B^2)}}{\sqrt{X}}dx=\frac{1}{\sqrt{(y_1-y_3)}}\text{sn}^{-1}\sqrt{\frac{y_1-y}{y_1}}$$

$$=\frac{1}{\sqrt{(y_1-y_3)}}\text{cn}^{-1}\sqrt{\frac{y}{y_1}}=\frac{1}{\sqrt{(y_1-y_3)}}\text{dn}^{-1}\sqrt{\frac{y-y_3}{y_1-y_3}}, \quad \text{...(110)}$$

with $\kappa^2=y_1/(y_1-y_3),\ \kappa'^2=-y_3/(y_1-y_3)$.

73. By another method of reduction we shall find

(Enneper, *Elliptische Functionen*, p. 23)

$$\int_\alpha^x\frac{dx}{\sqrt{\{x-\alpha\,.\,x-\beta\,.\,(x-m)^2+n^2\}}}$$

$$=\frac{1}{\sqrt{(HK)}}\text{cn}^{-1}\left\{\frac{H(x-\beta)-K(x-\alpha)}{H(x-\beta)+K(x-\alpha)},\ \kappa\right\}, \quad \text{......(111)}$$

$$\int_\beta^x\frac{dx}{\sqrt{\{\alpha-x\,.\,x-\beta\,.\,(x-m)^2+n^2\}}}$$

$$=\frac{1}{\sqrt{(HK)}}\text{cn}^{-1}\left\{\frac{K(\alpha-x)-H(x-\beta)}{K(\alpha-x)+H(x-\beta)},\ \kappa'\right\}, \quad \text{......(112)}$$

etc.; where $H^2=(\alpha-m)^2+n^2,\ K^2=(\beta-m)^2+n^2$;

and $\kappa^2=\frac{1}{2}-\frac{1}{4}\{(\alpha-\beta)^2-H^2-K^2\}/HK,$

$\kappa'^2=\frac{1}{2}+\frac{1}{4}\{(\alpha-\beta)^2-H^2-K^2\}/HK$;

so that $2\kappa\kappa'=n(\alpha-\beta)/HK.$

Degenerate forms occur when α and β are equal; and now

$$\int^\infty\frac{dx}{(x-\alpha)\sqrt{\{(x-m)^2+n^2\}}}$$

$$=\frac{1}{\sqrt{\{(\alpha-m)^2+n^2\}}}\cosh^{-1}\frac{\sqrt{\{(\alpha-m)^2+n^2\}}\sqrt{\{(x-m)^2+n^2\}}}{n(x-\alpha)},$$

$$\int_{-\infty}\frac{dx}{(\alpha-x)\sqrt{\{(x-m)^2+n^2\}}}$$

$$=\frac{1}{\sqrt{\{(\alpha-m)^2+n^2\}}}\cosh^{-1}\frac{\sqrt{\{(\alpha-m)^2+n^2\}}\sqrt{\{(x-m)^2+n^2\}}}{n(\alpha-x)}.$$

74. Replacing y by N/D in equations (102), then

$$Dy_1 - N = (Ay_1 - a)(x_1 - x)^2,$$
$$N - Dy_2 = (a - Ay_2)(x - x_2)^2;$$

so that we may write, according to Mr. R. Russell,

$$D = Ax^2 + 2Bx + C = P(x_1 - x)^2 + Q(x - x_2)^2,$$
$$N = ax^2 + 2bx + c = p(x_1 - x)^2 + q(x - x_2)^2; \quad\ldots\ldots\ldots(113)$$

where $P = (Ay_1 - a)/(y_1 - y_2)$, $Q = (a - Ay_2)/(y_1 - y_2)$;
and $p = Py_2$, $q = Qy_1$.

Interesting numerical examples can be constructed by giving arbitrary integral values to x_1, x_2, P, Q, p, q; and now the substitution

$$z = \frac{x - x_2}{x_1 - x},$$

will make, as in § 37,

$$\int \frac{dx}{\sqrt{X}} = \int \frac{(x_1 - x_2)dz}{\sqrt{(p + qz^2 \,.\, P + Qz^2)}}. \quad\ldots\ldots\ldots\ldots(114)$$

75. When the factors of the quartic X are unknown, we employ Weierstrass's function, and we shall show subsequently in Chap. IV. that the elliptic integral $\int dx/\sqrt{X}$ is reduced to Weierstrass's canonical form $\frac{1}{2}\int ds/\sqrt{S}$ (§ 50) by the substitution

$$s = -H/X,$$

H denoting the *Hessian* of the quartic X (Cayley, *Elliptic Functions*, p. 346); we may thus write

$$\int \frac{dx}{\sqrt{X}} = \tfrac{1}{2}\wp^{-1}\left(-\frac{H}{X};\ g_2, g_3\right), \quad\ldots\ldots\ldots\ldots(115)$$

where g_2, g_3 are the *quadrinvariant* and *cubinvariant* of the quartic X or $ax^4 + 4bx^3 + 6cx^2 + 4dx + e$,

so that

$$g_2 = ae - 4bd + 3c^2,$$
$$g_3 = ace + 2bcd - ad^2 - eb^2 - c^3,$$
$$H = (ac - b^2)x^4 + 2(ad - bc)x^3 + (ae + 2bd - 3c^2)x^2 + 2(be - cd)x + ce - d^2;$$

and the general reduction of the elliptic integral of the first kind $\int dx/\sqrt{X}$, where X is a cubic or quartic function of x, is now complete.

The application of this general method to the particular cases already discussed is left as an exercise for the student.

76. Systematic Tables of the integrals of the elliptic functions sn u, cn u, dn u, ns u, ds u, cs u, dc u, nc u, sc u, cd u, sd u, nd u, and of their powers have been given by Glaisher (*Messenger of Mathematics*, 1881).

Suppose $\int_0 \text{cn}\,u du$ is required; we may write it

$$\int_0 \frac{\text{cn}\,u\,\text{dn}\,u du}{\text{dn}\,u} = \int \frac{d\,\text{sn}\,u}{\sqrt{(1-\kappa^2\text{sn}^2 u)}} = \frac{1}{\kappa}\sin^{-1}(\kappa\,\text{sn}\,u) = \frac{1}{\kappa}\cos^{-1}(\text{dn}\,u),$$

etc.; so that

$$\int_0 \kappa\,\text{cn}\,u du = \cos^{-1}(\text{dn}\,u) = \sin^{-1}(\kappa\,\text{sn}\,u) = \tan^{-1}(\kappa\,\text{sn}\,u/\text{dn}\,u)$$

$$= \tfrac{1}{2}\sin^{-1}(2\kappa\,\text{sn}\,u\,\text{dn}\,u) = \text{am}(\kappa u,\ 1/\kappa),\ \text{etc.}$$

Similarly,

$$\int^K \kappa\,\text{sn}\,u du = \cosh^{-1}(\text{dn}\,u/\kappa') = \sinh^{-1}(\kappa\,\text{cn}\,u/\kappa') = \tanh^{-1}(\kappa\,\text{cn}\,u/\text{dn}\,u)$$

$$= \tfrac{1}{2}\log\frac{\text{dn}\,u+\kappa\,\text{cn}\,u}{\text{dn}\,u-\kappa\,\text{cn}\,u} = \log\frac{\text{dn}\,u+\kappa\,\text{cn}\,u}{\kappa'} = \log\frac{\kappa'}{\text{dn}\,u-\kappa\,\text{cn}\,u},\ \text{etc.},$$

while $\int_0 \text{dn}\,u du = \cos^{-1}(\text{cn}\,u) = \sin^{-1}(\text{sn}\,u) = \text{am}\,u$.........(116)

As an exercise the student may integrate $\text{ns}\,u$, $\text{ds}\,u$, ...; also $\text{sn}^3 u$, $\text{cn}^3 u$, $\text{dn}^3 u$, ...; and obtain formulas of reduction for the integrals of $(\text{sn}\,u)^n$, $(\text{cn}\,u)^n$, $(\text{dn}\,u)^n$, ...

As a general method, for $(\text{sn}\,u)^n$ for instance, we put $\text{sn}^2 u = s$; and now

$$\int(\text{sn}\,u)^n du = \tfrac{1}{2}\int\frac{s^{\frac{1}{2}(n-1)}ds}{\sqrt{(1-s\,.\,1-ks)}} = u_n,\ \text{suppose.}$$

By means of the well known *formula of reduction*,

$$(p+1)av_{p+1} + (2p+1)bv_p + pcv_{p-1} = x^p\sqrt{N},$$

for $\qquad v_p = \int x^p dx/\sqrt{N}$, where $N = ax^2 + 2bx + c$,

we have, on comparison,

$$a = k,\ b = -\tfrac{1}{2}(1+k),\ c = 1,\ p = \tfrac{1}{2}(n-1);$$

so that $\qquad v_p = 2u_n,\ v_{p+1} = 2u_{n+2},\ v_{p-1} = 2u_{n-2}$; and

$(n+1)ku_{n+2} - n(1+k)u_n + (n-1)u_{n-2} = \text{sn}^{n-1}u\,\text{cn}\,u\,\text{dn}\,u$,...(117)

the formula of reduction for $u_n = \int(\text{sn}\,u)^n du$.

When the limits are 0 and K, we obtain the recurring formula

$(n+1)ku_{n+2} - n(1+k)u_n + (n-1)u_{n-2} = 0$, ...(118)

analogous to Wallis's formulas for $\int_0^{\frac{1}{2}\pi}(\sin \text{ or } \cos\theta)^n d\theta$.

The same formulas hold for $u_n = (\text{cd}\,u)^n du$, since (§ 57)

$$\text{cd}\,u = \text{sn}(K-u).$$

Thus u_n is made to depend ultimately on u_1, already determined, or on u_2; and a similar procedure will hold for the integrals of $(\text{cn}\,u)^n$ or $(\text{sd}\,u)^n$, $(\text{dn}\,u)^n$ or $(\text{nd}\,u)^n$, etc.

77. *The Elliptic Integral of the Second Kind.*

We may mention here incidentally that the integrals of

$$\text{sn}^2u,\ \text{cn}^2u,\ \text{dn}^2u,\ \text{ns}^2u,\ \text{ds}^2u,\ \text{cs}^2u,\ \ldots$$

require for their expression new functions called *elliptic integrals of the second kind,* such as occur for instance in the rectification of the ellipse.

For if, in the ellipse $(x/a)^2+(y/b)^2=1$,

we put $$x=a\sin\phi,\quad y=b\cos\phi;$$

then $$\frac{ds^2}{d\phi^2}=\frac{dx^2}{d\phi^2}+\frac{dy^2}{d\phi^2}=a^2\cos^2\phi+b^2\sin^2\phi=a^2(1-e^2\sin^2\phi);$$

so that $$\frac{s}{a}=\int_0\sqrt{(1-e^2\sin^2\phi)}\,d\phi=\int_0\Delta(\phi,e)\,d\phi=\int_0\text{dn}^2u\,du, \tag{119}$$

on putting $\phi=\text{am}(u, e)$; and e, the excentricity of the ellipse, is now the modulus.

The integral $\int_0\sqrt{(1-\kappa^2\sin^2\phi)}\,d\phi$ or $\int\Delta(\phi,\kappa)\,d\phi$ is denoted by $E(\phi,\kappa)$ by Legendre, and called the *elliptic integral of the second kind;* and when the upper limit is $\frac{1}{2}\pi$, the integral is denoted by $E^1\kappa$, or by E simply, and called the *complete elliptic integral of the second kind.*

Examples.—The following examples are collected chiefly from Legendre's *Functions Elliptiques;* the results, being now expressed by the inverse elliptic functions, will serve as a guide to the substitutions required to reduce the integrals to the standard elliptic forms, and the correctness can be tested by differentiation as an exercise.

1. $\displaystyle\int_0(1+x^2)^{-\frac{3}{4}}dx=\sqrt{2}\,\text{cn}^{-1}\left\{(1+x^2)^{-\frac{1}{4}},\quad \tfrac{1}{2}\sqrt{2}\right\}.$

2. $\displaystyle\int_0(1-x^2)^{-\frac{3}{4}}dx=\sqrt{2}\,\text{cn}^{-1}\left\{(1-x^2)^{-\frac{1}{4}},\quad \tfrac{1}{2}\sqrt{2}\right\}.$

3. $\displaystyle\int_1(x^2-1)^{-\frac{3}{4}}dx=\quad\text{cn}^{-1}\left\{\frac{1-\sqrt{(x^2-1)}}{1+\sqrt{(x^2-1)}},\ \tfrac{1}{2}\sqrt{2}\right\}.$

4. $\displaystyle\int(x-\alpha\,.\,x-\beta)^{-\frac{3}{4}}dx$

$$=\sqrt{\left(\frac{2}{\alpha-\beta}\right)}\,\text{cn}^{-1}\left\{\frac{\alpha-\beta-2\sqrt{(x-\alpha\,.\,x-\beta)}}{\alpha-\beta+2\sqrt{(x-\alpha\,.\,x-\beta)}},\ \tfrac{1}{2}\sqrt{2}\right\}$$

5. $\displaystyle\int(\alpha-x\,.\,x-\beta)^{-\frac{3}{4}}dx=\frac{2}{\sqrt{(\alpha-\beta)}}\,\text{cn}^{-1}\left\{\frac{(\alpha-x\,.\,x-\beta)^{\frac{1}{4}}}{\sqrt{\frac{1}{2}(\alpha-\beta)}},\ \tfrac{1}{2}\sqrt{2}\right\}.$

6. $\int\{(x-m)^2+n^2\}^{-\frac{3}{4}}dx=\frac{2}{\sqrt{n}}\,\mathrm{cn}^{-1}\left\{\frac{\sqrt{n}}{\{(x-m)^2+n^2\}^{\frac{1}{4}}},\ \tfrac{1}{2}\sqrt{2}\right\}.$

7. Prove that, if $w^n=4x^n(1-x^n)$,

$$\int_{2^{-1/n}}^{x}(1-x^n)^{1-1/n}dx=\int_{x}^{2^{-1/n}}(1-x^n)^{1-1/n}dx=2^{-2/n}\int(1-w^n)^{-\frac{1}{2}}dw;$$

and express the result when $n=3$, 4, or 6.

8. Prove that, if $x-\alpha$ is a factor of the cubic X, so that

$$X=(x-\alpha)(ax^2+2bx+c);$$

$$\int_{\alpha}X^{-\frac{2}{3}}dx=\frac{3}{a\alpha^2+2b\alpha+c}\wp^{-1}\left(\frac{X^{\frac{1}{3}}}{x-\alpha};\ 0,\ 4\frac{ac-b^2}{a\alpha^2+2b\alpha+c}\right),$$

an integral occurring in the determination of the motion of a projectile in a resisting medium.

Evaluate the integral when $a\alpha^2+2b\alpha+c=0$, so that

$$X=(x-\alpha)^2(x-\gamma).$$

9. Prove that (i.) $\displaystyle\int_0^K\frac{\mathrm{cn}\,u\,du}{1+\mathrm{dn}\,u}=\sqrt{\frac{1-\mathrm{dn}\,u}{1+\mathrm{dn}\,u}}.$

(ii.) $\displaystyle\int_0^K\frac{\mathrm{sn}\,u\,du}{\mathrm{dn}\,u+\kappa'}=\frac{1}{\kappa'(1+\kappa')}.$

(iii.) $\displaystyle\int_0^{2K}u\,\mathrm{sn}^2u\,du=2K(K-E)/\kappa^2.$

(iv.) $\displaystyle\int_0^{\frac{1}{2}\pi}F(\phi,\kappa)\sin\phi\,d\phi=\frac{1}{\kappa}\sin^{-1}\kappa.$

(v.) $\displaystyle\int_0^1\frac{K\,d\kappa}{1+\kappa}=\tfrac{1}{4}\pi^2.$

10. Prove that

$$E/\kappa'^2>K>E>2K\kappa'^2/(1+\kappa'^2).$$

11. Denoting the integral $\int_0(\Delta\phi)^{-n}d\phi$ by u_n, establish the formula of reduction

$$n\kappa'^2u_{n+2}-(n-1)(1+\kappa'^2)u_n+(n-2)u_{n-2}=-\kappa^2\sin\phi\cos\phi(\Delta\phi)^{-n}.$$

Evaluate u_n for $n=2$, 3, 4, ...

CHAPTER III.

GEOMETRICAL AND MECHANICAL ILLUSTRATIONS OF THE ELLIPTIC FUNCTIONS.

78. *Graphs of the Elliptic Functions.*

Now that the Elliptic Functions have been defined and a few of their fundamental properties have been established in Chapter I. in connexion with the pendulum; while in Chapter II. the reductions of the elliptic integral to the standard form have been tabulated, let us consider some further applications, and first in connexion with the graphs of $\operatorname{am} u$, $\operatorname{cn} u$, $\operatorname{sn} u$, $\operatorname{dn} u$, represented by curves whose equations are of the form
$$y = \operatorname{am} x,\ \operatorname{cn} x,\ \operatorname{sn} x, \text{ or } \operatorname{dn} x.$$

The graphs of these equations are given in fig. 5, in curves (i.), (ii.), (iii.), (iv.); the modular angle employed is 45°, so that the curves can be plotted from the numerical values given in Table II., analogous to the graphs of the circular and hyperbolic functions, given in Chrystal's *Algebra*, Part II.; thus, for instance, the curve $y = \operatorname{am} x$ is the graph of the relation between ϕ and u in § 5.

We notice from the equations of § 57, Chap. II., that by sliding the curves along Ox through a distance $\pm K$, the curve $y = \operatorname{sn} x$ becomes changed into $y = \operatorname{sn}(K + x) = \operatorname{cn} x/\operatorname{dn} x$ or $\operatorname{cd} x$, and not into $y = \operatorname{cn} x$; while the curve $y = \operatorname{cn} x$ becomes changed into $y = \operatorname{cn}(x - K) = \kappa' \operatorname{sn} x/\operatorname{dn} x$ or $\kappa' \operatorname{sd} x$, and not into $y = \operatorname{sn} x$; so that the curves $y = \operatorname{sn} x$ and $y = \operatorname{cn} x$ are essentially distinct curves, and cannot be superposed, like $y = \cos x$ and $y = \sin x$.

The curve (i.), the graph of $\operatorname{am} x$, consists of a regular undulation, running along the straight line $y = \frac{1}{2}\pi x/K$; so that
$$\operatorname{am} x = \tfrac{1}{2}\pi x/K + \text{periodic terms} = \tfrac{1}{2}\pi x/K + \Sigma B_n \sin(n\pi x/K),$$

in a Fourier series, where the B's are to be determined subsequently; and then by differentiation,

$$\text{dn}\, x = (\tfrac{1}{2}\pi/K)\{1 + 2\Sigma n B_n \cos(n\pi x/K)\}.$$

So also the graph of $E\phi$ or $E\,\text{am}\, u$, the elliptic integral of the second kind (§ 77) consists, like (i.) the graph of $\text{am}\, x$, of an undulation running along the straight line $y = Ex/K$; so that we may write, in Jacobi's notation,

$$E\,\text{am}\, x = Ex/K + Zx,$$

where Zx is a periodic function of x, which can be expressed in a Fourier series

$$Zx = \Sigma C_n \sin n\pi x/K;$$

and then, by differentiation,

$$\text{dn}^2 x = E/K + (\pi/K)\Sigma n C_n \cos n\pi x/K;$$

whence also the expression for $\text{sn}^2 x$ and $\text{cn}^2 x$ in a Fourier series.

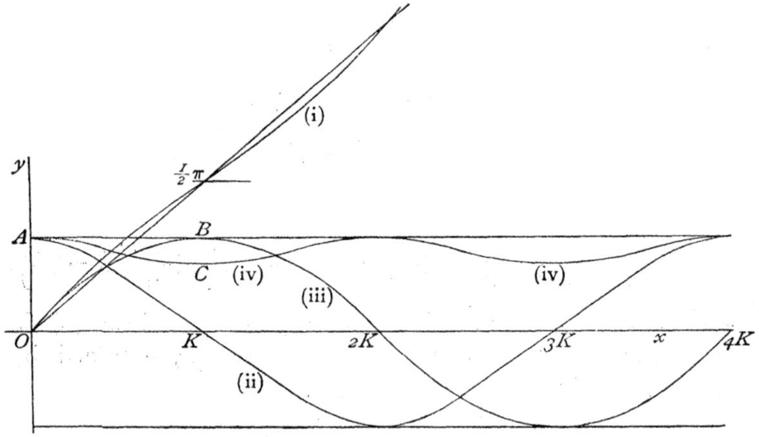

Fig. 5.

We proceed now to some mechanical and geometrical applications of these curves.

79. Problem I. *The curve assumed by a revolving chain* We shall prove that

$$y/b = \text{sn}\, Kx/a$$

(fig. 5, iii.) is the equation of the curve of a uniform chain, rotating steadily with constant angular velocity n about an axis Ox, to which the chain is fixed at two points, $2a$ feet apart, gravity being left out of account, *e.g.* a skipping rope.

Denote by t the tension in poundals of the chain at any point, and by w the weight in lb. per foot of the chain.

Then the equations to be satisfied are

$$\frac{d}{ds}\left(t\frac{dx}{ds}\right)=0,\quad \frac{d}{ds}\left(t\frac{dy}{ds}\right)+n^2wy=0.$$

Therefore $tdx/ds=T$, a constant, the thrust in poundals in the axis due to the pull of the chain; and therefore

$$\frac{d}{ds}\left(\frac{dy}{dx}\right)+\frac{n^2w}{T}y=0,\text{ or }\frac{d^2y}{dx^2}\frac{dx}{ds}+\frac{n^2w}{T}y=0,$$

the differential equation of the curve of the chain.

But $$1+\frac{dy^2}{dx^2}=\frac{ds^2}{dx^2},$$

so that $$\frac{dy}{dx}\frac{d^2y}{dx^2}=\frac{ds}{dx}\frac{d^2s}{dx^2};$$

and therefore $$\frac{d^2s}{dx^2}+\frac{n^2w}{T}y\frac{dy}{dx}=0.$$

Integrating, supposing $y=b$ when $dy/dx=0$ and $ds/dx=1$,

$$\frac{ds}{dx}=1+\frac{n^2w}{2T}(b^2-y^2);$$

so that $$t=Tds/dx=T+\tfrac{1}{2}n^2w(b^2-y^2).$$

Then $$\frac{dy^2}{dx^2}=\left(\frac{ds}{dx}-1\right)\left(\frac{ds}{dx}+1\right)=\frac{n^2w}{T}(b^2-y^2)\left\{1+\frac{n^2w}{4T}(b^2-y^2)\right\},$$

so that x is an elliptic integral of y, of the form (5) in Chap. II.; and y is an elliptic function of x, obtained by inverting the function of the integral.

To obtain this function, let $y=b\sin\phi$; then

$$\frac{dy^2}{dx^2}=b^2\cos^2\phi\frac{d\phi^2}{dx^2}=\frac{n^2w}{T}(b^2-y^2)\left\{1+\frac{n^2w}{4T}(b^2-y^2)\right\},$$

or $$\frac{d\phi^2}{dx^2}=\frac{n^2w}{T}\left(1+\frac{n^2wb^2}{4T}\right)(1-\kappa^2\sin^2\phi),\quad \kappa^2=\frac{n^2wb^2}{4T+n^2wb^2};$$

so that $$\phi=\operatorname{am}K\frac{x}{a},\text{ where }\frac{K\kappa}{a}=\frac{n^2wb}{2T};$$

and $$y/b=\operatorname{sn}Kx/a,$$

the equation of the curve formed by the chain; and now $2a$ denotes the distance between the ends of the chain.

We may denote $T/\frac{1}{4}n^2w$ by h^2; and now

$$\kappa^2=\frac{b^2}{h^2+b^2},\ \kappa'^2=\frac{h^2}{h^2+b^2},\ \frac{\kappa'^2}{\kappa^2}=\frac{h^2}{b^2},\ K\kappa=\frac{2ab}{h^2},\ K\kappa'=\frac{2a}{h};$$

whence the modulus κ and quarter period K can be determined when h and a are given; and

$$\frac{ds}{dx}=1+2\frac{b^2-y^2}{h^2}=1+\frac{2\kappa^2\cos^2\phi}{\kappa'^2}=2\frac{\Delta^2\phi}{\kappa'^2}-1,$$

while $$\frac{d\phi}{dx}=\frac{K}{a}\Delta\phi;$$

so that $$\frac{ds}{d\phi}=\frac{2a}{K\kappa'^2}\Delta\phi-\frac{a}{K}\frac{1}{\Delta\phi}=\frac{Kh^2}{2a}\Delta\phi-\frac{a}{K}\frac{1}{\Delta\phi};$$

and integrating, with the notation of §§ 5 and 77,

$$s=\frac{Kh^2}{2a}E(\phi,\kappa)-\frac{a}{K}F(\phi,\kappa).$$

If $2l$ denotes the length of the chain, then $s=l$ when $\phi=\frac{1}{2}\pi$, and $F(\phi,\kappa)=K$, $E(\phi,\kappa)=E$; and therefore

$$l+a=\tfrac{1}{2}EKh^2/a=bE/\kappa=2aE/K\kappa'^2,$$

from which κ, K, and E must be found by a tentative process, from Legendre's *F.E.*, II., Table II., when a and l are given.

For instance, if $\kappa=\kappa'=\frac{1}{2}\sqrt{2}$, as in Table II., page 11,

$$K=1{\cdot}85407,\ E=1{\cdot}35064;$$

and $$b/a=1{\cdot}5255,\ l/a=1{\cdot}9206.$$

80. When the chain is fixed at two points not in the axis, nor in the same plane through the axis, the chain when revolving in relative equilibrium will form a tortuous curve, which will sweep out a surface of revolution, of which the preceding curve $y/b=\operatorname{sn}Kx/a$ is a particular case of the meridian curve, while the general equation is of the form

$$y^2+z^2=b^2\operatorname{sn}^2(Kx/a)+c^2\operatorname{cn}^2(Kx/a).$$

For in this more general case the equations of relative equilibrium are now

$$\frac{d}{ds}\left(t\frac{dx}{ds}\right)=0,\ \frac{d}{ds}\left(t\frac{dy}{ds}\right)+n^2wy=0,\ \frac{d}{ds}\left(t\frac{dz}{ds}\right)+n^2wz=0.$$

Three first integrals of these equations are

$$t\frac{dx}{ds}=T;\ \dots\dots\dots(1)$$

$$t\left(y\frac{dz}{ds}-z\frac{dy}{ds}\right)=H,\text{ a constant};\ \dots\dots\dots(2)$$

and $$t+\tfrac{1}{2}n^2w(y^2+z^2)=\lambda,\text{ a constant.}\ \dots\dots\dots(3)$$

Putting $$y^2+z^2=r^2,$$

then $$y\frac{dy}{dx}+z\frac{dz}{dx}=\tfrac{1}{2}\frac{dr^2}{dx},$$

and from (1) and (2), $$y\frac{dz}{dx}-z\frac{dy}{dx}=\frac{H}{T};$$

therefore, squaring and adding,

$$r^2\left(\frac{dy^2}{dx^2}+\frac{dz^2}{dx^2}\right)=\tfrac{1}{4}\left(\frac{dr^2}{dx}\right)^2+\frac{H^2}{T^2},$$

or $$\left(\frac{dr^2}{dx}\right)^2=4r^2\left(\frac{ds^2}{dx^2}-1\right)-\frac{4H^2}{T^2}$$

$$=r^2\left(\frac{4t^2}{T^2}-4\right)-\frac{4H^2}{T^2}=\frac{r^2}{T^2}(2\lambda-n^2wr^2)^2-4r^2-\frac{4H^2}{T^2}$$

$$=\frac{n^4w^2}{T^2}(r^6-Ar^4+Br^2-C)=\frac{n^4w^2}{T^2}(r^2-b^2)(r^2-c^2)(r^2-d^2),$$

suppose; and for r^2 to lie between b^2 and c^2, we must suppose $d^2>b^2>r^2>c^2$, and as it is of the form (17), p. 37, we put

$$r^2=b^2\sin^2\phi+c^2\cos^2\phi,$$

$$b^2-r^2=(b^2-c^2)\cos^2\phi,\ r^2-c^2=(b^2-c^2)\sin^2\phi,$$

$$d^2-r^2=d^2-c^2-(b^2-c^2)\sin^2\phi=(d^2-c^2)\Delta^2\phi,$$

where $$\kappa^2=(b^2-c^2)/(d^2-c^2).$$

Then $$\left(\frac{dr^2}{dx}\right)^2=4(b^2-c^2)^2\cos^2\phi\,\sin^2\phi\frac{d\phi^2}{dx^2}$$

$$=\frac{n^4w^2}{T^2}(b^2-c^2)^2(d^2-c^2)\cos^2\phi\,\sin^2\phi\,\Delta^2\phi,$$

or $$\frac{d\phi^2}{dx^2}=\frac{n^4w^2}{4T^2}(d^2-c^2)\Delta^2\phi,$$

so that $$\phi=\text{am}\,Kx/a,$$

where $$K^2/a^2=n^4w^2(d^2-c^2)/4T^2=4(d^2-c^2)/h^4;$$

and then $$r^2=y^2+z^2=b^2\text{sn}^2Kx/a+c^2\text{cn}^2Kx/a,$$

the equation of the surface swept out by the chain, the meridian curve being similar to curve (iv.) in fig. 5.

81. The chain will obviously take up the form which, with given length between the two fixed ends, has the maximum moment of inertia about the axis of revolution; and we have thus investigated the solution of an interesting problem in the Calculus of Variations.

The form of the chain for a *minimum* moment of inertia is obtained by supposing that $r^2 > d^2$, as in (13), p. 35; and by putting

$$r^2 - d^2 = (d^2 - b^2)\tan^2\phi,$$
$$r^2 - b^2 = (d^2 - b^2)\sec^2\phi,$$
$$r^2 - c^2 = (d^2 - c^2)\Delta^2\phi\sec^2\phi,$$
$$\kappa^2 = (b^2 - c^2)/(d^2 - c^2), \text{ as before.}$$

Then
$$\left(\frac{dr^2}{dx}\right)^2 = 4(d^2 - b^2)^2\tan^2\phi\sec^4\phi\frac{d\phi^2}{dx^2},$$
$$= \frac{n^4w^2}{T^2}(d^2 - b^2)^2(d^2 - c^2)\tan^2\phi\sec^4\phi\Delta^2\phi,$$

or
$$\frac{d\phi^2}{dx^2} = \frac{n^4w^2}{4T^2}(d^2 - c^2)\Delta^2\phi,$$

so that
$$\phi = \text{am } Kx/a,$$

and then
$$y^2 + z^2 = d^2\sec^2\phi - b^2\tan^2\phi$$
$$= \frac{d^2 - b^2\text{sn}^2 Kx/a}{\text{cn}^2 Kx/a}$$
$$= d^2\text{nc}^2 Kx/a - b^2\text{sc}^2 Kx/a$$

is the equation of the surface of revolution upon which the chain lies, when its moment of inertia about the axis of x is a minimum.

The projection of the chain upon a plane perpendicular to the axis is to be investigated subsequently.

82. When the two points to which the ends of the chain are fastened lie in the axis, or in a plane through the axis, the chain takes the form of a plane curve, whose equation is

$$y/b = \text{sn } Kx/a$$

for a maximum moment of inertia, as already shown in § 79; and

$$y \text{ cn } Kx/a = d, \text{ or } y = d \text{ nc } Kx/a$$

for a minimum moment of inertia; which can be proved as a simple exercise in the Calculus of Variations, by considering the variation of the integral

$$\int (y^2 + \lambda)\sqrt{(1 + p^2)}ds.$$

83. Problem II. "The curve on which an ellipse, of semi-axes a and b, must roll for its centre to describe a straight line Ox is the curve whose equation is

$$y/a = \text{dn } x/b,$$

the modulus κ being the excentricity of the ellipse."

For if the centre M of the ellipse describes the horizontal straight line Ox (fig. 6), M must always lie vertically over P, the point of contact with the fixed curve, so that the ellipse rests in neutral equilibrium if its centre of gravity is at the centre M; teeth being cut in the curves, if requisite, to prevent slipping.

Therefore the polar subnormal

$$MG=-\frac{dr}{d\theta} \text{ in the ellipse } \frac{1}{r^2}=\frac{\cos^2\theta}{a^2}+\frac{\sin^2\theta}{b^2}$$

must be equal to the subnormal

$$MG=-y\frac{dy}{dx} \text{ in the fixed curve } AP, \text{ where } MP=r=y.$$

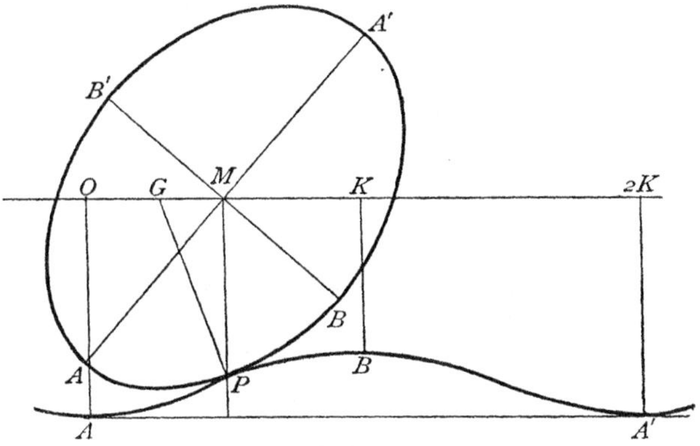

Fig. 6.

Now in the ellipse, differentiating,

$$-\frac{2}{r^3}\frac{dr}{d\theta}=\left(\frac{1}{b^2}-\frac{1}{a^2}\right)2\sin\theta\cos\theta=2\sqrt{\left(\frac{1}{r^2}-\frac{1}{a^2}\cdot\frac{1}{b^2}-\frac{1}{r^2}\right)},$$

since $\quad \frac{1}{r^2}-\frac{1}{a^2}=\left(\frac{1}{b^2}-\frac{1}{a^2}\right)\sin^2\theta,\ \frac{1}{b^2}-\frac{1}{r^2}=\left(\frac{1}{b^2}-\frac{1}{a^2}\right)\cos^2\theta;$

or $\quad -\frac{dr}{d\theta}=\frac{r\sqrt{(a^2-r^2\,.\,r^2-b^2)}}{ab};$

so that in the fixed curve AP

$$\frac{dy}{dx}=-\frac{\sqrt{(a^2-y^2\,.\,y^2-b^2)}}{ab},$$

$$x=\int^a\frac{ab\,.\,dy}{\sqrt{(a^2-y^2\,.\,y^2-b^2)}}=b\,\mathrm{dn}^{-1}\left\{\frac{y}{a},\ \sqrt{\left(1-\frac{b^2}{a^2}\right)}\right\},$$

by (9), p. 33; or, by inversion of the function,

$$y/a=\mathrm{dn}\,x/b.$$

The arc of the rolling curve is obviously the same function of r as the arc of the fixed curve is of y; and therefore the arcs are expressible by elliptic integrals of the second kind.

The curve AP can be described as a *roulette*, by a point P fixed to a certain curve which rolls on Ox, and therefore touches Ox at G, since G, the foot of the normal PG, is the centre of instantaneous rotation.

Since PM is the perpendicular from a pole P on the tangent of the rolling curve, and that the relative orbit of P and M is the ellipse, therefore the pedal of the rolling curve with respect to the pole P is an ellipse; or, in other words, the rolling curve is the *first negative pedal* of an ellipse with respect to its centre, that is, the envelope of lines drawn through each point on the ellipse perpendicular to the line joining the point to the centre of the ellipse.

The first negative pedal of an ellipse with respect to its centre is called *Talbot's curve;* its (p, ω) equation is

$$\frac{1}{p^2}=\frac{\cos^2\omega}{a^2}+\frac{\sin^2\omega}{b^2};$$

and it is of the sixth degree (Cayley, *Proc. R. S.*, 1857-9, p. 171).

84. For a rolling hyperbola, changing the sign of b^2, the fixed curve must be given by

$$x=\int_a \frac{ab\,dy}{\sqrt{(y^2-a^2\,.\,y^2+b^2)}}=\frac{ab}{\sqrt{(a^2+b^2)}}\operatorname{cn}^{-1}\left\{\frac{a}{y},\ \frac{b}{\sqrt{(a^2+b^2)}}\right\},$$

by (8), p. 33; so that, by inversion of the function,

$$a/y=\operatorname{cn} x/a\kappa, \text{ or } y/a=\operatorname{nc} x/a\kappa,$$

is the equation of the fixed curve for the hyperbola.

85. When the fixed curves are of the form of curves (ii.) and (iii.) in fig. 5, we shall find in a similar manner that the rolling curves which will rest upon them in neutral equilibrium are given by

$$\frac{1}{r^2}=\frac{\cosh^2\theta}{a^2}+\frac{\sinh^2\theta}{b^2}, \text{ or } \frac{1}{r^2}=\frac{\cosh^2\theta}{a^2}-\frac{\sinh^2\theta}{b^2}.$$

Taking the first of these two rolling curves,

$$\frac{1}{r^2}-\frac{1}{a^2}=\left(\frac{1}{a^2}+\frac{1}{b^2}\right)\sinh^2\theta,\ \frac{1}{r^2}+\frac{1}{b^2}=\left(\frac{1}{a^2}+\frac{1}{b^2}\right)\cosh^2\theta;$$

and $$-\frac{2}{r^2}\frac{dr}{d\theta}=\left(\frac{1}{a^2}+\frac{1}{b^2}\right)2\sinh\theta\cosh\theta=2\sqrt{\left(\frac{1}{r^2}-\frac{1}{a^2}\cdot\frac{1}{r^2}+\frac{1}{b^2}\right)},$$

or $$-\frac{dr}{d\theta}=\frac{r\surd(a^2-r^2 \,.\, b^2+r^2)}{ab};$$

so that in the corresponding fixed curve

$$\frac{dy}{dx}=-\frac{\surd(a^2-y^2 \,.\, b^2+y^2)}{ab},$$

$$x=\int_y^a \frac{abdy}{\surd(a^2-y^2 \,.\, b^2+y^2)}=\frac{ab}{\surd(a^2+b^2)}\text{cn}^{-1}\left\{\frac{y}{a}, \frac{a}{\surd(a^2+b^2)}\right\},$$

by (7), p. 33; so that, by inversion,

$$y/a=\text{cn}\, x/b\kappa, \text{ with mod. } \kappa=a/\surd(a^2+b^2).$$

Similarly it can be proved that the second rolling curve can rest in neutral equilibrium on the fixed curve (fig. 5, iii.)

$$y/a=\text{sn}\, x/a, \text{ with mod. } a/b.$$

86. PROBLEM III. *Dynamical Problem.* "The curve $r\,\text{cn}\,\theta=c$ is the relative orbit of the centres of gravity of a straight rod fitting into a smooth straight tube, resting on a smooth horizontal table, when struck by an impulsive couple, the centres of gravity of the rod and of the tube being initially c feet apart."

Suppose the rod to weigh m lb. and the tube to weigh M lb., and denote the moments of inertia about the centres of gravity by mk^2, MK^2 (lb. ft.^{2}).

Then, if P is the C.G. of the rod, Q of the tube ($PQ=r$), and O the (stationary) C.G. of the system,

$$OP=Mr/(m+M),\ OQ=mr/(m+M).$$

Denoting by n the initial angular velocity communicated to the system by the impulsive couple, then from the Principle of the Conservation of Angular Momentum,

$$\{m(k^2+OP^2)+M(K^2+OQ^2)\}(d\theta/dt),$$

or $$\left(mk^2+MK^2+\frac{mMr^2}{m+M}\right)\frac{d\theta}{dt}=\left(mk^2+MK^2+\frac{mMc^2}{m+M}\right)n. \dots (1)$$

Again, from the Principle of the Conservation of Energy,

$$\tfrac{1}{2}m\left(\frac{M}{m+M}\right)^2\frac{dr^2}{dt^2}+\tfrac{1}{2}m\left(\frac{M}{m+M}\right)^2 r^2\frac{d\theta^2}{dt^2}+\tfrac{1}{2}mk^2\frac{d\theta^2}{dt^2}$$

$$+\tfrac{1}{2}M\left(\frac{m}{m+M}\right)^2\frac{dr^2}{dt^2}+\tfrac{1}{2}M\left(\frac{m}{m+M}\right)^2 r^2\frac{d\theta^2}{dt^2}+\tfrac{1}{2}MK^2\frac{d\theta^2}{dt^2},$$

or, after reduction,

$$\frac{1}{2}\frac{mM}{m+M}\left(\frac{dr^2}{dt^2}+r^2\frac{d\theta^2}{dt^2}\right)+\tfrac{1}{2}(mk^2+MK^2)\frac{d\theta^2}{dt^2},$$

the kinetic energy in foot-poundals, is constant, and

$$=\frac{1}{2}\frac{mM}{m+M}c^2n^2+\tfrac{1}{2}(mk^2+MK^2)n^2. \qquad (2)$$

Therefore, employing the value of $d\theta/dt$ given by (1),

$$\frac{mM}{m+M}\left(\frac{dr^2}{d\theta^2}+r^2\right)+mk^2+MK^2=\frac{\left(mk^2+MK^2+\dfrac{mMr^2}{m+M}\right)^2}{mk^2+MK^2+\dfrac{mMc^2}{m+M}},$$

or, finally,

$$\frac{dr^2}{d\theta^2}=(r^2-c^2)\frac{mk^2+MK^2+mMr^2/(m+M)}{mk^2+MK^2+mMc^2/(m+M)}, \qquad (3)$$

so that r is an elliptic function of θ, given by (8), p. 33.

We therefore put $r=c\sec\phi$; and then find

$$\frac{d\phi^2}{d\theta^2}=1-\kappa^2\sin^2\phi=\Delta^2\phi,$$

where $$\kappa^2=\frac{mk^2+MK^2}{mk^2+MK^2+mMc^2/(m+M)};$$

so that $\phi=\operatorname{am}\theta$, $\cos\phi=\operatorname{cn}\theta$; and therefore

$$r\operatorname{cn}\theta=c.$$

87. When $c=0$, $\kappa=1$, and this method fails; but now

$$\frac{1}{r^2}\frac{dr^2}{d\theta^2}=1+\frac{mMr^2}{(mk^2+MK^2)(M+m)}=1+\frac{r^2}{a^2},$$

suppose, where $a^2=(m+M)(mk^2+MK^2)/mM$;

and now $$\theta=\int_r^\infty\frac{dr}{r\sqrt{(1+r^2/a^2)}}=\sinh^{-1}\frac{a}{r},$$

or $$r\sinh\theta=a,$$

the equation of one of Cotes's spirals, the relative orbit of the centres of gravity of the rod and tube, ultimately described after leaving the unstable position of coincidence.

The system of the rod and tube may be supposed started by any arbitrary impulse, not necessarily a couple, and the essential character of the relative motion is unaltered: but now the C.G. of the system is no longer at rest.

88. Other mechanical arrangements, leading to the same equations of motion, will readily suggest themselves; thus the tube may be supposed to be one of the hollow spokes of a wheel of weight M lb., moveable about a fixed vertical axis, while the rod is one of a number of equal rods, or balls, of collective weight m lb., one in each tube, and initially placed with the C.G. at a distance c from the axis of the wheel.

Now, if the wheel is started by an impulsive couple with angular velocity n, the path of the C.G. of each rod or ball in its spoke will be of the form

$$r \operatorname{cn} \theta = c.$$

89. PROBLEM IV. *Central Orbits and Catenaries expressed by Elliptic Functions.*

When a Central Orbit, expressed in the polar coordinates $(1/u, \theta)$, is described under an attraction to the pole, of magnitude P (dynes per gramme), then, as is proved in treatises on Dynamics, P is given by the equation

$$P = h^2u^2\left(\frac{d^2u}{d\theta^2} + u\right), \text{ where } h = r^2\frac{d\theta}{dt} = \frac{1}{u^2}\frac{d\theta}{dt},$$

and the constant h is twice the rate of area swept out by the radius vector; and v the velocity is given by

$$v^2 = \frac{h^2}{p^2} = h^2\left(\frac{du^2}{d\theta^2} + u^2\right).$$

Given the equation of the orbit as a relation between u and θ, the value of P as a function of u is thence easily determined by differentiation, as in § 30; let us then determine P for the orbits $\quad au = \operatorname{sn}, \operatorname{cn}, \operatorname{tn}, \text{ or } \operatorname{dn} m\theta;$

also for the inverse curves

$$au = \operatorname{ds}, \operatorname{nc}, \operatorname{cs}, \text{ or } \operatorname{nd} m\theta,$$

in Glaisher's notation; the remaining orbits

$$au = \operatorname{cd}, \operatorname{sd}, \operatorname{dc}, \operatorname{ds} m\theta;$$

are not distinct curves, being merely formed by reflexion in the line $\theta = \frac{1}{2}K/m$, since $\operatorname{cd} m\theta = \operatorname{sn}(K - m\theta)$ (§ 57), etc.

As in § 30, we shall find by differentiation that $(d^2u/d\theta^2) + u$ is always of the form $Au + Bu^3$, so that P is of the form $\mu u^3 + \nu u^5$; and conversely, given this form of P, we find by integration that $(du/d\theta)^2$ is of the form $C + Du^2 + Eu^4$; so that θ is an elliptic integral of u, and u an elliptic function of θ, of which the results are given in § 36.

When the orbit is given by

$$au = \operatorname{sn}^2 m\theta, \operatorname{cn}^2 m\theta, \operatorname{dn}^2 m\theta, \ldots,$$

we find by differentiation, as in § 30, that P is of the form $\lambda u^2 + \mu u^3 + \nu u^4$; and conversely, when P is of this form, $(du/d\theta)^2$ is a cubic form in u; and θ is given as an elliptic integral or inverse elliptic function of u, by the results of equations (12) to (45), Chap. II.

As an exercise the student may determine the value of P and v^2, as functions of u or r, in the orbit

$$\frac{1}{r^2}=\frac{\mathrm{cn}^2 m\theta}{a^2}+\frac{\mathrm{sn}^2 m\theta}{b^2}$$

and its inverse curve, whose equation is of the form

$$r^2=a^2\mathrm{cn}^2 m\theta+b^2\mathrm{sn}^2 m\theta.$$

Similarly the central forces required to make a chain assume the form of one of the preceding curves can also be determined (Biermann, *Problemata quaedam mechanica functionum ellipticarum ope soluta*, Berolini, 1865).

When a transverse force T is introduced into the field of force, then h is no longer constant, but, as demonstrated in treatises on Dynamics and the Lunar Theory,

$$\frac{dh^2}{d\theta}=\frac{2T}{u^3}, \text{ or } \frac{T}{h^2u^3}=\frac{d\log h}{d\theta};$$

while
$$\frac{d^2u}{d\theta^2}+u=\frac{P}{h^2u^2}-\frac{T}{h^2u^3}\frac{du}{d\theta},$$

so that
$$P=h^2u^2\left(\frac{d^2u}{d\theta^2}+u+\frac{d\log h}{d\theta}\frac{du}{d\theta}\right).$$

If we assume $P=h^2u^3$; then

$$\frac{d}{d\theta}\left(\log\frac{du}{d\theta}\right)+\frac{d\log h}{d\theta}=0, \text{ or } h\frac{du}{d\theta}=C, \text{ a constant.}$$

But $\dfrac{d\theta}{dt}=hu^2$, so that $\dfrac{du}{dt}=Cu^2$, or $\dfrac{dr}{dt}=-C$, which shows that the body approaches the centre with constant velocity C.

Suppose, for instance, we take an orbit given by

$$m\theta=\mathrm{am}\, au,$$

then
$$h=C\frac{d\theta}{du}=C\frac{a}{m}\mathrm{dn}\, au=C\frac{a}{m}\sqrt{(1-\kappa^2\sin^2 m\theta)};$$

and
$$P=h^2u^3 = C^2\frac{a^2u^3}{m^2}(1-\kappa^2\sin^2 m\theta),$$

$$T=\tfrac{1}{2}u^3\frac{dh^2}{d\theta}=-C^2\frac{a^2u^3}{m}\kappa^2\sin m\theta\cos m\theta;$$

so that V, the potential of the field of force, is given by

$$V=\frac{1}{2}\frac{C^2}{m^2}\frac{a^2}{r^2}(1-\kappa^2\sin^2 m\theta);$$

and then
$$P=-\frac{\partial V}{\partial r},\ T=-\frac{\partial V}{r\partial\theta}.$$

90. PROBLEM V. *The motion of Watt's Governor.*

"The oscillations of Watt's Governor between the inclinations α and β to the vertical, when constrained to revolve with constant angular velocity ω, are given by

$$\tan\tfrac{1}{2}\theta = \tan\tfrac{1}{2}\alpha \,\mathrm{dn}(nt, \kappa), \text{ with } \kappa' = \tan\tfrac{1}{2}\beta/\tan\tfrac{1}{2}\alpha,$$

where θ denotes the inclination of an arm to the vertical axis at the time t."

Consider the motion of either rod and ball, as if unconstrained by the other, and denote by C the moment of inertia of the rod and ball about its axis of figure, and by A the moment of inertia about the axis on which the rod turns at the upper joint O (fig. 7).

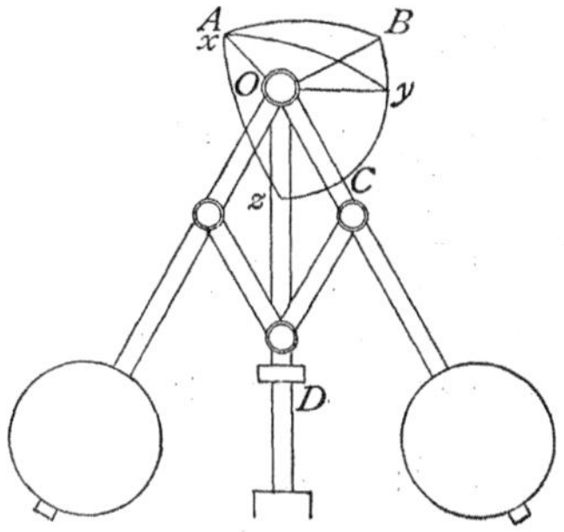

Fig. 7.

Drawing the three principal axes OA, OB, OC at O, and three moving coordinate axes Ox, Oy, Oz, such that Ox and OA are coincident, Oz is vertical, and yOz, BOC in the same vertical plane, then the components of angular velocity about OA, OB, OC are $-(d\theta/dt)$, $-\omega\sin\theta$, $\omega\cos\theta$; and the corresponding components of angular momentum are $-A(d\theta/dt)$, $-A\omega\sin\theta$, $C\omega\cos\theta$.

The components of angular momentum about Ox, Oy, Oz will therefore be

$h_1 = -A(d\theta/dt)$, $h_2 = (C-A)\omega\sin\theta\cos\theta$, $h_3 = (C\cos^2\theta + A\sin^2\theta)\omega$;

while the component angular velocities of the coordinate axes Ox, Oy, Oz are $\theta_1 = 0$, $\theta_2 = 0$, $\theta_3 = \omega$, with the notation of Routh's *Rigid Dynamics*.

Take the poundal as the unit of force, and denote by M the weight in lb. of either arm and ball, by h the distance in feet from O of the centre of gravity; the equation of motion obtained by taking moments about Ox or OA is

$$\frac{dh_1}{dt}-h_2\theta_3+h_3\theta_2=L,$$

or $$-A(d^2\theta/dt^2)+(A-C)\omega^2\sin\theta\cos\theta=Mgh\sin\theta\,;\ \ldots\ldots\ldots(1)$$

so that, if $A=C$, the motion reduces to simple pendulum motion.

Integrating, on the supposition that $\alpha>\theta>\beta$, and that $d\theta/dt=0$ when $\theta=\alpha$ and β,

$$\frac{d\theta^2}{dt^2}=\frac{A-C}{A}\omega^2(\cos\theta-\cos\alpha)(\cos\beta-\cos\theta).\ \ldots\ldots\ldots\ldots(2)$$

The position of relative equilibrium is given by $d^2\theta/dt^2=0$; and then, if $\theta=\gamma$,

$$\cos\gamma=Mgh/\{(A-C)\omega^2\}=\tfrac{1}{2}(\cos\alpha+\cos\beta),\ \ldots\ldots\ldots(3)$$

so that in these oscillations the point D, which controls the valve, makes equal excursions above and below its position of relative equilibrium.

The technical name for these oscillations is "Hunting"; and some kind of frictional constraint is required to prevent these oscillations from becoming established.

(Maxwell, *Proc. R. S.*, 1868.)

Denoting $\tan\frac{1}{2}\alpha$, $\tan\frac{1}{2}\beta$, $\tan\frac{1}{2}\theta$ by a, b, x respectively, then equation (2) may be written

$$\frac{4}{(1+x^2)^2}\frac{dx^2}{dt^2}=\frac{A-C}{A}\omega^2\left(\frac{1-x^2}{1+x^2}-\frac{1-a^2}{1+a^2}\right)\left(\frac{1-b^2}{1+b^2}-\frac{1-x^2}{1+x^2}\right),$$

or $$\frac{dx^2}{dt^2}=\frac{A-C}{A}\omega^2\cos^2\tfrac{1}{2}\alpha\cos^2\tfrac{1}{2}\beta(a^2-x^2)(x^2-b^2)\,;$$

and this, by equation (9), p. 33, gives

$$x=a\,\text{dn}(nt,\kappa),\ \text{or}\ \tan\tfrac{1}{2}\theta=\tan\tfrac{1}{2}\alpha\,\text{dn}\,nt,$$

where $\kappa'=b/a=\tan\frac{1}{2}\beta/\tan\frac{1}{2}\alpha$, and $n=\omega\sin\frac{1}{2}\alpha\cos\frac{1}{2}\beta\sqrt{(1-C/A)}$.

For a small oscillation, we put $\alpha=\beta$; and then $\kappa'=1$, $\kappa=0$; and now the period of an oscillation

$$\frac{2\pi}{n}=\frac{4\pi}{\omega\sin\alpha}\sqrt{\frac{A}{A-C}}.$$

91. If we suppose the whole weight of a rod and ball concentrated at the centre of gravity, we have $C=0$, $A=Mh^2$; and now the motion may be assimilated to that of a particle in a smooth circular tube, which is made to rotate about a vertical diameter with constant angular velocity ω.

(Prof. B. Price, *Analytical Mechanics*, § 403).

The equation of motion (1) now reduces to

$$h\frac{d^2\theta}{dt^2} - h\omega^2 \sin\theta\cos\theta = -g\sin\theta,$$

where h denotes the radius of the circle; and for oscillations on one side of the vertical between α and β, $\alpha > \theta > \beta$,

$$(d\theta/dt)^2 = \omega^2(\cos\theta - \cos\alpha)(\cos\beta - \cos\theta),$$

the solution of which is, as before,

$$\tan\tfrac{1}{2}\theta = \tan\tfrac{1}{2}\alpha \operatorname{dn} nt,$$

where $\kappa' = \tan\tfrac{1}{2}\beta/\tan\tfrac{1}{2}\alpha, \quad n = \omega\sin\tfrac{1}{2}\alpha\cos\tfrac{1}{2}\beta.$

If the particle in its oscillations just reaches the lowest point of the circle, $\beta=0$; and then $\kappa'=0$, $\kappa=1$; and now $\operatorname{dn} nt$ degenerates into $\operatorname{sech} nt$ (§ 16); so that

$$\tan\tfrac{1}{2}\theta = \tan\tfrac{1}{2}\alpha \operatorname{sech} nt, \text{ where } n = \omega\sin\tfrac{1}{2}\alpha;$$

the position of relative equilibrium being given by

$$\cos\gamma = g/\omega^2 h = \tfrac{1}{2}(1+\cos\alpha) = \cos^2\tfrac{1}{2}\alpha.$$

If the particle passes through the lowest point, it will come to rest again where $\theta = -\alpha$; and now

$$(d\theta/dt)^2 = \omega^2(\cos\theta - \cos\alpha)(2\cos\gamma - \cos\alpha - \cos\theta),$$

where $2\cos\gamma - \cos\alpha > 1$; and the solution of this equation is

$$\tan\tfrac{1}{2}\theta = \tan\tfrac{1}{2}\alpha \operatorname{cn} nt, \text{ where } n = \omega\sqrt{(\cos\gamma - \cos\alpha)}.$$

When $\alpha = \pi$, we shall find the motion given by

$$\tan\tfrac{1}{2}\theta = \frac{\sinh\sqrt{(\omega^2 + g/h)}t}{\sqrt{(1 + h\omega^2/g)}};$$

so that, after an infinite time, the particle just reaches the highest point of the circle, where it will be in unstable equilibrium.

A still greater velocity of the particle relative to the tube will make the particle perform complete revolutions, which will be expressed by

$$\tan\tfrac{1}{2}\theta = C \operatorname{tn} nt.$$

We have supposed the circular tube to be made to rotate with constant angular velocity about a vertical diameter; but the motion of the particle relatively to the tube will be found to depend on similar equations when the tube is attached in any other manner to the vertical axis.

92. Such will be the motion of a pendulum swinging about an axis fixed to the Earth, and now it is interesting to notice other cases of motion of bodies which can be directly compared and made to synchronize with the motion of an ordinary pendulum, swinging through a finite angle.

Thus the pendulum, if moveable about a smooth vertical axis, which is fixed to a wheel moveable about a fixed vertical axis, the inertia of the wheel being sufficiently great for the reaction of the pendulum to have no sensible effect on its angular velocity, will perform pendulum oscillations, with g replaced by $a\omega^2$, ω being the angular velocity of the wheel and a the distance between the axis of the wheel and of the pendulum.

Again a cylinder of radius a and radius of gyration k, rolling inside a fixed horizontal cylinder of radius b, will synchronize with a pendulum of length $l=(b-a)(1+k^2/a^2)$.

If the fixed horizontal cylinder is free to rotate about its axis, and has its centre of gravity in the axis, then the length of the equivalent pendulum is

$$l=(b-a)(1+n), \text{ where } n=\frac{h^2}{a^2}\Big/\left(1+\frac{b^2}{a^2}\frac{mk^2}{MK^2}\right),$$

mk^2, MK^2 denoting the moments of inertia about the axes of the rolling and fixed cylinders.

The rolling cylinder may be replaced by a waggon on wheels, and the motion can still be compared with that of a pendulum.

A circular cone, whose C.G. is in its axis of figure, and whose axis is a principal axis, performs pendulum oscillations when it rolls on an inclined plane, or inside or outside another fixed cone, whose axis is sloping, the vertices of the cones being coincident; the determination of l, the length of the equivalent pendulum, in these cases is left as an exercise to the student.

In those cases where the finite oscillations are not of the pendulum character, we suppose the motion indefinitely small; and now, in small oscillations under gravity, instead of giving the formula for the period of a small oscillation, it is in general simpler to give l, the length of the pendulum, whose small oscillations have the same period.

Thus for the vertical oscillation of a carriage on springs, l is equal to the permanent average vertical deflection of the springs, due to the weight of the body of the carriage.

For the small vertical oscillations of a ship, $l=V/A$, where V denotes the displacement of the ship (in cubic feet), and A the water line area (in square feet); and if the ship is floating in a dock of area B sq. feet, then it is easily proved that

$$l=V\left(\frac{1}{A}-\frac{1}{B}\right).$$

93. *The Reaction of the Axis of Suspension of a Pendulum.* It is important to know the magnitude of this reaction in the case of a large swinging body, like a bell in a church tower.

Denote by X and Y the horizontal and vertical components of this reaction, considered as acting on the swinging body; and take the gravitation unit of force, the force of a pound.

Then X, Y and W, applied at the centre of gravity G (fig. 1), will be the dynamical equivalents of the motion of the body, collected as a particle at G; and since the component accelerations of G are $h(d\theta/dt)^2$ in the direction GO,

and $h(d^2\theta/dt^2)$ perpendicular to GO,

therefore, resolving horizontally and vertically,

$$Wh(d^2\theta/dt^2)\cos\theta - Wh(d\theta/dt)^2\sin\theta = Xg,$$

$$Wh(d^2\theta/dt^2)\sin\theta + Wh(d\theta/dt)^2\cos\theta = Yg - Wg;$$

while, from the pendulum motion,

$$l(d^2\theta/dt^2) = -g\sin\theta,\ \tfrac{1}{2}l^2(d\theta/dt)^2 = g(2R - l\operatorname{vers}\theta).$$

From these equations we find

$$\frac{Y}{W} = 1 - \frac{h}{l}\sin^2\theta + \frac{4Rh}{l^2}\cos\theta - \frac{2h}{l}\cos\theta(1-\cos\theta),$$

or $$\frac{Y}{W} - 1 + \frac{h}{l} = -\left(\frac{2h}{l} - \frac{4Rh}{l^2}\right)\cos\theta + \frac{3h}{l}\cos^2\theta;$$

$$\frac{X}{W} = \left(\frac{2h}{l} - \frac{4Rh}{l^2}\right)\sin\theta - \frac{3h}{l}\sin\theta\cos\theta,$$

and therefore the resultant of X and $Y - W(1-h/l)$ is a force

$$T = W\left(-\frac{2h}{l} + \frac{4Rh}{l^2} + \frac{3h}{l}\cos\theta\right) = W\frac{3h}{l^2}(\tfrac{1}{3}l + \tfrac{4}{3}R - y)$$

in the direction GO; and T varies as the depth of P below the line $$y = \tfrac{1}{3}l + \tfrac{4}{3}R,$$
whence X and Y are easily constructed.

94. In the simple pendulum, $h=l$, and the tension T of the thread PO is given by

$$\frac{T}{W}=\frac{3}{l}(\tfrac{1}{3}l+\tfrac{4}{3}R-y),$$

At the end of a swing $y=2R$, and $T/W=1-2R/l$; so that, if $2R$ is less than l, T is always positive.

But if $2R$ is greater than l, so that the plummet swings through more than 180°, T changes sign, and the thread will become slack, unless replaced by a light stiff rod.

When $2R$ is greater than $2l$, the pendulum makes complete revolutions; and now, at the top of a revolution, $y=2l$, and $T/W=4R/l-5$; and when $2R$ is greater than $\frac{5}{2}l$, T is again always positive, and the plummet can be whirled round at the end of a thread, without the thread becoming slack.

95. When the axis of suspension of the pendulum is horizontal, and cut into a smooth screw of pitch p, the equation of energy gives

$$\tfrac{1}{2}W(h^2+k^2+p^2)(d\theta/dt)^2=Wg(H-h\,\text{vers}\,\theta),$$

if the centre of gravity descends from a height H above its lowest position; so that

$$(h^2+k^2+p^2)(d^2\theta/dt^2)=-gh\sin\theta,$$

and therefore $$l=h+(k^2+p^2)/h;$$

and now in addition to X and Y, the reaction of the axis exerts a horizontal longitudinal component Z and a couple pZ, given by

$$Z=\frac{W}{g}p\frac{d^2\theta}{dt^2}=\frac{-Wph\sin\theta}{h^2+k^2+p^2}.$$

Similarly the increase in l due to the pendulum being supported on friction wheels may be investigated.

As an exercise the student may investigate the small oscillations of a system of clockwork, in which the wheels are unbalanced about the axes, and prove that for small oscillations the length of the simple equivalent pendulum is given by

$$l=(\Sigma wk^2p^2)/(\Sigma whp^2\cos a),$$

where w denotes the weight, wh the moment, and wk^2 the moment of inertia of a wheel about its axis; a denoting the angle which the plane through the axis and centre of gravity makes with the vertical in the position of equilibrium; and p denoting the velocity ratio of the wheel.

96. *The Internal Stresses of a Swinging Body.*

These internal stresses are most forcibly realized on board a ship rolling in the sea, not only in their effects as producing sea-sickness, but also in causing the cargo to shift, if the cargo is grain, coal, or petroleum, in bulk.

It is usual to consider the ship as acted upon by two forces,

(i.) W tons, the weight or displacement of the ship, acting vertically downwards through the centre of gravity G,

(ii.) W tons, the buoyancy of the water, acting vertically upwards through M the metacentre (fig. 8).

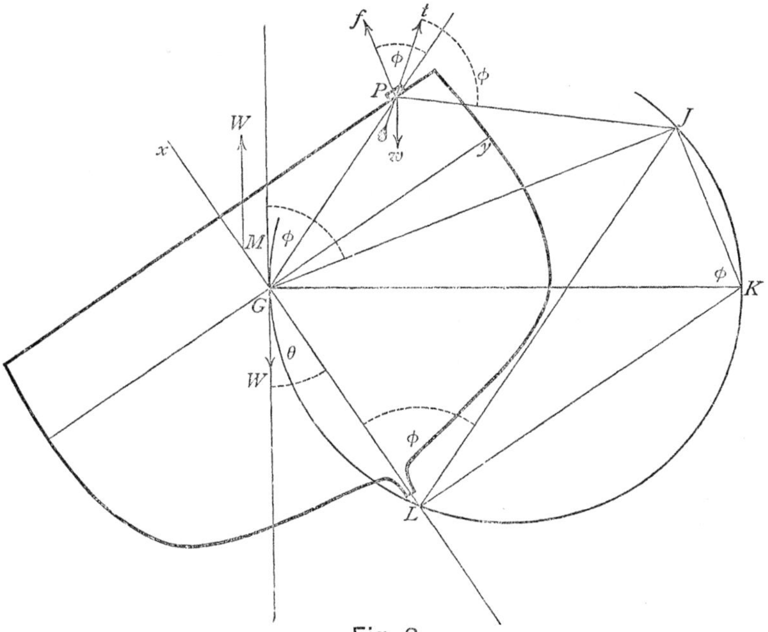

Fig. 8.

These two forces form a couple of moment $W \cdot GM \cdot \sin\theta$ (foot tons), so that the ship will roll about a horizontal longitudinal axis through G, like a pendulum of length $GL = k^2/GM$ feet, Wk^2 denoting the moment of inertia of the ship about this axis of rotation.

Now to find the force which acts upon w, any infinitesimal part at P of the ship, to give it its acceleration and to balance its weight, we refer the point P to axes Gx and Gy, drawn upwards through GM and perpendicular to GM.

This force will balance the reversed *effective force* of w at P and the effect of gravity on w; and therefore, in gravitation measure, will have components

$$\frac{w}{g}y\frac{d^2\theta}{dt^2}-\frac{w}{g}x\left(\frac{d\theta}{dt}\right)^2+w\cos\theta, \text{ parallel to } Gx,$$

$$-\frac{w}{g}x\frac{d^2\theta}{dt^2}-\frac{w}{g}y\left(\frac{d\theta}{dt}\right)^2+w\sin\theta, \text{ parallel to } Gy.$$

If w is suspended as a plummet by a very short thread, the thread will take the direction of this force, and will therefore make an angle with Gx

$$\tan^{-1}\frac{g\sin\theta-x(d^2\theta/dt^2)-y(d\theta/dt)^2}{g\cos\theta+y(d^2\theta/dt^2)-x(d\theta/dt)^2}.$$

Supposing the ship to roll like a pendulum of length l, through an angle 2α, then

$$l(d^2\theta/dt^2)=-g\sin\theta, \text{ and } \tfrac{1}{2}l(d\theta/dt)^2=g(\cos\theta-\cos\alpha);$$

and by § 8,

$$d^2\theta/dt^2=-n^2\sin\theta=-2n^2\sin\tfrac{1}{2}\theta\cos\tfrac{1}{2}\theta=-2n^2\kappa\operatorname{sn}nt\operatorname{dn}nt,$$

$$(d\theta/dt)^2=2n^2(\cos\theta-\cos\alpha)=4n^2(\sin^2\tfrac{1}{2}\alpha-\sin^2\tfrac{1}{2}\theta)=4n^2\kappa^2\operatorname{cn}^2nt.$$

At any instant the lines of reversed resultant acceleration will be equiangular spirals, of radial angle ϕ, round the centre of acceleration G as pole, the resultant acceleration at P being

$g\frac{r}{l}\sin\theta\operatorname{cosec}\phi$, and the resultant effective force $w\frac{r}{l}\sin\theta\operatorname{cosec}\phi$,

when we put $GP=r$, and $l(d\theta/dt)^2=g\sin\theta\cot\phi$; so that

$$\tan\phi=(\operatorname{sn}nt\operatorname{dn}nt)/(2\kappa\operatorname{cn}^2nt).$$

Superposing the effect of gravity, the resultant lines of force or internal stress will be equiangular spirals of the same radial angle ϕ, round a pole J, the position of which is obtained as follows (fig. 8):—Draw LK perpendicular to GL to meet the horizontal line GK in K; describe the circle on GK as diameter, and draw KJ making an angle $GKJ=\phi$ with GK; this will meet the circle in J.

For the resultant effective force of w at P, being

$$f=w\frac{r}{l}\sin\theta\operatorname{cosec}\phi=w\frac{PG}{GJ},$$

making an angle ϕ with GP, will, when compounded with w upwards, and taking the triangle PGJ turned through an angle ϕ as the triangle of forces, have a resultant

$$t=w\,.\,PJ/GJ, \text{ making an angle } \phi \text{ with } JP.$$

This will be the tension and in the direction of a short thread, from which w is suspended as a plummet at any point P; and the deflection of this plumb line from its original mean direction in the ship will be a measure of the tendency of a body to slide or of a grain cargo to shift; and to a certain extent of the tendency to sea-sickness at this point of the ship and at this instant of its motion.

The tendency will clearly have its maximum value at the end of a roll, when $d\theta/dt=0$, and $\phi=\frac{1}{2}\pi$, and then J coincides with K. (Prof. P. Jenkins, *On the Shifting of Cargoes*, Transactions of the Institute of Naval Architects, 1887.)

The plumb line at P will now set itself at right angles to KP, while the surface of water in a tumbler at P will pass through K; and a granular substance at P will begin to slip if KP makes with its surface an angle greater than the angle of repose of this grain.

Thus up the mast, at a distance a feet from G, water would be spilt out of a tumbler, or sand in a box would shift, by the rolling of the ship through an angle 2α, which would not spill or shift, if the ship heeled over steadily, until an inclination β (the angle of repose of the sand) was reached, given by

$$\tan\beta=(1+a/l)\tan\alpha.$$

At the centre of oscillation L, where $a=-l$, there is no tendency for the water to spill, and this shows that the motion of the ship is felt least by going down below as far as possible in the middle of the ship.

In a swing the body is very near the centre of oscillation, so that ordinary swinging is very little preparation for the motion of a vessel.

A swing to act properly as a preparation for a sea voyage should be constructed as in fig. 5, to imitate, in full size, the cross section of the ship, suspended at M; and now the varying effect of the motion can be experienced by taking up different positions on the deck, up the mast, and in the cabins, constructed in this swing.

Sir W. Thomson proposes to find the axis of rotation of a ship and the angle through which the ship rolls by noting the direction of the plumb lines of two such plummets, suspended at two given points across the ship; planes through the plummets perpendicular to the plumb lines at the extreme end of a roll would intersect in K; the horizontal plane through K would meet the median longitudinal plane of the ship in the axis G; while the plane through K perpendicular to the median plane would meet it in L, whence GL, the length of the equivalent pendulum, and therefore the period of small oscillations could be inferred, as a check on this construction.

Example. A rod AB, whose density varies in any manner, is swung in a vertical plane about a horizontal axis through A. Prove that the bending moment of the rod is a maximum at a point P, determined by the condition that the C.G. of the part PB is the centre of oscillation of the pendulum.

97. PROBLEM VI. *The Elastica or Lintearia.*

The Elastica is the name given to the curve assumed by a uniform elastic beam, wire, or spring, originally straight, when bent into a plane curve (fig. 9) by a stress composed of two equal opposite forces T, on the assumption that at a point P at a distance y from the line of the applied stress the bending moment Ty is equilibrated by a moment of resistance B/ρ, proportional to the curvature $1/\rho$; and the constant B is called the flexural rigidity of the spring (Thomson and Tait, *Natural Philosophy*, § 611).

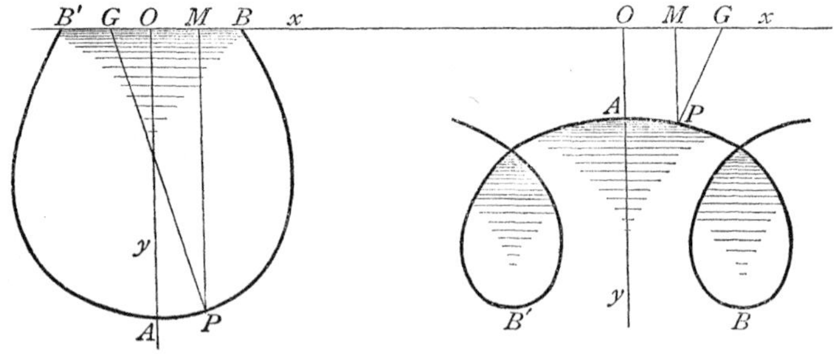

Fig. 9.

Then $\qquad Ty = B/\rho$, or $y\rho = B/T = c^2$, suppose;

and by Kirchhoff's Kinetic Analogue, the normal of the Elastica performs pendulum oscillations on each side of a perpendicular to the line of stress, as the point on the curve moves with a constant velocity.

For, when the normal has turned through an angle θ, the curvature

$$\frac{1}{\rho}=\frac{d\theta}{ds}=\frac{y}{c^2};$$

and by differentiation

$$\frac{d^2\theta}{ds^2}=\frac{1}{c^2}\frac{dy}{ds}=-\frac{1}{c^2}\sin\theta,$$

which agrees with the equation of pendulum motion

$$d^2\theta/dt^2=-n^2\sin\theta,\text{ if } s/c=nt.$$

Corresponding with the oscillating pendulum we have the undulating Elastica, intersecting the line of stress at an angle α; and thus, writing s/c for nt in § 8,

$$\sin\tfrac{1}{2}\theta=\kappa\,\mathrm{sn}\,s/c,\ \cos\tfrac{1}{2}\theta=\mathrm{dn}\,s/c,$$

$$\sin\theta=-dy/ds=2\kappa\,\mathrm{sn}\,s/c\,\mathrm{dn}\,s/c,$$

so that

$$y=2c\kappa\,\mathrm{cn}\,s/c,$$

measuring s from the point A, at a maximum distance from the line of thrust; and a graduated bow might thus be employed for giving mechanically the numerical values of the cn function.

In the nodal Elastica corresponding with the revolving pendulum,

$$\theta=2\,\mathrm{am}\,s/c\kappa,\ \sin\theta=2\,\mathrm{sn}\,s/c\kappa\,\mathrm{cn}\,s/c\kappa=-dy/ds;$$

so that

$$y=2(c/\kappa)\,\mathrm{dn}\,s/c\kappa.$$

In the separating case, $\kappa=1$, and $y=2c\,\mathrm{sech}\,s/c$; and

$$\tfrac{1}{2}\theta=\mathrm{amh}\,s/c,\ \sin\tfrac{1}{2}\theta=\tanh s/c,\ \tan\tfrac{1}{2}\theta=\sinh s/c,\text{ etc.}$$

In the undulating Elastica

$$\frac{dx}{ds}=\cos\theta=\surd(1-4\kappa^2\,\mathrm{sn}^2 s/c\,\mathrm{dn}^2 s/c)=1-2\kappa^2\mathrm{sn}^2 s/c;$$

and in the nodal Elastica

$$\frac{dx}{ds}=\cos\theta=\surd(1-4\,\mathrm{sn}^2 s/c\,\mathrm{cn}^2 s/c)\quad=1-2\,\mathrm{sn}^2 s/c;$$

so that x is given in terms of s by means of elliptic integrals of the second kind (§ 77).

A great simplification is introduced when $\kappa=\kappa'=\tfrac{1}{2}\surd 2$; the Elastica now cuts the line of thrust at right angles, and

$$\cos\theta=\mathrm{cn}^2 s/c=\tfrac{1}{2}y^2/c^2,$$

which shows that this Elastica is the roulette of the centre of a rectangular hyperbola, rolling on the line of thrust.

It is easily proved that in this curve the radius of curvature ρ is half the normal PG; also that a chain can hang in this curve as a catenary, provided the linear density is proportional to $(\mathrm{nc}\,s/c)^3$; this is left as an exercise for the student.

When $\kappa=0$, the undulating Elastica corresponds with small oscillations of the pendulum, and the Elastica is ultimately coincident with the line of thrust, the ordinate y varying as $\sin s/c$ or $\sin x/c$; and then the length of the beam, $\pi c=\pi\sqrt{(B/T)}$, is the extreme length at which the straight form of the beam begins to become unstable under the thrust T.

The nodal Elastica becomes practically a circle when $\kappa=0$, corresponding in Kirchhoff's Kinetic Analogue to the practically uniform revolutions of a pendulum when the velocity is indefinitely increased.

The Elastica is also called Bernoulli's *Lintearia*, being the cross section of a horizontal flexible watertight cylinder, when filled with water, the free surface of which lies in the line of thrust Ox; for if t denotes the constant circumferential tension,

$$t/\rho=wy, \text{ the pressure of the water,}$$

or

$$y\rho=t/w=c^2.$$

It is also the profile of the surface of water drawn up by Capillary Attraction between two parallel plates (Maxwell, Encyclopædia Britannica, *Capillary Action*).

The student may prove, as an exercise, as in § 80, that if the wire is bent into a tortuous curve by balancing forces and couples at its ends, it will assume the form of a curve in a surface of revolution defined by an equation of the form

$$y^2+z^2=a^2\text{cn}^2(s/c)+b^2\text{sn}^2(s/c).$$

(*Proc. London Math. Society*, vol. XVIII.)

98. PROBLEM VII. *Sumner Lines on Mercator's Chart.*

Sumner Lines, so called after Captain Sumner, of Boston, Massachusetts, are the projections on Mercator's chart of small circles on a sphere; if simultaneous observations are taken of the chronometer and of the altitude of the sun or a star, the observer knows that he must lie on a small circle having its pole where the Sun or star at that instant was in the zenith, and having an angular radius the complement of the observed altitude; and two such observations are employed in Sumner's Method for determining the ship's place.

According as the observed altitude of the Sun or the star is greater or less than the declination, the small circle on the

Earth does not or does enclose the polar axis; and the corresponding Sumner line will be a closed or open curve, whose equation may be thrown into the form

$$\cosh y/c = \sec \alpha \cos x/c, \quad \text{(i.)}$$

or

$$\sinh y/c = \tan \beta \cos x/c. \quad \text{(ii.)}$$

On Mercator's chart (§ 16) the latitude θ and the longitude ϕ of a point whose coordinates are x, y may be written

$$\phi = x/c, \ \theta = \text{amh}\, y/c,$$

where $\pi c/180$ is the length on the chart of a degree of longitude at the equator.

These relations are obtained by noticing that the bearing by compass of two adjacent points on the chart will be the same as on the terrestrial sphere, if

$$\frac{dy}{dx} = \frac{d\theta}{\cos \theta d\phi},$$

and now, if $x = c\phi$, so as to make the meridians of longitude equidistant parallel straight lines, then

$$dy/d\theta = c \sec \theta, \ y/c = \int \sec \theta d\theta,$$

or (§ 16)

$$\theta = \text{amh}\, y/c.$$

Now let δ denote the declination of the Sun or star, γ the observed altitude, ϕ the difference of longitude of the observer and of the object; then in the spherical triangle SPZ

$$PS = \tfrac{1}{2}\pi - \delta, \ SZ = \tfrac{1}{2}\pi - \alpha, \ PZ = \tfrac{1}{2}\pi - \theta, \ SPZ = \phi,$$

S denoting the Sun or star, Z the zenith of the observer, and P the pole of the Earth's axis.

Since $\cos SZ = \cos PS \cos PZ + \sin PS \sin PZ \cos SPZ$,

therefore $\sin \alpha = \sin \delta \sin \theta + \cos \delta \cos \theta \cos \phi$,

or

$$\cos \delta \cos \phi = \sin \alpha \sec \theta - \sin \delta \tan \theta$$
$$= \sin \alpha \cosh y/c - \sin \delta \sinh y/c;$$

and according as α is greater or less than δ, this is reducible to the form $A \cosh(y-b)/c$ or $-B \sinh(y-b)/c$; and this again by a change of axes to the form of (i.) or (ii.).

(Crelle, XI., Gudermann, on the *Loxodrome*; *Messenger of Mathematics*, XVI. and XX., *Sumner Lines.*)

Differentiating equation (i.) with respect to x,

$$\frac{dy}{dx} = \frac{-\sec \alpha \sin x/c}{\sinh y/c} = \frac{-\sec \alpha \sin x/c}{\sqrt{(\sec^2 \alpha \cos^2 x/c - 1)}},$$

$$\frac{ds}{dx} = \frac{\tan \alpha}{\sqrt{(\sec^2 \alpha \cos^2 x/c - 1)}} = \frac{\sin \alpha}{\sqrt{(\sin^2 \alpha - \sin^2 x/c)}};$$

so that, as in §§ 3, 4, and 8,

$$\sin x/c = \kappa \operatorname{sn} s/c, \qquad \cos x/c = \operatorname{dn} s/c,$$

$$\cosh y/c = \sin \alpha \operatorname{dn} s/c, \quad \sinh y/c = \tan \alpha \operatorname{cn} s/c,$$

the modular angle being α.

This shows that s/c in the closed Sumner Line (i.) may be equated to nt in the oscillating pendulum, and then x/c will be half the angle made by the pendulum with the vertical; also in the Sumner Line

$$\cos \psi = \frac{dx}{ds} = \operatorname{cn} s/c, \text{ or } \psi = \operatorname{am} s/c,$$

the intrinsic equation; and $\rho = c \sin \alpha \sec x/c$.

The differentiation of equation (ii.) gives in a similar manner

$$\frac{ds}{dx} = \frac{1}{\sqrt{(1 - \sin^2\beta \sin^2 x/c)}},$$

so that $x/c = \operatorname{am} s/c$, with mod. angle β;

and now, in the corresponding undulating Sumner Line, x/c is half the angle made with the vertical by a revolving pendulum, if we put $s/c = \kappa nt$.

Also $$\cos \psi = \frac{dx}{ds} = \operatorname{dn} s/c = (\operatorname{cn} \kappa s/c, 1/\kappa)$$

by § 29; so that $\psi = \operatorname{am}(\kappa s/c, 1/\kappa)$,

the intrinsic equation; and $\rho = c \operatorname{cosec} \beta \sec x/c$.

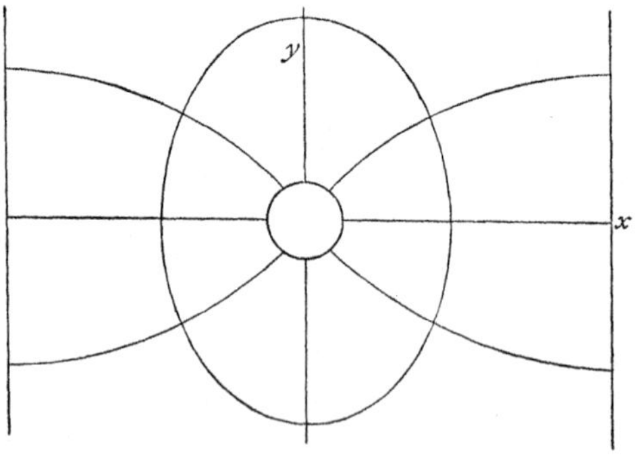

Fig. 10.

The second curve, by a shift of origin a distance $\frac{1}{2}\pi c$ to the right, becomes $\sinh y/c = \tan \beta \sin x/c$,

and then it cuts at right angles the first curve (fig. 10)

$$\cosh y/c = \sec \alpha \cos x/c.$$

For, differentiating these equations logarithmically,

$$\coth\frac{y}{c}\frac{dy}{dx}=\quad\cot\frac{x}{c},$$

$$\tanh\frac{y}{c}\frac{dy}{dx}=-\tan\frac{x}{c};$$

and therefore the product of the $\frac{dy}{dx}$'s is -1.

In fact putting $\sec\alpha=\coth\alpha'$, the curves are derivable as conjugate functions from the equation

$$x+iy=c\operatorname{amh}(\alpha'+i\beta).$$

99. Problem VIII. *Catenaries.*

"The catenary for a line density proportional to $\cosh s/a$, where s is the length of the arc measured from the lowest point, is of the form

$$\tanh y/b=\operatorname{dn} x/a,\text{ or }\operatorname{dn} x/b,$$

according as a, the ratio of the tension in pounds to the density in lb. per foot at the lowest point of the catenary is greater or less than b; the *Catenary of Uniform Strength* being the curve in the separating case of $a=b$."

The equation of the *Catenary of Uniform Strength*, in which the linear density or cross section is so arranged as to be proportional to the tension, is well known (Thomson and Tait, *Natural Philosophy*, § 583) being

$$e^{y/b}\cos x/b=1,\text{ or }e^{y/b}=\sec x/b;$$

or as it may be written

$$\tanh\tfrac{1}{2}y/b=\tan^2\tfrac{1}{2}x/b.$$

For if σ_0 denotes the density in lb. per foot, and $\sigma_0 b$ the tension in pounds at the lowest point A, σ the density and σb the tension at any other point P, at a distance s from A, measured along the curve, the equations of equilibrium of AP are

$$\sigma b\cos\psi=\sigma_0 b,\ \sigma b\sin\psi=\int\sigma ds.$$

Thence $\quad\sigma=\sigma_0\sec\psi$, and $\int\sigma ds=\sigma_0 b\tan\psi$;

so that $\quad\sigma=\sigma_0 b\sec^2\psi d\psi/ds=\sigma_0\sec\psi,$

or $\quad ds/d\psi=b\sec\psi,$

$$s=\int_0 b\sec\psi d\psi=b\cosh^{-1}\sec\psi=b\cosh^{-1}\sigma/\sigma_0,$$

$$\sigma=\sigma_0\cosh s/b.$$

We might therefore take a piece of uniform flexible and inextensible material, cut out from a plane piece by two catenaries, or modified catenaries, say $y/c = \pm\cosh x/b$, and hang it up in a catenary of equal strength.

Also $$x = \int \cos\psi ds = \int b d\psi = b\psi,$$

$$y = \int \sin\psi ds = \int b \tan\psi d\psi = b \log\sec\psi;$$

so that $$y/b = \log\sec x/b, \text{ or } e^{y/b} = \sec x/b,$$

the equation of the Catenary of Uniform Strength.

But now suppose two supports at the same level to be made to approach or recede from each other; the piece of cloth or the chain will hang in a different catenary.

Denoting by $\sigma_0 a$ the tension in pounds at the lowest point A, and by t the tension at P, then

$$t\cos\psi = \sigma_0 a,\ t\sin\psi = \int \sigma ds = \sigma_0 b \sinh s/b;$$

so that $$p \text{ or } \frac{dy}{dx} = \tan\psi = \frac{b}{a}\sinh\frac{s}{b},$$

the intrinsic equation of the curve.

Then $$\frac{dp}{dx} = \frac{1}{a}\cosh\frac{s}{b}\frac{ds}{dx} = \sqrt{\left(\frac{1}{a^2}+\frac{p^2}{b^2}\right)}\sqrt{(1+p^2)},$$

or $$x = \int \frac{abdp}{\sqrt{(b^2+a^2p^2 \,.\, 1+p^2)}},$$

an elliptic integral, of the form (10), p. 33; and putting $p = \tan\psi$,

$$\frac{d\psi}{dx} = \sqrt{\left(\frac{\cos^2\psi}{a^2}+\frac{\sin^2\psi}{b^2}\right)}.$$

In the separating case, $a = b$; and then $x = b\psi$, as in the Catenary of Uniform Strength; the greatest possible span of a catenary of given material is therefore $\pi b = \pi\tau/w$, where τ denotes the tenacity of the material, in pounds per sq. foot, and w the density or heaviness, in lb. per cubic foot.

But with $a > b$,

$$\frac{d\psi}{dx} = \frac{1}{b}\sqrt{(1-\kappa^2\cos^2\psi)} = \frac{1}{b}\Delta(\tfrac{1}{2}\pi+\psi, \kappa), \text{ where } \kappa' = b/a;$$

so that $$\tfrac{1}{2}\pi + \psi = \operatorname{am} x/b,$$

and $$\frac{dy}{dx} = \tan\psi = -\frac{\operatorname{cn} x/b}{\operatorname{sn} x/b},$$

$$y = \int_0 \frac{-\operatorname{cn} x/b \operatorname{sn} x/b}{\operatorname{sn}^2 x/b}dx = \int \frac{-\kappa^2 \operatorname{sn} x/b \operatorname{sn} x/b}{1-\operatorname{dn}^2 x/b}dx = b\tanh^{-1}\operatorname{dn} x/b,$$

or $$\tanh y/b = \operatorname{dn} x/b.$$

With $a<b$,

$$\frac{d\psi}{dx}=\frac{1}{a}\sqrt{(1-\kappa^2\sin^2\psi)}=\frac{1}{a}\Delta(\psi,\kappa),\text{ where }\kappa'=a/b;$$

so that $\psi=\text{am}\,x/a$,

and $$\frac{dy}{dx}=\tan\psi=\frac{\text{sn}\,x/a}{\text{cn}\,x/a},$$

$$y=\int_0\frac{\text{sn}\,x/a\,\text{cn}\,x/a}{\text{cn}^2x/a}dx=\int\frac{\kappa^2\text{sn}\,x/a\,\text{cn}\,x/a}{\text{dn}^2x/a-\kappa'^2}dx$$

$$=\frac{a}{2\kappa'}\log\frac{\text{dn}\,x/a+\kappa'}{\text{dn}\,x/a-\kappa'}=b\coth^{-1}\frac{\text{dn}\,x/a}{\kappa'},$$

or $$\tanh y/b=\frac{\kappa'}{\text{dn}\,x/a}=\text{dn}(K-x/a),$$

by § 57; so that by a change of origin, taking the axis of y in a vertical asymptote of the curve, its equation may be written

$$\tanh y/b=\text{dn}\,x/a.$$

(Compare Cayley, on *A Torse depending on Elliptic Functions*, Q. J. M., XIV., p. 241.)

100. In the catenary formed by an elastic rope or flexible wire, obeying Hooke's Law "*ut tensio sic vis*," we may still have $p=\sinh u$; but u is no longer proportional to the arc s.

We use σ_0 to denote the uniform density of the rope when unstretched, and s_0 to denote the length of rope which stretches in AP to length s, $\sigma_0 b$ denotes as before the tension in pounds of the rope at the lowest point A, and $\sigma_0 c$ is used to denote the modulus of elasticity of the rope in pounds; so that, by Hooke's law,

$$\frac{ds}{ds_0}=1+\frac{t}{\sigma_0 c}.$$

Then, as before, for the equilibrium of AP,

$$t\cos\psi=\sigma_0 a,\quad t\sin\psi=\int\sigma ds=\sigma_0 s_0,$$

so that $$p=\frac{dy}{dx}=\frac{s_0}{b}=\sinh u,$$

if we put $$s_0=a\sinh u;$$

and then $$t=\sigma_0\sqrt{(a^2+s_0{}^2)}=\sigma_0 a\cosh u.$$

Then $$\frac{ds}{du}=\left(1+\frac{t}{\sigma_0 c}\right)\frac{ds_0}{du}=a\cosh u+\frac{a^2}{c}\cosh^2 u,$$

and $$\frac{ds}{dx}=\sqrt{(1+p^2)}=\cosh u,$$

so that $$\frac{dx}{du}=a+\frac{a^2}{c}\cosh u,$$

$$\frac{dy}{du}=a\sinh u+\frac{a^2}{c}\cosh u\sinh u.$$

Integrating, putting $a/c=h$,

$$s/a=\sinh u+\tfrac{1}{2}h(u+\cosh u\sinh u),$$

$$x/a=\quad u+\ h\sinh u,$$

$$y/a=\cosh u+\tfrac{1}{2}h\sinh^2 u.$$

For the corresponding points on the rope, when it is supposed inextensible, putting $c=\infty$, and $h=0$,

$$s_0/a=\sinh u,\ x_0/a=u,\ y_0/a=\cosh u,$$

giving an ordinary catenary; so that the tangents are parallel at corresponding points of the catenaries of the elastic and of the inextensible rope.

The terms depending on h, considered separately, define an ordinary parabola; so that the catenary formed by an elastic rope is something intermediate to a parabola and a common catenary.

101. Problem IX. *Geodesics.*

"Investigation of the geodesics on the *Catenoid*, the surface formed by the revolution of a catenary round its directrix, and on the *Helicoid*, into which it can be developed; also of the geodesics on the *Unduloid* and *Nodoid*, the capillary surfaces of revolution, of which the meridian curves are the roulette of the focus of a conic section, an ellipse or hyperbola, rolling upon the axis of revolution."

The simplest mode of determining a geodesic on a surface of revolution is to treat it as the path of a particle moving under no forces on the surface, considered as smooth, so that ds/dt is constant; and then, since the reaction of the surface passes through the axis, $r^2 d\theta/dt$ is constant; and therefore

$$r^2\frac{d\theta}{ds}=b, \text{ a constant,}$$

r and θ denoting the polar coordinates of any point of the projection on a plane perpendicular to the axis Ox; and thus

$$\frac{ds^2}{d\theta^2}=\frac{dx^2}{d\theta^2}+\frac{dr^2}{d\theta^2}+r^2=\frac{r^4}{b^2}.$$

In the catenoid $\quad r/a = \cosh x/a$,

so that
$$\frac{dr}{dx} = \sinh x/a = \frac{\sqrt{(r^2-a^2)}}{a};$$

and therefore, in the geodesic,
$$\frac{r^2-a^2}{a^2}\frac{dr^2}{d\theta^2} + \frac{dr^2}{d\theta^2} + r^2 = \frac{r^4}{b^2},$$

or
$$\frac{dr^2}{d\theta^2} = \frac{(r^2-a^2)(r^2-b^2)}{b^2}.$$

We must distinguish the two cases according as $b^2 \gtrless a^2$.

When $b^2 > a^2$, then $r^2 > b^2$; the geodesic osculates the circular cross section of radius b; and we have
$$r \operatorname{sn} \theta = b, \text{ with } \kappa = a/b,$$
as the polar equation of the projection of the geodesic.

When $b^2 < a^2$, then $r^2 > a^2$; the geodesic crosses the circular section of minimum radius a; and supposing it cuts the meridian here at an angle α, $b = a \sin \alpha$; and now
$$r \operatorname{sn}(\theta/\kappa) = a, \text{ the modular angle being } \alpha.$$

In the separating case, $b = a$ and $\kappa = 1$; and then $\operatorname{sn}\theta = \tanh\theta$; so that
$$r \tanh \theta = a$$
is now the polar equation of the projection of the geodesic, a curve having $r = a$ as an asymptotic circle.

Generally in any geodesic on a surface of revolution, which cuts the meridian curve at a distance r from the axis at an angle χ,
$$\sin \chi = r\frac{d\theta}{ds} = \frac{b}{r};$$
so that $\sin \chi$ varies inversely as r.

102. Now suppose the catenoid is divided along a meridian curve AP, and again along the smallest circular section AA', and that this section AA' is drawn out into a straight line, of length $2\pi a$; the rest of the surface, if flexible and inextensible, will assume the form of a *Helicoid*, or uniform screw surface of pitch a, such that its equation is
$$z = a\phi,$$
taking the axis of z along the axis of the surface, and ρ, ϕ the polar coordinates of the projection of a point on a plane perpendicular to the axis; and AP will become a generating line of the Helicoid; this is proved geometrically, by noticing that the length of the helix PP' on the Helicoid is equal to the length of the circle PP' on the Catenoid.

The surface being inextensible, and a circular cross section of the Catenoid becoming a helix on the Helicoid, it follows that

$$r^2 d\theta^2 = \rho^2 d\phi^2 + dz^2 = (\rho^2 + a^2) d\phi^2 ;$$

and since $$r^2 = \rho^2 + a^2, \text{ therefore } \theta = \phi.$$

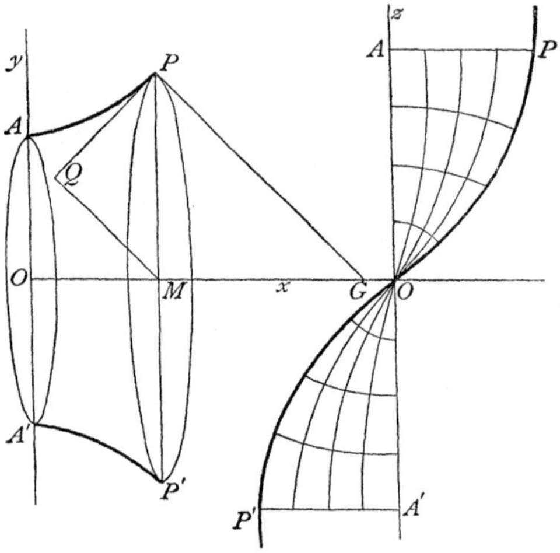

Fig. II.

Therefore the equation of the projection of a geodesic on the helicoid is either of the forms

$$(\rho^2 + a^2)\text{sn}^2(\phi/\kappa) = a^2,$$

$$\rho \text{ tn } (\phi/\kappa) = a ;$$

or $$(\rho^2 + a^2)\text{sn}^2\phi = b^2 = a^2/\kappa^2,$$

$$\rho = \frac{a \text{ dn } \phi}{\kappa \text{ sn } \phi},$$

$$\rho \text{ cn}(K - \phi) = a\kappa'/\kappa.$$

The Catenoid is the surface of revolution formed by a capillary soap bubble film, when the pressure of the air is the same on both sides of the film. The surface is easily formed practically by dipping a circular wire into soapy water and raising it vertically; and it is evident from mechanical considerations that the surface is a *minimum surface* (§ 31).

The Helicoid, into which the Catenoid can be deformed, can be produced in the same manner by a film between two coaxial helical wires of the same pitch (C. V. Boys, *Soap Bubbles*).

These surfaces are particular cases of *Scherk's minimum surface,* whose equation is

$$z = a \tan^{-1}\frac{y}{x} + a \tan^{-1}\frac{a\sqrt{(x^2+y^2-b^2)}}{b\sqrt{(x^2+y^2+a^2)}} + b \tanh^{-1}\frac{\sqrt{(x^2+y^2-b^2)}}{\sqrt{(x^2+y^2+a^2)}},$$

or

$$z = a \cos^{-1}\frac{bx\sqrt{(x^2+y^2+a^2)} - ay\sqrt{(x^2+y^2-b^2)}}{\sqrt{(a^2+b^2)(x^2+y^2)}}$$

$$+ b \cosh^{-1}\frac{\sqrt{(x^2+y^2+a^2)}}{\sqrt{(a^2+b^2)}};$$

reducing to the Catenoid when $a=0$, and to the Helicoid when $b=0$.

The verification in the manner of § 32 is left as an exercise for the student.

103. The meridian curve of the Catenoid is the roulette AP of the focus of a parabola aG, the pressure of the air being the same on both sides of the film (fig. 12).

But when the pressure of the air inside the film is increased or diminished, we find that the surface of revolution formed by the capillary film has as meridian curve BP or CP, the roulette of the focus of an ellipse or hyperbola, the first surface being called the *Unduloid* and the second the *Nodoid.*

(Maxwell, *Capillary Attraction, Encyclopædia Britannica.*)

Denoting by y, y' the perpendiculars from the foci P, P' on the axis Ox on which the conic rolls, then in the Unduloid BP, generated by the focus P of a rolling ellipse bQ,

$$y + y' = (PQ + QP')\cos\psi = 2a\cos\psi,$$

and $$yy' = b^2;$$

so that $$b^2 + y^2 = 2ay\cos\psi.$$

If in the meridian curve BP of the Unduloid, we denote the radius of curvature by ρ, and the normal PG by n, then,

since $$b^2 + y^2 = 2ay\cos\psi = 2ay^2/n,$$

therefore $$\frac{1}{n} = \frac{b^2}{2ay^2} + \frac{1}{2a};$$

and since $$\cos\psi = \frac{b^2}{2ay} + \frac{y}{2a},$$

differentiating,

$$\sin\psi\frac{d\psi}{ds} = \left(\frac{b^2}{2ay^2} - \frac{1}{2a}\right)\frac{dy}{ds},$$

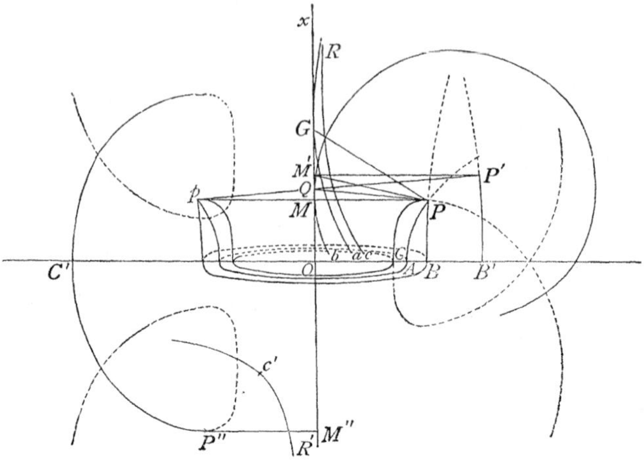

Fig. 12.

or $$\frac{1}{\rho}=\frac{b^2}{2ay^2}-\frac{1}{2a};$$

so that $$\frac{1}{n}-\frac{1}{\rho}=\frac{1}{a}.$$

Then, if p denotes the excess over the atmospheric pressure of the air inside a capillary film, in the shape of an Unduloid, and t the tension of the film,

$$p=t\left(\frac{1}{n}-\frac{1}{\rho}\right)=\frac{t}{a};$$

so that, if inside a Catenoid, the pressure is increased, the surface is changed into an Unduloid.

If the pressure is slightly diminished by p, the surface becomes a portion of a Nodoid CP; for now

$$p=t\left(\frac{1}{\rho}-\frac{1}{n}\right),$$

and in the meridian curve CP of the Nodoid, the roulette of the focus P of a hyperbola cR with foci P and P'',

$$y''-y=(P''R-RP)\cos\psi=2a\cos\psi, \text{ and } yy'=b^2;$$

so that $$b^2-y^2=2ay\cos\psi=2ay^2/n;$$

$$\frac{1}{n}=\frac{b^2}{2ay^2}-\frac{1}{2a},$$

$$\frac{1}{\rho}=\frac{b^2}{2ay^2}+\frac{1}{2a},$$

and $$p=t/a.$$

In the geodesic on the Unduloid,

$$y^2 d\theta/ds = \alpha \sin\gamma,$$

supposing the geodesic cuts the meridian curve at an angle γ at its maximum distance α from the axis; also $\alpha = a(1+e)$, and the minimum distance $\beta = a(1-e)$, so that $\alpha\beta = b^2$, $\alpha+\beta = 2a$; and y lies between α and β.

Now, in the projection of the geodesic on a plane perpendicular to Ox, writing r for y, so that $\tan\psi = dy/dx = dr/dx$,

$$\frac{ds^2}{d\theta^2} = \frac{dx^2}{d\theta^2} + \frac{dr^2}{d\theta^2} + r^2 = \frac{dr^2}{d\theta^2}\operatorname{cosec}^2\psi + r^2 = \frac{r^4}{\alpha^2\sin^2\gamma};$$

or

$$\frac{dr^2}{d\theta^2} = r^2\sin^2\psi\left(\frac{r^2}{\alpha^2\sin^2\gamma} - 1\right);$$

and $r\cos\psi = (b^2+r^2)/2a$; so that

$$\frac{dr^2}{d\theta^2} = \left\{r^2 - \frac{(b^2+r^2)^2}{4a^2}\right\}\left(\frac{r^2}{\alpha^2\sin^2\gamma} - 1\right)$$

$$= \frac{(\alpha^2 - r^2)(r^2 - \beta^2)(r^2 - \alpha^2\sin^2\gamma)}{\alpha^2(\alpha+\beta)^2\sin^2\gamma};$$

leading to integrals of the form (72) and (73), p. 52.

We suppose first that $\beta > \alpha\sin\gamma$, so that the geodesic crosses the minimum section of the surface, and therefore all the sections if produced; and now with $\alpha > r > \beta > \alpha\sin\gamma$, we have, according to equation (72),

$$m\theta = \int_r^\alpha \frac{m\alpha(\alpha+\beta)\sin\gamma\, dr}{\sqrt{(\alpha^2 - r^2 \,.\, r^2 - \beta^2 \,.\, r^2 - \alpha^2\sin^2\gamma)}} = \operatorname{sn}^{-1}\sqrt{\frac{\beta^2 \,.\, \alpha^2 - r^2}{\alpha^2 - \beta^2 \,.\, r^2}};$$

or

$$\frac{1}{r^2} = \frac{\operatorname{cn}^2 m\theta}{\alpha^2} + \frac{\operatorname{sn}^2 m\theta}{\beta^2}.$$

Secondly, if $\alpha > r > \alpha\sin\gamma > \beta$, then the geodesic osculates the circle of radius $\alpha\sin\gamma$, and is limited by the convex part of the surface between two such circles; and the equation of the projection of the geodesic is obtained from the above merely by interchanging $\alpha\sin\gamma$ and β.

In the separating case $\alpha\sin\gamma = \beta$; and then $\kappa = 1$, $m = \tan\frac{1}{2}\gamma$; and the polar equation of the projection of the geodesic is

$$\frac{1}{r^2} = \frac{\operatorname{sech}^2 m\theta}{\alpha^2} + \frac{\tanh^2 m\theta}{\beta^2},$$

a curve having an asymptotic circle $\gamma = \beta$.

The formulas are similar for the geodesics on the Nodoid.

104. *Euler's Equations resumed. Poinsot's Geometrical Representation of the Motion of a Body under No Forces.*

We now resume these equations of motion, of which the solution by elliptic functions has been indicated in § 32. p. 78

By the *Principle of the Conservation of Angular Momentum* (Routh, *Rigid Dynamics*, Chap. IX.) the axis OC of the resultant angular momentum G will be fixed in space; and the direction cosines of this axis with respect to the principal axes of the body being

$$Ap/G,\ Bq/G,\ Cr/G,$$

the component angular velocity about OC will be

$$\frac{Ap^2+Bq^2+Cr^2}{G}=\frac{T}{G},\ \text{a constant},$$

where, as before, T denotes twice the kinetic energy of the body.

It is convenient to denote this component of angular velocity about OC by a single letter, say μ; and also to replace G and T by $D\mu$ and $D\mu^2$, making $T/G=\mu$ and $G^2/T=D$; and then D will be a constant quantity, of the same dimensions as A, B, C.

If I denotes the moment of inertia about the instantaneous axis of rotation OP, and if OP denotes the vector of the momental ellipsoid at O, then I varies as OP^{-2}, so that we may put $I=Dh^2/OP^2$, where h is a new constant length.

Now, if ω denotes the resultant angular velocity about OP,

$$T=I\omega^2, \text{ or } D\mu^2=Dh^2\omega^2/OP^2,$$

so that the angular velocity ω varies as OP; and

$$\frac{h}{\mu}=\frac{OP}{\omega}=\frac{x}{p}=\frac{y}{q}=\frac{z}{r}.$$

The direction cosines of the normal of the momental ellipsoid at P being proportional to Ax, By, Cz, or Ap, Bq, Cr, are therefore Ap/G, Bq/G, Cr/G; so that OC, the axis of G, is perpendicular to the tangent plane at P; and if OC meets this tangent plane in C, it follows that $OC=h$, so that the tangent plane at P is a fixed plane; and during the motion the momental ellipsoid rolls on this fixed plane, called the *invariable plane*, with angular velocity proportional to OP.

The curve traced out by the point of contact P on the momental ellipsoid is called the *polhode*, and the curve traced out by P on the invariable plane is called the *herpolhode;*

these names are due to Poinsot, as well as this geometrical representation of the motion.

(*Théorie nouvelle de la rotation des corps*, Paris, 1852.)

The equation of the momental ellipsoid may now be written

$$Ax^2+By^2+Cz^2=Dh^2;$$

while Ax/Dh, By/Dh, Cz/Dh are the direction cosines of the *invariable line* OC; so that

$$A^2x^2+B^2y^2+C^2z^2=D^2h^2.$$

The polhode is therefore the curve of intersection of these two coaxial quadric surfaces, and therefore lies on the cone

$$A(A-D)x^2+B(B-D)y^2+C(C-D)z^2=0,$$

called the *polhode* cone; and the projections of the polhode on the principal planes are therefore

$$(A-B)By^2+(A-C)Cz^2=(A-D)Dh^2, \ldots.$$

105. Denoting by ν the component angular velocity of the body about the axis OH, where OH is equal and parallel to CP,

$$p^2+q^2+r^2=\omega^2=\mu^2+\nu^2,$$
$$Ap^2+Bq^2+Cr^2=T=D\mu^2,$$
$$A^2p^2+B^2q^2+C^2r^2=G^2=D^2\mu^2;$$

and, by solution of these equations,

$$\frac{A-B\,.\,A-C}{BC}p^2=\omega^2-\left(\frac{1}{B}+\frac{1}{C}\right)T-\frac{G^2}{BC}=\omega^2-\omega_a{}^2, \text{ suppose,}$$

or

$$=\nu^2+\left(1-\frac{D}{B}\right)\left(1-\frac{D}{C}\right)\mu^2=\nu^2-\nu_a{}^2, \text{ suppose;}$$

$$\frac{B-C\,.\,B-A}{CA}q^2=\nu^2+\left(1-\frac{D}{C}\right)\left(1-\frac{D}{A}\right)\mu^2=\nu^2-\nu_b{}^2, \ldots,$$

$$\frac{C-A\,.\,C-B}{AB}r^2=\nu^2+\left(1-\frac{D}{A}\right)\left(1-\frac{D}{B}\right)\mu^2=\nu^2-\nu_c{}^2, \ldots;$$

and in these equations we may replace p, q, r, ω, μ, ν by x, y, z, OP, h, ρ, respectively, where $\rho^2=OP^2-h^2$.

Example.—Prove that

$$\left(A\frac{dp}{dt}\right)^2+\left(B\frac{dq}{dt}\right)^2+\left(C\frac{dr}{dt}\right)^2=(D\mu\nu)^2,$$

$$A\left(\frac{dp}{dt}\right)^2+B\left(\frac{dq}{dt}\right)^2+C\left(\frac{dr}{dt}\right)^2=\mu^2\nu^2+\frac{A-D\,.\,B-D\,.\,C-D}{ABC}\mu^4,$$

and simplify

$$\left(\frac{dp}{dt}\right)^2+\left(\frac{dq}{dt}\right)^2+\left(\frac{dr}{dt}\right)^2.$$

106. On the supposition that

$$AT > BT > G^2 > CT, \text{ or } A > B > D > C,$$

r never vanishes, and the polhode encloses the principal axis C; but p and q alternately vanish, so that ν^2 oscillates in value between $\left(1-\frac{D}{B}\right)\left(\frac{D}{C}-1\right)\mu^2$ and $\left(1-\frac{D}{A}\right)\left(\frac{D}{C}-1\right)\mu^2$.

If we put $\frac{\nu^2}{\mu^2}=\left(\frac{D}{C}-1\right)\left\{\left(1-\frac{D}{A}\right)\cos^2\theta+\left(1-\frac{D}{B}\right)\sin^2\theta\right\}$,

then
$$Ap^2=D\mu^2\frac{D-C}{A-C}\cos^2\theta,$$

$$Bq^2=D\mu^2\frac{D-C}{B-C}\sin^2\theta,$$

$$Cr^2=D\mu^2\left(\frac{A-D}{A-C}\cos^2\theta+\frac{B-D}{B-C}\sin^2\theta\right).$$

We now find, on substituting in one of Euler's equations,

$$\frac{d\theta^2}{dt^2}=D\mu^2\frac{A-C\,.\,B-C}{ABC}\left(\frac{A-D}{A-C}\cos^2\theta+\frac{B-D}{B-C}\sin^2\theta\right)$$

and
$$\frac{d^2\theta}{dt^2}=-D\mu^2\frac{(A-B)(D-C)}{ABC}\sin\theta\cos\theta,$$

the solution of which is of the form, as before in §§ 18 and 32,

$$\theta=\operatorname{am}(nt,\kappa),$$

where
$$n^2=D\mu^2\frac{A-D\,.\,B-C}{ABC}, \text{ and } \kappa^2=\frac{A-B\,.\,D-C}{A-D\,.\,B-C},$$

the anharmonic ratio of A, B, D, C; while

$$Ap^2=D\mu^2\frac{D-C}{A-C}\operatorname{cn}^2 nt,$$

$$Bq^2=D\mu^2\frac{D-C}{B-C}\operatorname{sn}^2 nt,$$

$$Cr^2=D\mu^2\frac{A-D}{A-C}\operatorname{dn}^2 nt;$$

giving (§ 32)

$$P^2=\frac{D\,.\,D-C}{A\,.\,A-C}\mu^2,\ Q^2=\frac{D\,.\,D-C}{B\,.\,B-C}\mu^2,\ R^2=\frac{D\,.\,A-D}{C\,.\,A-C}\mu^2.$$

107. *Quadrantal Oscillations.*

The oscillations given by a differential equation of the form

$$d^2\theta/dt^2=-m^2\sin\theta\cos\theta$$

are called *quadrantal oscillations* (Thomson and Tait, *Natural Philosophy*, § 322), the system having two positions of stable equilibrium given by $\theta=0$ and $\theta=\pi$, and two unstable positions in the remaining quadrants, given by $\theta=\pm\frac{1}{2}\pi$; for instance, an elongated piece of soft iron in a uniform magnetic field, or an elliptic cylinder moveable about its axis in a current of liquid performs quadrantal oscillations. (*Q. J. M.*, xvi.)

When the system performs complete revolutions, the solution is (§ 18) $\theta=\text{am}(mt/\kappa, \kappa)$;

but if it oscillates about the positions of stable equilibrium, given by $\theta=0$, the solution is (§ 29)

$$\theta= \text{am}(mt, 1/\kappa),$$

or

$$\cos\theta= \text{dn}(mt/\kappa, \kappa),$$

$$\sin\theta=\kappa\,\text{sn}(mt/\kappa, \kappa),$$

where κ is less than unity.

The second solution will apply to the second state of motion in § 32, where $AT>G^2>BT>CT$, or $A>D>B>C$, and where p never vanishes, and the polhode encloses the principal axis A.

108. Differentiating the equations of § 105 with respect to t,

$$\omega\frac{d\omega}{dt}=\nu\frac{d\nu}{dt}=\frac{A-B.A-C}{BC}p\frac{dp}{dt}=\frac{B-C.B-A}{CA}q\frac{dq}{dt}=\frac{C-A.C-B}{AB}r\frac{dr}{dt}$$

$$=-\frac{B-C.C-A.A-B}{ABC}pqr;$$

or

$$\frac{d\omega^2}{dt}=-\surd(4.\omega_a^2-\omega^2.\omega_b^2-\omega^2.\omega_c^2-\omega^2),$$

$$\frac{d\nu^2}{dt}=-\surd(4.\nu_a^2-\nu^2.\nu_b^2-\nu^2.\nu_c^2-\nu^2);$$

so that ω^2 and ν^2 are elliptic functions of t, of the form given by equation (15), p. 36.

But, on reference to equation (A), p. 43, we see that

$$\wp'u=-\surd(4\wp^3u^3-g_2\wp u-g_3)=-\surd(4.\wp u-e_a.\wp u-e_b.\wp u-e_c),$$

if e_a, e_b, e_c denote the roots of $4s^3-g_2s-g_3=0$; so that on comparison we may make

$$\omega_a^2-\omega^2,\ \omega_b^2-\omega^2,\ \omega_c^2-\omega^2, \text{ or } \nu_a^2-\nu^2,\ \nu_b^2-\nu^2,\ \nu_c^2-\nu^2,$$

proportional to $\wp u-e_a$, $\wp u-e_b$, $\wp u-e_c$;

or, symmetrically, we can put

$$Ap^2=-m^2(B-C)(\wp u-e_a),$$

$$Bq^2=-m^2(C-A)(\wp u-e_b),$$

$$Cr^2=-m^2(A-B)(\wp u-e_c);$$

where the factor $-m^2$ is introduced for the sake of *homogeneity*,

m being of the dimensions of an angular velocity, such as p, q, r, ω, μ, ν; and now, on substitution in Euler's equations,

$$\frac{du^2}{dt^2} = -\frac{B-C \,.\, C-A \,.\, A-B}{ABC} m^2 = \left(\frac{B-C}{A} + \frac{C-A}{B} + \frac{A-B}{C}\right) m^2 = n^2,$$

suppose; so that $\qquad u = \text{a constant} \pm nt.$

109. As in § 32, we take $A > B > C$; and then

(i.) when $\quad AT > BT > G^2 > CT$, or $A > B > D > C$,

r never vanishes, and we must take

$$e_c > e_a > \wp u > e_b;$$

so that $\qquad e_1 = e_c,\ e_2 = e_a,\ e_3 = e_b;$

(ii.) when $\quad AT > G^2 > BT > CT$, or $A > D > B > C$,

p never vanishes; and then

$$e_a > e_c > \wp u > e_b;$$

and we must take $\quad e_1 = e_a,\ e_2 = e_c,\ e_3 = e_b.$

Since $\wp u$ oscillates between e_2 and e_3, and is taken initially equal to e_3, we find, on reference to equation (42), p. 45, that we must put

$$u = 2\omega_1 + \omega_3 - nt,$$

so that the constant of integration for u in § 108 is $2\omega_1 + \omega_3$.

Now, at the cost of symmetry, to get rid of the imaginary ω_3, and to make the argument of the elliptic functions a real quantity nt, equation (42), expressed in the direct notation,

gives
$$\wp u - e_3 = \frac{e_1 - e_3 \,.\, e_2 - e_3}{\wp nt - e_3},$$

$$\wp u - e_2 = \frac{e_1 - \wp nt \,.\, e_2 - e_3}{\wp nt - e_3},$$

$$\wp u - e_1 = \frac{e_1 - e_3 \,.\, e_2 - \wp nt}{\wp nt - e_3};$$

and e_b always replaces e_3, while e_a replaces e_1, e_c replaces e_2, or *vice versa*, according as the polhode encloses A or C.

110. For the determination of e_a, e_b, e_c, we have the equations

$$e_a + e_b + e_c = 0,$$

$$(B-C)e_a + (C-A)e_b + (A-B)e_c = T/m^2 = D\mu^2/m^2,$$

$$A(B-C)e_a + B(C-A)e_b + C(A-B)e_c = G^2/m^2 = D^2\mu^2/m^2,$$

whence
$$AT - G^2 = m^2(C-A)(A-B)(e_b - e_c),$$

$$BT - G^2 = m^2(A-B)(B-C)(e_c - e_a),$$

$$CT - G^2 = m^2(B-C)(C-A)(e_a - e_b);$$

or
$$e_b - e_c = \frac{D\mu^2}{m^2}\frac{A-D}{C-A\,.\,A-B},$$
$$e_c - e_a = \frac{D\mu^2}{m^2}\frac{B-D}{A-B\,.\,B-C},$$
$$e_a - e_b = \frac{D\mu^2}{m^2}\frac{C-D}{B-C\,.\,C-A};$$
so that $e_c - e_a$ is taken positive or negative, according as $BT - G^2$ or $B - D$ is positive or negative; while $e_b - e_c$ and $e_b - e_a$ are always negative, as explained above.

Also $$(e_a - e_b) - (e_c - e_a) = 3e_a, \ldots,$$
whence the values of e_a, e_b, e_c.

Then $$g_2 = \tfrac{2}{3}\{(e_b - e_c)^2 + (e_c - e_a)^2 + (e_a - e_b)^2\}$$
can be found; and the *discriminant* (§ 53)
$$\Delta = 16(e_b - e_c)^2(e_c - e_a)^2(e_a - e_b)^2$$
$$= 16\frac{D^6\mu^{12}}{m^{12}}\frac{(A-D)^2(B-D)^2(C-D)^2}{(B-C)^4(C-A)^4(A-B)^4};$$
$$J = \frac{g_2{}^3}{\Delta} = \frac{\{(B-C)^2(A-D)^2 + (C-A)^2(B-D)^2 + (A-B)^2(C-D)^2\}^3}{108(B-C)^2(C-A)^2(A-B)^2(A-D)^2(B-D)^2(C-D)^2}.$$

111. We have supposed no forces to act; but the case in which the impressed couple is always parallel and proportional to the resultant angular momentum leads to equations which can be solved in a similar manner; in this way we imitate the motion of a body, like the Earth, which is cooling and contracting uniformly.

Now, the component impressed couples about the principal axes being of the form λAp, λBq, λCr,
$$A(dp/dt) - (B-C)qr = \lambda Ap, \ldots,$$
which, on putting $p = e^{-\lambda t}p'$, and $\lambda t' = 1 - e^{-\lambda t}$, reduce to
$$A\frac{dp'}{dt'} - (B-C)q'r' = 0, \ldots,$$
so that p', q', r' are the same functions of t', which p, q, r would be of t, in the case where no forces act.

In the case of the cooling and contracting body, we put $A = e^{-\lambda t}A_0$, $B = e^{-\lambda t}B_0$, $C = e^{-\lambda t}C_0$; and the equations become
$$A_0\frac{dp}{dt'} - (B_0 - C_0)qr = 0, \ldots,$$
which are solved as before; and Poinsot's geometrical representation of the motion still holds, with slight modification.

A similar procedure will solve the following theorem:

"A rigid body is moving under the action of a force whose direction and magnitude are constant, always passing through the centre of inertia (*e.g.* gravity), and of an absolutely constant couple.

"If p, q, r denote the component angular velocities about the principal axes at the centre of inertia, and if u, v, w denote the compound velocities of the centre of inertia along the principal axes at the time t; then the determination of

$$p/t,\ q/t,\ r/t,\ u/t,\ v/t,\ w/t,$$

in terms of $\frac{1}{2}t^2$ is the same as that of p, q, r, u, v, w, in terms of t, when no forces act; t being reckoned from the commencement of the motion." (W. Burnside, *Math. Tripos*, 1881.)

112. To obtain the equation of the herpolhode, we notice that during the motion the polhode cone, fixed in the body, rolls on the herpolhode cone, fixed in space, O being the common vertex; corresponding areas of these cones are therefore equal, as also their projections on any fixed plane, for instance the invariable plane.

Therefore if ρ, ϕ denote with respect to C the polar coordinates of P on the herpolhode,

$$\rho^2\frac{d\phi}{dt} = \frac{Ax}{Dh}\left(y\frac{dz}{dt} - z\frac{dy}{dt}\right) + \frac{By}{Dh}\left(z\frac{dx}{dt} - x\frac{dz}{dt}\right) + \frac{Cz}{Dh}\left(x\frac{dy}{dt} - y\frac{dx}{dt}\right).$$

Since
$$\frac{x}{p} = \frac{y}{q} = \frac{z}{r} = \frac{\rho}{\nu} = \frac{h}{\mu},$$

therefore
$$\frac{dx}{dt} = \frac{\mu}{h}\frac{B-C}{A}yz,$$

and (§ 104)
$$y\frac{dz}{dt} - z\frac{dy}{dt} = \frac{\mu}{h}\left(\frac{A-B}{C}xy^2 - \frac{C-A}{B}xz^2\right) = \frac{D(A-D)}{BC}\mu hx,$$

so that
$$\rho^2\frac{d\phi}{dt} = \left(\frac{A-D}{BC}Ax^2 + \frac{B-D}{CA}By^2 + \frac{C-D}{AB}Cz^2\right)\mu$$

$$= \frac{(A-D)A^2x^2 + (B-D)B^2y^2 + (C-D)C^2z^2}{ABC}\mu$$

$$= \rho^2\mu + \frac{A-D\,.\,B-D\,.\,C-D}{ABC}h^2\mu;$$

which, combined with the value of $d\nu^2/dt$ or $d\rho^2/dt$ of § 108,

$$\frac{d\rho^2}{dt} = -\frac{\mu}{h}\sqrt{(4\,.\,\rho_a^2 - \rho^2\,.\,\rho_b^2 - \rho^2\,.\,\rho_c^2 - \rho^2)},$$

will determine the equation of the herpolhode.

113. Using Weierstrass's functions of § 108,

$$\frac{\rho^2}{h^2}=\frac{\nu^2}{\mu^2}=\frac{p^2+q^2+r^2}{\mu^2}-1$$

$$=-\frac{m^2}{\mu^2}\left\{\frac{B-C}{A}(\wp u-e_a)+\frac{C-A}{B}(\wp u-e_b)+\frac{A-B}{C}(\wp u-e_c)+\frac{\mu^2}{m^2}\right\}$$

$$=\frac{n^2}{\mu^2}(\wp v-\wp u),$$

with
$$\wp v=\frac{\frac{B-C}{A}e_a+\frac{C-A}{B}e_b+\frac{A-B}{C}e_c-\frac{\mu^2}{m^2}}{\frac{B-C}{A}+\frac{C-A}{B}+\frac{A-B}{C}};$$

and then
$$\wp v-e_a=\frac{\mu^2}{n^2}\left(1-\frac{D}{B}\right)\left(\frac{D}{C}-1\right),\ \text{(positive)},$$

$$\wp v-e_b=\frac{\mu^2}{n^2}\left(1-\frac{D}{C}\right)\left(\frac{D}{A}-1\right),\ \text{(positive)},$$

$$\wp v-e_c=\frac{\mu^2}{n^2}\left(1-\frac{D}{A}\right)\left(\frac{D}{B}-1\right),\ \text{(negative)},$$

$$\wp'^2 v=\ 4(\wp v-e_a)(\wp v-e_b)(\wp v-e_c)$$

$$=-4\frac{\mu^6}{n^6}\frac{(A-D)^2(B-D)^2(C-D)^2}{A^2B^2C^2};$$

and, since e_1 (or e_c) $>\wp v>e_2$ (or e_a),
we must, by (39), § 54, where t' is a proper fraction, take

$$v=\omega_1+t'\omega_3.$$

Therefore
$$\frac{d\phi}{dt}=\mu+n\frac{\frac{1}{2}i\wp' v}{\wp v-\wp u},$$

or
$$\frac{d\phi}{du}=\frac{\mu}{n}+\frac{\frac{1}{2}i\wp' v}{\wp v-\wp u},$$

and, integrating,
$$\phi=\mu t+\tfrac{1}{2}i\int\frac{\wp' v\,du}{\wp v-\wp u},$$

and we are thus introduced to a new integral, called an *elliptic integral of the third kind.*

The cone described in the body by OH (§ 105) is called by Poinsot the *rolling and sliding cone*; during the motion this cone rolls on an invariable plane through O, while at the same time this plane turns with constant angular velocity μ about OC; so that, if ρ, ϕ' denote with respect to O the polar coordinates of H on this plane,

$$\phi'=\phi-\mu t=\tfrac{1}{2}i\int\frac{\wp' v\,du}{\wp v-\wp u}.$$

114. With the notation of the elliptic functions of Jacobi, as in § 106,

$$\frac{\rho^2}{h^2}=\frac{D}{A}\frac{D-C}{A-C}\operatorname{cn}^2nt+\frac{D}{B}\frac{D-C}{B-C}\operatorname{sn}^2nt+\frac{D}{C}\frac{A-D}{A-C}\operatorname{dn}^2nt$$

$$=\frac{A-D\,.\,D-C}{AC}-\frac{D\,.\,A-B\,.\,D-C}{ABC}\operatorname{sn}^2nt,$$

which can be thrown into the form

$$\frac{\rho^2}{h^2}=\frac{A-D\,.\,D-C}{AC}(1-\kappa^2\operatorname{sn}^2a\operatorname{sn}^2nt)$$

on putting $$\kappa^2\operatorname{sn}^2a=\frac{D}{D}\frac{A-B}{A-D},$$

$$\operatorname{sn}^2a=\frac{D}{B}\frac{B-C}{D-C},\quad \operatorname{cn}^2a=-\frac{C}{B}\frac{B-D}{D-C},\quad \operatorname{dn}^2a=\frac{A}{B}\frac{B-D}{A-D}.$$

With $e_a=e_2$, $e_b=e_3$, $e_c=e_1$, and $v=\omega_1+t'\omega_3$, then by (32), p. 44,

$$\surd(e_1-e_3)v=K+t'iK';$$

and $$\operatorname{dn}^2(K+t'iK')=\frac{\wp v-e_a}{\wp v-e_b}=\frac{A}{B}\frac{B-D}{A-D}=\operatorname{dn}^2a;$$

so that $$a=K+t'iK'.$$

Then $$\frac{d\phi}{dt}=\mu-\frac{B-D}{B}\frac{\mu}{1-\kappa^2\operatorname{sn}^2a\operatorname{sn}^2nt}$$

$$=\mu-\frac{i\operatorname{cn}a\operatorname{dn}a}{\operatorname{sn}a}\frac{n}{1-\kappa^2\operatorname{sn}^2a\operatorname{sn}^2nt},$$

and, writing u for nt,

$$\phi=\mu t-\frac{i\operatorname{sn}a\operatorname{dn}a}{\operatorname{sn}a}\int_0\frac{du}{1-\kappa^2\operatorname{sn}^2a\operatorname{sn}^2u}$$

$$=\mu t-\frac{i\operatorname{cn}a\operatorname{dn}a}{\operatorname{sn}a}u-i\int_0\frac{\kappa^2\operatorname{sn}a\operatorname{cn}a\operatorname{dn}a\operatorname{sn}^2u}{1-\kappa^2\operatorname{sn}^2a\operatorname{sn}^2u}du,$$

the last term an *elliptic integral of the third kind*, in the form employed by Jacobi.

On putting $\operatorname{sn}u=\sin\theta$, and $\operatorname{sn}a=\sin\alpha$, $\kappa^2\operatorname{sn}^2a=-m$, then

$$\phi=\mu t-i\frac{\cos\alpha\Delta\alpha}{\sin\alpha}\int_0\frac{d\theta}{(1+m\sin^2\theta)\surd(1-\kappa^2\sin^2\theta)},$$

the third elliptic integral, as employed by Legendre; the further discussion of this integral must be reserved for a subsequent chapter.

EXAMPLES.

1. Prove that, if the excentric anomaly in an undisturbed planetary orbit of excentricity e is represented by $2\,\mathrm{am}(u, e)$, the mean anomaly is

$$2\,\mathrm{am}\,u + 2\frac{d^2\mathrm{am}\,u}{edu^2}.$$

2. Prove that the envelope of the straight line rays

$$\kappa'^2 x\,\mathrm{sn}\,u + (\mathrm{cn}\,u + \kappa\,\mathrm{dn}\,u)y = \kappa\,\mathrm{sn}\,u(\mathrm{dn}\,u + \kappa\,\mathrm{cn}\,u)$$

where u is the variable parameter, is the curve

$$\kappa'^2 x = \kappa^2(1 - \kappa^{-\frac{4}{3}} y^{\frac{2}{3}})^{\frac{3}{2}} + \kappa(1 - \kappa^{\frac{2}{3}} y^{\frac{2}{3}})^{\frac{3}{2}};$$

the caustic of parallel rays, after refraction at a circle, of refractive index $1/\kappa$; and find the order of this curve.

(Cayley, *Phil. Trans.*, 1857, "Caustics.")

3. Prove that a portion of a flexible inextensible spherical surface of radius a, bounded by two meridians (a *lune*, or gore of a spherical balloon) can be bent into the surface of revolution given by

$$x = a\cos\theta\cos(\phi/\kappa),\quad y = a\cos\theta\sin(\phi/\kappa),\quad z = aE(\theta, \kappa);$$

θ, ϕ denoting the latitude and longitude of the point on the sphere.

Explain the geometrical theory, distinguishing the cases of $\kappa < 1$, and $\kappa > 1$.

4. Denoting by ω the solid angle subtended by a circle of radius a at a point whose cylindrical coordinates are r, z with respect to the axis of the circle, prove that

$$\frac{d\omega}{da} = \frac{az}{2(ar)^{\frac{3}{2}}}\frac{\kappa^3}{\kappa'^2}E,$$

where $$\kappa^2 = \frac{4ar}{z^2 + (a+r)^2},\quad \kappa'^2 = \frac{z^2 + (a-r)^2}{z^2 + (a+r)^2}.$$

Show how to determine the illumination at any point of the surface of the water at the bottom of a deep well, due to the light from the sky.

5. A uniform circular wire, charged with $-e$ coulombs, is presented symmetrically to a fixed insulated sphere of radius a centimetres, so that every point of the wire is at a distance f cm from the centre of the sphere, the radius of the wire subtending an angle α at the centre of the sphere.

Prove that the electricity, in coulombs per cm^2, induced at a point of the sphere whose angular distance from the axis of symmetry is θ, is given by

$$e\frac{f^2-a^2}{2\pi^2 a}\frac{E}{\{a^2-2af\cos(\theta-\alpha)+f^2\}\sqrt{\{a^2-2af\cos(\theta+\alpha)+f^2\}}},$$

where $\kappa^2=\dfrac{4af\sin\alpha\sin\theta}{a^2-2af\cos(\theta+\alpha)+f^2}$, $\kappa'^2=\dfrac{a^2-2af\cos(\theta-\alpha)+f^2}{a^2-2af\cos(\theta+\alpha)+f^2}$.

6. Prove that if this sphere and wire gravitate to each other, and if the wire is free to turn about a fixed diameter perpendicular to the line joining the centres, the wire will be in stable equilibrium when its plane passes through the centre of the sphere; and prove that the oscillations of the wire due to the gravitation will synchronize with a pendulum of length

$$\frac{\pi b^2(b+c)}{2CMF}g \text{ cm},$$

where b denotes the radius of the wire, c the distance between the centres of the sphere and wire in cm, M the weight of the sphere in g, C the gravitation constant; and

$$F=\frac{1+\kappa'^2}{2\kappa'^2}E-K=\tfrac{1}{2}\kappa(1+\kappa'^2)\frac{d}{d\kappa}\{(1+\kappa'^2)^{\frac{1}{2}}K\},$$

where $\qquad \kappa^2=4bc/(b+c)^2.$

Determine the position of stable equilibrium and the length of the equivalent pendulum, when the attraction is changed to repulsion.

7. Two uniform concentric circular wires of radii b and c cm, weighing M and M'g, are freely moveable about a common fixed diameter. Prove that in consequence of their gravitation, the oscillations will synchronize with a pendulum of length

$$\frac{\pi b^2c^2(b+c)}{CF(Mb^2+M'c^2)}g \text{ cm},$$

where F and κ have the same values as before.

CHAPTER IV.

THE ADDITION THEOREM FOR ELLIPTIC FUNCTIONS.

115. So far we have considered the elliptic functions of a single argument u; but now we have to determine the formulas which give the elliptic functions of the sum or difference, $u \pm v$, of two arguments u and v, in terms of the elliptic functions of u and v; and thence generally the formulas for the elliptic functions of the sum of any number of arguments $u+v+w+\ldots$; and the formulas for the duplication, triplication, etc., of the argument.

The Addition Theorem for Circular and Hyperbolic Functions.

The analogous formulas in Trigonometry for the Circular Functions are well known, namely,

$$\sin(u \pm v) = \sin u \cos v \pm \cos u \sin v,$$

$$\cos(u \pm v) = \cos u \cos v \mp \sin u \sin v;$$

or, as they may be written,

$$\sin(u \pm v) = \sin u \sin' v \pm \sin' u \sin v,$$

$$\cos(u \pm v) = \cos u \cos v \mp \cos' u \cos' v;$$

the accents denoting differentiation; and to these may be added

$$\tan(u \pm v) = \frac{\tan u \pm \tan v}{1 \mp \tan u \tan v};$$

these formulas constituting the Addition Theorem for the Circular Functions.

For the Hyperbolic Functions, the formulas are

$$\cosh(u \pm v) = \cosh u \cosh v \pm \sinh u \sinh v,$$

$$\sinh(u \pm v) = \sinh u \cosh v \pm \cosh u \sinh v;$$

or, as they may be written,

$$\cosh(u\pm v)=\cosh u\cosh v\pm\cosh' u\cosh' v,$$
$$\sinh(u\pm v)=\sinh u\sinh' v\pm\sinh' u\sinh v;$$

and to these may be added

$$\tanh(u\pm v)=\frac{\tanh u\pm\tanh v}{1\pm\tanh u\tanh v};$$

constituting the Addition Theorem for the Hyperbolic Functions.

116. *The Addition Theorem for the Elliptic Functions.*

For the Elliptic Functions the analogous formulas of the Addition Theorem are found to be

$$\mathrm{sn}(u\pm v)=(\mathrm{sn}\,u\,\mathrm{sn}'v\pm\quad\mathrm{sn}'u\,\mathrm{sn}\,v)/D,$$
$$\mathrm{cn}(u\pm v)=(\mathrm{cn}\,u\,\mathrm{cn}\,v\mp\quad\mathrm{cn}'u\,\mathrm{cn}'v)/D,$$
$$\mathrm{dn}(u\pm v)=(\mathrm{dn}\,u\,\mathrm{dn}\,v\pm\kappa^{-2}\mathrm{dn}'u\,\mathrm{dn}'v)/D,$$

where $D=1-\kappa^2\mathrm{sn}^2u\,\mathrm{sn}^2v$;

or, performing the differentiations, and dropping the double signs,

$$\mathrm{sn}(u+v)=\frac{\mathrm{sn}\,u\,\mathrm{cn}\,v\,\mathrm{dn}\,v+\mathrm{cn}\,u\,\mathrm{dn}\,u\,\mathrm{sn}\,v}{1-\kappa^2\mathrm{sn}^2u\,\mathrm{sn}^2v},\qquad\text{..........(1)}$$

$$\mathrm{cn}(u+v)=\frac{\mathrm{cn}\,u\,\mathrm{cn}\,v-\mathrm{sn}\,u\,\mathrm{dn}\,u\,\mathrm{sn}\,v\,\mathrm{dn}\,v}{1-\kappa^2\mathrm{sn}^2u\,\mathrm{sn}^2v},\qquad\text{..........(2)}$$

$$\mathrm{dn}(u+v)=\frac{\mathrm{dn}\,u\,\mathrm{dn}\,v-\kappa^2\mathrm{sn}\,u\,\mathrm{cn}\,u\,\mathrm{sn}\,v\,\mathrm{cn}\,v}{1-\kappa^2\mathrm{sn}^2u\,\mathrm{sn}^2v}.\qquad\text{.........(3)}$$

Putting $\kappa=0$, we obtain the formulas for the Circular Functions, $\sin(u+v)$ and $\cos(u+v)$, the denominator D reducing to unity.

Putting $\kappa=1$, remembering that then (§ 16) $\mathrm{sn}\,u$ becomes $\tanh u$, $\mathrm{cn}\,u$ or $\mathrm{dn}\,u$ becomes $\mathrm{sech}\,u$, we obtain from (1)

$$\tanh(u+v)=\frac{\tanh u\,\mathrm{sech}^2v+\mathrm{sech}^2u\tanh v}{1-\tanh^2u\tanh^2v}$$

$$=\frac{\tanh u(1-\tanh^2v)+(1-\tanh^2u)\tanh v}{1-\tanh^2u\tanh^2v}=\frac{\tanh u+\tanh v}{1+\tanh u\tanh v},$$

as before; with the corresponding formula for $\mathrm{sech}(u+v)$ or $\cosh(u+v)$, the formulas for the Hyperbolic Functions.

117. To establish these formulas of the Addition Theorem for Elliptic Functions, let us employ the geometry invented by Jacobi (*Crelle*, Band 3; *Gesammelte Werke*, I., p. 279), at the same time interpreting the geometry in connexion with Pendulum Motion.

To do this, let us suppose that P' would be the position of P in fig. 2 at the time t, if it had started τ seconds later, and put $t-\tau=t'$; then (§ 6)

$AN'=AD\,\text{sn}^2 nt'$, $N'D=AD\,\text{cn}^2 nt'$, $N'E=AE\,\text{dn}^2 nt'$, etc.;
and we shall prove that PP' touches a fixed circle through B and B' during the motion (fig. 13).

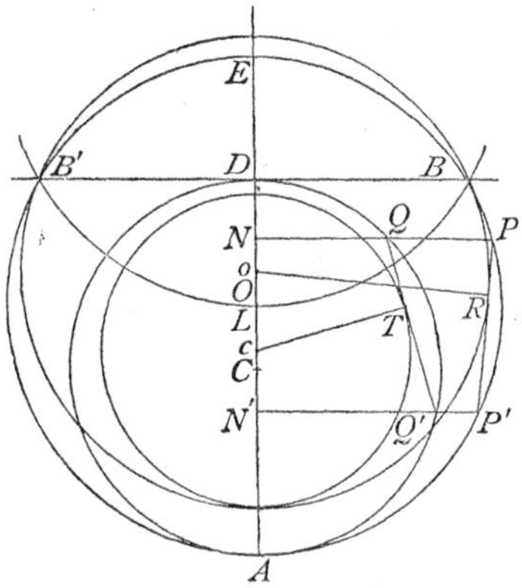

Fig. 13.

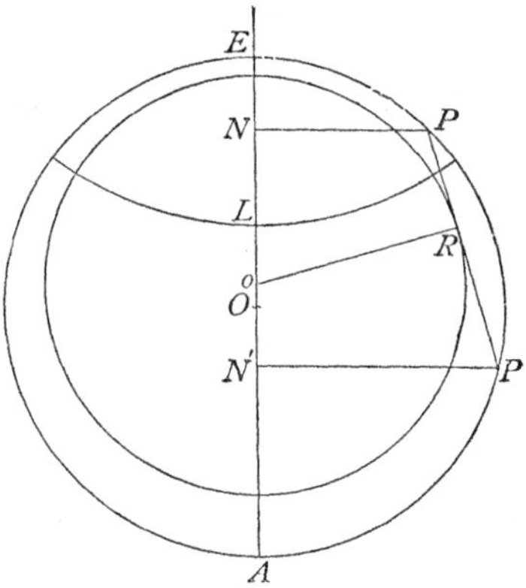

Fig. 14.

For suppose that, in the small element of time dt, P has moved to an adjacent point p and P' to p'; and let PP', pp' intersect in R, so that R is ultimately the point of contact on the envelope of PP'.

Then since, by a property of the circle, PP' cuts the circle $AP'P$ at equal angles at P and P',

$$\frac{PR}{RP'}=\text{lt}\,\frac{Pp}{P'p'}=\frac{\text{velocity of } P}{\text{velocity of } P'}=\sqrt{\frac{ND}{ND'}}.$$

Now describe a circle with centre o on AE, passing through B and B', and touching PP' at a point which we shall denote by R'; then

$$\begin{aligned} PR'^2=Po^2-oR'^2 &= PO^2+Oo^2-2Oo\,.\,ON-oR'^2 \\ &= OB^2+Oo^2-2Oo\,.\,ON-Bo^2 \\ &= OD^2-Do^2+Oo^2-2Oo\,.\,ON \\ &= Oo(OD+Do+oO-2ON) \\ &= Oo(2OD-2ON)=2Oo\,.\,ND. \end{aligned}$$

Similarly, $\qquad R'P'^2=2Oo\,.\,N'D,$

so that
$$\frac{PR'}{R'P'}=\sqrt{\frac{ND}{N'D}}=\frac{PR}{RP'},$$

and therefore R and R' coincide; and we have thus verified that PP' touches at R the circle oR (using the notation oR to mean a circle of centre o, and radius oR).

Putting $Oo=a$, and denoting the angles AOP, AOP' by θ, θ', and ADQ, ADQ' by ϕ, ψ, then

$$PR^2=2a\,.\,ND=4aR\cos^2\psi=4al\kappa^2\cos^2\phi,\quad RP'^2=4al\kappa^2\cos^2\psi;$$

so that $$P'R+RP=2\sqrt{(al)}\kappa(\cos\psi+\cos\phi),$$

while $$P'P=2l\sin\tfrac{1}{2}(\theta-\theta'),$$

and therefore $\sin\frac{1}{2}(\theta-\theta')=\sqrt{(a/l)}\kappa(\cos\psi+\cos\phi)$.

Putting $nt=u$, $nt'=v$, $n\tau=u-v=w$; then since (§ 8)

$$\phi=\operatorname{am}u,\quad \sin\tfrac{1}{2}\theta=\kappa\sin\phi=\kappa\operatorname{sn}u,\quad \cos\tfrac{1}{2}\theta=\operatorname{dn}u;$$

$$\psi=\operatorname{am}v,\quad \sin\tfrac{1}{2}\theta'=\kappa\sin\psi=\kappa\operatorname{sn}v,\quad \cos\tfrac{1}{2}\theta'=\operatorname{dn}v;$$

$$\sqrt{\frac{a}{l}}=\frac{\sin\frac{1}{2}(\theta-\theta')}{\kappa(\cos\psi+\cos\phi)}=\frac{\operatorname{sn}u\operatorname{dn}v-\operatorname{dn}u\operatorname{sn}v}{\operatorname{cn}v+\operatorname{cn}u},\text{ a constant.}$$

Putting $t'=0$, $v=0$, and therefore $u=n\tau=w$, we find

$$\sqrt{\frac{a}{l}}=\frac{\operatorname{sn}w}{1+\operatorname{cn}w}=\frac{1-\operatorname{cn}w}{\operatorname{sn}w}=\sqrt{\frac{1-\operatorname{cn}w}{1+\operatorname{cn}w}};$$

so that

$$\sqrt{\frac{1-\operatorname{cn}(u-v)}{1+\operatorname{cn}(u-v)}}=\frac{\operatorname{sn}u\operatorname{dn}v-\operatorname{dn}u\operatorname{sn}v}{\operatorname{cn}v+\operatorname{cn}u}=\frac{\operatorname{cn}v-\operatorname{cn}u}{\operatorname{sn}u\operatorname{dn}v+\operatorname{dn}u\operatorname{sn}v},$$

one form of the Addition Theorem, which by algebraical transformation can be reduced to one of the preceding forms of § 116.

118. Representing, as in § 31, $\operatorname{sn}u$ by s_1, $\operatorname{cn}u$ by c_1, $\operatorname{dn}u$ by d_1, and the corresponding functions of v by s_2, c_2, d_2; then

$$\sqrt{\frac{1-\operatorname{cn}(u-v)}{1+\operatorname{cn}(u-v)}}=\frac{s_1d_2-s_2d_1}{c_2+c_1}=\frac{c_2-c_1}{s_1d_2+s_2d_1},$$

so that $$\frac{1-\operatorname{cn}(u-v)}{1+\operatorname{cn}(u-v)}=\frac{(c_2-c_1)(s_1d_2-s_2d_1)}{(c_2+c_1)(s_1d_2+s_2d_1)},$$

or $$\operatorname{cn}(u-v)=\frac{s_1c_1d_2+s_2c_2d_1}{s_1c_2d_2+s_2c_1d_1};$$

and changing the sign of v,

$$\operatorname{cn}(u+v)=\frac{s_1c_1d_2-s_2c_2d_1}{s_1c_2d_2-s_2c_1d_1},$$

another form of the Addition Equation.

Again $$\frac{1-\operatorname{cn}(u-v)}{1+\operatorname{cn}(u-v)}=\left(\frac{s_1d_2-s_2d_1}{c_2+c_1}\right)^2,\text{ or }=\left(\frac{c_2-c_1}{s_1d_2+s_2d_1}\right)^2,$$

$$\operatorname{cn}(u-v)=\frac{(c_2+c_1)^2-(s_1d_2-s_2d_1)}{(c_2+c_1)^2+(s_1d_2-s_2d_1)},\text{ or }=\frac{(s_1d_2+s_2d_1)^2-(c_2-c_1)^2}{(s_1d_2+s_2d_1)^2+(c_2-c_1)^2};$$

and, adding numerators and denominators (*componendo*),

$$\text{cn}(u-v)=\frac{2(c_1c_2+s_1d_1s_2d_2)}{c_1^2+c_2^2+s_1^2d_2^2+s_2^2d_1^2}$$

$$=\frac{c_1c_2+s_1d_1s_2d_2}{1-\kappa^2s_1^2s_2^2},$$

$$\text{cn}(u+v)=\frac{c_1c_2-s_1d_1s_2d_2}{1-\kappa^2s_1^2s_2^2},\quad\ldots\ldots\ldots\ldots\ldots\ldots\ldots(2)$$

the usual form (2) of the Addition Theorem for the cn function.

But, subtracting numerators and denominators (*dividendo*),

$$\text{cn}(u-v)=\frac{c_1^2+c_2^2-s_1^2d_2^2-s_2^2d_1^2}{2(c_1c_2-s_1d_1s_2d_2)}$$

$$=\frac{1-s_1^2-s_2^2+\kappa^2s_1^2s_2^2}{c_1c_2-s_1d_1s_2d_2};$$

$$\text{cn}(u+v)=\frac{1-s_1^2-s_2^2+\kappa^2s_1^2s_2^2}{c_1c_2+s_1d_1s_2d_2};$$

and another form can be easily established in the same way,

$$\text{cn}(u+v)=\frac{c_1d_1c_2d_2-\kappa^2s_1s_2}{d_1d_2+\kappa^2s_1c_1s_2c_2}.$$

(Glaisher, *Messenger of Mathematics*, vol. x., p. 106;
M. M. U. Wilkinson, *Proc. London Math. Soc.*, vol. xiii., p. 109;
Woolsey Johnson, *Messenger of Mathematics*, vol. xi., p. 138.)

119. Expressed again in Legendre's trigonometrical form, with $\phi=\text{am}\,u$, $\psi=\text{am}\,v$, $\gamma=\text{am}(u-v)$,

$$\sqrt{\frac{a}{l}}=\frac{1-\cos\gamma}{\sin\gamma}=\frac{\sin\phi\Delta\psi-\sin\psi\Delta\phi}{\cos\psi+\cos\phi},$$

$$\sqrt{\frac{l}{a}}=\frac{1+\cos\gamma}{\sin\gamma}=\frac{\sin\phi\Delta\psi+\sin\psi\Delta\phi}{\cos\psi-\cos\phi}.$$

Therefore, eliminating $\Delta\psi$,

$$2\sin\psi\sin\gamma\Delta\phi=(\cos\psi-\cos\phi)(1+\cos\gamma)-(\cos\psi+\cos\phi)(1-\cos\gamma)$$

$$=-2\cos\phi+2\cos\psi\cos\gamma,$$

or $$\cos\phi=\cos\psi\cos\gamma-\sin\psi\sin\gamma\Delta\phi.$$

Expressed in Jacobi's notation, since $u=v+w$,

$$\text{cn}(v+w)=\text{cn}\,v\,\text{cn}\,w-\text{sn}\,v\,\text{sn}\,w\,\text{dn}(v+w).$$

Changing $v+w$ into $u-v$, this becomes

$$\text{cn}(u-v)=\text{cn}\,u\,\text{cn}\,v+\text{sn}\,u\,\text{sn}\,v\,\text{dn}(u-v),$$

or $$\cos\gamma=\cos\phi\cos\psi+\sin\phi\sin\psi\Delta\gamma.$$

Conversely, these relations, treating γ as constant, lead to the differential relations $du-dv=0$,

or $$d\phi/\Delta\phi-d\psi/\Delta\psi=0,$$

or $$(d\phi)^2(1-\kappa^2\sin^2\psi)-(d\psi)^2(1-\kappa^2\sin^2\phi)=0.$$

Writing x for $\sin\phi\sin\psi$, y for $\cos\phi\cos\psi$, and m for $\Delta\gamma$, then $\cos\gamma=\sqrt{(m^2-\kappa'^2)}/\kappa$ (§ 17); and the integral relation becomes

$$y+mx=\sqrt{(m^2-\kappa'^2)}/\kappa,$$

leading to the differential equation, of Clairaut's form,

$$y-xp=\sqrt{(p^2-\kappa'^2)}/\kappa,$$

denoting dy/dx by p; this is the form of the differential equation when we change to these new variables x and y.

120. We have begun in § 117 by supposing the points P and P' to oscillate on a circle with velocity due to the level of the horizontal line BDB', cutting the circle in B and B' (figs. 2, 13); but if they are performing complete revolutions with velocity due to the level of a horizontal line BB' through D not cutting the circle, but lying above it (figs. 3, 14), a similar proof will show that PP' touches a fixed circle having with the circle PP' the common radical axis BB', the two circles not intersecting; and the Landen point L (§ 28) will be a limiting point of these two circles.

But this motion of P and P' in fig. 14 is imitated by the circulating motion of Q and Q' on the circle AQ in fig. 13; so that QQ' touches at T a fixed circle, centre c; and the horizontal line through E is the common radical axis of this circle and the circle CQ, the Landen point L being a limiting point; and thus the Addition Theorem for Elliptic Functions can be deduced from the motion of P and P' in fig. 14, or of Q and Q' in fig. 13, as given by Durège, *Elliptische Functionen*, X.

For if in fig. 14 a circle is drawn with centre o and radius oR, such that BDB' (fig. 3) is the common radical axis of this circle and of the circle AP, then, since the tangents to these circles from D are equal in length,

$$DO^2-OP^2=Do^2-oR^2;$$

and now, if the tangent to the inner circle at R cuts the outer circle in P and P',

$$PR^2=Po^2-oR^2=PO^2+Oo^2-2Oo\,.\,ON-PO^2+OD^2-Do^2$$
$$=OD^2-Do^2+Oo^2-2Oo\,.\,ON=2Oo\,.\,ND,$$

as in § 117; and similarly $RP'^2=2Oo\,.\,N'D$; so that

$$\frac{PR}{RP'}=\sqrt{\frac{N'D}{ND}}=\frac{\text{velocity of } P}{\text{velocity of } P'};$$

and therefore PP' will continue to touch the circle R, during the subsequent motion of P and P'.

Similarly, in fig. 13, QQ' during the motion touches a fixed circle, centre c and radius cT; and putting $Oc=c$,

$$QT^2=2c\,.\,NE=4cl\,\text{dn}^2nt,\ TQ'^2=4cl\,\text{dn}^2nt'.$$

We notice, on reference to § 28, that

$$LQ^2=2LC\,.\,EN=2LC\,.\,EA\,\text{dn}^2nt=4l^2(1-\kappa')^2\text{dn}^2nt=LA^2\text{dn}^2nt,$$

so that $$LQ=LA\,\text{dn}\,nt;$$

and therefore $$\frac{LQ}{QT}=\frac{LQ'}{Q'T'}$$

or LT bisects the angle QLQ' in fig. 13; while LR bisects the angle PLP' in fig. 14; we may state this theorem geometrically, "the segments of a tangent to one circle, cut off by another circle, subtend equal angles at a limiting point of the two circles."

Then, with the notation of § 117,

$$Q'T+TQ=2\surd(cl)(\Delta\psi+\Delta\phi),$$

and $$Q'Q=2R\sin(\phi-\psi)=2\kappa^2l\sin(\phi-\psi);$$

so that, in Legendre's trigonometrical form,

$$\frac{\kappa\sin(\phi-\psi)}{\Delta\psi+\Delta\phi}=\sqrt{\frac{c}{\kappa^2l}},\text{ or }\sqrt{\frac{c}{R}},\text{ a constant,}$$

Putting $\psi=0$, then $\phi=\gamma$; so that

$$\sqrt{\frac{c}{R}}=\frac{\kappa\sin(\phi-\psi)}{\Delta\psi+\Delta\phi}=\frac{\kappa\sin\gamma}{1+\Delta\gamma},\text{ or }\frac{1-\Delta\gamma}{\kappa\sin\gamma},$$

$$\sqrt{\frac{R}{c}}=\frac{\kappa\sin(\phi+\psi)}{\Delta\psi-\Delta\phi}=\frac{\kappa\sin\gamma}{1-\Delta\gamma},\text{ or }\frac{1+\Delta\gamma}{\kappa\sin\gamma},$$

the product of the two equations being unity.

Conversely, the relation

$$\sin(\phi\pm\psi)=C(\Delta\psi+\Delta\phi),$$

where C is an arbitrary constant, leads to the differential relation

$$d\phi/\Delta\phi\pm d\psi/\Delta\psi=0.$$

121. Taking the equations

$$\frac{1+\Delta\gamma}{\sin\gamma}=\frac{\kappa^2\sin(\phi+\psi)}{\Delta\psi-\Delta\phi},\ \frac{1-\Delta\gamma}{\sin\gamma}=\frac{\kappa^2\sin(\phi-\psi)}{\Delta\psi+\Delta\phi},$$

we find, on eliminating $\sin\phi$,

$$2\kappa^2\cos\phi\sin\psi\sin\gamma=(1+\Delta\gamma)(\Delta\psi-\Delta\phi)-(1-\Delta\gamma)(\Delta\psi+\Delta\phi)$$
$$=-2\Delta\phi+\Delta\gamma\Delta\psi,$$
$$\Delta\phi=\Delta\gamma\Delta\psi-\kappa^2\cos\phi\sin\psi\sin\gamma,$$

or $$\text{dn}\,u=\text{dn}\,v\,\text{dn}\,w-\kappa^2\text{cn}\,u\,\text{sn}\,v\,\text{sn}\,w,$$

with $$u=v+w.$$

By eliminating $\cos\phi$,

$$2\kappa^2\sin\phi\cos\psi\sin\gamma=2\Delta\psi-2\Delta\gamma\Delta\phi,$$

$$\Delta\psi=\Delta\phi\Delta\gamma+\kappa^2\sin\phi\cos\psi\sin\gamma,$$

or
$$\text{dn}(u-w)=\text{dn}\,u\,\text{dn}\,w+\kappa^2\text{sn}\,u\,\text{sn}\,w\,\text{cn}(u-w).$$

Changing w into v,

$$\text{dn}(u-v)=\text{dn}\,u\,\text{dn}\,v+\kappa^2\text{sn}\,u\,\text{sn}\,v\,\text{cn}(u-v),$$

or
$$\Delta\gamma=\Delta\phi\Delta\psi+\kappa^2\sin\phi\sin\psi\cos\gamma.$$

Writing x for $\kappa^2\sin\phi\sin\psi$, y for $\Delta\phi\Delta\psi$, and m for $\text{cn}\,\gamma$, then

$$y+mx=\sqrt{(\kappa'^2+\kappa^2m^2)},$$

the integral relation of Clairaut's differential equation

$$y-xp=\sqrt{(\kappa'^2+\kappa^2p^2)},$$

which is therefore the transformation of

$$d\phi/\Delta\phi-d\psi/\Delta\psi=0,$$

when we change to these new variables x and y.

Taking the two trigonometrical expressions from § 119, 120, for the Addition Theorem,

$$\frac{1-\cos\gamma}{\sin\gamma}=\frac{\sin\phi\Delta\psi-\sin\psi\Delta\phi}{\cos\psi+\cos\phi},\quad \frac{1-\Delta\gamma}{\sin\gamma}=\frac{\kappa^2\sin(\phi-\psi)}{\Delta\psi+\Delta\phi},$$

we obtain, by subtraction and reduction,

$$\frac{\Delta\gamma-\cos}{\sin\gamma}=\frac{\cos\psi\Delta\phi-\cos\phi\Delta\psi}{\sin\phi+\sin\psi},$$

or
$$\frac{\text{dn}(u-v)-\text{cn}(u-v)}{\text{sn}(u-v)}=\frac{\text{dn}\,u\,\text{cn}\,v-\text{cn}\,u\,\text{dn}\,v}{\text{sn}\,u+\text{sn}\,v},$$

the form of the Addition Theorem given by J. J. Thomson (*Messenger of Mathematics*, vol. IX., p. 53).

122. With the notation of the elliptic functions,

$$\frac{1+\text{dn}(u-v)}{\kappa\,\text{sn}(u-v)}=\frac{\kappa(\text{sn}\,u\,\text{cn}\,v+\text{sn}\,v\,\text{cn}\,u)}{\text{dn}\,v-\text{dn}\,u},$$

$$\frac{1-\text{dn}(u-v)}{\kappa\,\text{sn}(u-v)}=\frac{\kappa(\text{sn}\,u\,\text{cn}\,v-\text{sn}\,v\,\text{cn}\,u)}{\text{dn}\,v+\text{dn}\,u}.$$

Therefore, as before, with Glaisher's abbreviations,

$$\frac{1-\text{dn}(u-v)}{1+\text{dn}(u-v)}=\frac{(d_2-d_1)(s_1c_2-s_2c_1)}{(d_2+d_1)(s_1c_2+s_2c_1)},$$

$$\text{dn}(u-v)=\frac{s_1d_1c_2+s_2d_2c_1}{s_1d_2c_2+s_2d_1c_1}.$$

Similar algebraical reductions to those given above for cn$(u-v)$ will establish the formulas for dn$(u-v)$ and dn$(u+v)$, given by Glaisher (*Messenger*, X., p. 106),

$$\operatorname{dn}(u+v)=\frac{s_1d_1c_2-s_2d_2c_1}{s_1d_2c_2-s_2d_1c_1}=\frac{c_1c_2d_1d_2+\kappa'^2s_1s_2}{c_1c_1+ss_1d_1d_2}$$
$$=\frac{1-\kappa^2s_1{}^2-\kappa^2s_2{}^2+\kappa^2s_1{}^2s_2{}^2}{d_1d_2+\kappa^2s_1s_2c_1c_2}=\frac{d_1d_2-\kappa^2s_1s_2c_1c_2}{1-\kappa^2s_1{}^2s_2{}^2},$$

the last of form (3), § 116.

123. *The Duplication, Triplication, etc., Formulas.*

Putting $v=u$ in formulas (1), (2), (3) of § 116, and writing s, c, d for sn u, cn u, dn u, we find

$$\operatorname{sn} 2u=\frac{2scd}{1-\kappa^2s^4},$$
$$\operatorname{cn} 2u=\frac{1-2s^2+\kappa^2s^4}{1-\kappa^2s^4}=\frac{-\kappa'^2+2\kappa'^2c^2+\kappa^2c^4}{\kappa'^2+2\kappa'^2c^2-\kappa^2c^4},$$
$$\operatorname{dn} 2u=\frac{1-2\kappa^2s^2+\kappa^2s^4}{1-\kappa^2s^4}=\frac{\kappa'^2-2\kappa'^2d^2+d^4}{-\kappa'^2+2d^2-d^4}.$$

Writing S, C, D for sn $2u$, cn $2u$, dn $2u$, we find

$$\frac{1-C}{1+C}=\frac{s^2d^2}{c^2},\quad \frac{1-D}{1+D}=\frac{\kappa^2s^2c^2}{d^2},\quad \frac{D-C}{D+C}=\frac{\kappa'^2s^2}{c^2d^2};$$
$$s^2=\frac{1-C}{1+D}=\frac{1}{\kappa^2}\frac{1-D}{1+C}=\text{etc.},$$
$$c^2=\frac{D+C}{1+D}=\frac{\kappa'^2}{\kappa^2}\frac{1-D}{D-C}=\text{etc.},$$
$$d^2=\frac{D+C}{1+C}=\kappa'^2\frac{1-C}{D-C}=\text{etc.}$$

Putting $u=\tfrac{1}{2}K$, then $S=1$, $C=0$, $D=\kappa'$; and

$$\operatorname{sn}\tfrac{1}{2}K=\sqrt{\left(\frac{1}{1+\kappa'}\right)},\quad \operatorname{cn}\tfrac{1}{2}K=\sqrt{\left(\frac{\kappa'}{1+\kappa'}\right)},\quad \operatorname{dn}\tfrac{1}{2}K=\sqrt{\kappa'}.$$

Again, in § 67,

$$\operatorname{sn}(v,\lambda)=\frac{(1+\kappa')\operatorname{sn}(u,\kappa)\operatorname{cn}(u,\kappa)}{\operatorname{dn}(u,\kappa)}=\frac{1+\kappa'}{\kappa}\sqrt{\frac{1-\operatorname{dn}(2u,\kappa)}{1+\operatorname{dn}(2u,\kappa)}},$$

and $\qquad 2u=(1+\lambda)v,\quad \lambda=(1-\kappa')/(1+\kappa'),$

$$\operatorname{dn}(1+\lambda\,.\,v,\kappa)=\frac{1-\lambda\operatorname{sn}^2(v,\lambda)}{1+\lambda\operatorname{sn}^2(v,\lambda)},$$
$$\operatorname{sn}(1+\lambda\,.\,v,\kappa)=\frac{(1+\lambda)\operatorname{sn}(v,\lambda)}{1+\lambda\operatorname{sn}^2(v,\lambda)},$$
$$\operatorname{cn}(1+\lambda\,.\,v,\kappa)=\frac{\operatorname{cn}(v,\lambda)\operatorname{dn}(v,\lambda)}{1+\lambda\operatorname{sn}^2(v,\lambda)},$$

which is called *Landen's second transformation.*

Again, putting $v=2u$, and making use of the above formulas, we shall find

$$\operatorname{sn} 3u=\frac{3s-4(1+\kappa^2)s^3+6\kappa^2s^5-\kappa^4s^9}{1-6\kappa^2s^4+4(1+\kappa^2)\kappa^2s^6-3\kappa^4s^8},$$

$$\frac{1-\operatorname{sn} 3u}{1+\operatorname{sn} 3u}=\frac{1+s}{1-s}\left(\frac{1-2s+2\kappa^2s^3-\kappa^2s^4}{1+2s-2\kappa^2s^3-\kappa^2s^4}\right)^2,$$

$$\frac{1-\kappa\operatorname{sn} 3u}{1+\kappa\operatorname{sn} 3u}=\frac{1+\kappa s}{1-\kappa s}\left(\frac{1-2\kappa s+2\kappa s^3-\kappa^2s^4}{1+2\kappa s-2\kappa s^3-\kappa^2s^4}\right)^2;$$

with similar expressions for $\operatorname{cn} 3u$ and $\operatorname{dn} 3u$, leading to

$$\frac{1-\operatorname{cn} 3u}{1+\operatorname{cn} 3u}=\frac{1-c}{1+c}\left(\frac{\kappa'^2+2\kappa'^2c+2\kappa^2c^3+\kappa^2c^4}{\kappa'^2-2\kappa'^2c-2\kappa^2c^3+\kappa^2c^4}\right)^2,$$

$$\frac{1-\operatorname{dn} 3u}{1+\operatorname{dn} 3u}=\frac{1-d}{1+d}\left(\frac{\kappa'^2+2\kappa'^2d-2d^3-d^4}{\kappa'^2-2\kappa'^2d+2d^3-d^4}\right)^2,$$

$$\frac{\operatorname{dn} 3u-\kappa'}{\operatorname{dn} 3u+\kappa'}=\frac{d-\kappa'}{d+\kappa'}\left(\frac{d^4+2\kappa'd^3-2\kappa'd-\kappa'^2}{d^4-2\kappa'd^3+2\kappa'd-\kappa'^2}\right)^2;$$

the algebraical work is left as an exercise for the student.

124. *Poristic Polygons of Poncelet, with respect to two Circles.* Starting from the point A in fig. 13, and drawing the successive tangents AQ_1, Q_1Q_2, Q_2Q_3, ... to the inner circle, centre c, from the points Q_1, Q_2, Q_3, ... on the circle CQ; or starting from A in fig. 14, and drawing the tangents AP_1, P_1P_2, P_2P_3, ... to the inner circle, centre o, from P_1, P_2, P_3, ... on the circle OP; then, if we denote the first angle ADQ_1 or AEP_1 by $\operatorname{am} w$, it follows from this construction that

$$ADQ_2=AEP_2=\operatorname{am} 2w,\ ADQ_3=AEP_3=\operatorname{am} 3w, \ldots;$$

and we have thus a geometrical construction for the elliptic functions of the duplicated, triplicated, ... argument.

When w is an aliquot part, one n^{th}, of the half period $2K$, or τ of the half period $2T$ seconds, then after n such operations the polygon $AQ_1Q_2Q_3$, ..., or $AP_1P_2P_3$, ..., will close on itself at the starting point A; and the preceding investigations show that during the subsequent motion of these points, the polygon formed by them will continue to be a closed polygon, inscribed in the circle CQ and circumscribed to the circle cT, or inscribed in the circle OP and circumscribed to the circle oR; and thus we have a mechanical proof of Poncelet's Poristic Theorem for two circles, a problem discussed by Fuss, Steiner, Jacobi, Richelot, and Minding.

(Cayley, *Philosophical Magazine*, 1853, 1854, 1861.)

Let us consider the particular cases of w equal to $\frac{1}{2}$, $\frac{1}{3}$, $\frac{1}{4}$, $\frac{1}{5}$, ... of the half period $2K$.

(i.) When $w=2K$, PP' is horizontal in fig. 13; and P and P' coincide in fig. 14.

(ii.) When $w=K$, the circle oR in fig. 14 and the circle cT in fig. 13 shrink up into the limiting point L, Landen's point (§ 28); and now any straight line through L will divide these circles OP or CQ into two parts described in equal times, $\frac{1}{2}T$; while in fig. 13 the line PP' will touch the circle described with centre E through B, L, and B', subtending an angle 4α at O; and any arc PP' will be described in time $\frac{1}{4}T$, half the time of describing BAB'; hence the following theorem—

"Two segments of circles are described on the under side of the same horizontal straight line, one subtending twice as many degrees at the centre as the other; if a particle oscillates on the lower segmental arc under gravity, any tangent to the upper arc will cut off from the lower an arc described in half the time of oscillation." (Maxwell, *Math. Tripos*, 1866.)

As P' is passing through A in fig. 15, P is instantaneously at rest at B or B'; and AB, AB' are obviously tangents at B and B' to the circle BLB', drawn with centre E; while PP' is one side of a crossed quadrilateral, escribed to this circle BLB', and inscribed in the circle BAB'.

When the circle cT shrinks up into the limiting point L, then, as in § 120,

$$QL^2=2CL\,.\,EN,\quad LQ'^2=2CL\,.\,EN';$$

and since $QL\,.\,LQ'$ is constant in the circle CQ, therefore $EN\,.\,EN'$ is constant, and equal to LE^2, the value it assumes when N and N' pass each other at the point L.

Since $$EN\,.\,EN'=EL^2=EB^2,$$

a circle can be drawn passing through N, N', and touching EB at B; and the triangles ENB, EBN' are therefore similar, so that $$ENB=EBN',\quad EN'B=EBN.$$

(Landen, *Phil. Trans.*, 1771, p. 308.)

Translated into a theorem of elliptic functions,

$$EN\,.\,EN'=EA^2\mathrm{dn}^2u\,\mathrm{dn}^2v,\text{ and } EB^2=\kappa'^2\,.\,EA^2,$$

so that, as in (59), § 57,

$$\mathrm{dn}\,u\,\mathrm{dn}\,v=\kappa',\text{ when } u-v=K,$$

Otherwise, since (§ 28)

$$QL = AL \operatorname{dn} u, \quad LQ' = AL \operatorname{dn} v,$$

and

$$QL \,.\, LQ' = AL \,.\, LD,$$

therefore

$$\operatorname{dn} u \operatorname{dn} v = LD/AL = \kappa'.$$

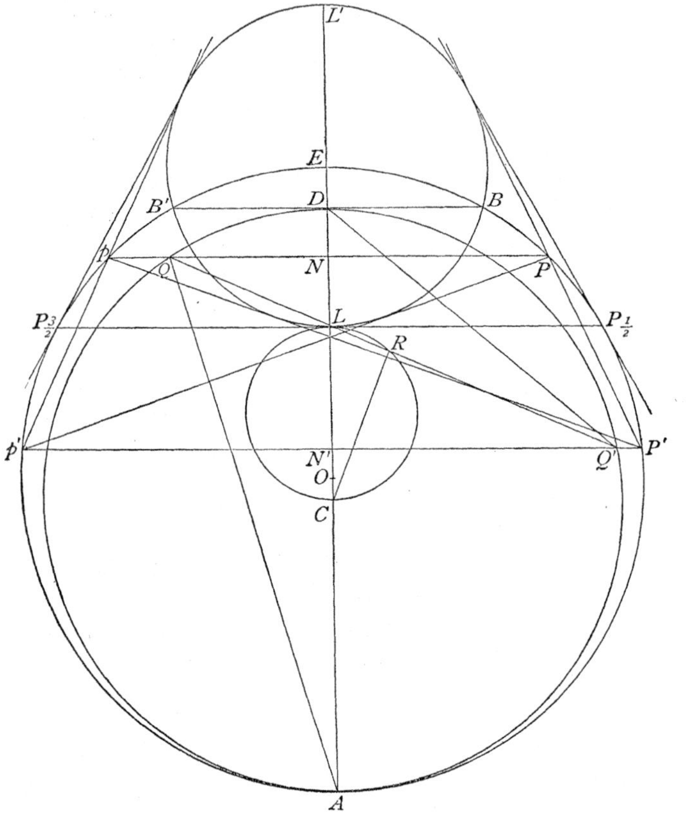

Fig. 15.

The similarity of the triangles AQL, LDQ' shows that

$$AQ/AL = DQ'/LQ';$$

and since (§ 10) $AQ = AD \operatorname{sn} u$, $DQ' = AD \operatorname{cn} v$,
therefore, as in (57), § 57,

$$\operatorname{sn} u = \operatorname{cn} v/\operatorname{dn} v \text{ or } \operatorname{cd} v, \text{ when } u = v + K.$$

Again, since

$$DQ'/DL = AQ/LQ,$$

therefore

$$\operatorname{cn} v = \frac{DL}{AL}\frac{\operatorname{sn} u}{\operatorname{dn} u} = \frac{\kappa' \operatorname{sn} u}{\operatorname{dn} u},$$

as in (58), 57, when $v = u - K$.

Conversely, if the straight line QLQ', passing through L, moves into the adjacent position qLq', then

$$\text{lt}\frac{qQ}{q'Q'}=\frac{QL}{LQ'}=\sqrt{\frac{EN}{EN'}}=\frac{\text{velocity of } Q}{\text{velocity of } Q'},$$

if Q and Q' move under gravity, or diluted gravity, on the circle CQ with velocity due to the level of E; so that QLQ' will continue to pass through L, and will divide the circle CQ into two parts described in the same time $\frac{1}{4}T$ (§ 28).

If in fig. 13 we denote the radius of the circle cT by r, then

$$\cos\gamma=r/(R+c),$$

γ or am w denoting the angle ADQ_1; while, from § 120,

$$\frac{1-\Delta\gamma}{1+\Delta\gamma}=\frac{c}{R}, \text{ or } \Delta\gamma=\frac{R-c}{R+c};$$

and thence $$\kappa^2=\frac{4cR}{(R+c)^2-r^2},\ \kappa'^2=\frac{(R-c)^2-r^2}{(R+c)^2-r^2}.$$

Again, if Dq is drawn from D to touch the circle cT, and the angle ADq is denoted by γ' or am w', then

$$\sin\gamma'=\frac{r}{R-c}=\frac{\cos\gamma}{\Delta\gamma}, \text{ or } \operatorname{sn} w'=\frac{\operatorname{cn} w}{\operatorname{dn} w},$$

so that (§ 57) $w+w'=K$.

125. *Poristic Triangles.*

(iii.) When $w=\frac{2}{3}K$ or $\frac{4}{3}K$, triangles $Q_1Q_2Q_3$ can be inscribed in the circle CQ and circumscribed to the circle cT, while at the same time triangles $P_1P_2P_3$ (or hexagons) can be inscribed in the circle OP and escribed to the circle oR (fig. 16).

The well known relations of Trigonometry

$$c^2=R^2-2Rr, \text{ or } a^2=R^2+2Rr',$$

where $Cc=c$, $Oo=a$, $cT=r$, $oR=r'$, are now easily deduced.

We may write these relations, more symmetrically,

$$\frac{r}{R-c}+\frac{r}{R+c}=1, \text{ or } \frac{r'}{a-R}-\frac{r'}{a+R}=1.$$

In fig. 16, $ADQ_{\frac{2}{3}}=\gamma=\text{am}\frac{2}{3}K$, $ADQ_{\frac{1}{3}}=\gamma'=\text{am}\frac{1}{3}K$;
and since $cQ_{\frac{2}{3}}$ bisects the angle $N_{\frac{2}{3}}Q_{\frac{2}{3}}A$, which is equal to γ, therefore $DcQ_{\frac{2}{3}}=\frac{1}{2}(\pi-\gamma)$; and $DcQ_{\frac{2}{3}}=DQ_{\frac{2}{3}}c$, or $DQ_{\frac{2}{3}}=Dc$.

Similarly $AQ_{\frac{1}{3}}=Ac$; so that

$$AQ_{\frac{1}{3}}+DQ_{\frac{2}{3}}=AD.$$

Therefore $$\sin\gamma'+\cos\gamma=1,$$

or $$\operatorname{sn}\tfrac{1}{3}K+\operatorname{cn}\tfrac{2}{3}K=1,$$

or $$\frac{r}{R-c}+\frac{r}{R+c}=1.$$

We shall employ this suffix notation for the points N, P, Q to signify points corresponding to aliquot parts of K.

Corresponding to $w=\frac{4}{3}K$, the circle oR becomes the circle through B, $N_{\frac{2}{3}}$, B'; and now $P_{\frac{2}{3}}AP_{\frac{4}{3}}$ is a triangle escribed to this circle, and inscribed in the circle OP.

For $w=\frac{2}{3}K$, the circle oR becomes the circle through B, $N_{\frac{1}{3}}$, B'; and now we shall find that hexagons can be escribed to this circle, and inscribed in the circle OP.

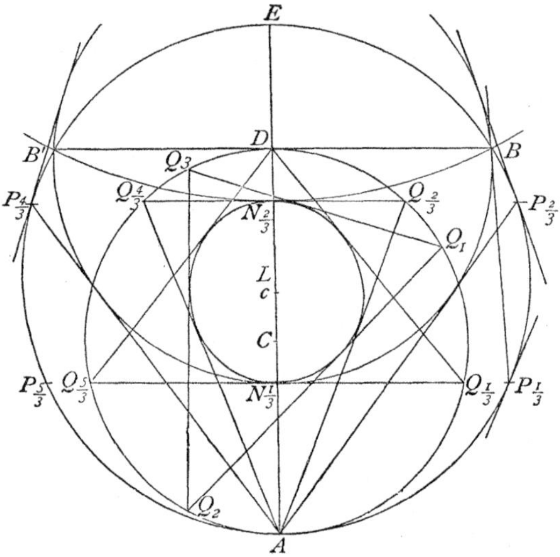

Fig. 16.

The tangents at $P_{\frac{2}{3}}$, $P_{\frac{4}{3}}$ touch the circle $BN_{\frac{1}{3}}B'$, and the tangents at $P_{\frac{1}{3}}$, $P_{\frac{5}{3}}$ touch the circle $BN_{\frac{2}{3}}B'$; while $AP_{\frac{2}{3}}$, $AP_{\frac{4}{3}}$ are the common tangents of the circles $BN_{\frac{1}{3}}B'$, $BN_{\frac{2}{3}}B'$.

Denoting the sides of the triangle $Q_1Q_2Q_3$ by q_1, q_2, q_3, then

$$Rr=\frac{q_1q_2q_3}{2(q_1+q_2+q_3)}.$$

But u_1, u_2, u_3 denoting the value of u corresponding to the points Q_1, Q_2, Q_3, and d_1, d_2, d_3 denoting the corresponding values of dn u, then (§ 120)

$$q_1=Q_2Q_3=2\sqrt{(cl)(d_2d_3)}, \ldots;$$

so that
$$\frac{(d_2+d_3)(d_3+d_1)(d_1+d_2)}{d_1+d_2+d_3}=\frac{Rr}{cl},$$

a constant, a relation connecting d_1, d_2, d_3, when

$$u_1-u_2=u_2-u_3=\tfrac{2}{3}K.$$

126. *Poristic Quadrilaterals.*

(iv.) When $w=\frac{1}{2}K$, quadrilaterals $Q_1Q_2Q_3Q_4$ can be inscribed in the circle CQ which are circumscribed to the circle cT, and now the corresponding relation is found to be

$$\left(\frac{r}{R-c}\right)^2+\left(\frac{r}{R+c}\right)^2=1,$$

while T_1T_3, T_2T_4 intersect at right angles in L, being the bisectors of the angles between Q_1LQ_2, Q_2LQ_3 (fig. 17).

This relation is proved immediately by taking the quadrilateral in the position $AQ_{\frac{1}{2}}DR_{\frac{3}{2}}$; and now $\gamma=\gamma'=\text{am}\,\frac{1}{2}K$,

$$\text{sn}\,\tfrac{1}{2}K=\frac{r}{R-c},\quad \text{cn}\,\tfrac{1}{2}K=\frac{r}{R+c};$$

so that squaring and adding leads to the desired relation.

As in (ii.), quadrilaterals can be escribed to the circle BLB', which are inscribed in the circle OP, since $N_{\frac{1}{2}}$ coincides with L.

But the circles $BN_{\frac{1}{4}}B'$ and $BN_{\frac{3}{4}}B'$ are related to the circle OP with regard to poristic octagons; and the common tangents of these circles are easily recognised at the points $P_{\frac{1}{4}}$, $P_{\frac{1}{2}}$, $P_{\frac{3}{4}}$.

Conversely, starting with the circle cT and the internal point L, and drawing T_1LT_3, T_2LT_4 through L at right angles to each other, the tangents to the circle cT at T_1, T_2, T_3, T_4 will form a quadrilateral $Q_1Q_2Q_3Q_4$ which is inscribed in a circle CQ, the diagonals Q_1Q_3, Q_2Q_4 passing through L, and being equally inclined to T_1T_3 and T_2T_4.

If Q_1c, Q_2c, Q_3c, Q_4c are produced to meet the circle CQ again in q_1, q_2, q_3, q_4, then q_1q_3 and q_2q_4 are diameters of the circle CQ; for Q_1q_1 bisects the angle $Q_2Q_1Q_4$, so that the arc $Q_2q_1=$ arc q_1Q_4, and similarly the arc $Q_2q_3=$ arc q_3Q_4, so that the arc $q_1Q_2q_3=$ arc $q_1Q_4q_3$, and each is therefore a semi-circle.

It follows, from elementary geometrical considerations, that

$$LT_1^2+LT_2^2+LT_3^2+LT_4^2=4r^2,$$

or $$T_1T_2^2+T_3T_4^2=T_2T_3^2+T_1T_4^2=4r^2;$$

and $$\frac{1}{cQ_1^2}+\frac{1}{cQ_3^2}=\frac{1}{cQ_2^2}+\frac{1}{cQ_4^2}=\frac{1}{r^2};$$

so that $$cq_1^2+cq_3^2=cq_2^2+cq_4^2=(R^2-c^2)^2/r^2,$$

leading to $$2(R^2+c^2)=(R^2-c^2)^2/r^2,$$

or, as before, $$\left(\frac{r}{R-c}\right)^2+\left(\frac{r}{R+c}\right)^2=1.$$

Denoting by u_1, u_2, u_3, u_4 the values of u at Q_1, Q_2, Q_3, Q_4,
so that $$u_1-u_2=u_2-u_3=u_3-u_4=\tfrac{1}{2}K;$$
and denoting by d_1, d_2, d_3, d_4 the corresponding values of dn u,
then (§ 57) $$d_1d_3=d_2d_4=\kappa';$$
and (§ 120) $$LQ=2l(1-\kappa')\text{dn}\,u,$$
so that $$Q_1Q_3=2l(1-\kappa')(d_1+d_3),\ Q_2Q_4=2l(1-\kappa')(d_2d_4);$$
while $$Q_1Q_2=2\surd(cl)(d_1+d_2), \text{ etc.}$$

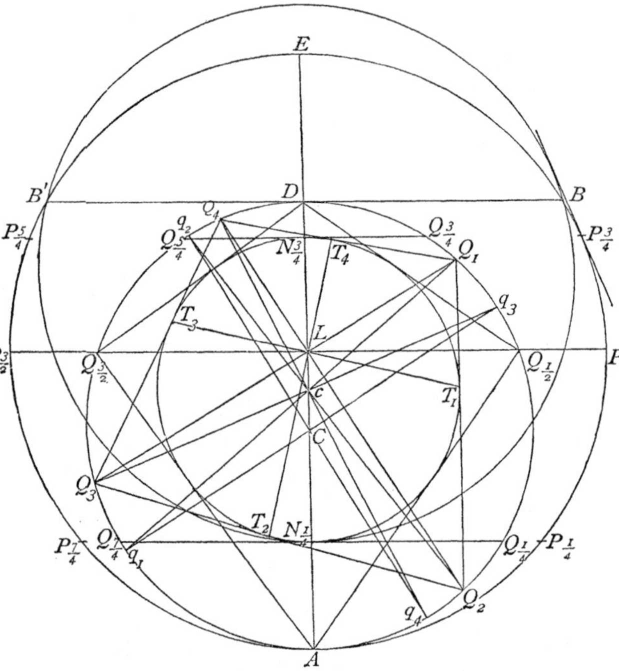

Fig. 17.

Now by a property of the circle (Euclid VI. D)
$$Q_1Q_3\,.\,Q_2Q_4=Q_1Q_2\,.\,Q_3Q_4+Q_1Q_4\,.\,Q_2Q_3;$$
so that $$l^2(1-\kappa')^2(d_1+d_3)(d_2+d_4)$$
$$=cl\{(d_1+d_2)(d_3+d_4)+(d_1+d_4)(d_2+d_3)\}$$
$$=cl\{(d_1+d_3)(d_2+d_4)+4\kappa'\},$$
or $(d_1+d_3)(d_2+d_4)$ is constant, and $=2\surd\kappa'(1+'),\kappa$
the value obtained by putting $u_4=0$, when
$$u_1=\tfrac{3}{2}K,\ u_2=K,\ u_3=\tfrac{3}{2}K;$$
and $d_1=d_3=\surd\kappa'$, $d_2=\kappa'$, $d_4=1$.

Then $\left(\operatorname{dn} u_1+\frac{\kappa'}{\operatorname{dn} u_1}\right)\left(\operatorname{dn} u_2+\frac{\kappa'}{\operatorname{dn} u_2}\right)=2\sqrt{\kappa'(1+\kappa')}$,

when $u_1-u_2=\frac{1}{2}K.$

Thus
$$\operatorname{dn}(u+\tfrac{1}{2}K)+\frac{\kappa'}{\operatorname{dn}(u+\frac{1}{2}K)}=\frac{2\sqrt{\kappa'(1+\kappa')}\operatorname{dn} u}{\operatorname{dn}^2 u+\kappa'},$$
$$\operatorname{dn}(u+\tfrac{1}{2}K)-\frac{\kappa'}{\operatorname{dn}(u+\frac{1}{2}K)}=\frac{-2\sqrt{\kappa'(1-\kappa'^2)}\operatorname{sn} u \operatorname{cn} u}{\operatorname{dn}^2 u+\kappa'};$$

so that
$$\operatorname{dn}(u+\tfrac{1}{2}K)=\sqrt{\kappa'(1+\kappa')}\frac{\operatorname{dn} u-(1-\kappa')\operatorname{sn} u \operatorname{cn} u}{\operatorname{dn}^2 u+\kappa'},$$
$$\operatorname{sn}(u+\tfrac{1}{2}K)=\sqrt{(1+\kappa')}\frac{\kappa'\operatorname{sn} u+\operatorname{cn} u \operatorname{dn} u}{\operatorname{dn}^2 u+\kappa'},$$
$$\operatorname{cn}(u+\tfrac{1}{2}K)=\sqrt{\kappa'}\sqrt{(1+\kappa')}\frac{\operatorname{cn} u-\operatorname{sn} u \operatorname{dn} u}{\operatorname{dn}^2 u+\kappa'}.$$

127. *Poristic Pentagons, etc.*

(v.) When $v=\frac{2}{5}K$, or $\frac{4}{5}K$, the poristic polygons are pentagons (fig. 18), and the relation to be satisfied is of the form
$$1+p+q-(p+q)^2-(p+q)(p-q)^2=0,$$
or
$$(p-q)^2=p+q-1-1/(p+q),$$
where p and q are used to denote $r/(R-c)$ and $r/(R+c)$.

We notice that the relation for pentagons leads to a cubic equation, when two of the three quantities R, r, c are given; but the equation reduces to a quadratic when $c=0$ or the circles are concentric, the case considered by Euclid.

The reader is referred to the articles of Cayley (*Phil. Mag.*, Series IV., Vol. 7, and *Collected Works*) and to Halphen's *Fonctions Elliptiques*, t. II., chap. X., for the proof of this relation and the similar relations for other polygons.

We shall find that Halphen's α and γ (t. II., p. 375) are connected with our R, r, c, κ, and w by the relations
$$\alpha=\kappa^2=\frac{4Rc}{(R+c)^2-r^2},\ \gamma=\operatorname{dn}^2 w=\left(\frac{R-c}{R+c}\right)^2;$$
and thence Halphen's x and y can be formed.

By the use of Legendre's Table IX. for $F(\phi, \kappa)$ (*F. E.*, t. II.) we are able to construct geometrically, to any required degree of accuracy, figures of circles related to each other for poristic polygons of any given number n of sides.

Having selected an arbitrary modulus κ or modular angle $\frac{1}{2}\alpha$, we look out the value of K, and then determine, by proportional parts, the value of ϕ in degrees corresponding to an amplitude of K/n, $2K/n$, ...; and these values of ϕ will mark the position of the points Q_1, Q_2, ...

Thus, in drawing figs. 13, 14, 16, 17, we have selected $\kappa=\sin 60°$, when $K=2{\cdot}1565$; and in drawing fig. 16 for poristic triangles, we find, from Legendre's Table IX.,

$$\text{am}\,\tfrac{1}{3}K=\text{c.m. of } 38°49', \quad \text{am}\,\tfrac{2}{3}K=\text{c.m. of } 68°5'.$$

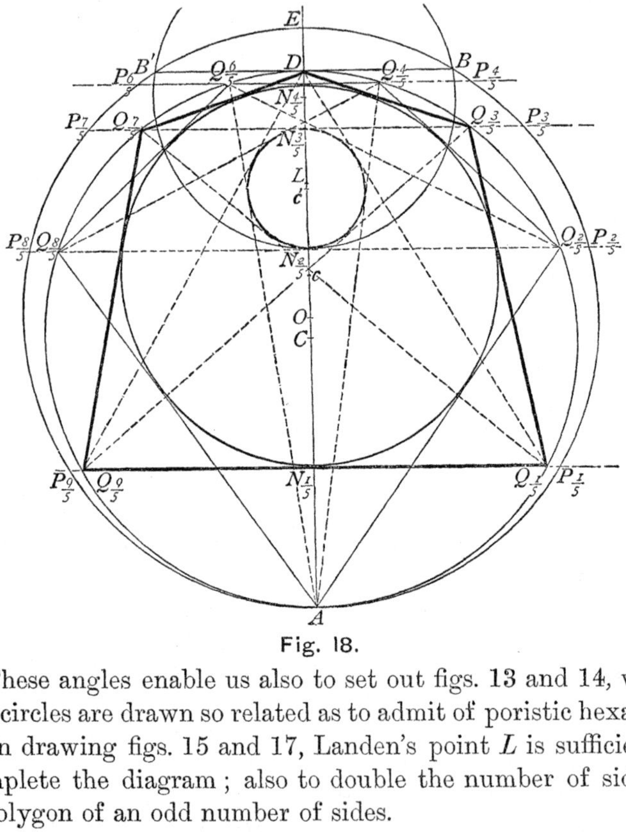

Fig. 18.

These angles enable us also to set out figs. 13 and 14, where the circles are drawn so related as to admit of poristic hexagons.

In drawing figs. 15 and 17, Landen's point L is sufficient to complete the diagram; also to double the number of sides of a polygon of an odd number of sides.

In fig. 18, κ has been taken as $\sin 75°$, as in figs. 1, 2, 3; and now $K=2{\cdot}76806$; and from Legendre's Table IX.,

$$\text{am}\,\tfrac{1}{5}K=\text{c.m. of } 30°18', \quad \text{am}\,\tfrac{3}{5}K=\text{c.m. of } 70°20',$$

by means of which the figures can be drawn.

Fig. 19 shows poristic heptagons, to the same modular angle of 75°, laid out by means of the relations

$$\phi_1=\text{am}\,\tfrac{1}{7}K=\text{c.m. of } 22°8', \quad \phi_3=\text{am}\,\tfrac{3}{7}K=\text{c.m. of } 56°49',$$

$$\phi_5=\text{am}\,\tfrac{5}{7}K=\text{c.m. of } 77°6'.$$

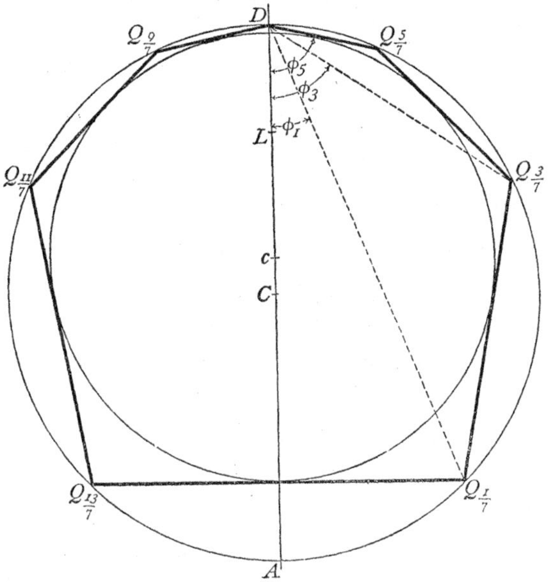

Fig. 19.

128. The poristic relation between the quantities R, r, c has been obtained by placing the polygon in a symmetrical position; but another method is employed by Wolstenholme (*Proceedings London Math. Society*, vol. VIII., p. 136; also by Halphen, *F.E.*, II., chap. X.), where the polygon on the circle OP is considered in its limiting form, when passing through one or both of the common points B and B'.

Thus with triangles, the tangent to the circle oR at B must meet the circle OP again at a point $P_{\frac{1}{3}}$, the point of contact of a common tangent of the two circles P and R, the degenerate triangle being BPP.

For quadrilaterals, the tangents to R at B, B' must meet at A on the circle P, $BACAB$ being the degenerate quadrilateral.

For pentagons we obtain the degenerate form $BP_{\frac{1}{5}}P_{\frac{7}{5}}P_{\frac{7}{5}}P_{\frac{1}{5}}B$, where $BP_{\frac{1}{5}}$ is the tangent at B to oR, the circle through B, $N_{\frac{2}{5}}$, B', and $P_{\frac{7}{5}}$ is the point of contact of a common tangent of the circles OP and oR (fig. 18).

For hexagons (fig. 16) the limiting form is $BP_{\frac{1}{3}}P_{\frac{5}{3}}B'P_{\frac{5}{3}}P_{\frac{1}{3}}B$, where $BP_{\frac{1}{3}}$, $P_{\frac{5}{3}}B'$ are tangents at B, B' to the circle through B, $N_{\frac{1}{3}}$, B'; and so on.

129. *Geometrical Applications of Elliptic Functions to Spherical Trigonometry.*

Taking the fundamental formulas of Spherical Trigonometry

$$\cos c = \cos a \cos b + \sin a \sin b \cos C,$$

$$\frac{\sin A}{\sin a} = \frac{\sin B}{\sin b} = \frac{\sin C}{\sin c} = \kappa, \text{ suppose};$$

then $$\cos C = \surd(1 - \kappa^2 \sin^2 c) = \Delta c,$$

so that $$\cos c = \cos a \cos b + \sin a \sin b \Delta c,$$

a formula like that of § 119, with a, b, c for ϕ, ψ, γ; so that if, keeping C, c, and therefore κ constant, we vary a and b, then

$$\cos B \,.\, da + \cos A \,.\, db = 0,$$

or $$da/\Delta a - db/\Delta b = 0;$$

and, conversely, the integral of this differential relation is the formula above.

(Lagrange, *Théorie des fonctions*, p. 85, §§ 81, 82; Legendre, *Fonctions elliptiques*, t. I., p. 20.)

If, in Jacobi's notation, we put

$$a = \text{am}(u, \kappa), \quad b = \text{am}(v, \kappa), \quad c = \text{am}(w, \kappa),$$

then the differential relation becomes

$$du - dv = 0,$$

so that $$u - v = \text{a constant} = w,$$

since $$a = c, \text{ or } u = w, \text{ when } b = 0 \text{ and } v = 0.$$

Supposing κ is less than unity, and the angle C is acute, then $c > C$, and of the other angles, one, A, must be obtuse, and the other, B, acute.

But by changing to the colunar triangle on the side BC, we may convert the triangle ABC into one in which all three angles are obtuse; and in such a triangle we may put

$$a = \text{am}\, u, \quad b = \pi - \text{am}\, v = \text{am}(2K - v), \quad c = \text{am}(2K - w);$$

so that if the triangle ABC has three obtuse angles, we may put

$$a = \text{am}\, u_1, \quad b = \text{am}\, u_2, \quad c = \text{am}\, u_3,$$

where $$u_1 + u_2 + u_3 = u + 2K - v + 2K - w = 4K;$$

and now

$$\cos A = -\text{dn}\, u_1, \quad \cos B = -\text{dn}\, u_2, \quad \cos C = -\text{dn}\, u_3,$$

so that, by § 29, we may write

$$A = \pi - \text{am}(\kappa u_1, 1/\kappa), \quad B = \pi - \text{am}(\kappa u_2, 1/\kappa), \quad C = \pi - \text{am}(\kappa u_3, 1/\kappa),$$

where κ is less than unity.

For instance, if ABC is the spherical triangle formed by three summits of a regular tetrahedron,

$$A = B = C = \tfrac{2}{3}\pi,$$

and
$$\cos a = \cos b = \cos c = -\tfrac{1}{3},$$
$$\sin a = \sin b = \sin c = \tfrac{2}{3}\sqrt{2},$$
$$\kappa = \frac{\sin A}{\sin a} = \frac{3\sqrt{3}}{4\sqrt{2}} = \frac{3\sqrt{6}}{8},\ \kappa' = \frac{\sqrt{10}}{8},\ 2\kappa\kappa' = \frac{3\sqrt{15}}{16};$$

while
$$u_1 = u_2 = u_3 = \tfrac{4}{3}K,$$

so that
$$\text{cn}\,\tfrac{4}{3}K = -\tfrac{1}{3},\ \text{sn}\,\tfrac{4}{3}K = \tfrac{2}{3}\sqrt{2},\ \text{dn}\,\tfrac{4}{3}K = \tfrac{1}{2}.$$

When $\kappa = 0$, $K = \frac{1}{2}\pi$, and the triangle ABC is coincident with a great circle; and now

$$a = u_1,\ b = u_2,\ c = u_3, \text{ and } a + b + c = 2\pi;$$

while
$$\cos A = \cos B = \cos C = -1,\ A = B = C = \pi.$$

When $\kappa = 1$, $K = \infty$; and therefore of u_1, u_2, u_3, two of them, say u_1 and u_2, are infinite; so that

$$\cos a = \text{sech}\, u_1 = 0, \text{ or } a = \tfrac{1}{2}\pi; \text{ and similarly } b = \tfrac{1}{2}\pi;$$

the triangle ABC now has two quadrantal sides and therefore two right angles, the third side c and angle C being equal, and taken greater than a right angle.

130. For values of κ which would be greater than unity, we change the notation by considering the polar triangle; and now if ABC is such a polar triangle, having three acute sides, instead of three obtuse angles, we put

$$\frac{\sin a}{\sin A} = \frac{\sin b}{\sin B} = \frac{\sin c}{\sin C} = \kappa;$$

and
$$A = \text{am}\, v_1,\ B = \text{am}\, v_2,\ C = \text{am}\, v_3,$$

where
$$v_1 = 2K - u_1,\ v_2 = 2K - u_2,\ v_3 = 2K - u_3,$$

so that
$$v_1 + v_2 + v_3 = 2K.$$

Now
$$\sin a = \kappa\, \text{sn}\, v_1,\ \sin b = \kappa\, \text{sn}\, v_2,\ \sin c = \kappa\, \text{sn}\, v_3;$$
$$\cos a = \text{dn}\, v_1,\ \cos b = \text{dn}\, v_2,\ \cos c = \text{dn}\, v_3;$$

so that
$$a = \text{am}(\kappa v_1, 1/\kappa),\ b = \text{am}(\kappa v_2, 1/\kappa),\ c = \text{am}(\kappa v_3, 1/\kappa).$$

The fundamental formula

$$\cos c = \cos a \cos b + \sin a \sin b \cos c$$

now leads to the formula of § 121,

$$\text{dn}\, v_3 = \text{dn}\, v_1 \text{dn}\, v_2 + \kappa^2 \text{sn}\, v_1 \text{sn}\, v_2\, \text{cn}\, v_3,$$

or
$$\text{dn}(v_1 + v_2) = \text{dn}\, v_1 \text{dn}\, v_2 - \kappa^2 \text{sn}\, v_1 \text{sn}\, v_2 \text{cn}(v_1 + v_2).$$

In the degenerate case of $\kappa = 0$, $K = \frac{1}{2}\pi$, and

$$v_1 + v_2 + v_3 = \pi, \text{ or } A + B + C = \pi;$$

and now $a = 0$, $b = 0$, $c = 0$, so that the spherical triangle is indefinitely small, and may be considered a plane triangle; and we can thus deduce the formulas of Plane Trigonometry.

131. A spherical triangle thus falls into one of two Classes, I. or II.; in Class I. the triangle, or a colunar triangle, has three obtuse angles; in Class II. the triangle, or a colunar triangle, has three acute sides; the quadrantal triangle falling into Class I., and the right-angled triangle into Class II.

In Class I. we put

$$\frac{\sin A}{\sin a}=\frac{\sin B}{\sin b}=\frac{\sin C}{\sin c}=\kappa,$$

and then κ is less than unity; and we put

$$a=\operatorname{am} u_1,\ b=\operatorname{am} u_2,\ c=\operatorname{am} u_3,$$

where
$$u_1+u_2+u_3=4K,$$

and then

$$A=\pi-\operatorname{am}(\kappa u_1,\ 1/\kappa),\ B=\pi-\operatorname{am}(\kappa u_2,\ 1/\kappa),\ C=\pi-\operatorname{am}(\kappa u_3,\ 1/\kappa).$$

In Class II. we put

$$\frac{\sin a}{\sin A}=\frac{\sin b}{\sin B}=\frac{\sin c}{\sin C}=\kappa,$$

and then κ is less than unity; and we put

$$A=\operatorname{am} v_1,\ B=\operatorname{am} v_2,\ C=\operatorname{am} v_3,$$

where
$$v_1+v_2+v_3=2K,$$

and then $a=\operatorname{am}(\kappa v_1,\ 1/\kappa),\ b=\operatorname{am}(\kappa v_2,\ 1/\kappa),\ c=\operatorname{am}(\kappa v_3,\ 1/\kappa)$.

When this triangle of Class II. is the polar of the triangle in Class I., $\quad u_1+v_1=u_2+v_2=u_3+v_3=2K.$

The change from one Class to the other affords an illustration of the change from one modulus to the reciprocal modulus (§ 29).

The spherical triangles employed originally by Lagrange and Legendre fall into Class I.; and a full discussion of the connexion between Elliptic Functions and Spherical Trigonometry will be found in the *Quarterly Journal of Mathematics*, vols. 17, 18, 19, in articles by Glaisher and Woolsey Johnson.

But it is preferable in some respects to work with the spherical triangles of Class II., as growing out on the sphere more naturally from the infinitesimal plane triangle; so it is proposed to develop here the relations with Elliptic Functions by means of a typical triangle of Class II., having three acute sides, and to refer to the articles of Glaisher and Woolsey Johnson for the corresponding relations of Class I.

132. Writing c_1, s_1, d_1 for $\text{cn}\, v_1$, $\text{sn}\, v_1$, $\text{dn}\, v_1$, etc.; then with

$$v_1+v_2+v_3=2K,$$

we may put, in Class II.,

$$A=\text{am}\, v_1,\ B=\text{am}\, v_2,\ C=\text{am}\, v_3;$$

so that $\cos A=c_1,\ \sin A=s_1$, etc.;

and now $\sin a=\kappa \sin A=\kappa s_1,\ \cos a=d_1$, etc.

From the fundamental formulas

$$\cos c=\cos a \cos b+\sin a \sin b \cos C,$$

$$-\cos C=\cos A \cos B-\sin A \sin B \cos c,$$

we obtain

$$d_3=d_1d_2+\kappa^2 s_1 s_2 c_3,$$

$$-c_3=c_1c_2-s_1s_2d_3,$$

where $d_3=\text{dn}\, v_3=\text{dn}(v_1+v_2),\ c_3=\text{cn}\, v_3=-\text{cn}(v_1+v_2)$.

Again, from these two formulas of spherical trigonometry,

$$-\cos C=\cos A \cos B-\sin A \sin B(\cos a \cos b+\sin a \sin b \cos C),$$

or

$$-\cos C=\frac{\cos A \cos B-\sin A \sin B \cos a \cos b}{1-\sin A \sin B \sin a \sin b},$$

so that

$$-c_3=\text{cn}(v_1+v_2)=\frac{c_1c_2-s_1s_2d_1d_2}{1-\kappa^2 s_1^2 s_2^2}.$$

Similarly,

$$\cos c=\frac{\cos a \cos b-\sin a \sin b \cos A \cos B}{1-\sin A \sin B \sin a \sin b},$$

leading to

$$d_3=\text{dn}(v_1+v_2)=\frac{d_1d_2-\kappa^2 s_1s_2c_1c_2}{1-\kappa^2 s_1^2 s_2^2}.$$

As a specimen of Class II., take the spherical triangle formed by three adjacent summits of a regular icosahedron; then

$$A=B=C=\tfrac{2}{5}\pi;$$

and

$$\cos c=\frac{\cos C+\cos A \cos B}{\sin A \sin B}=\frac{\cos C}{1-\cos C}=\frac{1}{\sqrt{5}},$$

so that $\kappa=\sin c/\sin C=\tfrac{2}{5}\sqrt{(10-2\sqrt{5})}$;

and then $v_1=v_2=v_3=\tfrac{2}{3}K$,

so that

$$\text{cn}\, \tfrac{2}{3}K=\cos C=\tfrac{1}{4}(\sqrt{5}-1),$$

$$\text{dn}\, \tfrac{2}{3}K=\cos c=\tfrac{1}{5}\sqrt{5}.$$

133. To prove that in a triangle of Class II. we obtain the differential relation

$$\cos b\, .\, dA+\cos b\, .\, dB=0,\ \text{or}\ dA/\Delta A+dB/\Delta B=0,$$

when we change A and B, keeping c and C constant, displace the triangle ABC into the consecutive position ABC', keeping the points A, B fixed and the angle $AC'B$ unchanged in magnitude (fig. 20).

Then, if CA and CB produced on the sphere meet the great circle of which C is the pole in P and Q, the arc $PQ=C$; and if $C'A$ and $C'B$ produced meet this great circle in P' and Q', the arc $P'Q'$ is ultimately equal to the arc PQ, or

$$\text{lt}(PP'/QQ')=1.$$

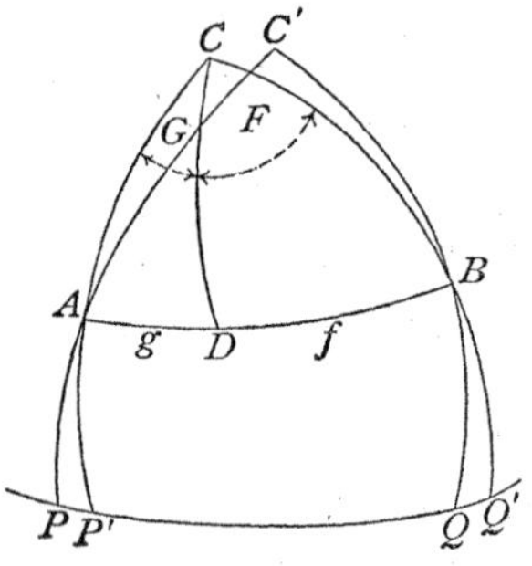

Fig. 20.

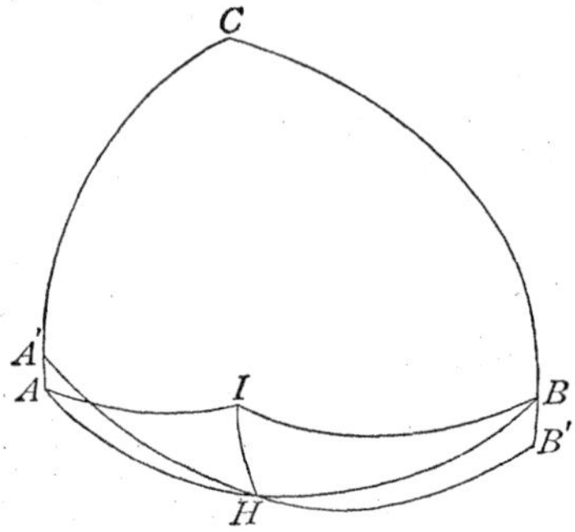

Fig. 21.

But $PAP'=-dA$, $QBQ'=dB$; while ultimately

$$PP'=-\sin AP \,.\, dA=-\cos b \,.\, dA, \quad QQ'=\cos a \,.\, dB;$$

so that $$\cos b \,.\, dA+\cos a \,.\, dB=0,$$

or $$dA/\Delta A+dB/\Delta B=0,$$

since $$\sin a=\kappa \sin A, \quad \cos a=\Delta A.$$

With $A=\text{am}\, v_1$, $B=\text{am}\, v_2$, this becomes

$$dv_1+dv_2=0,$$

so that $v_1+v_2=\text{constant}=2K-v_3$, where $C=\text{am}\, v_3$; since $B+C=\pi$, or $v_2+v_3=2K$, when $A=0$, $v_1=0$.

Conversely, this differential relation, interpreted with respect to the triangle ABC, of which the side AB is fixed, expresses the constancy of the opposite angle C.

134. If, as is customary, we deduce the differential relation

$$\cos B \,.\, da+\cos A \,.\, db=0, \quad \text{or} \quad da/\Delta a+db/\Delta b=0,$$

from a spherical triangle ABC of Class I., in which

$$\sin A=\kappa \sin a, \quad \cos A=\Delta a,$$

we keep the angle C fixed, and displace the side AB into its consecutive position $A'B'$, without change of length, through an infinitesimal angle θ about the centre of instantaneous rotation I, the point of intersection of the arcs AI, BI, drawn perpendicular to CA, CB respectively (fig. 21).

Then $$\frac{db}{da}=-\text{lt}\frac{AA'}{BB'}=-\frac{\sin IA}{\sin IB}=-\frac{\sin IBH}{\sin IAH}=-\frac{\cos B}{\cos A}.$$

135. To obtain immediately the addition formulas (1), (2), (3) of § 116 for the elliptic functions, Mr. Kummell draws the arc CD perpendicular to AB (fig. 20), and denotes the perpendicular CD by p, the segments BCD, ACD of the angle C by F, G, and the segments BD, DA of the base C by f, g; so that

$$F+G=C,\ f+g=c.$$

(Kummell, *Analyst*, vol. V., 1878.)

Now, from the right-angled spherical triangles ACD, BCD,

$$\cos G=\sin A\cos b/\cos p,\quad \sin G=\cos A/\cos p;$$
$$\cos F=\sin B\cos a/\cos p,\quad \sin F=\cos B/\cos p;$$

or with $\sin A=s_1$, $\cos A=c_1$, $\sin a=\kappa s$, $\cos a=d_1$, etc., and writing M for $\cos p$,

$$\cos G=s_1d_2/M,\quad \sin G=c_1/M;$$
$$\cos F=s_2d_1/M,\quad \sin F=c_2/M.$$

Also $\quad \sin p=\sin A\sin b=\sin a\sin B=\kappa s_1s_2,$

so that $\quad M^2=\cos^2 p=1-\kappa^2s_1{}^2s_2{}^2,$

a quantity which we have found it convenient to denote by D.

Now, $\quad \cos C=\cos F\cos G-\sin F\sin G,$

or $\quad c_3=(s_1s_2d_1d_2-c_1c_2)/D,$

or $\quad \operatorname{cn}(v_1+v_2)=-\operatorname{cn}v_3=(c_1c_2-s_1s_2d_1d_2)/D,$

formula (2).

Again, $\quad \sin C=\sin(F+G)$
$$=\sin F\cos G+\cos F\sin G,$$

or $\quad s_3=(s_1c_2d_2+s_2c_1d_1)/D,$

where $\quad s_3=\operatorname{sn}v_3=\operatorname{sn}(v_1+v_2)$, as in formula (1).

Changing the sign of v_2,

$$\operatorname{sn}(v_1-v_2)=\sin(F-G),$$

or $\quad F-G=\operatorname{am}(v_1-v_2),$

while $\quad F+G=\operatorname{am}v_3=\operatorname{am}(2K-v_1-v_2)$
$$=\pi-\operatorname{am}(v_1+v_2),$$

so that $\quad F=\tfrac{1}{2}\pi-\tfrac{1}{2}\operatorname{am}(v_1+v_2)+\tfrac{1}{2}\operatorname{am}(v_1-v_2),$
$$G=\tfrac{1}{2}\pi-\tfrac{1}{2}\operatorname{am}(v_1+v_2)-\tfrac{1}{2}\operatorname{am}(v_1-v_2).$$

Thus, for instance,

$$\tan\{\tfrac{1}{2}\operatorname{am}(v_1+v_2)+\tfrac{1}{2}\operatorname{am}(v_1-v_2)\}=\cot G=\tan A\cos b=s_1d_2/c_1,$$
$$\tan\{\tfrac{1}{2}\operatorname{am}(v_1+v_2)-\tfrac{1}{2}\operatorname{am}(v_1-v_2)\}=\cot F=\tan B\cos a=s_2d_1/c_2.$$

Again, from the right-angled spherical triangles BCD, ACD,

$$\cos f=\cos a/\cos p=d_1/M,\quad \sin f=\sin a\cos B/\cos p=\kappa s_1c_2/M;$$
$$\cos g=\cos b/\cos p=d_2/M,\quad \sin g=\sin b\cos A/\cos p=\kappa s_2c_1/M;$$

and therefore

$$\begin{aligned}\text{dn}(v_1+v_2)&=\text{dn}\,v_3=\cos c=\cos(f+g)\\&=\cos f\cos g-\sin f\sin g\\&=(d_1d_2-\kappa^2s_1s_2c_1c_2)/D\\&=\frac{d_1d_2-\kappa^2s_1s_2c_1c_2}{1-\kappa^2s_1{}^2s_2{}^2},\end{aligned}$$

as before, in (3), § 116.

Also $\sin(f+g)=\kappa\,\text{sn}(v_1+v_2)$, $\sin(f-g)=\kappa\,\text{sn}(v_1-v_2)$; whence f and g can be found as functions of v_1+v_2 and v_1-v_2.

136. The formula employed by Morgan Jenkins in the *Messenger of Mathematics*, vol. XVII., p. 30, as fundamental in Spherical Trigonometry, is

$$\frac{\sin(A+B)}{\cos b+\cos a}=\frac{\sin C}{1+\cos c}, \qquad\ldots\ldots(\alpha)$$

and this now leads to

$$\frac{s_1c_2+s_2c_1}{d_2+d_1}=\frac{s_3}{1+d_3},$$

or, in the Legendrian form

$$\frac{\sin(A+B)}{\Delta B+\Delta A}=\frac{\sin C}{1+\Delta C},$$

a formula already obtained from pendulum motion in § 120.

Then the formula

$$\frac{s_1c_2-s_2c_1}{d_2-d_1}=\frac{s_3}{1-d_3},$$

or

$$\frac{\sin(A-B)}{\Delta B-\Delta A}=\frac{\sin C}{1-\Delta C},$$

gives

$$\frac{\sin(A-B)}{\cos b-\cos a}=\frac{\sin C}{1-\cos c}. \qquad\ldots\ldots(\beta)$$

The formulas of § 120, in the form

$$\frac{s_1d_2+s_2d_1}{c_2+c_1}=\frac{s_3}{1+c_3},\quad \frac{s_1d_2-s_2d_1}{c_2-c_1}=\frac{s_3}{1-c_3},$$

lead to the relations

$$\frac{\sin(a+b)}{\cos B+\cos A}=\frac{\sin c}{1-\cos C}, \qquad\ldots\ldots(\gamma)$$

$$\frac{\sin(a-b)}{\cos B-\cos A}=\frac{\sin c}{1+\cos C}, \qquad\ldots\ldots(\delta)$$

and from these four formulas of Spherical Trigonometry Mr. Morgan Jenkins deduces the analogies of Napier, Delambre, and Gauss.

137. Write, as before, in § 135,

$$A = \text{am}\, u,\quad B = \text{am}\, v;$$
$$F = \tfrac{1}{2}\pi - \tfrac{1}{2}\,\text{am}(u+v) + \tfrac{1}{2}\,\text{am}(u-v),$$
$$G = \tfrac{1}{2}\pi - \tfrac{1}{2}\,\text{am}(u+v) - \tfrac{1}{2}\,\text{am}(u-v).$$

Then, since

$$\sin(F+G) + \sin(F-G) = 2\sin F\cos G,$$

therefore, writing c_1, s_1, d_1 for cn u, sn u, dn u, and c_2, s_2, d_2 for cn v, sn v, dn v, and D for $\cos^2 p$ or $1-\kappa^2 s_1^2 s_2^2$,

$$\text{sn}(u+v) + \text{sn}(u-v) = 2\,s_1c_2d_2/D;\ \ldots\ldots(1)$$
$$\cos(F-G) - \cos(F+G) = 2\sin F\sin G,$$
$$\text{cn}(u-v) + \text{cn}(u+v) = 2\,c_1c_2/D;\ \ldots\ldots(2)$$
$$\cos(f-g) + \cos(f+g) = 2\cos f\cos g,$$
$$\text{dn}(u-v) + \text{dn}(u+v) = 2\,d_1d_2/D;\ \ldots\ldots(3)$$
$$\sin(F+G) - \sin(F-G) = 2\cos F\sin G,$$
$$\text{sn}(u+v) - \sin(u-v) = 2\,s_2c_1d_1/D;\ \ldots\ldots(4)$$
$$\cos(F-G) + \cos(F+G) = 2\cos F\cos G,$$
$$\text{cn}(u-v) - \text{cn}(u+v) = 2\,s_1d_1s_2d_2/D;\ \ldots\ldots(5)$$
$$\cos(f-g) - \cos(f+g) = 2\sin f\sin g,$$
$$\text{dn}(u-v) - \text{dn}(u+v) = 2\,\kappa^2 s_1c_1s_2c_2/D;\ \ldots\ldots(6)$$
$$\sin(F+G)\sin(F-G) = \sin^2 F - \sin^2 G,$$
$$\text{sn}(u+v)\ \text{sn}(u-v) = (c_2^2 - c_1^2)D = (s_1^2 - s_2^2)/D.\ \ldots(7)$$

Again, since

$$1 + \sin(f+g)\sin(f-g) = \cos^2 g + \sin^2 f,$$

and $\sin(f+g) = \kappa\,\text{sn}(u+v)$, $\sin(f-g) = \kappa\,\text{sn}(u-v)$,

$$1 + \kappa^2\text{sn}(u+v)\ \text{sn}(u-v) = (d_2^2 + \kappa^2 s_1^2 c_2^2)/D;\ \ldots\ldots(8)$$
$$1 + \sin(F+G)\sin(F-G) = \sin^2 F + \cos^2 G,$$
$$1 + \text{sn}(u+v)\ \text{sn}(u-v) = (c_2^2 + s_1^2 d_2^2)/D;\ \ldots\ldots(9)$$
$$1 - \cos(F+G)\cos(F-G) = \sin^2 G + \sin^2 F,$$
$$1 + \text{cn}(u+v)\ \text{cn}(u-v) = (c_1^2 + c_2^2)/D;\ \ldots\ldots(10)$$
$$1 + \cos(f+g)\ \cos(f-g) = \cos^2 f\cos^2 g,$$
$$1 + \text{dn}(u+v)\ \text{dn}(u-v) = (d_1^2 + d_2^2)/D;\ \ldots\ldots(11)$$
$$1 - \sin(f+g)\ \sin(f-g) = \cos^2 f + \sin^2 g),$$
$$1 - \kappa^2\text{sn}(u+v)\ \text{sn}(u-v) = (d_1^2 + \kappa^2 s_2^2 c_1^2)/D;\ \ldots\ldots(12)$$
$$1 - \sin(F+G)\sin(F-G) = \sin^2 G + \cos^2 F,$$
$$1 - \text{sn}(u+v)\ \text{sn}(u-v) = (c_1^2 + s_2^2 d_1^2)/D;\ \ldots\ldots(13)$$
$$1 + \cos(F+G)\sin(F-G) = \cos^2 G + \cos^2 F,$$
$$1 - \text{cn}(u+v)\ \text{cn}(u-v) = (s_1^2 d_2^2 + s_2^2 d_1^2)/D;\ \ldots\ldots(14)$$
$$1 - \cos(f+g)\ \cos(f-g) = \sin^2 f + \sin^2 g,$$
$$1 - \text{dn}(u+v)\ \text{dn}(u-v) = \kappa^2(s_1^2 c_2^2 + s_2^2 c_1^2)/D;\ \ldots\ldots(15)$$

$$\{1\pm\sin(F+G)\}\{1\pm\sin(F-G)\}=(\sin F\pm\cos G)^2,$$
$$\{1\pm\ \mathrm{sn}(u+v)\}\{1\pm\ \mathrm{sn}(u-v)\}=(c_2\pm s_1d_2)^2/D\,;\ \ldots\ldots\ldots\ldots(16)$$
$$\{1\pm\sin(F+G)\}\{1\mp\sin(F-G)\}=(\sin G\pm\cos F)^2,$$
$$\{1\pm\ \mathrm{sn}(u+v)\}\{1\mp\ \mathrm{sn}(u-v)\}=(c_1\pm s_2d_1)^2/D\,;\ \ldots\ldots\ldots\ldots(17)$$
$$\{1\pm\ \sin(f+g)\}\{1\pm\ \sin(f-g)\}=(\cos g\pm\sin f)^2,$$
$$\{1\pm\kappa\,\mathrm{sn}(u+v)\}\{1\pm\kappa\,\mathrm{sn}(u-v)\}=(d_2\pm\kappa s_1c_2)^2/D\,;\ \ldots\ldots\ldots\ldots(18)$$
$$\{1\pm\ \sin(f+g)\}\{1\mp\ \sin(f-g)\}=(\cos f\pm\sin g)^2,$$
$$\{1\pm\kappa\,\mathrm{sn}(u+v)\}\{1\mp\kappa\,\mathrm{sn}(u-v)\}=(d_1\pm\kappa s_2c_1)^2/D\,;\ \ldots\ldots\ldots\ldots(19)$$
$$\{1\mp\cos(F+G)\}\{1\pm\cos(F-G)\}=(\sin F\pm\sin G)^2,$$
$$\{1\pm\ \mathrm{cn}(u+v)\}\{1\pm\ \mathrm{cn}(u-v)\}=(c_1\pm c_2)^2/D\,;\ \ldots\ldots\ldots\ldots(20)$$
$$\{1\pm\cos(F+G)\}\{1\pm\cos(F-G)\}=(\cos G\mp\cos F)^2,$$
$$\{1\mp\ \mathrm{cn}(u+v)\}\{1\pm\ \mathrm{cn}(u-v)\}=(s_1d_2\mp s_2d_1)^2/D\,;\ \ldots\ldots\ldots\ldots(21)$$
$$\{1\pm\ \cos(f+g)\}\{1\pm\ \cos(f-g)\}=(\cos f\pm\cos g)^2,$$
$$\{1\pm\ \mathrm{dn}(u+v)\}\{1\pm\ \mathrm{dn}(u-v)\}=(d_1\pm d_2)^2/D\,;\ \ldots\ldots\ldots\ldots(22)$$
$$\{1\pm\ \cos(f+g)\}\{1\mp\ \cos(f-g)\}=(\sin f\mp\sin g)^2,$$
$$\{1\pm\ \mathrm{dn}(u+v)\}\{1\mp\ \mathrm{dn}(u-v)\}=\kappa^2(s_1c_2\mp s_2c_1)^2/D\,;\ \ldots\ldots\ldots\ldots(23)$$
$$\sin(F+G)\cos(F-G)=\sin G\cos G+\sin F\cos F,$$
$$\mathrm{sn}(u+v)\ \ \mathrm{cn}(u-v)=(s_1c_1d_2+s_2c_2d_1)/D\,;\ \ldots\ldots\ldots(24)$$
$$-\sin(F-G)\cos(F+G)=\sin G\cos G-\sin F\cos F,$$
$$\mathrm{sn}(u-v)\ \ \mathrm{cn}(u+v)=(s_1c_1d_2-s_2c_2d_1)/D\,;\ \ldots\ldots\ldots(25)$$
$$\sin(f+g)\ \cos(f-g)=\sin f\cos f+\sin g\cos g,$$
$$\mathrm{sn}(u+v)\ \ \mathrm{dn}(u-v)=(s_1d_1c_2+s_2d_2c_1)/D\,;\ \ldots\ldots\ldots(26)$$
$$\sin(f-g)\ \cos(f+g)=\sin f\cos f-\sin g\cos g,$$
$$\mathrm{sn}(u-v)\ \ \mathrm{dn}(u+v)=(s_1d_1c_2-s_2d_2s_1/D\,;\ \ldots\ldots\ldots(27)$$
$$-\cos(F+G)\cos(f-g)=\{\cos A\cos B-\sin A\sin B\cos(f+g)\}\cos(f-g),$$
$$\mathrm{cn}(u+v)\ \ \mathrm{dn}(u-v)=(c_1c_2d_1d_2-\kappa'^2s_1s_2)/D\,;\ \ldots\ldots(28)$$
$$\cos(F-G)\cos(f+g)=\cos(F-G)\{\cos a\cos b+\sin a\sin b\cos(F+G)\},$$
$$\mathrm{cn}(u-v)\ \ \mathrm{dn}(u+v)=(c_1c_2d_1d_2+\kappa'^2s_1s_2)/D\,;\ \ldots\ldots(29)$$
$$\sin 2G=2\sin G\cos G,$$
$$\sin\{\mathrm{am}(u+v)+\mathrm{am}(u-v)\}=2\,s_1c_1d_2/D\,;\ \ldots\ldots\ldots\ldots(30)$$
$$\sin 2F=2\sin F\cos F,$$
$$\sin\{\mathrm{am}(u+v)-\mathrm{am}(u-v)\}=2\,s_2c_2d_1/D\,;\ \ldots\ldots\ldots\ldots(31)$$
$$-\cos 2G=\sin^2G-\cos^2G,$$
$$\cos\{\mathrm{am}(u+v)+\mathrm{am}(u-v)\}=(c_1^2-s_1^2d_2^2)/D\,;\ \ldots\ldots\ldots\ldots(32)$$
$$-\cos 2F=\sin^2F-\cos^2F,$$
$$\cos\{\mathrm{am}(u+v)-\mathrm{am}(u-v)\}=(c_2^2-s_2^2d_1^2)/D\,;\ \ldots\ldots\ldots\ldots(33)$$

the thirty-three formulas of Jacobi, given in his *Fundamenta Nova*, 18, and reproduced in Cayley's *Elliptic Functions*.

Similarly any other formula in Spherical Trigonometry is converted into a form of the Addition Theorem of the Elliptic Functions, and conversely; by writing c_1, s_1 for $\cos A$, $\sin A$, and d_1, κs_1 for $\cos a$, $\sin a$, etc., with

$$v_1+v_2+v_3=2K.$$

Thus the six four-part formulas, of which

$$\cot a \sin c=\cot A \sin B+\cos c \cos B$$

is the type, obtained by eliminating $\cos b$ between (α) and (β), lead to

$$s_3 d_1=s_2 c_1+s_1 c_2 d_3,$$

with five other similar relations.

By means of these and the preceding relations we can prove the following examples on the formulas of Elliptic Functions.

EXAMPLES.

1. Prove that, if $u+v+w+x=0$,

(i.) $$\frac{\text{cn}\, u\, \text{dn}\, v-\text{dn}\, u\, \text{cn}\, v}{\text{sn}\, u-\text{sn}\, v}+\frac{\text{cn}\, w\, \text{dn}\, x-\text{dn}\, w\, \text{cn}\, x}{\text{sn}\, w-\text{sn}\, x}=0.$$

(ii.) $$\kappa'^2-\kappa^2\kappa'^2\text{sn}\, u\, \text{sn}\, v\, \text{sn}\, w\, \text{sn}\, x+\kappa^2\text{cn}\, u\, \text{cn}\, v\, \text{cn}\, w\, \text{cn}\, x$$
$$-\text{dn}\, u\, \text{dn}\, v\, \text{dn}\, w\, \text{dn}\, x=0.$$

2. Prove that

(i.) $$\text{ns}(u-v)+\text{sn}(u+v)=\frac{2\kappa^2\text{sn}\, u\, \text{cn}\, v\, \text{dn}\, v}{\text{dn}^2 v-\text{dn}^2 u};$$

(ii.) $$1-\kappa^2\text{sn}^2(u+v)\text{sn}^2(u-v)=(1-\kappa^2\text{sn}^4 u)(1-\kappa^2\text{sn}^4 v)/D^2;$$

(iii.) $$\kappa^2\text{sn}(u+v)\text{sn}(u-v)\text{sn}(u+w)\text{sn}(u-w)$$
$$+\frac{(1-\kappa^2\text{sn}^4 u)(1-\kappa^2\text{sn}^2 v\, \text{sn}^2 w)}{(1-\kappa^2\text{sn}^2 u\, \text{sn}^2 v)(1-\kappa^2\text{sn}^2 u\, \text{sn}^2 w)}=1;$$

(iv.) $$\frac{1-\kappa^2\text{cd}^2(u+v)\text{cd}^2(u-v)}{1-\kappa^2\text{sn}^2(u+v)\text{sn}^2(u-v)}=\kappa'^2\left(\frac{1-\kappa^2\text{sn}^2 u\, \text{sn}^2 v}{\kappa'^2+\kappa^2\text{cn}^2 u\, \text{cn}^2 v}\right)^2.$$

3. (i.) $$\frac{1-\text{sn}\, u}{1+\text{sn}\, u}=\frac{\text{cn}^2\tfrac{1}{2}(u+K)\text{dn}^2\tfrac{1}{2}(u+K)}{\kappa'^2\text{sn}^2\tfrac{1}{2}(u+K)};$$

(ii.) $$\frac{1-\kappa'\text{dn}\, u+\kappa^2\text{sn}\, u}{1+\kappa'\text{dn}\, u+\kappa^2\text{sn}\, u}=\kappa^2\text{sn}^4\tfrac{1}{2}(u+K).$$

4. Prove that

$$1\pm\kappa\, \text{sn}\, u\, \text{sn}\, v=\frac{\{1\pm\kappa\, \text{sn}^2\tfrac{1}{2}(u+v)\}\{1\mp\kappa\, \text{sn}^2\tfrac{1}{2}(u-v)\}}{1-\kappa^2\text{sn}^2\tfrac{1}{2}(u+v)\text{sn}^2\tfrac{1}{2}(u-v)};$$

and hence prove that the expression

$$\frac{1-\kappa\, \text{sn}\, x\, \text{sn}\, y}{1+\kappa\, \text{sn}\, x\, \text{sn}\, y}\cdot\frac{1+\kappa\, \text{sn}\, z\, \text{sn}\, w}{1-\kappa\, \text{sn}\, z\, \text{sn}\, w}$$

remains unaltered when for x, y, z, w we substitute respectively

$$\tfrac{1}{2}(x+y+z+w),\ \tfrac{1}{2}(x+y-z-w),\ \tfrac{1}{2}(x-y+z-w),\ \tfrac{1}{2}(x-y-z+w).$$

5. Prove that, if $\tanh A=\kappa\,\mathrm{sn}^2\alpha$, $\tanh B=\kappa\,\mathrm{sn}^2\beta$,

$$\tanh(A-B)=\kappa\,\mathrm{sn}(\alpha+\beta)\mathrm{sn}(\alpha-\beta).$$

Deduce Jacobi's relations,

$$\mathrm{sn}(\beta+\gamma)\mathrm{sn}(\beta-\gamma)+\mathrm{sn}(\gamma+\alpha)\mathrm{sn}(\gamma-\alpha)+\mathrm{sn}(\alpha+\beta)\mathrm{sn}(\alpha-\beta)$$
$$+\kappa^2\mathrm{sn}(\beta+\gamma)\mathrm{sn}(\gamma+\alpha)\mathrm{sn}(\alpha+\beta)\mathrm{sn}(\beta-\gamma)\mathrm{sn}(\gamma-\alpha)\mathrm{sn}(\alpha-\beta)=0;$$

or

$$\frac{1-\kappa\,\mathrm{sn}(\beta+\gamma)\mathrm{sn}(\beta-\gamma)}{1+\kappa\,\mathrm{sn}(\beta+\gamma)\mathrm{sn}(\beta-\gamma)}\cdot\frac{1-\kappa\,\mathrm{sn}(\gamma+\alpha)\mathrm{sn}(\gamma-\alpha)}{1+\kappa\,\mathrm{sn}(\gamma+\alpha)\mathrm{sn}(\gamma-\alpha)}\cdot\frac{1-\kappa\,\mathrm{sn}(\alpha+\beta)\mathrm{sn}(\alpha-\beta)}{1+\kappa\,\mathrm{sn}(\alpha+\beta)\mathrm{sn}(\alpha-\beta)}=1;$$

or

$$\frac{1-\kappa\,\mathrm{sn}(t-x)\mathrm{sn}(y-z)}{1+\kappa\,\mathrm{sn}(t-x)\mathrm{sn}(y-z)}\cdot\frac{1-\kappa\,\mathrm{sn}(t-y)\mathrm{sn}(z-x)}{1+\kappa\,\mathrm{sn}(t-y)\mathrm{sn}(z-x)}\cdot\frac{1-\kappa\,\mathrm{sn}(t-z)\mathrm{sn}(x-y)}{1+\kappa\,\mathrm{sn}(t-z)\mathrm{sn}(x-y)}=1;$$

or

$$\frac{1-\kappa\,\mathrm{sn}\,u\,\mathrm{sn}\,v}{1+\kappa\,\mathrm{sn}\,u\,\mathrm{sn}\,v}\cdot\frac{1+\kappa\,\mathrm{sn}(u+w)\mathrm{sn}(v+w)}{1-\kappa\,\mathrm{sn}(u+w)\mathrm{sn}(v+w)}\cdot\frac{1-\kappa\,\mathrm{sn}(u+v+w)\mathrm{sn}\,w}{1+\kappa\,\mathrm{sn}(u+v+w)\mathrm{sn}\,w}=1.$$

(Glaisher, *Q. J. M.*, vol. XIX., p. 22.)

6. Prove that the tangents at the points on an ellipse of excentricity e whose excentric angles are

$$\phi=\tfrac{1}{2}\pi-\mathrm{am}(u,\,e),\quad \psi=\tfrac{1}{2}\pi-\mathrm{am}(v,\,e),$$

will meet on a confocal ellipse when $u-v$ is constant, and on a confocal hyperbola when $u+v$ is constant.

Hence show that the general integral of

$$d\phi/\sqrt{(1-e^2\sin^2\phi)}-d\psi/\sqrt{(1-e^2\sin^2\psi)}=0$$

may be written

$$\frac{a^2}{a^2+\lambda}\sin^2\tfrac{1}{2}(\phi+\psi)+\frac{b^2}{b^2+\lambda}\cos^2\tfrac{1}{2}(\phi+\psi)=\cos^2\tfrac{1}{2}(\phi-\psi);$$

and convert this into the form

$$\cos\gamma=\cos\phi\cos\psi+\sin\phi\sin\psi\sqrt{(1-e^2\sin^2\gamma)},$$

proving that $$\tan^2\tfrac{1}{2}\gamma=\frac{\lambda(a^2+\lambda)}{a^2(b^2+\lambda)}.$$

7. Prove that the straight line joining the points

$$c\,\mathrm{cn}(u+v),\ c\,\mathrm{sn}(u+v)\ \text{and}\ c\,\mathrm{cn}(u-v),\ c\,\mathrm{sn}(u-v),$$

on a given circle of radius c, will touch an ellipse whose semi-axes are $c\,\mathrm{sn}(K-v)$, $c\,\mathrm{cn}\,v$, when u is constant and v is variable; and determine the envelope when u is variable and v is constant.

CHAPTER V.

THE ALGEBRAICAL FORM OF THE ADDITION THEOREM.

138. The first demonstration of the existence of an Addition Theorem for Elliptic Functions is due to Euler (*Acta Petropolitana*, 1761; *Institutiones Calculi Integralis*), who showed that the differential relation

$$dx/\sqrt{X}+dy/\sqrt{Y}=0,$$

connecting $\quad X=ax^4+4bx^3+6cx^2+4dx+e,$

or $\quad (a, b, c, d, e)(x, 1)^4,$

the most general quartic function of a variable x, and Y the same function of another variable y, leads to an algebraical relation between x and y, X and Y.

This algebraical relation is

$$\left(\frac{\sqrt{X}-\sqrt{Y}}{x-y}\right)^2=a(x+y)^2+4b(x+y)+C,$$

where C is the arbitrary constant of integration; and this relation when rationalized leads to a symmetrical quadri-quadric function of x and y, of the form (§ 148)

$$\alpha x^2y^2+2\beta xy(x+y)+\gamma(x^2+4xy+y^2)+2\delta(x+y)+\epsilon=0,$$

or $\quad (\alpha x^2+2\beta x+\gamma)y^2+2(\beta x^2+2\gamma x+\delta)y+\gamma x^2+2\delta x+\epsilon=0,$

or $\quad (\alpha y^2+2\beta y+\gamma)x^2+2(\beta y^2+2\gamma y+\delta)x+\gamma y^2+2\delta y+\epsilon=0.$

(Cayley, *Elliptic Functions*, chap. XIV.)

With $a=0$ and $b=0$, X and Y reduce to quadratic functions of x and y; and then

$$\frac{\sqrt{X}-\sqrt{Y}}{x-y}=\text{a constant}$$

is the general integral of $dx/\sqrt{X}+dy/\sqrt{Y}=0$.

139. By writing $(lx'+m)/(l'x'+m')$ for x, which is called a *linear substitution*, this symmetrical quadri-quadric function becomes unsymmetrical, the five constants α, β, γ, δ, ϵ being thereby raised in number to nine; and then

$dx/\sqrt{X}$ becomes changed to $(lm'-l'm)dx'/\sqrt{X'}$,

where $$X'=(a, b, c, d, e)(lx+m, l'x+m')^4.$$

The *invariants* g_2 and g_3 of the quartic X have been defined in § 75, and in § 53 the *discriminant* $\Delta=g_2^3-27g_3^2$, and the *absolute invariant* $J=g_2^3/\Delta$; and now, if g_2', g_3', Δ', J' denote the same invariants of X', we find

$$g_2'=(lm'-l'm)^4 g_2,\ g_3'=(l'm-lm')^6 g_3,\ \Delta'=(lm'-l'm)^{12}\Delta;$$

while the absolute invariants J and J' are equal.

Conversely, any unsymmetrical quadri-quadric function whatever of x and y may be written

$$G(x,y)=(ax^2+2\beta x+\gamma)y^2+2(\beta'x^2+2\gamma'x+\delta')y+\gamma''x^2+2\delta''x+\epsilon''$$
$$=Ly^2+2My+N=0;$$
$$G(x,y)=(\alpha y^2+2\beta'y+\gamma'')x^2+2(\beta y^2+2\gamma'y+\delta'')x+\gamma y^2+2\delta'y+\epsilon''$$
$$=Px^2+2Qx+R=0;$$

L, M, N being quadratic functions of x, and P, Q, R being quadratic functions of y.

Then by differentiation

$$(Px+Q)dx+(Ly+M)dy=0;$$

and by solution of quadratic equations

$$Ly+M=\sqrt{(M^2-LN)}=\sqrt{X}, \text{ suppose};$$
$$Px+Q=\sqrt{(Q^2-PR)}=\sqrt{Y}, \text{ suppose};$$

and thus we are led to the differential relation

$$dx/\sqrt{X}+dy/\sqrt{Y}=0,$$

where X and Y are quartic functions of X, not necessarily of the same form, but having the same g_2 and g_3.

A linear transformation, such as that given by

$$y=(ly'+m)/(l'y'+m'),$$

can however always be found, which will transform

$$dy/\sqrt{Y} \text{ into } dy'/\sqrt{Y'},$$

where Y' is a quartic having the same coefficients as the quartic X; in other words, the quartics X and Y have the same invariants; so that we may, without loss of generality, consider X and Y as of the same form, and therefore drop the accents in the expression for $G(x, y)$.

Now $\sqrt{X}=Ly+M=(\alpha x^2+2\beta x+\gamma)y+\beta x^2+2\gamma x+\delta,$
$\sqrt{Y}=Px+Q=(\alpha y^2+2\beta y+\gamma)x+\beta y^2+2\gamma y+\delta;$

so that $$\frac{\sqrt{X}-\sqrt{Y}}{x-y}=\alpha xy+\beta(x+y)+\gamma,$$
a form of the integral relation, in which the coefficients a, b, c, d, e in X and Y are functions of $\alpha, \beta, \gamma, \delta, \epsilon$, determined by
$$ax^4+4bx^3+6cx^2+4dx+e$$
$$\equiv(\beta x^2+2\gamma x+\delta)^2-(\alpha x^2+2\beta x+\gamma)(\gamma x^2+2\delta x+\epsilon),$$
the Hessian, with changed sign, of $(\alpha, \beta, \gamma, \delta, \epsilon)(x, 1)^4$; and
$$a(x+y)^2+4b(x+y)+C$$
$$\equiv\{\alpha xy+\beta(x+y)+\gamma\}^2$$
$$\equiv(\beta^2-\alpha\gamma)(x+y)^2+2(\beta\gamma-\alpha\delta)(x+y)+\gamma^2-\alpha\epsilon.$$

140. Lagrange proves Euler's Addition Equation as follows:—

Put $dx/dt=\sqrt{X}$, and therefore $dy/dt=-\sqrt{Y}$; then
$$\frac{d^2x}{dt^2}=2(ax^3+3bx^2+3cx+d)=2X_1,$$
$$\frac{d^2y}{dt^2}=2(ay^3+3by^2+3cy+d)=2Y_1,$$
suppose; so that putting $x+y=p$, $x-y=q$, then
$$\frac{dp}{dt}=\sqrt{X}-\sqrt{Y},\quad \frac{dq}{dt}=\sqrt{X}+\sqrt{Y};$$
$$\frac{d^2p}{dt^2}=2(X_1+Y_1)$$
$$=\tfrac{1}{2}a(p^3+3pq^2)+3b(p^2+q^2)+6cp+4d,$$
$$\frac{dp}{dt}\frac{dq}{dt}=X-Y$$
$$=\tfrac{1}{2}apq(p^2+q^2)+bq(3p^2+q^2)+6cpq+4dq;$$
whence $$q\frac{d^2p}{dt^2}-\frac{dp}{dt}\frac{dq}{dt}=apq^3+2bq$$
or $$\frac{2}{q^2}\frac{dp}{dt}\frac{d^2p}{dt^2}-\frac{2}{q^3}\frac{dq}{dt}\left(\frac{dp}{dt}\right)^2=2ap\frac{dp}{dt}+4b\frac{dp}{dt}.$$

Both sides of this equation are now integrable, so that
$$\left(\frac{1}{q}\frac{dp}{dt}\right)^2=ap^2+4bp+C,$$
or $$\left(\frac{\sqrt{X}-\sqrt{Y}}{x-y}\right)^2=a(x+y)^2+4b(x+y)+C.$$

We notice here that, if $C=4b^2/a$,
$$\frac{\sqrt{X}-\sqrt{Y}}{x-y}=\frac{a(x+y)+2b}{\sqrt{a}}.$$

141. In the canonical form considered by Legendre, with

$$x=\operatorname{sn} u, \qquad dx/du=\surd(1-x^2 \,.\, 1-\kappa^2x^2),$$

$$y=\operatorname{sn} v, \qquad dy/dv=\surd(1-y^2 \,.\, 1-\kappa^2y^2),$$

then $\quad X=1-x^2 \,.\, 1-\kappa^2x^2, \qquad Y=1-y^2 \,.\, 1-\kappa^2y^2.$

Therefore $\quad dx/\surd X+dy/\surd Y=0,$

leads to $\quad du+ \; dv \; =0,$

or $\quad u+ \; v \; =\text{constant};$

which, in Clifford's notation, may be written

$$\operatorname{sn}^{-1}x+\operatorname{sn}^{-1}y=\text{constant}.$$

Euler's Addition Theorem of § 138 now gives

$$C=\left(\frac{\surd X-\surd Y}{x-y}\right)^2-\kappa^2(x+y)^2$$

$$=\frac{(\operatorname{cn} u \operatorname{dn} u-\operatorname{cn} v \operatorname{dn} v)^2-\kappa^2(\operatorname{sn}^2u-\operatorname{sn}^2v)^2}{(\operatorname{sn} u-\operatorname{sn} v)^2}$$

$$=\left(\frac{\operatorname{dn} u \operatorname{cn} v-\operatorname{cn} u \operatorname{dn} v}{\operatorname{sn} u-\operatorname{sn} v}\right)^2=\left\{\frac{\operatorname{dn}(u+v)-\operatorname{cn}(u+v)}{\operatorname{sn}(u+v)}\right\}^2,$$

by J. J. Thomson's formula of § 121.

142. But the Addition Theorem (1) for $\operatorname{sn}(u+v)$ of § 116,

$$\operatorname{sn}(u+v)=\frac{\operatorname{sn} u \operatorname{cn} v \operatorname{dn} v+\operatorname{sn} v \operatorname{cn} u \operatorname{dn} u}{1-\kappa^2\operatorname{sn}^2u \operatorname{sn}^2v},$$

when translated into the inverse function notation, gives

$$\operatorname{sn}^{-1}x+\operatorname{sn}^{-1}y=\operatorname{sn}^{-1}\frac{x\surd(1-y^2.1-\kappa^2y^2)+y\surd(1-x^2.1-\kappa^2x^2)}{1-\kappa^2x^2y^2}$$

This reduces, for $\kappa=0$, to the trigonometrical formula

$$\sin^{-1}x+\sin^{-1}y=\sin^{-1}\{x\surd(1-y^2)+y\surd(1-x^2)\},$$

the integral of $\quad dx/\surd(1-x^2)+dy/\surd(1-y^2)=0;$

and for $\kappa=1$, to

$$\tanh^{-1}x+\tanh^{-1}y=\tanh^{-1}\frac{x+y}{1+xy},$$

the integral of $\quad dx/(1-x^2)+dy/(1-y^2)=0.$

Similarly, equations (2) and (3) of § 116 may be written

$$\operatorname{cn}^{-1}x+\operatorname{cn}^{-1}y=\operatorname{cn}^{-1}\frac{xy-\surd(1-x^2.\kappa'^2+\kappa^2x^2)\surd(1-y^2.\kappa'^2+\kappa^2y^2)}{1-\kappa^2x^2y^2},$$

$$\operatorname{dn}^{-1}x+\operatorname{dn}^{-1}y=\operatorname{dn}^{-1}\frac{xy-\kappa^{-2}\surd(1-x^2.x^2-\kappa'^2)\surd(1-y^2.y^2-\kappa'^2)}{1-\kappa^2x^2y^2}.$$

We can now see why so little progress was made with the Theory of Elliptic Functions, so long as the Elliptic Integrals alone were studied, and also why Abel's idea of the inversion of the integral has revolutionised the subject.

143. A slight change of notation in the canonical integral (11) of § 38, suggested by Kronecker (*Berlin Sitz.*, July, 1886), introduces a further simplification, on writing

$$x = \kappa\,\mathrm{sn}^2(\tfrac{1}{2}u/\surd\kappa);$$

then $$dx/du = \surd\kappa\,\mathrm{sn}(\tfrac{1}{2}u/\surd\kappa)\mathrm{cn}(\tfrac{1}{2}u/\surd\kappa)\mathrm{dn}(\tfrac{1}{2}u/\surd\kappa),$$

$$\frac{dx^2}{du^2} = \kappa\frac{x}{\kappa}\left(1-\frac{x}{\kappa}\right)(1-\kappa x)$$

$$= x(1-\rho x + x^2),$$

with $$\rho = \kappa^{-1} + \kappa;$$

and now $$u = \int_0 dx/\surd X,$$

with $$X = x(1-\rho x + x^2).$$

Now

$$\tfrac{1}{2}(u+v)/\surd\kappa = \mathrm{sn}^{-1}\surd(x/\kappa) + \mathrm{sn}^{-1}\surd(y/\kappa)$$

$$= \mathrm{sn}^{-1}\frac{\surd x\surd(1-\rho y+y^2) + \surd y\surd(1-\rho x + x^2)}{1-xy}.$$

144. In Weierstrass's notation, we take

$$X = 4x^3 - g_2 x - g_3,$$

so that, in the general expression of the quartic X,

$$a=0,\ b=1,\ c=0,\ d=-\tfrac{1}{4}g_2,\ e=-g_3;$$

and now Euler's form of the Addition Theorem becomes, with z for C the arbitrary constant,

$$z = \tfrac{1}{4}\left(\frac{\surd X - \surd Y}{x-y}\right)^2 - x - y.$$

Now if $x = \wp u$, $y = \wp v$, so that $\surd X = -\wp' u$, $\surd Y = -\wp' v$, then we shall find (§ 147) that $z = \wp(u+v)$; so that

$$\wp(u+v) = \tfrac{1}{4}\left(\frac{\wp' u - \wp' v}{\wp u - \wp v}\right)^2 - \wp u - \wp v; \quad \text{(F)}$$

or, in the inverse notation,

$$\wp^{-1}x + \wp^{-1}y = \wp^{-1}\left\{\tfrac{1}{4}\left(\frac{\surd X - \surd Y}{x-y}\right)^2 - x - y\right\}.$$

Put $u+v=-w$, so that

$$\wp(u+v) = \wp w,\quad \wp'(u+v) = -\wp' w,$$

since (§ 51) $\wp w$ is an even function, and $\wp' w$ an odd function of w; then, with

$$u+v+w=0,$$

$$\wp u + \wp v + \wp w = \tfrac{1}{4}\left(\frac{\wp' u - \wp' v}{\wp u - \wp v}\right)^2,$$

and therefore also, by symmetry,

$$= \tfrac{1}{4}\left(\frac{\wp' v - \wp' w}{\wp v - \wp w}\right)^2 = \tfrac{1}{4}\left(\frac{\wp' w - \wp' u}{\wp w - \wp u}\right)^2 \quad \text{(F)*}$$

Thus $$\frac{\wp'v-\wp'w}{\wp v-\wp w}=\frac{\wp'w-\wp'u}{\wp w-\wp u}=\frac{\wp'u-\wp'v}{\wp u-\wp v},$$

or $$(\wp v-\wp w)\wp'u+(\wp w-\wp u)\wp'v+(\wp u-\wp v)\wp'w=0,$$

or $$(\wp'v-\wp'w)\wp u+(\wp'w-\wp'u)\wp v+(\wp'u-\wp'v)\wp w=0,$$

or $$\begin{vmatrix} 1, & \wp u, & \wp'u \\ 1, & \wp v, & \wp'v \\ 1, & \wp w, & \wp'w \end{vmatrix}=0 \text{.......................... (G)}$$

Weierstrass thus replaces the three elliptic functions sn u, cn u, dn u by a single function $\wp u$, and its derivative $\wp'u$.

145. Take for example the integral of ex. 8, p. 65,

$$\int X^{-\frac{2}{3}}dx, \text{ where } X=(x-a)(ax^2+2bx+c),$$

a cubic function of x, having a factor $x-a$.

This example shows that we may put

$$\wp u=\frac{X^{\frac{1}{3}}}{x-a}, \text{ with } g_2=0,\ g_3=4\frac{ac-b^2}{aa^2+2ba+c};$$

and then $$\wp'^2u=4\frac{ax^2+2bx+c}{(x-a)^2}-4\frac{ac-b^2}{aa^2+2ba+c}$$

$$=4\frac{\{(aa+b)(x-a)+aa^2+2ba+c\}^2}{(aa^2+2ba+c)(x-a)^2}.$$

Now, if y and z are the values of x corresponding to the values v and w of u, and if

$$u+v+w=0, \text{ or } \int_a X^{-\frac{2}{3}}dx+\int_a Y^{-\frac{2}{3}}dy+\int_a Z^{-\frac{2}{3}}dz=0,$$

then the integral relation (G) of § 144 connecting x, y, z becomes

$$(y-z)X^{\frac{1}{3}}+(z-x)Y^{\frac{1}{3}}+(x-y)Z^{\frac{1}{3}}=0. \text{.................(1)}$$

We notice that the integral relation does not require the knowledge of the factor $x-a$ of X; so that, writing

$$X=Ax^3+3Bx^2+3Cx+D,$$

we have, on rationalizing the relation (1),

$$3(y-z)(z-x)(x-y)(XYZ)^{\frac{1}{3}}=(y-z)^3X+(z-x)^3Y+(x-y)^3Z$$
$$=3(y-z)(z-x)(x-y)\{Axyz+B(yz+zx+xy)+C(x+y+z)+D\};$$

or $$XYZ=\{Axyz+B(yz+zx+xy)+C(x+y+z)+D\}^3. \text{ ...(2)}$$

(MacMahon, *Comptes Rendus*, 1882; *Q. J. M.*, XIX., p. 158.)

Then $$X^{\frac{1}{3}}Y^{\frac{1}{3}}\{(y-z)X^{\frac{1}{3}}+(z-x)Y^{\frac{1}{3}}\}$$
$$+(x-y)\{Axyz+B(yz+zx+xy)+C(x+y+z)+D\}=0,$$

so that $$z=\frac{X^{\frac{1}{3}}Y^{\frac{1}{3}}(yX^{\frac{1}{3}}-xY^{\frac{1}{3}})+(x-y)\{Bxy+C(x+y)+D\}}{X^{\frac{1}{3}}Y^{\frac{1}{3}}(X^{\frac{1}{3}}-Y^{\frac{1}{3}})-(x-y)\{Axy+B(x+y)+C\}},$$

equivalent to Allégret's result (*Comptes Rendus*, 66).

146. We shall find it convenient to replace the constant C in Euler's integral relation by $4c+4s$, and to consider s as the arbitrary constant, the meaning of which is to be interpreted; and then

$$s=\tfrac{1}{4}\left(\frac{\sqrt{X}-\sqrt{Y}}{x-y}\right)^2-\tfrac{1}{4}a(x+y)^2-b(x+y)-c,$$

or
$$s=\frac{F(x,\,y)-\sqrt{X}\sqrt{Y}}{2(x-y)^2},$$

where

$$\begin{aligned}F(x,\,y)&=ax^2y^2+2bxy(x+y)+c(x^2+4xy+y^2)+2d(x+y)+e\\&=(ax^2+2bx+c)y^2+2(bx^2+2cx+d)y+cx^2+2dx+e\\&=(ay^2+2by+c)x^2+2(by^2+2cy+d)x+cy^2+2dy+e,\end{aligned}$$

a symmetrical quadri-quadric function of x and y.

Treating s as a function of the independent variables x and y, we shall find

$$\begin{aligned}\sqrt{X}\frac{\partial s}{\partial x}&=\frac{\frac{1}{2}\frac{\partial F}{\partial x}\sqrt{X}-\frac{1}{4}\frac{dX}{dx}\sqrt{Y}}{(x-y)^2}-\frac{F\sqrt{X}-X\sqrt{Y}}{(x-y)^3}\\&=-\frac{(ay^3+3by^2+3cy+d)x+by^3+3cy^2+3dy+e}{(x-y)^3}\sqrt{X}\\&\quad+\frac{(ax^3+3bx^2+3cx+d)y+bx^3+3cx^2+3dx+e}{(x-y)^3}\sqrt{Y}\\&=-\frac{Y_1x+Y_2}{(x-y)^3}\sqrt{X}+\frac{X_1y+X_2}{(x-y)^3}\sqrt{Y},\text{ suppose;}\end{aligned}$$

and similarly we shall find that $\sqrt{Y}\dfrac{\partial s}{\partial y}$ has the same value.

But if s is taken as constant, then

$$\frac{\partial s}{\partial x}dx+\frac{\partial s}{\partial y}dy=0,$$

or
$$dx/\sqrt{X}+dy/\sqrt{Y}=0,$$

so that the differential relation which leads to Euler's integral relation is thus verified.

147. But now denote

$$4s^3-g_2s-g_3 \text{ by } S,$$

where $g_2=ae-4bd+3c^2$, $g_3=ace+2bcd-ad^2-eb^2-c^3$, so that (§ 75) g_2 and g_3 are the *quadrivariant* and *cubicvariant* of the quartic X (Burnside and Panton, *Theory of Equations*; Salmon, *Higher Algebra*).

We shall find, after considerable algebraical reduction, that

$$\sqrt{S}=\frac{(Y_1x+Y_2)\sqrt{X}-(X_1y+X_2)\sqrt{Y}}{(x-y)^3},$$

so that

$$\frac{1}{\sqrt{X}}\frac{dx}{dt}+\frac{1}{\sqrt{Y}}\frac{dy}{dt}=-\frac{1}{\sqrt{S}}\frac{ds}{dt},$$

and the elliptic elements $dx/\sqrt{X}$ and $dy/\sqrt{Y}$ are now reduced by this substitution to Weierstrass's canonical form $ds/\sqrt{S}$ of § 50.

Mr. R. Russell points out a concise way of performing this algebraical reduction, by means of the linear substitution $t=(\tau x+y)/(\tau+1)$ in the quartic $(a, b, c, d, e)(t, 1)^4$; which then becomes of the form

$$X\tau^4+4(X_1y+X_2)\tau^3+6F(x, y)\tau^2+4(Y_1x+Y_2)\tau+Y,$$

or $$A\tau^4+4B\tau^3+6C\tau^2+4D\tau+E, \text{ suppose.}$$

If the invariants of this new quartic are denoted by G_2, G_3,

then $$G_2=(x-y)^4g_2,\ G_3=(x-y)^6G_3;$$

and $$S=4s^3-g_2s-g_3$$

$$=\frac{(C-\sqrt{A}\sqrt{E})^3}{2(x-y)^6}-g_2\frac{C-\sqrt{A}\sqrt{E}}{2(x-y)^2}-g_3$$

$$=\frac{(C-\sqrt{A}\sqrt{E})^3-G_2(C-\sqrt{A}\sqrt{E})-2G_3}{2(x-y)^6}$$

$$=\frac{(D\sqrt{A}-B\sqrt{E})^2}{(x-y)^6}$$

$$=\frac{\{(Y_1x+Y_2)\sqrt{X}-(X_1y+X_2)\sqrt{Y}\}^2}{(x-y)^6}.$$

148. Rationalizing the integral relation of § 146,

$$\{2s(x-y)^2-F(x, y)\}^2=XY,$$

or $$s^2(x-y)^2-sF(x, y)-E(x, y)=0,$$

where $$E(x, y)=\{(ac-b^2)y^2+(ad-bc)y+\tfrac{1}{4}(ae-c^2)\}x^2$$
$$+\{(ad-bc)y^2+(\tfrac{1}{2}ae+2bd-\tfrac{5}{2}c^2)y+be-cd\}x$$
$$+\tfrac{1}{4}(ae-c^2)y^2+(be-cd)y+ce-d^2;$$

or $$(s^2-\tfrac{1}{12}g_2)(x-y)^2-sF(x, y)-H(x, y)=0,$$

where $$H(x, y)=(ac-b^2)x^2y^2+(ad-bc)xy(x+y)$$
$$+\tfrac{1}{6}(ae+2bd-3c^2)(x^2+4xy+y^2)+(be-cd)(x+y)+(ce-d),$$

a symmetrical quadri-quadric function of x and y.

149. When $x=y$, $F(x, x)=X$, and

$$E(x, x)=H(x, x)=(ac-b^2)x^4+2(ad-bc)x^3+(ae+2bd-3c^2)x^2$$
$$+2(be-cd)x+ce-d^2,$$

the *Hessian* H of the quartic X.

One value of s is now infinite, and the other

$$t = -H/X,$$

as in § 75; for, when $x=y$,

$$t = \frac{F(x, y) - \surd X \surd Y}{2(x-y)^2} = \frac{0}{0}$$

$$= \text{lt} \frac{\{F(x, y)\}^2 - XY}{2(x-y)^2\{F(x, y) + \surd X \surd Y\}} = \text{lt} \frac{-2E(x, y)}{F(x, y) + \surd X \surd Y} = -\frac{H}{X},$$

a substitution due originally to Hermite (*Crelle*, LII., 1856).

Now, since $t=\infty$, when $X=0$, or $x=a$,

$$\int_a dx/\surd X = \tfrac{1}{2}\int^\infty dt/\surd T = \tfrac{1}{2}\wp^{-1}(-H/X),$$

a denoting a root of the quartic $X=0$; and here

$$T = \surd(4t^3 - g_2 t - g_3)$$

$$= \text{lt} \frac{(Y_1 x + Y_2)\surd X - (X_1 y + X_2)\surd Y}{(x-y)^3} = \frac{0}{0}$$

$$= \text{lt} \frac{(Y_1 x + Y_2)^2 X - (X_1 y + X_2)^2 Y}{(x-y)^3\{(Y_1 x + Y_2)\surd X + (X_1 y + X_2)\surd Y\}} = \frac{G}{X^{\frac{3}{2}}},$$

where G is a certain rational integral function of x of the *sixth* degree, called the *sextic covariant* of the quartic X; the preceding algebra showing that

$$T^2 X^3 = G^2, \text{ or } 4H^3 - g_2 H X^2 + g_3 X^3 + G^2 = 0, \ldots\ldots\ldots\text{(H)}$$

this is called a *syzygy* between X, H, and G.

(Burnside and Panton, *Theory of Equations*, p. 346.)

For instance, if X is already in Weierstrass's canonical form, so that, if

$$x = \wp u,$$

$$X = \wp'^2 u = 4x^3 - g_2 x - g_3,$$

then

$$H = -(x^2 + \tfrac{1}{4}g_2)^2 - 2g_3 x;$$

and now

$$t = \wp 2u,$$

so that

$$\wp 2u = \frac{(\wp^2 u + \frac{1}{4}g_2)^2 + 2g_3\wp u}{4\wp^3 u - g_2\wp u - g_3}.$$

This may also be written

$$\wp 2u = \wp u - \frac{1}{4}\frac{d^2}{du^2}\log \wp' u.$$

150. With $y=\infty$,

$$2s = ax^2 + 2bx + c - \surd a \surd X,$$

or $s^2 - (ax^2 + 2bx + c)s - (ac - b^2)x^2 - (ad - bc)x - \frac{1}{4}(ae - c^2) = 0.$

With $y=0$,

$$2s = (cx^2 + 2dx + e - \surd e \surd X)/x^2,$$

or $x^2 s^2 - (cx^2 + 2dx + e)s - \frac{1}{4}(ae - c^2)x^2 - (bc - cd)x - ce + d^2 = 0.$

Writing $F(x, y)$ in the first equation of § 146 in the form

$$Y+\tfrac{1}{2}Y'(x-y)+\tfrac{1}{12}Y''(x-y)^2,$$

we can find x as a function of s and y by the solution of a quadratic, in the form

$$x-y=\frac{\sqrt{Y}\sqrt{S}+\tfrac{1}{2}Y'(s-\tfrac{1}{24}Y'')+\tfrac{1}{24}YY'''}{2(s-\tfrac{1}{24}Y'')^2-\tfrac{1}{2}aY}.$$

This method of the reduction of the general elliptic element $dx/\sqrt{X}$ to Weierstrass's canonical form $ds/\sqrt{S}$ is taken from a tract "*Problemata quædam mechanica functionum ellipticarum ope soluta.—Dissertatio inauguralis*," 1865, by G. G. A. Biermann, where the formulas are quoted as derived from Weierstrass's lectures.

151. Changing the sign of $\sqrt{Y}$, we find that

$$s=\frac{F(x, y)+\sqrt{X}\sqrt{Y}}{2(x-y)^2}$$

leads to the differential relation

$$\frac{1}{\sqrt{X}}\frac{dx}{dt}-\frac{1}{\sqrt{Y}}\frac{dy}{dt}=-\frac{1}{\sqrt{S}}\frac{ds}{dt};$$

so that, putting $\int^x dx/\sqrt{X}=u, \int^y dy/\sqrt{Y}=v,$

$$u-v=\int_y^x dx/\sqrt{X}=\int_s^\infty ds/\sqrt{S},$$

implying that $u-v=0$ when $x=y$, since $s=\infty$ when $x=y$; and now, in Weierstrass's notation,

$$s=\wp(u-v)=\frac{F(x, y)+\sqrt{X}\sqrt{Y}}{2(x-y)^2}.$$

Changing the sign of v, and therefore again of Y,

$$\wp(u+v)=\frac{F(x, y)-\sqrt{X}\sqrt{Y}}{2(x-y)^2};$$

so that $\qquad \wp 2u=-H_x/X, \ \wp 2v=-H_y/Y,$

implying that $u=0$ when $X=0$, $v=0$ when $Y=0$; so that

$$u=\int_a dx/\sqrt{X}, \ v=\int_a dy/\sqrt{Y},$$

where a denotes a root of the equation $X=0$.

Then

$$\wp(u-v)+\wp(u+v)=\frac{F(x, y)}{(x-y)^2},$$

$$\wp(u-v)-\wp(u+v)=\frac{\sqrt{X}\sqrt{Y}}{(x-y)^2}.$$

Mr. R. Russell finds, as is easily verified algebraically, that

$$\frac{F(x, y)}{(x-y)^2}-\frac{H_x}{X}=\frac{(X_1y+X_2)^2}{(x-y)^2X}, \quad \frac{F(x, y)}{(x-y)^2}-\frac{H_y}{Y}=\frac{(Y_1x+Y_2)^2}{(x-y)^2Y}.$$

But, from the Addition Theorem (F) of § 144,

$$\wp(u-v)+\wp(u+v)+\wp 2u=\frac{1}{4}\left\{\frac{\wp'(u-v)-\wp'(u+v)}{\wp(u-v)-\wp(u+v)}\right\}^2,$$

$$\wp(u-v)+\wp(u+v)+\wp 2v=\frac{1}{4}\left\{\frac{\wp'(u-v)+\wp'(u+v)}{\wp(u-v)-\wp(u+v)}\right\}^2;$$

and therefore

$$\frac{X_1y+X_2}{(x-y)\surd X}=-\frac{1}{2}\frac{\wp'(u-v)-\wp'(u+v)}{\wp(u-v)-\wp(u+v)},$$

$$\frac{Y_1x+Y_2}{(x-y)\surd Y}=-\frac{1}{2}\frac{\wp'(u-v)+\wp'(u+v)}{\wp(u-v)-\wp(u+v)};$$

the sign being determined by taking v small, when $y=a$, nearly.

Now, $$\wp'(u-v)-\wp'(u+v)=-2\frac{X_1y+X_2}{(x-y)^3}\surd Y,$$

$$\wp'(u-v)+\wp'(u+v)=-2\frac{Y_1x+Y_2}{(x-y)^3}\surd X;$$

so that, as in § 147,

$$\wp'(u-v)=\frac{-(Y_1x+Y_2)\surd X-(X_1y+X_2)\surd Y}{(x-y)^3},$$

$$\wp'(u+v)=\frac{-(Y_1x+Y_2)\surd X+(X_1y+X_2)\surd Y}{(x-y)^3}.$$

152. When $y=\infty$,

$$\wp 2v=-\text{lt}\, H_y/Y=(b^2-ac)/a,$$

and $$\wp' 2v=-\text{lt}\, G_y/Y^{\frac{3}{2}}=(a^2d-3abc+2b^3)/a^{\frac{3}{2}};$$

$$\frac{1}{2}\frac{\wp'(u-v)+\wp'(u+v)}{\wp(u-v)-\wp(u+v)}=-\text{lt}\frac{Y_1x+Y_2}{(x-y)\surd Y}=\frac{ax+b}{\surd a}.$$

Again, from equations (F)* and (G) of § 144,

$$\frac{1}{2}\frac{\wp'(u-v)-\wp' 2v}{\wp(u-v)-\wp 2v}=\frac{1}{2}\frac{\wp'(u-v)+\wp'(u+v)}{\wp(u-v)-\wp(u+v)}=-\frac{Y_1x+Y_2}{(x-y)\surd Y},$$

and putting $u=0$, and therefore $x=a$, we find

$$\frac{aa+b}{\surd a}=\frac{\wp' v+\wp' 2v}{\wp v-\wp 2v},$$

so that the quartic can be solved, when $\wp v$ and $\wp' v$ are known.

(*Solution of the Cubic and Quartic Equation*, *Proc. London Math. Soc.*, vol. XVIII., 1886.)

Otherwise, with $t=-H/X$,

$$\frac{dt}{dx}=-\frac{H'X-HX'}{X^2}=-\frac{2G}{X^2},$$

while $T^3=4t^3-g_2t-g_3=G^2/X^3$,

so that $dt/\sqrt{T}=-2dx/\sqrt{X}$,

and $$u=\int_a dx/\sqrt{X}=\tfrac{1}{2}\int_s^{\infty}dt/\sqrt{T}=\tfrac{1}{2}\wp^{-1}(-H/X),$$

a denoting a root of the quartic $X=0$.

Then $\wp 2u=t=-H/X$, $\wp' 2u=-T=-G/X^{\frac{3}{2}}$;

while $v=0$ when $y=a$, and $Y=0$;

so that $$\wp u=s=\frac{F(x,a)}{2(x-a)^2},$$

$$\wp' u=-\sqrt{S}=-\frac{(aa^3+3ba^2+3ca+d)x+ba^3+3ca^2+3da+e}{(x-a)^3}\sqrt{X}.$$

If v, k, K denote the values of u, s, S, when $x=\infty$,

$k=\tfrac{1}{2}(aa^2+2ba+c)=\wp v$, $K=(aa^3+3ba^2+3ca+d)\sqrt{a}=-\wp' v$;

$$s-k=\frac{aa^3+3ba^2+3ca+d}{x-a},$$

so that $$x-a=\frac{K}{(s-k)\sqrt{a}}=\frac{-\wp' v}{(\wp u-\wp v)\sqrt{a}};$$

and now $\wp 2v=(b^2-ac)/a$, $\wp' 2v=(a^2d-3abc+2b^3)/a^{\frac{3}{2}}$.

Conversely, given these values of $\wp 2v$ and $\wp' 2v$, and supposing the bisection of the argument of the elliptic functions to be carried out, we can determine $\wp v$ and $\wp' v$, and thence solve the quartic equation $X=0$.

153. Since $F(x,a)$ vanishes when $x=a$, a root of $X=0$, it is divisible by $x-a$; so that

$$s=\frac{(aa^2+2ba+c)x^2+2(ba^2+2ca+d)x+ca^2+2da+e}{2(x-a)^2}$$

$$=\tfrac{1}{2}(aa^2+2ba+c)\frac{x-a'}{x-a},\text{ suppose,}$$

a typical linear transformation, which converts $dx/\sqrt{X}$ into $ds/\sqrt{S}$, the canonical form of Weierstrass.

Denoting the four roots of $X=0$ by $\alpha, \beta, \gamma, \delta$, then since

$b/a=-\tfrac{1}{4}(\alpha+\beta+\gamma+\delta)$, $c/a=\tfrac{1}{6}(\alpha\beta+\alpha\gamma+\alpha\delta+\gamma\delta+\delta\beta+\beta\gamma)$,

we may write

$$s=\tfrac{1}{12}a\frac{\alpha-\beta\,.\,\alpha-\gamma\,.\,\alpha-\delta}{x-\alpha}\left(\frac{x-\beta}{\alpha-\beta}+\frac{x-\gamma}{\alpha-\gamma}+\frac{x-\delta}{\alpha-\delta}\right),$$

and now

$$\alpha' = \frac{\beta(\alpha-\gamma)(\alpha-\delta)+\gamma(\alpha-\delta)(\alpha-\beta)+\delta(\alpha-\beta)(\alpha-\gamma)}{(\alpha-\gamma)(\alpha-\delta)+(\alpha-\delta)(\alpha-\beta)+(\alpha-\beta)(\alpha-\gamma)},$$

with three other values β', γ', δ' corresponding to β, γ, δ.

Now $$\sqrt{S} = \frac{(a\alpha^3+3b\alpha^2+3c\alpha+d)x+b\alpha^3+3c\alpha^2+3d\alpha+e}{(x-\alpha)^3}\sqrt{X}$$

$$=(a\alpha^3+3b\alpha^2+3c\alpha+d)\frac{\sqrt{X}}{(x-\alpha)^2}$$

$$=\tfrac{1}{4}a(\alpha-\beta)(\alpha-\gamma)(\alpha-\delta)\sqrt{\left\{\frac{a(x-\beta)(x-\gamma)(x-\delta)}{(x-\alpha)^3}\right\}}.$$

Denoting by e_1, e_2, e_3, the roots of the *discriminating cubic*

$$4e^3-g_2e-g_3=0,$$

so that $$S=4(s-e_1)(s-e_2)(s-e_3),$$

then we may write

$$s-e_1=\tfrac{1}{4}a(\alpha-\gamma)(\alpha-\delta)\frac{x-\beta}{x-\alpha},$$

$$s-e_2=\tfrac{1}{4}a(\alpha-\delta)(\alpha-\beta)\frac{x-\gamma}{x-\alpha},$$

$$s-e_3=\tfrac{1}{4}a(\alpha-\beta)(\alpha-\gamma)\frac{x-\delta}{x-\alpha};$$

so that, to $x=\alpha$, β, γ, δ, corresponds $s=\infty$, e_1, e_2, e_3; and then

$$e_1=\tfrac{1}{12}a\{(\alpha-\gamma)(\delta-\beta)-(\alpha-\delta)(\beta-\gamma)\},$$
$$e_2=\tfrac{1}{12}a\{(\alpha-\delta)(\beta-\gamma)-(\alpha-\beta)(\gamma-\delta)\},$$
$$e_3=\tfrac{1}{12}a\{(\alpha-\beta)(\gamma-\delta)-(\alpha-\gamma)(\delta-\beta)\}.$$

If we interchange α and β, and put

$$s_1=\tfrac{1}{12}a\frac{\beta-\gamma\,.\,\beta-\delta\,.\,\beta-\alpha}{z-\beta}\left(\frac{z-\gamma}{\beta-\gamma}+\frac{z-\delta}{\beta-\delta}+\frac{z-\alpha}{\beta-\alpha}\right),$$

then to $z=\beta$, γ, δ, α, corresponds $s_1=\infty$, e_3, e_2, e_1;
so that $s=s_1$ gives a linear substitution converting

$$dx/\sqrt{X} \text{ into } dz/\sqrt{Z},$$

in which $x=\alpha$, β, γ, δ, corresponds to $z=\beta$, γ, δ, α.

If s is replaced by $\wp u$, and the same function of z by $\wp v$, then we find from § 54 that

$$v=u,\ u+\omega_1,\ u+\omega_1+\omega_3,\ u+2\omega_1+\omega_3,$$

gives the four linear transformations which leave $dx/\sqrt{X}$ unaltered; and corresponding to the values $(\alpha, \beta, \gamma, \delta)$ of x we find $(\alpha, \beta, \gamma, \delta)$, $(\beta, \gamma, \delta, \alpha)$, $(\gamma, \delta, \alpha, \beta)$, $(\delta, \alpha, \beta, \gamma)$ of z; the first transformation being merely $z=x$, not a distinct transformation.

154. When, as at first,

$$s=\frac{F(x, y)-\surd X\surd Y}{2(x-y)^2},$$

and when e is a root of the discriminating cubic, then $s-e$ is a perfect square; and we find

$$\surd(s-e)=\frac{\surd N_x\surd D_y-\surd N_y\surd D_x}{2(x-y)},$$

where, as in § 70, the quartic X is resolved into the quadratic factors N_x and D_x, and Y into the corresponding factors N_y and D_y; this can be done in three ways, corresponding to the three roots of the discriminating cubic.

Thus the integral relation

$$\frac{\surd N_x\surd D_y-\surd N_y\surd D_x}{x-y}=\text{constant}$$

leads to the differential relation

$$dx/\surd(N_xD_x)+dy/\surd(N_yD_y)=0,$$

as is easily verified algebraically, N and D being quadratics.

155. A more elegant expression can be given to these relations if we follow Klein (*Math. Ann.*, XIV., p. 112; Klein and Fricke, *Elliptische Modulfunctionen*, 1890) in employing *homogeneous* variables x_1 and x_2, by writing x_1/x_2 for x, and y_1/y_2 for y; and now

$$\int\frac{dx}{\surd X}=\int\frac{x_2dx_1-x_1dx_2}{\surd(ax_1^4+4bx_1^3x_2+6cx_1^2x_2^2+4dx_1x_2^3+ex_2^4)}.$$

Conversely, by writing x for x_1, and 1 for x_2, we return to our original non-homogeneous variable x.

Klein employs the abbreviations

(xdx) for $x_2dx_1-x_1dx_2$, and (xy) for $x_1y_2-x_2y_1$;

also $\mathrm{f}x$ for $(a, b, c, d, e)(x_1, x_2)^4$; and now with

$$w=u-v=\int_y^x dx/\surd X,$$

$$s=\wp w=\frac{\mathrm{F}(x, y)+\surd \mathrm{f}x\surd \mathrm{f}y}{2(xy)^2},$$

where $$\mathrm{F}(x, y)=\frac{1}{12}\left(\frac{\partial^2\mathrm{f}x}{\partial x_1^2}y_1^2+2\frac{\partial^2\mathrm{f}}{\partial x_1\partial x_2}y_1y_2+\frac{\partial^2\mathrm{f}}{\partial x_2^2}y_2^2\right)$$

$$=\frac{1}{12}\left(\frac{\partial^2\mathrm{f}y}{\partial y_1^2}x_1^2+2\frac{\partial^2\mathrm{f}}{\partial y_1\partial y_2}x_1x_2+\frac{\partial^2\mathrm{f}}{\partial y_2^2}x_2^2\right),$$

$$\text{and}\quad \sqrt{S} = -\wp' w = \frac{\left(\frac{\partial \mathrm{f}x}{\partial x_1} y_1 + \frac{\partial \mathrm{f}x}{\partial x_2} y_2\right)\sqrt{\mathrm{f}y} + \left(\frac{\partial \mathrm{f}y}{\partial y_1} x_1 + \frac{\partial \mathrm{f}y}{\partial y_2} x_2\right)\sqrt{\mathrm{f}x}}{4(xy)^3};$$

reducing to the above in § 153, when $\mathrm{f}y = 0$.

The Hessian H or $H(x_1, x_2)$ of X or $\mathrm{f}(x_1, x_2)$ is now given by

$$144H = \begin{vmatrix} \frac{\partial^2 \mathrm{f}}{\partial x_1^2}, & \frac{\partial^2 \mathrm{f}}{\partial x_2 \partial x_2} \\ \frac{\partial^2 \mathrm{f}}{\partial x_2 \partial x_2}, & \frac{\partial^2 \mathrm{f}}{\partial x_2^2} \end{vmatrix},$$

and the sextic covariant G or $G(x_1, x_2)$ by

$$8G = \begin{vmatrix} \frac{\partial \mathrm{f}}{\partial x_1}, & \frac{\partial \mathrm{f}}{\partial x_2} \\ \frac{\partial H}{\partial x_1}, & \frac{\partial H}{\partial x_2} \end{vmatrix}.$$

We may also use x and y as the homogeneous variables in the quantities, instead of x_1 and x_2.

Thus, for example, the integral $\int \mathrm{f}^{-\frac{1}{6}}(xdy)$, where

$$\mathrm{f} = x^{11}y + 11x^6y^6 - xy^{11} \quad \text{(the *icosahedron* form)}$$

is shown to be elliptic by means of the substitution

$$z = -H\mathrm{f}^{-\frac{5}{3}},$$

where

$$H = \tfrac{1}{121}\begin{vmatrix} \frac{\partial^2 \mathrm{f}}{\partial x^2}, & \frac{\partial^2 \mathrm{f}}{\partial x \partial y} \\ \frac{\partial^2 \mathrm{f}}{\partial x \partial y}, & \frac{\partial^2 \mathrm{f}}{\partial y^2} \end{vmatrix}$$

$$= -x^{20} + 228x^{15}y^5 - 494x^{10}y^{10} - 228x^5y^{15} - y^{20}.$$

Then we can verify the syzygy

$$-H^3 + 1728\mathrm{f}^5 = T^2,$$

where

$$T = -\tfrac{1}{20}\begin{vmatrix} \frac{\partial \mathrm{f}}{\partial x}, & \frac{\partial \mathrm{f}}{\partial y} \\ \frac{\partial H}{\partial x}, & \frac{\partial H}{dy} \end{vmatrix}$$

$$= x^{30} + y^{30} + 522(x^{25}y^5 - x^5y^{25}) - 10005(x^{20}y^{10} + x^{10}y^{20}).$$

Now
$$\frac{dz}{z(xdy)} = \frac{3\mathrm{f}H' - 5\mathrm{f}'H}{3\mathrm{f}H} = \frac{-5T}{3\mathrm{f}H},$$

so that
$$\frac{dz}{\sqrt{(4z^3 - g_3)}} = \frac{-5Tz}{3\mathrm{f}H}\frac{\mathrm{f}^{\frac{5}{2}}}{2T}(xdy) = \frac{5}{6}\frac{(xdy)}{\mathrm{f}^{\frac{1}{6}}};$$

since
$$4z^3 - g_3 = 4T^2\mathrm{f}^{-5}, \text{ provided } g_3 = -6912;$$

and $$\int^1 \frac{xdy}{\sqrt[6]{\mathrm{f}}} = \frac{6}{5}\int^\infty \frac{dz}{\sqrt{(4z^3-g_3)}} = \frac{6}{5}\wp^{-1}z = \frac{6}{5}\wp^{-1}(-H\mathrm{f}^{-\frac{5}{3}}).$$

Similar reductions will show that the integrals

$$\int H^{-\frac{1}{10}}(xdy) \text{ and } \int T^{-\frac{1}{15}}(xdy)$$

are also elliptic; also the integrals

$$\int (x^5y-xy^5)^{-\frac{1}{3}}(xdy) \text{ and } \int (x^8+14x^4y^4+y^8)^{-\frac{1}{4}}(xdy),$$

depending on the *octahedron* form, $x^8+14x^4y^4+y^8$.
(Schwarz, *Werke*, II., p. 252; Klein, *Lectures on the Icosahedron.*)

156. The further development introduces the theorems of Higher Algebra on the quartic and cubic, for the treatment of which the reader is referred to Salmon's *Higher Algebra* and Burnside and Panton's *Theory of Equations.*

Thus, H denoting the Hessian of a quartic X, and e_1, e_2, e_3 the roots of the discriminating cubic

$$4e^3-g_2e-g_3=0,$$

then $4(H+e_1X)(H+e_2X)(H+e_3X)=4H^3-g_2HX^2+g_3X^3=-G^2$, where G denotes the sextic covariant (§ 149); so that $H+eX$ is the square of a quadratic factor of G.

Following Burnside and Panton (p. 345) we shall find it convenient to put $16(H+eX)=-P^2$; and then

$$P_1P_2P_3=32G,$$

P_1, P_2, P_3 denoting the quadratic factors of the sextic covariant G.

Then $$P_1^2+P_2^2+P_3^2=-48H,$$

since $$e_1+e_2+e_3=0;$$

while $$(e_2-e_3)P_1^2+(e_3-e_1)P_2^2+(e_1-e_2)P_3^2=0;$$

and $$e_1P_1^2+e_2P_2^2+e_3P_3^2=-16(e_1^2+e_2^2+e_3^2)X=-8g_2X.$$

Since $$(e_2-e_3)P_1^2=(e_1-e_3)P_2^2-(e_1-e_2)P_3^2$$
$$=\{\sqrt{(e_1-e_3)}P_2+\sqrt{(e_1-e_2)}P_3\}\{\sqrt{(e_1-e_3)}P_2-\sqrt{(e_1-e_2)}P_3\},$$

therefore each of these factors must be the square of a linear factor, and we may therefore put

$$\sqrt{(e_1-e_3)}P_2+\sqrt{(e_1-e_2)}P_3=2u_1^2,$$
$$\sqrt{(e_1-e_3)}P_2-\sqrt{(e_1-e_2)}P_3=2u_2^2,$$

so that u_1 and u_2 are linear; and now

$$\sqrt{(e_2-e_3)}P_1=2u_1u_2,$$
$$\sqrt{(e_1-e_3)}P_2=u_1^2+u_2^2,$$
$$\sqrt{(e_1-e_2)}P_3=u_1^2-u_2^2.$$

157. Mr. R. Russell points out (*Q. J. M.*, XX., p. 183) that Hermite's substitution of $t=-H/X$ reduces the integral

$$\int G^{-\frac{1}{3}}dx \text{ to } \tfrac{1}{2}\int(4t^3-g_2t-g_3)^{-\frac{2}{3}}dt. \quad\text{.............(1)}$$

For $\qquad \dfrac{dt}{dx}=-\dfrac{2G}{X^2}, \text{ and } 4t^3-g_2t-g_3=\dfrac{G^2}{X^3},$

so that $\qquad G^{-\frac{1}{3}}dx=-\tfrac{1}{2}(4t^3-g_2t-g_3)^{-\frac{2}{3}}dt.$

Again the integral $\int(4t^3-g_2t-g_3)^{-\frac{2}{3}}dt$, as well as the general integral

$$\int U^{-\frac{2}{3}}dx, \quad\text{.........................(2)}$$

where U or $U(x, 1)$ denotes the cubic $(a, b, c, d)(x, 1)^3$, is again proved to be elliptic by the substitution

$$s^3=-K^3/U^2, \quad\text{.........................(3)}$$

where K or $K(x, y)$ denotes the Hessian of the cubic $U(x, y)$, given by

$$9K(x, y)=\begin{vmatrix}\dfrac{\partial^2 U}{\partial x^2}, & \dfrac{\partial^2 U}{\partial x\partial y}\\ \dfrac{\partial^2 U}{\partial x\partial y}, & \dfrac{\partial^2 U}{\partial y^2}\end{vmatrix}=\frac{\partial^2 U}{\partial x^2}\frac{\partial^2 U}{\partial y^2}-\left(\frac{\partial^2 U}{\partial x\partial y}\right)^2. \quad (4)$$

The *cubicovariant* J of the cubic U is given by

$$3J(x, y)=\begin{vmatrix}\dfrac{\partial U}{\partial x}, & \dfrac{\partial U}{\partial y}\\ \dfrac{\partial K}{\partial x}, & \dfrac{\partial K}{\partial y}\end{vmatrix}, \quad\text{.........................(5)}$$

and the discriminant Δ by

$$\Delta=a^2d^2+4ac^3-6abcd+4db^3-3b^2c^2; \quad\text{.........(6)}$$

and now we have the *syzygy*

$$J^2=-4K^3+\Delta U^2. \quad\text{..............................(7)}$$

(Salmon, *Higher Algebra*, § 192; Burnside and Panton, *Theory of Equations*, § 159.)

Differentiating (3) logarithmically

$$\frac{3}{s}\frac{ds}{dx}=\frac{3K'}{K}-\frac{2U'}{U}=-\frac{3J}{KU};$$

while $\qquad \sqrt{(4s^3+\Delta)}=\dfrac{J}{U};$

so that $\qquad \dfrac{dx}{U^{\frac{2}{3}}}=-\dfrac{sdx}{K}=\dfrac{Uds}{J}=\dfrac{ds}{\sqrt{(4s^3+\Delta)}},$

and $\qquad \int U^{-\frac{2}{3}}dx=\wp^{-1}(s;\ 0,\ -\Delta)=\wp^{-1}(-KU^{-\frac{2}{3}}). \quad\text{......(8)}$

When we know a factor, $x-a$, of U, then we may employ, as in ex. 8, p. 65, the substitution

$$z=U^{\frac{1}{3}}/(x-a). \quad\text{.........................(9)}$$

Putting $\quad U=(x-a)(ax^2+2b'x+c')$

$$=(x-a)\{ax^2+(aa+3b)x+aa^2+3ba+3c\},$$

then $4z^3-g_3$ is a perfect square, when

$$g_3=4\frac{ac'-b'^2}{aa^2+2b'a+c'}=\frac{(aa+b)^2+4(ac-b^2)}{aa^2+2ba+c};$$

and now

$$z-\frac{g_3}{z^2}=\frac{ax^2+2b'x+c'-g_3(x-a)^2}{U^{\frac{2}{3}}}$$

$$=\frac{-3K}{(aa^2+2ba+c)U^{\frac{2}{3}}}=\frac{3s}{aa^2+2ba+c};$$

$$\left(1+\frac{2g_3}{z^3}\right)dz=\frac{3ds}{aa^2+2ba+c};$$

while

$$(4z^3-g_3)\left(1+\frac{2g_3}{z^3}\right)^2=\frac{3\{(aa+b)(x-a)+2(aa^2+2ba+c)\}^2}{(aa^2+2ba+c)(x-a)^2}$$

$$\times\frac{3(x-a)^2\{(a^2a^2+2aba-2b^2+3ac)x+\dots\}^2}{(aa^2+2ba+c)^2U^2}$$

$$=\frac{9J^2}{(aa^2+2ba+c)^3U^2}=\frac{9(4s^3+\Delta)}{(aa^2+2ba+c)^3};$$

so that

$$\frac{dz}{\surd(4z^3-g_3)}=\surd(aa^2+2ba+c)\frac{ds}{\surd(4s^3+\Delta)}, \quad\text{............(10)}$$

a transformation equivalent to that of § 47.

158. Mr. R. Russell also shows (*Proc. L. M. S.*, XVIII., p. 57), that

$$\int\frac{lx^2+2mx+n}{\surd(aX+\beta H\,.\,a'X+\beta' H)}dx,$$

where X denotes a quartic and H its Hessian, can be reduced to the sum of three elliptic integrals by Hermite's substitution

$$t=-H/X.$$

For we may replace (§ 156)

$$lx^2+2mx+n \text{ by } pP_1+qP_2+rP_3$$

or by $4p\surd(-H-e_1X)+4q\surd(-H-e_2X)+4r\surd(-H-e_3X)$, where p, q, r are determined by equating coefficients; while

$$dx/\surd X=\tfrac{1}{2}dt/\surd T=\tfrac{1}{4}dt/\surd(t-e_1\,.\,t-e_2\,.\,t-e_3);$$

so that the integral becomes

$$\int\frac{p\surd(-H-e_1X)+q\surd(-H-e_2X)+r\surd(-H-e_3X)}{\surd(aX+\beta H\,.\,a'X+\beta' H)}\frac{\surd X dt}{\surd(t-e_1.t-e_2.t-e_3)}$$

$$=\int \frac{p\sqrt{(t-e_1)}+q\sqrt{(t-e_2)}+r\sqrt{(t-e_3)}}{\sqrt{(a-\beta t \,.\, a'-\beta t)}} \frac{dt}{\sqrt{(t-e_1 \,.\, t-e_2 \,.\, t-e_3)}}$$

$$=\int \left\{\frac{p}{\sqrt{(t-e_2 \,.\, t-e_3)}}+\frac{q}{\sqrt{(t-e_3 \,.\, t-e_1)}}+\frac{r}{\sqrt{(t-e_1 \,.\, t-e_2)}}\right\}\frac{dt}{\sqrt{(a-\beta t \,.\, a'-\beta' t)}},$$

the sum of three elliptic integrals.

Particular cases may be constructed by making β and β' zero, or a and a' zero; when we obtain

$$\int (lx^2+2mx+n)dx/X, \text{ or } \int (lx^2+2mx+n)dx/H.$$

159. Mr. Russell remarks that the reduction of the well-known *hyperelliptic* integral

$$\int \frac{(lx^2+2mx+n)dx}{\sqrt{(1-x^2 \,.\, 1+\kappa x^2 \,.\, 1+\lambda x^2 \,.\, 1-\kappa\lambda x^2)}}$$

to the sum of elliptic integrals is a particular case of this theorem, since the quartics

$$1-x^2 \,.\, 1-\kappa\lambda x^2 \text{ and } 1+\kappa x^2 \,.\, 1+\lambda x^2$$

can be expressed in the forms $aX+\beta H$ and $a'X+\beta' H$, by taking $X=1+\kappa\lambda x^4$, and therefore $H=\kappa\lambda x^2$; and now $a=1$, $a'=1$, $\beta=-(1+\kappa\lambda)/\kappa\lambda$, $\beta'=(\kappa+\lambda)/\kappa\lambda$.

These integrals are considered in Cayley's *Elliptic Functions*, chap. XVI., where x^2 is replaced by x; they arise in the expression of Legendre's elliptic integral

$$\int d\phi/\Delta(\phi, b) \text{ in the form } E+iF,$$

when the modulus b is complex, so that $b^2=e+if$.

(Jacobi, *Werke*, I., p. 380; Pringsheim, *Math. Ann.*, IX., p. 475.)

Writing P for $x(1-x)(1+\kappa x)(1+\lambda x)(1-\kappa\lambda x)$, Jacobi finds

$$\int dx/\sqrt{P}=\tfrac{1}{2}(b'+c')\{F(\phi, c)+F(\phi, b)\},$$

$$\int x\,dx/\sqrt{P}=\tfrac{1}{2}\frac{(b'+c')^2}{b'-c'}\{F(\phi, c)-F(\phi, b)\},$$

where
$$\kappa=\left(\frac{b'-c'}{b-c}\right)^2, \qquad \lambda=\left(\frac{b'-c'}{b+c}\right)^2,$$

or
$$b=\frac{\sqrt{\kappa}+\sqrt{\lambda}}{\sqrt{(1+\kappa \,.\, 1+\lambda)}}, \qquad b'=\frac{1-\sqrt{\kappa\lambda}}{\sqrt{(1+\kappa \,.\, 1+\lambda)}};$$

$$c=\frac{\sqrt{\kappa}-\sqrt{\lambda}}{\sqrt{(1+\kappa \,.\, 1+\lambda)}}, \qquad c'=\frac{1+\sqrt{\kappa\lambda}}{\sqrt{(1+\kappa \,.\, 1+\lambda)}};$$

and
$$\sin^2\phi=\frac{(1+\kappa)(1+\lambda)x}{(1+\kappa x)(1+\lambda x)}, \qquad \cos^2\phi=\frac{(1-x)(1-\kappa\lambda x)}{(1+\kappa x)(1+\lambda x)};$$

$$\Delta^2(\phi, b)=\frac{(1-x\sqrt{\kappa\lambda})^2}{(1+\kappa x)(1+\lambda x)}, \quad \Delta(\phi, c)=\frac{(1+x\sqrt{\kappa\lambda})^2}{(1+\kappa x)(1+\lambda x)}.$$

Then employing the inverse function notation,

$$\int \frac{dx}{\surd P} =$$

$$\frac{1}{\surd(1+\kappa.1+\lambda)}\left\{\mathrm{sn}^{-1}\left(\sqrt{\frac{1+\kappa.1+\lambda.x}{1+\kappa x.1+\lambda x}}, b\right)+\mathrm{sn}^{-1}\left(\sqrt{\frac{1+\kappa.1+\lambda.x}{1+\kappa x.1+\lambda x}}, c\right)\right\},$$

$$\int \frac{xdx}{\surd P} =$$

$$\frac{1}{\surd(\kappa\lambda.1+\kappa.1+\lambda)}\left\{\mathrm{sn}^{-1}\left(\sqrt{\frac{1+\kappa.1+\lambda.x}{1+\kappa x.1+\lambda x}}, b\right)-\mathrm{sn}^{-1}\left(\sqrt{\frac{1+\kappa.1+\lambda.x}{1+\kappa x.1+\lambda x}}, c\right)\right\}.$$

When λ is negative, then b and c are conjugate imaginaries; so that we can now express $F(\phi, b)$ in the form $E+iF$, when b^2 is of the form $e+if$.

For, writing $-\lambda$ for λ, and now writing

$$P \text{ for } x(1-x)(1+\kappa x)(1-\lambda x)(1+\kappa\lambda x),$$

then $$\int \frac{dx}{\surd P} = \frac{2E}{\surd(1+\kappa.1-\lambda)}, \int \frac{xdx}{\surd P} = \frac{2F}{\surd(\kappa\lambda.1+\kappa.1-\lambda)}.$$

In the particular case considered by Legendre, $\lambda=1$, and now

$$P=x(1-x^2)(1-\kappa^2x^2),$$

on replacing κ by κ^2; so that

$$\int x^{\pm\frac{1}{2}}dx/\surd(1-x^2.1-\kappa^2x^2)$$

can be expressed by elliptic integrals.

Mr. R. Russell employs the substitution

$$y=Ax/(1+Bx)^2,$$

and now

$$\int \frac{dy}{\surd(y.1-y.1-\sigma y)} = \int \frac{A(1-Bx)dx}{\surd[Ax\{(1+Bx)^2-Ax\}\{(1+Bx)^2-\sigma Ax\}]},$$

so that, putting

$$x\{(1+Bx)^2-Ax\}\{(1+Bx)^2-\sigma Ax\}=P,$$

therefore $$B^4=\kappa^2\lambda^2,\ B=\pm\surd(\kappa\lambda).$$

Taking $B=\surd(\kappa\lambda)$, and

$$(1+Bx)^2-\ \ Ax=(1-x)(1-\kappa\lambda x),$$
$$(1+Bx)^2-\sigma Ax=(1+\kappa x)(1+\lambda x),$$

then $$2\surd\kappa\lambda-A=-1-\kappa\lambda,$$
$$2\surd\kappa\lambda-\sigma A=\kappa+\lambda,$$

or $$A=(1+\surd\kappa\lambda)^2,\quad \sigma A=-(\surd\kappa+\surd\lambda)^2;$$

and taking $$B=-\surd(\kappa\lambda),$$

then $$A=(1-\surd\kappa\lambda),\quad \sigma A=-(\surd\kappa-\surd\lambda)^2.$$

160. Mr. Roberts's integrals (*Tract on the Addition of the Elliptic and Hyperelliptic Integrals*, p. 53)

$$\int(A+Bx^2)dx/\surd Q,$$

where Q is a reciprocal quartic in x^2, say

$$Q=ax^8+4bx^6+6cx^4+4bx^2+a$$

or $$aQ=(ax^4+2bx^2+a)^2-(2a^2+4b^2-6c)x^4,$$

furnish another particular case of Mr. Russell's theorem, since Q can be expressed in the form

$$(aX+\beta H)(a'X+\beta' H),$$

where X and H are in their canonical forms,

$$X=x^4+6mx^2+1,\quad H=mx^4+(1-3m^2)x^2+m.$$

Or we may put $x+x^{-1}=u$, $x-x^{-1}=v$, when the integral becomes $$\tfrac{1}{2}A(U+V)+\tfrac{1}{2}B(U-V),$$

where $$U=\int\frac{du}{\surd\{au^4-4(a-b)u^2+2a-8b+6c\}},$$

$$V=\int\frac{dv}{\surd\{av^4+4(a+b)v^2+2a+8b+6c\}}.$$

Thus $$1+x^8=(1+\surd 2x^2+x^4)(1-\surd 2x^2+x^4)$$
$$=(X+\surd 2H)(X-\surd 2H),$$

where $$X=1+x^4,\quad H=x^2.$$

Therefore the integral $\displaystyle\int\frac{A+Bx^2}{\surd(1+x^8)}dx$

is reduced to elliptic integrals by a substitution, such as

$$y=(1+x^4)/x^2;$$

and then becomes

$$\tfrac{1}{4}(B-A)\int\frac{dy}{\surd(y+2\,.\,y^2-2)}-\tfrac{1}{4}(B+A)\int\frac{dy}{\surd(y-2\,.\,y^2-2)}.$$

Another particular case of the general theorem occurs in the reduction of the integral

$$\int(lx+m)dx/\surd R,$$

where R is a *sextic* function, the roots of which form an *involution*, and whose invariant E therefore vanishes (Salmon, *Higher Algebra*, 1866, p. 210).

This invariant E is the one tabulated in the Appendix, p. 253, *Higher Algebra*, where it occupies thirteen pages.

The sextic covariant G of a quartic X is a specimen of a sextic of which the roots form an involution; and writing $32G$ or

$$P_1P_2P_3=(a_1x^2+2b_1x+c_1)(a_2x^2+2b_2x+c_2)(a_3x^2+2b_3x+c_3)$$
$$=a_1(x-\theta_1\,.\,x-\phi_1)a_2(x-\theta_2\,.\,x-\phi_2)a_3(x-\theta_3\,.\,x-\phi_3),$$

then since the squares of P_1, P_2, P_3 are linearly connected by the relation of § 156, therefore P_1, P_2, P_3 are mutually harmonic, and any one is therefore the Jacobian of the remaining two; this leads to the three relations

$$a_2c_3+a_3c_2-2b_2b_3=a_3c_1+a_1c_3-2b_3b_1=a_1c_2+a_2c_1-2b_1b_2=0.$$

Now $\dfrac{z}{p}=\dfrac{x-\theta_1}{x-\phi_1},\ \dfrac{x-\phi_1}{x-\theta_1},\ \dfrac{x-\theta_2}{x-\phi_2},\ \dfrac{x-\phi_2}{x-\theta_2},\ \dfrac{x-\theta_3}{x-\phi_3},\ \dfrac{x-\phi_3}{x-\theta_3},$

are the six linear transformations which reduce

$\displaystyle\int\frac{dx}{\surd X}$ to Legendre's canonical form $\displaystyle\int\frac{dz}{\surd(Az^4+6Cz^2+E)},$

as in § 74; so that if the quartic X is resolved into the quadratic factors N and D, we may write

$$N=p(x-\theta)^2+q(x-\phi)^2,$$
$$D=P(x-\theta)^2+Q(x-\phi)^2.$$

Now N/D is maximum or minimum when $x=\theta$, or ϕ.

Making P_1, P_2, P_3 homogeneous by the introduction of y, which is afterwards replaced by unity, so that

$$P=(a_1,\ b_1,\ c_1)(x,\ y)^2,\ \ldots,$$

then the three distinct linear transformations of § 153, which leave $dx/\surd X$ unaltered, are found to be

$$z=-\frac{\partial P_1}{\partial y}\Big/\frac{\partial P_1}{\partial x},\quad -\frac{\partial P_2}{\partial y}\Big/\frac{\partial P_2}{\partial x},\quad -\frac{\partial P_3}{\partial y}\Big/\frac{\partial P_3}{dx}.$$

(R. Russell, *Proc. L. M. S.*, XVIII., p. 48.)

Now $\displaystyle\int\frac{lx+m}{\surd G}dx$, or $\displaystyle\int\frac{(Au_1+Bu_2)(u_2du_1-u_1du_2)}{\surd\{u_1u_2(u_1^4-u_2^4)\}},$

where u_1, u_2 are defined in § 155, is reduced by the substitution

$$y^2=u_2/u_1,\ \text{or}\ p(x-\phi)/(x-\theta),$$

to the form $\displaystyle\int\frac{A+By^2}{\surd(1-y^8)}dy.$

This integral has been considered by Richelot (*Crelle*, XXXII., p. 213); and by differentiation we find

$$\frac{d}{dy}\,\mathrm{sn}^{-1}(\sqrt{2}+1)y\sqrt{\frac{1-y^2}{1+y^2}}=\frac{1+(\sqrt{2}-1)y^2}{\sqrt{(1-y^8)}},$$

$$\frac{d}{dy}\,\mathrm{sn}^{-1}y\sqrt{\frac{1-y^2}{1+y^2}}=\frac{1-(\sqrt{2}+1)y^2}{\sqrt{(1-y^8)}},\ \text{or}\ \frac{(\sqrt{2}+1)y^2-1}{\sqrt{(1-y^8)}},$$

according as y^2 is less or greater than $\sqrt{2}-1$; and thence the integration can be inferred; the value of κ to be taken is $\sqrt{2}-1$ or $\tan 22\frac{1}{2}°$, when it will be found that $K'/K=\sqrt{2}$.

161. As further applications, consider the integrals

$$\int(\Delta\phi)^{-\frac{1}{2}}d\phi,\ \int(\Delta\phi)^{-\frac{3}{2}}d\phi,\ \int(\Delta\phi)^{-\frac{2}{3}}d\phi,\ \int(\Delta\phi)^{-\frac{4}{3}}d\phi,$$

where $$\Delta\phi=\sqrt{(1-b^2\sin^2\phi)}.$$

(Legendre, *Fonctions elliptiques*, I., p. 178.)

Putting $\Delta\phi=x^2$, and $1-b^2=c^2$, then

$$\int(\Delta\phi)^{-\frac{1}{2}}d\phi=\int^1\frac{2x^2dx}{\sqrt{(1-x^4\,.\,x^4-c^2)}},$$

the integration required in the rectification of the Cassinian oval, given by

$$r_1r_2=\beta^2,\ \text{or}\ r^4-2a^2r^2\cos 2\theta+a^4=\beta^4,$$

where r_1, r_2 are the distances from the foci $(\pm a, 0)$.

The expression $1-x^4\,.\,x^4-c^2$ can be expressed by H^2-X^2, where $$X=x^4+c,\ H=(1+c)x^2;$$

and now the substitution $y=X/H$ gives

$$x+\frac{\sqrt{c}}{x}=\sqrt{\{(1+c)y+2\sqrt{c}\}},\ x-\frac{\sqrt{c}}{x}=\sqrt{\{(1+c)y-2\sqrt{c}\}};$$

so that $\int(\Delta\phi)^{-\frac{1}{2}}d\phi$

$$=\frac{1}{2}\int\frac{dy}{\sqrt{\{(1+c)y-2\sqrt{c}\}}\sqrt{(1-y^2)}}+\frac{1}{2}\int\frac{dy}{\sqrt{\{(1+c)y+2\sqrt{c}\}}\sqrt{(1-y^2)}}$$

$$=\frac{1}{\sqrt{(2+2c)}}\left[\mathrm{cn}^{-1}\left\{\frac{x^2+\sqrt{c}}{(1+\sqrt{c})x},\frac{1+\sqrt{c}}{\sqrt{(2+2c)}}\right\}+\mathrm{cn}^{-1}\left\{\frac{x^2-\sqrt{c}}{(1-\sqrt{c})x},\frac{1-\sqrt{c}}{\sqrt{(2+2c)}}\right\}\right]$$

by means of the results of §§ 39-41.

In the Cassinian

$$\theta=\tfrac{1}{2}\cos^{-1}\frac{r^4+a^4-\beta^4}{2a^2r^2}$$

$$=\cos^{-1}\frac{\sqrt{\{(r^2+a^2)^2-\beta^4\}}}{2ar}=\sin^{-1}\frac{\sqrt{\{\beta^4-(r^2-a^2)^2\}}}{2ar},$$

$$r\frac{d\theta}{dr}=-\frac{r^4-a^4+\beta^4}{\sqrt{\{4a^4r^4-(r^4+a^4-\beta^4)^2\}}}.$$

$$\frac{ds}{dr} = -\frac{2a^2r^2}{\sqrt{\{(a^2+\beta^2)^2-r^4\}}\sqrt{\{r^4-(a^2-\beta^2)^2\}}},$$

$$s = \int_r^{\sqrt{(a^2+\beta^2)}} \frac{2a^2r^2dr}{\sqrt{\{(a^2+\beta^2)^2-r^4\}}\sqrt{\{r^4-(a^2-\beta^2)^2\}}}.$$

Now, if we put

$$r^4 = (a^2+\beta^2)^2\cos^2\phi + (a^2-\beta^2)^2\sin^2\phi,$$

then $$s = a^2\int_0 \{a^2+\beta^2)^2\cos^2\phi + (a^2-\beta^2)^2\sin^2\phi\}^{-\frac{1}{4}}d\phi$$

$$= \frac{a^2}{\sqrt{(a^2+\beta^2)}}\int_0 \left\{1-\frac{4a^2\beta^2}{(a^2+\beta^2)^2}\sin^2\phi\right\}^{-\frac{1}{4}}d\phi.$$

Similarly $$\int_0 (\Delta\phi)^{-\frac{3}{2}}d\phi = \int^1 \frac{2dx}{\sqrt{(1-x^4\,.\,x^4-c^2)}}$$

$$= \frac{1}{2\sqrt{c}}\int\frac{dy}{\sqrt{\{(1+c)y-2\sqrt{c}\}}\sqrt{(1-y^2)}} - \frac{1}{2\sqrt{c}}\int\frac{dy}{\sqrt{\{(1+c)y+2\sqrt{c}\}}\sqrt{(1-y^2)}},$$

which can be expressed in a similar manner.

Again, substituting $\Delta^2\phi = x^3$, then

$$\int_0 (\Delta\phi)^{-\frac{4}{3}}d\phi = \frac{3}{2}\int^1\frac{dx}{\sqrt{(1-x^3\,.\,x^3-c^2)}},$$

$$\int_0 (\Delta\phi)^{-\frac{2}{3}}d\phi = \frac{3}{2}\int^1\frac{xdx}{\sqrt{(1-x^3\,.\,x^3-c^2)}};$$

particular cases of the preceding general integrals.

Mr. R. A. Roberts (*Proc. L. M. S.*, XXII., p. 33) has shown that $$\int (lx+m)(ax^6+2bx^3+c)^{-\frac{2}{3}\text{ or }-\frac{1}{3}}dx$$
can be expressed as the sum of elliptic integrals, not always however in a real form.

Mr. Russell shows that if $x-\theta_1$, $x-\theta_2$ are the factors of P_1, a quadratic factor of the sextic covariant, then

$$\int\frac{lx+m}{\sqrt{P_1}\sqrt{X}}dx$$

is reduced by the substitution

$$y^2 = p(x-\theta_1)/(x-\theta_2)$$

to the form $$\int\frac{Ay^2+B}{\sqrt{(ay^8+2by^4+c)}}dy,$$

and this again by the substitution

$$z = y^2\sqrt{a} + y^{-2}\sqrt{c}$$

to the form $$\int\frac{p\sqrt{(z-2\sqrt[4]{ac})}+q\sqrt{(z+2\sqrt[4]{ac})}}{\sqrt{(z^2-4\sqrt{ac})}\sqrt{(z^2+2b-2\sqrt{ac})}}dz,$$

two elliptic integrals, not necessarily however in a real form.

Abel's Theorem applied to the Addition Equation.

162. Euler's Addition Theorem is now found to be a very special case of a Theorem of great generality, due to Abel, the method of which we shall employ here, in the very limited form required for the Addition of the First Elliptic Integrals.

Consider the points of intersection of the fixed quartic curve whose equation is

$$y^2 = X, \quad \text{.............................(1)}$$

with any arbitrary algebraical curve whose equation in a rational form may be written

$$f(x, y) = 0. \quad \text{.............................(2)}$$

By continually writing X for y^2, we can reduce equation (2) to the form

$$P + Qy = 0 \; ; \quad \text{..........................(3)}$$

and now the abscissas of the points of intersection of (1) and (2) are given by the equation

$$P + Q\sqrt{X} = 0, \quad \text{..........................(4)}$$

or, in a rational form,

$$P^2 - Q^2 X = 0. \quad \text{..........................(5)}$$

Denoting the degree of this equation (5) by μ, and its roots by $x_1, x_2, \ldots x_\mu$, Abel puts

$$\psi x = P^2 - Q^2 X = C(x - x_1)(x - x_2) \ldots (x - x_\mu), \quad \text{........(6)}$$

and now he supposes the roots of this equation to vary in consequence of arbitrary variations in the coefficients of the terms in equation (2), corresponding to arbitrary changes in the shape and position of this curve; the coefficients in equation (1) are however kept unchanged.

If ∂P, ∂Q denote small changes in P and Q due to the changes in the coefficients, and if dx_r denotes the corresponding change in any root x_r of equation (5), then

$$\psi' x_r \,.\, dx_r + 2P\delta P - 2Q\delta Q X_r = 0,$$

or, making use of equation (4),

$$\psi' x_r \,.\, dx_r - 2(Q\partial P - P\delta Q)\sqrt{X_r} = 0,$$

$$\frac{dx_r}{\sqrt{X_r}} = 2\frac{Q\partial P - P\delta Q}{\psi' x_r} = \frac{\theta x_r}{\psi' x_r}, \quad \text{..............(7)}$$

suppose.

Now, if the degrees of P and Q are denoted by p and q, then the degree of θx is $p + q$; and we shall find this is always at least one less than $\mu - 1$, the degree of $\psi' x$, or two less than μ, the degree of ψx.

For if in equation (3), P^2 and Q^2X are of equal degree, then $q=p-2$, and $\mu=2p$; so that $\mu-p-q=2$; and $\mu-p-q$ is greater than 2, if q is less than $p-2$.

But if q is greater than $p-2$, then the order of ψx is given by that of Q^2X, and therefore $\mu=2q+4$, while $p=q+1$ at most; so that $\mu-p-q=3$ at least.

Since $x\theta x$ is thus of lower degree than ψx, we can split the fraction $x\theta x/\psi x$ into a series of partial fractions, such that

$$\frac{x\theta x}{\psi x}=\sum_{r=1}^{r=\mu}\frac{x_r\theta x_r}{\psi' x_r(x-x_r)};$$

and now, if we make $x=0$, we find that

$$\sum\frac{\theta x_r}{\psi' x_r}=0, \qquad (8)$$

a theorem in Algebra due to Euler; otherwise stated as

$$\sum_{r=1}^{r=\mu}\frac{x_r{}^m}{(x_r-x_1)(x_r-x_2)\ldots * \ldots(x_r-x_\mu)}=0; \qquad (9)$$

provided m is less than $\mu-1$, the $*$ marking the position of the missing factor x_r-x_r.

Applying this theorem to equation (7), we find

$$\sum_{r=1}^{r=\mu} dx_r/\sqrt{X_r}=0, \qquad (10)$$

so that, if, in consequence of any finite alteration of the coefficients in equation (2) or (3), the roots of equation (5) become changed to $x'_1, x'_2, \ldots, x'_\mu$, then

$$\int_{x'_1}^{x_1} dx_1/\sqrt{X_1}+\int_{x'_2}^{x_2} dx_2/\sqrt{X_2}+\ldots+\int_{x'_\mu}^{x_\mu} dx_\mu=0, \qquad (11)$$

the Theorem of Abel, as required for present purposes.

It is the combination of the theory of Integrals and of the theory of Algebra which furnishes the key of Abel's Theorem; the algebraical laws are expressed very concisely by a single equation (5), of which the variables are the roots, and whose coefficients are not independent, but are connected by a number of relations.

Thus, if we take P of the pth order, and Q of the order $p-2$, we have a *plexus* of μ or $2p$ equations of the form (4)

$$\alpha x_r{}^p+\beta x_r{}^{h-1}+\gamma x_r{}^{h-2}+\ldots+(\gamma' x_r{}^{h-2}+\ldots)\sqrt{X_r}=0;$$

and the elimination of $\alpha, \beta, \gamma, \ldots, \gamma', \ldots$ leads to a determinant of $2p$ rows, each row of the form

$$x_r{}^p,\ x_r{}^{h-1},\ x_r{}^{h-2},\ \ldots,\ x_r{}^{h-2}\sqrt{X_r},\ x_r{}^{h-3}\sqrt{X_r},\ \ldots.$$

163. Suppose for instance that (2) is the parabola

$$y = ax^2 + 2\beta x + \gamma, \quad\text{..................(2) or (3)}$$

then equation (4) becomes

$$ax^2 + 2\beta x + \gamma - \sqrt{X} = 0, \quad\text{.....................(4)}$$

and (5) becomes the quartic equation

$$(ax^2 + 2\beta x + \gamma)^2 - X = 0, \quad\text{.....................(5)}$$

Denoting the roots by x_1, x_2, x_3, x_4, then the elimination of a, β, γ leads to the determinant

$$\begin{vmatrix} x_1^2, & x_1, & 1, & \sqrt{X_1} \\ x_2^2, & x_2, & 1, & \sqrt{X_2} \\ x_3^2, & x_3, & 1, & \sqrt{X_3} \\ x_4^2, & x_4, & 1, & \sqrt{X_4} \end{vmatrix} = 0,$$

as the integral relation, corresponding to $(\mu = 4)$,

$$\frac{dx_1}{\sqrt{X_1}} + \frac{dx_2}{\sqrt{X_2}} + \frac{dx_3}{\sqrt{X_3}} + \frac{dx_4}{\sqrt{X_4}} = 0.$$

By making $a = \sqrt{a}$, so that the parabolas are of constant size, or by writing equation (5) in the form

$$(ax^2 + 2\beta x + \gamma)^2 - aX = 0,$$

one root, x_4 suppose, becomes infinite; and now

$$4a(\beta - b)x^3 + (4\beta^2 + 2a\gamma - 6ac)x^2 + 4(\beta\gamma - ad)x + \gamma^2 - ac = 0,$$

so that

$$4(\beta - b)(x_1 + x_2 + x_3) = 6c - 2\gamma - 4\beta^2/a$$

$$= 2ax_3^2 + 4\beta x_3 + 6c - 2\sqrt{a}\sqrt{X_3} - 4\beta^2/a,$$

or $\quad 4(\beta - b)(x_1 + x_2) = 2ax_3^2 + 4bx_3 + 6c - 2\sqrt{a}\sqrt{X_3} - 4\beta^2/a.$

Now the two relations

$$ax_1^2 + 2\beta x_1 + \gamma - \sqrt{a}\sqrt{X_1} = 0,$$

$$ax_2^2 + 2\beta x_2 + \gamma - \sqrt{a}\sqrt{X_2} = 0,$$

give by subtraction

$$(x_1 - x_2)\{a(x_1 + x_2) + 2\beta\} = \sqrt{a}(\sqrt{X_1} - X_2),$$

so that $\left(\dfrac{\sqrt{X_1} - \sqrt{X_2}}{x_1 - x_2}\right)^2 = a(x_1 + x_2)^2 + 4\beta(x_1 + x_2) + \dfrac{4\beta^2}{a}$

$$= a(x_1 + x_2)^2 + 4b(x_1 + x_2) + C,$$

where $\quad C = 2ax_3^2 + 4bx_3 + 6c - 2\sqrt{a}\sqrt{X_3};$

and we thus obtain Euler's original integral relation, the general integral of the differential relation

$$dx_1/\sqrt{X_1} + dx_2/\sqrt{X_2} = 0,$$

when C is constant; and a particular integral of

$$dx_1/\sqrt{X_1} + dx_2/\sqrt{X_2} + dx_3/\sqrt{X_3} = 0,$$

when x_3 is considered as variable.

164. When X is in Legendre's canonical form $1-x^2 \,.\, 1-\kappa^2 x^2$, then Abel takes $$P=ax+x^3,\ Q=b;$$
and now equation (6) becomes
$$\begin{aligned}\psi x &= (ax+x^3)^2-b^2(1-x^2)(1-\kappa^2x^2)\\ &= x^6-(b^2\kappa^2-2a)x^4+(b^2+b^2\kappa^2+a^2)x^2-b^2\\ &= (x^2-x_1^2)(x^2-x_2^2)(x^2-x_3^2),\end{aligned}$$
where
$$x_1^2+x_2^2+x_3^2=b^2\kappa^2-2a,$$
$$x_2^2x_3^2+x_3^2x_1^2+x_1^2x_2^2=b^2+b^2\kappa^2+a^2,$$
$$x_1^2x_2^2x_3^2=b^2.$$

But a and b are determined by the equations
$$ax_1+x_1^3+bX_1=0,\ ax_2+x_2^3+bX_2=0;$$
so that
$$b=\frac{x_1x_2(x_1^2-x_2^2)}{x_1X_2-x_2X_1},$$
and therefore, as in formula (1), § 116,
$$x_3=\frac{x_1^2-x_2^2}{x_1X_2-x_2X_1}=\frac{x_1X_2+x_2X_1}{1-\kappa^2x_1^2x_2^2}.$$

Also $$1-x_1^2 \,.\, 1-x_2^2 \,.\, 1-x_3^2=1-b^2\kappa^2+2a+b^2+b^2\kappa^2+a^2-b^2$$
$$=(1+a)^2,$$
while $$x_1^2+x_2^2+x_3^2-\kappa^2x_1^2x_2^2x_3^2=-2a,$$
so that
$$\begin{aligned}2-x_1^2-x_2^2-x_3^2+\kappa^2x_1^2x_2^2x_3^2 &= 2(1+a)\\ &= 2\surd(1-x_1^2 \,.\, 1-x_2^2 \,.\, 1-x_3^2),\end{aligned}$$
or $(2-x_1^2-x_2^2-x_3^2+\kappa^2x_1^2x_2^2x_3^2)^2=4(1-x_1^2)(1-x_2^2)(1-x_3^2)$,
which may also be written
$$\surd(1-x_3^2)=\surd(1-x_1^2 \,.\, 1-x_2^2)\pm x_1x_2\surd(1-\kappa^2x_3^2),$$
as in § 119, with $x_1=\operatorname{sn}u$, $x_2=\operatorname{sn}v$, $x_3=\operatorname{sn}(u\pm v)$.

This, with $x_1=\operatorname{sn}u_1$, $x_2=\operatorname{sn}u_2$, $x_3=\operatorname{sn}u_3$, may be written
$$1-\operatorname{cn}^2u_1-\operatorname{cn}^2u_2-\operatorname{cn}^2u_3+2\operatorname{cn}u_1\operatorname{cn}u_2\operatorname{cn}u_3=\kappa^2\operatorname{sn}^2u_1\operatorname{sn}^2u_2\operatorname{sn}^2u_3;$$
where $$u_1+u_2+u_3=4K,$$
(§ 131); and, with a triangle of Class I., is equivalent to the formulas in Spherical Trigonometry
$$\begin{aligned}1-\cos^2a-\cos^2b-\cos^2c+2\cos a\cos b\cos c &= \kappa^2\sin^2a\sin^2b\sin^2c\\ =\sin^2A\sin^2b\sin^2c=\sin^2a\sin^2B\sin^2c &= \sin^2a\sin^2b\sin^2C.\end{aligned}$$

165. To obtain the Addition Theorem for Weierstrass's functions, we consider the intersections of the cubic curve
$$y^2=4x^3-g_2x-g_3, \text{ or } X, \quad\text{.....(1)}$$
with an arbitrary straight line
$$y=ax+\beta; \quad\text{.....(2)}$$

Now, if x_1, x_2, x_3 denote the roots of the equation

$$4x^3 - g_2 x - g_3 - (\alpha x + \beta)^2 = 0, \qquad \text{.....(5)}$$

then

$$\alpha x_1 + \beta + \sqrt{X_1} = 0,$$
$$\alpha x_2 + \beta + \sqrt{X_2} = 0,$$

so that

$$\alpha = \frac{\sqrt{X_1} - \sqrt{X_2}}{x_1 - x_2}, \quad \beta = \frac{x_1\sqrt{X_2} - x_2\sqrt{X_1}}{x_1 - x_2},$$

and (§ 144)

$$x_1 + x_2 + x_3 = \tfrac{1}{4}\alpha^2 = \tfrac{1}{4}\left(\frac{\sqrt{X_1} - \sqrt{X_2}}{x_1 - x_2}\right)^2.$$

The elimination of α and β between these two equations and

$$\alpha x_3 + \beta + \sqrt{X_3} = 0$$

leads, as in § 144, to the determinant (G)

$$\begin{vmatrix} 1, & x_1, & \sqrt{X_1} \\ 1, & x_2, & \sqrt{X_2} \\ 1, & x_3, & \sqrt{X_3} \end{vmatrix} = 0, \text{ or } \begin{vmatrix} 1, & \wp u, & \wp' u \\ 1, & \wp v, & \wp' v \\ 1, & \wp w, & \wp' w \end{vmatrix} = 0,$$

where

$$u + v + w = 0.$$

In addition, from (5),

$$x_2 x_3 + x_3 x_1 + x_1 x_2 = -\tfrac{1}{4}g_2 - \tfrac{1}{2}\alpha\beta,$$
$$x_1 x_2 x_3 = \tfrac{1}{4}g_3 + \tfrac{1}{4}\beta^2;$$

so that

$$(x_1 + x_2 + x_3)(4x_1 x_2 x_3 - g_3) = (x_2 x_3 + x_3 x_1 + x_1 x_2 + \tfrac{1}{4}g_2)^2 \qquad \text{........(I)}$$

166. Consider the intersections of the fixed cubic curve

$$y^3 = Ax^3 + 3Bx^2 + 3Cx + D, \qquad \text{.................(1)}$$

with a variable straight line

$$y = \alpha x + \beta. \qquad \text{.....................................(2)}$$

Then

$$\psi x = (\alpha x + \beta)^3 - (Ax^3 + 3Bx^2 + 3Cx + D)$$
$$= (\alpha^3 - A)(x - x_1)(x - x_2)(x - x_3), \qquad \text{.................(6)}$$

and

$$x_1 + x_2 + x_3 = -3\frac{\alpha^2\beta - B}{\alpha^3 - A},$$

$$x_2 x_3 + x_3 x_1 + x_1 x_2 = 3\frac{\alpha\beta^2 - C}{\alpha^3 - A},$$

$$x_1 x_2 x_3 = -\frac{\beta^3 - D}{\alpha^3 - A}.$$

Denoting by y_1, y_2, y_3 the corresponding values of y, then

$$\begin{aligned} y_1 y_2 y_3 &= (\alpha x_1 + \beta)(\alpha x_2 + \beta)(\alpha x_3 + \beta) \\ &= \alpha^3 x_1 x_2 x_3 + \{B - \tfrac{1}{3}(\alpha^3 - A)(x_1 + x_2 + x_3)\}(x_2 x_3 + x_3 x_1 + x_1 x_2) \\ &\quad + \{C + \tfrac{1}{3}(\alpha^3 - A)(x_2 x_3 + x_3 x_1 + x_1 x_2)\}(x_1 + x_2 + x_3) \\ &\quad + D - (\alpha^3 - A)x_1 x_2 x_3 \\ &= Ax_1 x_2 x_3 + B(x_2 x_3 + x_3 x_1 + x_1 x_2) + C(x_1 + x_2 + x_3) + D, \end{aligned}$$

as in § 145.

Now, if the constants α and β receive small increments $\delta\alpha$ and $\delta\beta$, then

$$\psi' x_1 dx_1 + 3(\alpha x_1+\beta)^2(x_1\delta\alpha+\delta\beta)=0,$$

and
$$\psi' x_1 = (\alpha^3 - A)(x_1-x_2)(x_1-x_3),$$

so that
$$\frac{dx_1}{y_1^{\,2}} = 3\frac{x_1\delta\alpha+\delta\beta}{(\alpha^3-A)(x_3-x_1)(x_1-x_2)}, \quad\text{.................(7)}$$

and

$$\frac{dx_1}{y_1^{\,2}}+\frac{dx_2}{y_2^{\,2}}+\frac{dx_3}{y_3^{\,2}} = 3\left(\frac{x_1}{x_3-x_1 . x_1-x_2}+\frac{x_2}{x_1-x_2 . x_2-x_3}+\frac{x_3}{x_2-x_3 . x_3-x_1}\right)\frac{\delta\alpha}{\alpha^3-A}$$
$$+3\left(\frac{1}{x_3-x_1 . x_1-x_2}+\frac{1}{x_1-x_2 . x_2-x_3}+\frac{1}{x_2-x_3 . x_3-x_1}\right)\frac{\delta\beta}{\alpha^3-A}$$
$$=0, \quad\text{...(10)}$$

and the sum of the three integrals is a constant, which can be made to vanish by taking for the lower limits a root of the equation $y=0$.

In the particular case of the cubic curve

$$x^3+y^3=1,$$

the relation expressing the collinearity of the three points is

$$x_1x_2x_3+y_1y_2y_3=1.$$

Now, as in § 145, with $g_2=0$, $g_3=1$, and

$$\wp u = \frac{(1-x^3)^{\frac{1}{3}}}{1-x},\ \wp' u = -\sqrt{3}\frac{1+x}{1-x},$$

and, by symmetry, with

$$\wp v = \frac{(1-y^3)^{\frac{1}{3}}}{1-y},\ \wp' v = -\sqrt{3}\frac{1+y}{1-y},$$

we find from (F) § 144, after reduction,

$$\wp(u+v) = \tfrac{1}{4}\left(\frac{\wp' u - \wp' v}{\wp u - \wp v}\right)^2 - \wp u - \wp v = 1,$$

so that
$$u+v=a, \text{ a constant.}$$

With $\wp a = 1$, then (§ 149) $\wp 2a = 1$; so that (§ 62)

$$\wp 2a = \wp(2\omega_2 - a), \text{ or } a = \tfrac{2}{3}\omega_2.$$

We may therefore put

$$u = \tfrac{1}{3}\omega_2 + t,\quad v = \tfrac{1}{3}\omega_2 - t,$$

and express x and y by functions of t.

For any other arbitrary value of a, the integral relation connecting x and y will be, by § 145,

$$(1-x^3)(1-y^3)(1-z^3) = (1-xyz)^3;$$

and treating z as constant, this leads to the differential relation

$$(1-x^3)^{-\frac{2}{3}}dx + (1-y^3)^{-\frac{2}{3}}dy = 0.$$

We can put

$$\wp u=\frac{(1-x^3)^{\frac{1}{3}}}{1-x},\ \wp v=\frac{(1-y^3)^{\frac{1}{3}}}{1-y},\ \wp w=\frac{(1-z^3)^{\frac{1}{3}}}{1-z},$$

where $$u+v+w=0;$$

and $\wp w=1$, for the value $z=\infty$; and then

$$x^3+y^3=1.$$

167. When the quartic X is resolved into two quadratic factors N and D, we may replace (1) by the quartic curve

$$y^2=N/D;\ \dots\dots\dots\dots\dots(1)$$

and now equation (4) is replaced by

$$P\sqrt{D}+Q\sqrt{N}=0;\ \dots\dots\dots\dots\dots(4)$$

so that equation (5) becomes

$$P^2D-Q^2N=0.\ \dots\dots\dots\dots\dots(5)$$

The elimination of the constants from the *plexus* of equations determined by the roots of this last equation (4) leads to determinants, whose rows are of the form

$$x_r^p\sqrt{D_r},\ x_r^{p-1}\sqrt{D_r},\ \dots,\ x_r^q\sqrt{N_r},\ x_r^{q-1}\sqrt{N_r},\ \dots.$$

For instance, by taking P and Q linear, so that the variable curve (2) or (3) in § 162 is a hyperbola, we can obtain the integral relation of § 154 in the form

$$\frac{\sqrt{N_1}\sqrt{D_2}-\sqrt{N_2}\sqrt{D_1}}{x_1-x_2}\cdot\frac{\sqrt{N_3}\sqrt{D_4}-\sqrt{N_4}\sqrt{D_3}}{x_3-x_4}=\text{constant}.$$

(W. Burnside, *Messenger of Mathematics.*)

We have taken X as a quartic function of x, so as to apply to the elliptic functions, but Abel's theorem holds for any higher degree of X, the method of proof being exactly the same; and, according to Klein, we resolve X, supposed of even degree, into factors N and D, differing in degree by 0 or a multiple of 4, when we wish to make use of the fixed curve

$$y^2=N/D.$$

168. The reader is referred to the treatises of Salmon or of Burnside and Panton for the proof of the Theorems in Higher Algebra quoted here; they are easily verified, however, if we work with the quartic in its canonical form

$$U=x^4-6m\,x^2y^2+y^4;$$

when

$$H=-mx^4+(1-3m^2)x^2y^2-my^4,$$

$$G=\tfrac{1}{2}(1-9m^2)xy(x^4-y^4).$$

The following examples, taken from recent examination papers, will illustrate the character of the algebraical work.

EXAMPLES.

1. Denoting by U the binary quartic, reduced to its canonical form, $x^4 - 6mx^2y^2 + y^4$, its quadrinvariant and cubinvariant by g_2 and g_3, and its Hessian and sextic covariant by H and G, prove that

(i.) $4m^3 - g_2 m - g_3 = 0$;

(ii.) $H + mU$ is a perfect square;

(iii.) $4H^3 - g_2 HU^2 + g_3 U^3 + G^2 = 0$;

(iv.) $H\left(\frac{\partial H}{\partial y}, -\frac{\partial H}{\partial x}\right) = 16H(gH - g_3 U)(g_2 H + 3g_3 U)$;

(v.) $\frac{\partial^2 H}{\partial x^2}\left(\frac{\partial U}{\partial y}\right)^2 - 2\frac{\partial^2 H}{\partial x \partial y}\frac{\partial U}{\partial x}\frac{\partial U}{\partial y} + \frac{\partial^2 H}{\partial y^2}\left(\frac{\partial U}{\partial x}\right)^2 = 32(g_2 U^2 - 6H^2)$;

(vi.) $\frac{\partial^2 U}{\partial x^2}\left(\frac{\partial H}{\partial y}\right)^2 - 2\frac{\partial^2 U}{\partial x \partial y}\frac{\partial H}{\partial x}\frac{\partial H}{\partial y} + \frac{\partial^2 U}{\partial y^2}\left(\frac{\partial H}{\partial x}\right)^2 = 48U(g_2 H - g_3 U)$;

(vii.) the Hessian of $\lambda U + \mu H$ is

$$(\lambda^2 - \tfrac{1}{12} g_2 \mu^2)H + (\tfrac{1}{6} g_2 \lambda\mu + \tfrac{1}{4} g_3 \mu^2)U,$$

and the sextic covariant is

$$\tfrac{1}{4}(4\lambda^3 - g_2 \lambda\mu^2 - g_3 \mu^3)G.$$

2. Denoting the roots of $4e^3 - g_2 e - g_2 = 0$ by e_1, e_2, e_3, prove that the roots of $(x^2 + \tfrac{1}{4} g_2)^2 \pm 2g_3 x = 0$ are of the form $\sqrt{(e_2 e_3)} + \sqrt{(e_3 e_1)} + \sqrt{(e_1 e_2)}$.

3. Denoting the discriminant, Hessian, and cubicovariant of a cubic U by Δ, K, and J, prove that

$$\Delta U^2 = J^2 + 4K^3.$$

(Work with the canonical form $U = ax^3 + by^3$.)

Denoting the same functions of $\lambda U + \mu G$ by Δ', K', J', prove that

$$\Delta' = (\lambda^2 - \mu^2 \Delta)^2 \Delta,$$

$$K' = (\lambda^2 - \mu^2 \Delta)K,$$

$$J' = (\lambda^2 - \mu^2 \Delta)(\lambda J + \mu \Delta U).$$

4. Prove that X and Y in § 139 have the same invariants g_2 and g_3 (Burnside and Panton, 1886, p. 418).

5. Prove that, in § 156,

$$\sqrt{(e_2 - e_3)}P_1 + \sqrt{(e_3 - e_1)}P_2 + \sqrt{(e_1 - e_2)}P_3$$

is the square of a linear factor of X.

6. Discuss the properties of the quartic X' in § 153, whose roots are α', β', γ', δ'.

7. Prove that (§ 160) θ_1, ϕ_1; θ_2, ϕ_3; θ_3, ϕ_2; define an involution of the roots of the sextic covariant G (R. Russell).

8. Prove that the cubic substitution

$$y = -(bx^3+3cx^2+3dx+e)/(ax^3+3bx^2+3cx+d) = -X_2/X_1$$

makes
$$\frac{dy}{\sqrt{(g_2H_y - g_3U_y)}} = \frac{3dx}{\sqrt{(g_2H_x+3g_3U_x)}},$$

where
$$U_x = (a, b, c, d, e)(x, 1)^3.$$

(Hermite; Crelle, LX., p. 304; R. Russell, *Proc. L. M. S.*, XVIII., p. 52.)

9. Integrate $\displaystyle\int \frac{P_2P_3}{H}\frac{dx}{\sqrt{X}}$.

10. Prove that, with $s = \wp u$,

$$\wp' 2u = (2s^6 - \tfrac{5}{2}g_2s^4 - 10g_3s^3 - \tfrac{5}{8}g_2{}^2s^2 - \tfrac{1}{2}g_2g_3s - g_3{}^2 + \tfrac{1}{32}g_2{}^3)/\wp'^3u;$$

$$\sqrt{(\wp 2u - e)} = -(s^2 - 2es - 2e^2 + \tfrac{1}{4}g_2)/\wp' u;$$

$$\sqrt{(\wp 2u - e_\alpha)} + \sqrt{(\wp 2u - e_\beta)} = -2(s-e_\alpha)(s-e_\beta)/\wp' u;$$

$$4\ 2u = s + \frac{e_1 - e_2 \,.\, e_1 - e_3}{s - e_1} + \frac{e_2 - e_3 \,.\, e_2 - e_1}{s - e_2} + \frac{e_3 - e_1 \,.\, e_1 - e_2}{s - e_2}.$$

11. Prove that, if

(i.) $\wp(v; -20, -40) = 5$, then $\wp 2v = 0$, $\wp 3v = -\frac{7}{5}$, $\wp 4v = \frac{5}{8}$, ...

(ii.) $\wp(v; -60, -10) = 5$, ... 0, ... $\frac{7}{5}$, ... $\frac{45}{2}$, ...

(iii.) $\wp(v; -15, \quad 19) = \frac{5}{2}$, ... $\frac{5}{4}$, ...$\frac{456}{25}$, ...

12. Prove that

(i.) $\int (A+Bx)dx/y$ is elliptic, if $y^2 = (1-x^2)(a+3x-4x^3)$;

(ii.) $\displaystyle\int (A+Bx+Cy)dx \Big/ \frac{\partial f}{\partial y}$ is elliptic, if

$$f(x, y) = (a, b, c, f, g, h)(x^2, y^2, 1).$$

(W. Burnside).

CHAPTER VI.

THE ELLIPTIC INTEGRALS OF THE SECOND AND THIRD KIND.

169. The Elliptic Integrals, and thence the Elliptic Functions, derive their name Elliptic from the early attempts of mathematicians at the rectification of the Ellipse.

It was some time before mathematicians perceived that the simple integral to begin considering is

$$F\phi = \int d\phi/\Delta\phi,$$

which has not originally such a special connexion with the ellipse; but the name Elliptic Integral has nevertheless been retained generally for all integrals of this nature.

To a certain extent this is a disadvantage; not only because we employ the name *hyperbolic* function to denote $\cosh u$, $\sinh u$, $\tanh u$, ..., by analogy with which the *elliptic* functions would be merely the *circular* functions $\cos\phi$, $\sin\phi$, $\tan\phi$, ...; but also because it is found that the elliptic functions are a particular case of a large class, called *hyperelliptic* functions, but included in a larger class, called Abelian functions after Abel, which, beginning with the algebraical, circular, hyperbolic, and elliptic functions of a single argument u $(p=1)$ are in the general case the functions of p arguments which are met with when we consider the integrals

$$\int (1, x, x^2, \ldots, x^{p-1})\, dx/\sqrt{X},$$

arising in the linear transformations of $\int dx/\sqrt{X}$, in which X is a rational integral function of x of the degree $2p+2$; for now the linear transformation $(lx+m)/(l'x+m')$ converts

$$\int dx/\sqrt{X} \text{ into } (lm'-l'm)\int (l'x+m')^{p-1} dx/\sqrt{X}.$$

170. Legendre's elliptic integral of the second kind has already been defined in § 77; and denoting it by $E\phi$, then the length of the arc BP of an ellipse is given by $aE\phi$, where the arc BP and the excentric angle of the point P are both measured from the minor axes OB, and now the modulus is the excentricity of the ellipse.

The quadrant of the ellipse BA is given by aE, where, as in § 77, E denotes $\int_0^{\frac{1}{2}\pi}\Delta\phi d\phi$, the complete elliptic integral of the second kind, in which $\phi=\frac{1}{2}\pi$.

The perimeter of the ellipse is therefore $4aE$, the same as that of a circle of radius $aE/\frac{1}{2}\pi$.

The periodicity of $\sin\phi$ and $\Delta\phi$ shows that, as in § 14,

$$E(\pi+\phi)=\int_0^{\pi+\phi}\Delta\phi d\phi=\int_0^{\pi}+\int_0^{\phi}=2E+E\phi,$$

and generally $$E(m\pi+\phi)=2mE+E\phi,$$
when m is an integer.

Expanded in ascending powers of the modulus κ,

$$\Delta\phi=(1-\kappa^2\sin^2\phi)^{\frac{1}{2}}=1-\sum_{n=1}^{n=\infty}\frac{1.3.5\,\ldots\,2n-1}{2.4.6\,\ldots\,2n}\frac{(\kappa\sin\phi)^{2n}}{2n-1};$$

so that, employing Wallis's theorems of integration, as in § 11,

$$E=\int_0^{\frac{1}{2}\pi}\Delta\phi d\phi=\tfrac{1}{2}\pi\left[1-\sum_{n=1}^{n=\infty}\left(\frac{1.3.5\,\ldots\,2n-1}{2.4.6\,\ldots\,2n}\right)^2\frac{\kappa^{2n}}{2n-1}\right],$$

whence the numerical value of E can be calculated.

Tables of the numerical values of $E\phi$ for every degree of ϕ and of the modular angle are given in Legendre's *F. E.*, II., Table IX.; while the values of $\log E$ are given in his Table I. for every tenth of a degree in the modular angle.

We reproduce this Table of $\log E$, and of $\log E'$, corresponding to the complementary modulus κ', to 7 decimals, and to every half degree in the modular angle $\frac{1}{2}a$, corresponding to the values of $\log K$ in Table I., p. 10.

171. By differentiation and integration, we prove that

$$\frac{d}{d\kappa}\left(\frac{E\phi}{\kappa}\right)=-\frac{F\phi}{\kappa^2},\quad \frac{d}{d\kappa}(\kappa F\phi)=\int\frac{d\phi}{\Delta^3\phi}=\frac{E\phi}{\kappa'^2}-\frac{\kappa^2}{\kappa'^2}\frac{\sin\phi\cos\phi}{\Delta\phi};$$

and therefore, with $\phi=\frac{1}{2}\pi$,

$$\frac{d}{d\kappa}\left(\frac{E}{\kappa}\right)=-\frac{K}{\kappa^2},\quad \frac{d}{d\kappa}(\kappa K)=\frac{E}{\kappa'^2}.$$

TABLE IV.

$\frac{1}{2}\alpha$	log E.	log E'.	$\frac{1}{2}(\pi - \alpha)$	$\frac{1}{2}\alpha$	log E.	log E'.	$\frac{1}{2}(\pi - \alpha)$	$\frac{1}{2}\alpha$	log E.	log E'.	$(\frac{1}{2}\pi - \alpha)$
0·0	0·1961199	0·0000000	90·0	15·5	0·1881869	0·0335799	74·5	31·0	0·1645829	0·0865694	59·0
0·5	1116	·0000931	89·5	16·0	·1876678	·0351996	74·0	31·5	·1635676	·0882630	58·5
1·0	0868	·0003263	89·0	16·5	1323	·0368337	73·5	32·0	·1625369	·0899500	58·0
1·5	0455	·0006736	88·5	17·0	·1865806	·0384812	73·0	32·5	·1614907	·0916301	57·5
2·0	·1959876	·0011208	88·0	17·5	0125	·0401409	72·5	33·0	·1604293	·0933030	57·0
2·5	9132	·0016581	87·5	18·0	·1854281	·0418118	72·0	33·5	·1593525	·0949683	56·5
3·0	8222	·0022778	87·0	18·5	·1848274	·0434930	71·5	34·0	·1582606	·0966256	56·0
3·5	7148	·0029734	86·5	19·0	2104	·0451834	71·0	34·5	·1571535	·0982747	55·5
4·0	5908	·0037396	86·0	19·5	·1835772	·0468822	70·5	35·0	·1560313	·0999152	55·0
4·5	4502	·0045716	85·5	20·0	·1829277	·0485885	70·0	35·5	·1548940	·1015469	54·5
5·0	2932	·0054652	85·0	20·5	2621	·0503014	69·5	36·0	·1537418	·1031694	54·0
5·5	1196	·0064165	84·5	21·0	·1815802	·0520201	69·0	36·5	·1525747	·1047826	53·5
6·0	·1949295	·0074221	84·0	21·5	·1808822	·0537438	68·5	37·0	·1513928	·1063860	53·0
6·5	7229	·0084788	83·5	22·0	1680	·0554717	68·0	37·5	·1501962	·1079796	52·5
7·0	4998	·0095837	83·0	22·5	·1794377	·0572032	67·5	38·0	·1489849	·1095629	52·0
7·5	2602	·0107340	82·5	23·0	·1786913	·0589374	67·0	38·5	·1477590	·1111359	51·5
8·0	0041	·0119270	82·0	23·5	·1779288	·0606737	66·5	39·0	·1465186	·1126982	51·0
8·5	·1937314	·0131605	81·5	24·0	·1771503	·0624115	66·0	39·5	·1452639	·1142496	50·5
9·0	4423	·0144321	81·0	24·5	·1763557	·0641501	65·5	40·0	·1439948	·1157899	50·0
9·5	1367	·0157396	80·5	25·0	·1755451	·0658888	65·0	40·5	·1427116	·1173189	49·5
10·0	·1928147	·0170811	80·0	25·5	·1747186	·0676271	64·5	41·0	·1414142	·1188364	49·0
10·5	4762	·0184545	79·5	26·0	·1738761	·0693644	64·0	41·5	·1401028	·1203423	48·5
11·0	1212	·0198581	79·0	26·5	·1730178	·0711002	63·5	42·0	·1387776	·1218362	48·0
11·5	·1917497	·0212901	78·5	27·0	·1721435	·0728339	63·0	42·5	·1374385	·1233181	47·5
12·0	3618	·0227487	78·0	27·5	·1712535	·0745649	62·5	43·0	·1360858	·1247878	47·0
12·5	·1909575	·0242324	77·5	28·0	·1703476	·0762929	62·0	43·5	·1347196	·1262450	46·5
13·0	5367	·0257397	77·0	28·5	·1694260	·0780173	61·5	44·0	·1333399	·1276897	46·0
13·5	0996	·0272690	76·5	29·0	·1684886	·0797376	61·0	44·5	·1319470	·1291217	45·5
14·0	·1896460	·0288190	76·0	29·5	·1675356	·0814534	60·5	45·0	·1305409	·1305409	45·0
14·5	1760	·0303884	75·5	30·0	·1665669	·0831642	60·0	45·5	·1291217	·1319470	44·5
15·0	·1886896	·0319758	75·0	30·5	·1655827	·0848697	59·5	46·0	·1276987	·1333399	44·0

We can now prove *Legendre's relation*, that

$$EK'+E'K-KK' \text{ is constant, and } =\tfrac{1}{2}\pi;$$

for denoting it by A, we find that $dA/d\kappa=0$, so that A is independent of κ; and taking $\kappa=0$, then

$$A=\text{lt}\int_0^{\frac{1}{2}\pi}\int_0^{\frac{1}{2}\pi}\frac{\Delta^2(\phi,\kappa)+\Delta^2(\psi,\kappa')-1}{\Delta(\phi,\kappa)\Delta(\psi,\kappa')}d\phi d\psi=\iint\cos\psi d\phi d\psi=\tfrac{1}{2}\pi.$$

172. In Jacobi's notation, with $\phi=\text{am}\,u$,

$$E\phi=E\,\text{am}\,u=\int_0\text{dn}^2u du;$$

and now, from the quasi-periodicity of am u (§ 14),

$$E(m\pi+\phi)=E\,\text{am}(2mK+u)=2mE+E\,\text{am}\,u,$$

where m is an integer.

We may therefore, as in § 78, separate $E\,\text{am}\,u$ into two parts, one the *secular* part, increasing uniformly with u, at a rate $2E$ per increase $2K$ of u, and the other a *periodic* part, denoted by Zu in Jacobi's notation, and called the *Zeta function;* so that

$$E\,\text{am}\,u=Eu/K+\text{Z}u,$$

or

$$\text{Z}u=\int_0(\text{dn}^2u-E/K)du.$$

The Addition Theorem for the Second Elliptic Integral.

173. A well-known theorem, due to Graves and Chasles, asserts that if an endless thread, placed round a fixed ellipse, is kept stretched by a pencil, the pencil will trace out a confocal ellipse (fig. 22). (Salmon, *Conic Sections*, § 399.)

If the excentric angles (measured from the minor axis of the ellipse) of the points of contact P, Q of the straight parts of the thread PR, RQ are denoted by ϕ, ψ, so that the

$$\text{arc } BP=aE\phi,\quad \text{arc } BQ=aE\psi;$$

and if we put $\phi=\text{am}\,u$, $\psi=\text{am}\,v$, the modulus κ being the excentricity of the ellipse, then, as asserted in ex. 6, at the end of Chap. IV., R moves on a confocal ellipse, when $u-v$ is constant, and conversely.

For the coordinates of R being given by

$$x=a\frac{\cos\psi-\cos\phi}{\sin(\phi-\psi)},\quad y=b\frac{\sin\phi-\sin\psi}{\sin(\phi-\psi)},$$

we find from Jacobi's formulas (4), (5), and (31), § 137, replacing u and v by $\tfrac{1}{2}(u+v)$ and $\tfrac{1}{2}(u-v)$,

$$x = a\frac{\operatorname{cn} v - \operatorname{cn} u}{\sin(\operatorname{am} u - \operatorname{am} v)} = a\frac{s_1 s_2 d_1 d_2}{s_2 c_2 d_1} = a\frac{s_1 d_2}{c_2} = a\frac{\operatorname{sn}\frac{1}{2}(u+v)\operatorname{dn}\frac{1}{2}(u-v)}{\operatorname{cn}\frac{1}{2}(u-v)},$$

$$y = b\frac{\operatorname{sn} u - \operatorname{sn} v}{\sin(\operatorname{am} u - \operatorname{am} v)} = b\frac{s_2 c_1 d_1}{s_2 c_2 d_1} = b\frac{c_1}{c_2} = a\frac{\kappa' \operatorname{cn}\frac{1}{2}(u+v)}{\operatorname{cn}\frac{1}{2}(u-v)}.$$

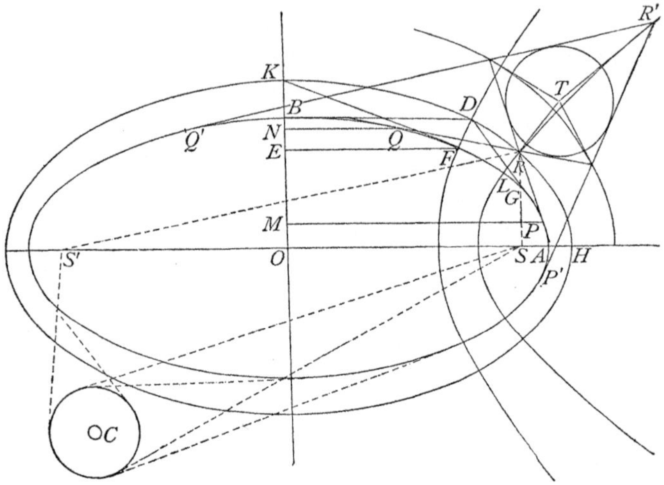

Fig. 22.

Therefore

$$\frac{x}{a}\operatorname{cd}\tfrac{1}{2}(u-v) = \operatorname{sn}\tfrac{1}{2}(u+v),\ \frac{y}{b}\operatorname{cn}\tfrac{1}{2}(u-v) = \operatorname{cn}\tfrac{1}{2}(u+v);$$

and $$(x/\alpha)^2 + (y/\beta)^2 = 1,$$

where $$\alpha = a\operatorname{dc}\tfrac{1}{2}(u-v),\ \beta = b\operatorname{nc}\tfrac{1}{2}(u-v);$$

so that $$\alpha^2 - \beta^2 = a^2 - b^2,$$

and therefore R describes a confocal ellipse, if $u-v$ is constant.

If $$u+v \text{ is constant},$$

we find $$(x/\alpha')^2 - (y/\beta')^2 = 1,$$

where $$\alpha' = a\kappa\operatorname{sn}\tfrac{1}{2}(u+v),\ \beta' = a\kappa\operatorname{cn}\tfrac{1}{2}(u+v),$$

so that $$\alpha'^2 + \beta'^2 = a^2\kappa^2 = a^2 - b^2,$$

and R therefore describes a confocal hyperbola (MacCullagh).

To realise mechanically this motion of R on the hyperbola, the threads RP, RQ must pass round the ellipse, and be led, in the *same* direction, round a reel moveable about a fixed axis at C; so that, as the reel revolves, equal lengths of thread are wound up or unwound.

If the hyperbola starts from the ellipse at L, then

$$PR - \operatorname{arc} PL = QR - \operatorname{arc} QL.$$

If the threads are wound in *opposite* directions on the reel, then R will describe a confocal ellipse, as at first; but in this case the reel may be suppressed, and the thread merely made to slide round the ellipse, as in the theorems of Graves and Chasles.

Moreover, it is not necessary that the tangents RP, RQ should proceed to the same ellipse, but to any two fixed confocals, and the same theorems hold.

If tangents $R'P'$, $R'Q'$, are drawn to the ellipse from any other point R' on the confocal hyperbola RR', forming with RP, RQ the quadrilateral $RrR'r'$, then r, r' lie on a confocal ellipse, by the preceding theorems; and now a circle can be inscribed in this quadrilateral whose centre is at T, the point of concourse of the tangents to the confocals at R, r, R', r'; for TR, Tr, TR', Tr' bisect the angles of the quadrilateral; (Salmon, *Conic Sections*, § 189).

If R is brought up to L, the circle touches the ellipse at L; so that the point of contact of the circle inscribed in the area bounded by two tangents and the ellipse is at the point where the confocal hyperbola through the point of intersection of the tangents cuts the ellipse.

174. Putting $\quad u-v=w$, or $F\phi-F\psi=F\gamma$,
then when $v=0$ and Q is at B, $u=w$ and P is at G where $\phi=\gamma$ suppose; while R will come to D on the ellipse RD, where it is cut by the tangent at B.

Now, since

$$PR+RQ-\text{arc } PQ=BD+DG-\text{arc } BG,$$

or

$$\text{arc } PQ-\text{arc } BG=PR+RQ-BD-DG;$$

therefore $\quad E\phi-E\psi-E\gamma=$ a certain trigonometrical function of ϕ, ψ, γ, which is found to be $-\kappa^2\sin\phi\sin\psi\sin\gamma$; this is the Addition Theorem for the Second Elliptic Integral.

For
$$PR^2=a^2\left\{\sin\phi\frac{\cos\psi-\cos\phi}{\sin(\phi-\psi)}\right\}^2+b^2\left\{\frac{\sin\phi-\sin\psi}{\sin(\phi-\psi)}-\cos\phi\right\}^2$$
$$=\frac{(a^2\cos^2\phi+b^2\sin^2\phi)\{1-\cos(\phi-\psi)\}^2}{\sin^2(\phi-\psi)},$$

so that
$$PR=a\Delta\phi\frac{1-\cos(\phi-\psi)}{\sin(\phi-\psi)},\quad RQ=a\Delta\psi\frac{1-\cos(\phi-\psi)}{\sin(\phi-\psi)},$$

while
$$BD=a\frac{1-\cos\gamma}{\sin\gamma},\quad DG=a\Delta\gamma\frac{1-\cos\gamma}{\sin\gamma}.$$

Therefore, by § 121,

$$PR+RQ=a\frac{\Delta\phi+\Delta\psi}{\sin(\phi-\psi)}\{1-\cos(\phi-\psi)\}=a\frac{1+\Delta\gamma}{\sin\gamma}\{1-\cos(\phi-\psi)\}$$

$$PR+RQ-BD-DG=a\frac{1+\Delta\gamma}{\sin\gamma}\{\cos\gamma-\cos(\phi-\psi)\};$$

$$=a\frac{1+\Delta\gamma}{\sin\gamma}\{\cos\phi\cos\psi+\sin\phi\sin\psi\Delta\gamma-\cos(\phi-\psi)\}$$

$$=-a\frac{1-\Delta^2\gamma}{\sin\gamma}\sin\phi\sin\psi$$

$$=-a\kappa^2\sin\phi\sin\psi\sin\gamma.$$

In Jacobi's notation this is written

$$E\text{ am }u-E\text{ am }v-E\text{ am}(u-v),\text{ or }Zu-Zv-Z(u-v)$$
$$=-\kappa^2\text{sn }u\text{ sn }v\text{ sn}(u-v).$$

175. Putting $v=w$, and therefore $u=2w$, then

$$E\text{ am }2w-2\,E\text{ am }w=-\kappa^2\text{sn }2w\text{ sn}^2w,$$

or changing w into $\frac{1}{2}w$,

$$E\text{ am }w-2E\text{ am}\tfrac{1}{2}w=-\kappa^2\text{sn }w\text{ sn}^2\tfrac{1}{2}w=-\text{sn }w\frac{1-\text{dn }w}{1+\text{cn }w}\quad(\S\ 123).$$

Then $\quad PR+RQ-\text{arc }PQ=BD+DG-\text{arc }BG$

$$=a(1+\text{dn }w)\frac{1-\text{cn }w}{\text{sn }w}-aE\text{ am }w$$

$$=a(1+\text{dn }w)\frac{\text{sn }w}{1+\text{cn }w}+a\text{ sn }w\,\frac{1-\text{dn }w}{1+\text{cn }w}-2a\,E\text{ am }\tfrac{1}{2}w$$

$$=2a\left(\frac{\text{sn }w}{1+\text{cn }w}-E\text{ am }\tfrac{1}{2}w\right)=2a\left(\frac{\text{sn }\frac{1}{2}w\text{ dn }\frac{1}{2}w}{\text{cn }\frac{1}{2}w}-E\text{ am }\tfrac{1}{2}w\right);$$

and now $\quad\text{cn }\frac{1}{2}w$, or $\text{cn }\frac{1}{2}(u-v)=b/\beta$, where $\beta=OK$.

176. A ready way of proving the Addition Theorem is to take the spherical triangle of Class II., in which

$$A=\text{am }v_1,\ B=\text{am }v_2,\ C=\text{am }v_3,$$

where $$v_1+v_2+v_3=2K,$$

and to vary all the sides and angles, keeping κ constant.

Then $$dv_1+dv_2+dv_3=0,$$

or $$dA/\cos a+dB/\cos b+dC/\cos c=0,$$

or $$\cos b\cos c\,.\,dA+\cos c\cos a\,.\,dB+\cos a\cos b\,.\,dC=0,$$

or $$(\cos a-\sin b\sin c\cos A)dA+(\cos b-\sin c\sin a\cos B)dB$$
$$+(\cos c-\sin a\sin b\cos C)dC=0,$$

or $$\cos a dA+\cos b dB+\cos c dC$$
$$=\kappa^2(\sin B\sin C\cos A dA+\sin C\sin A\cos B dB+\sin A\sin B\cos C dC)$$
$$=\kappa^2 d(\sin A\sin B\sin C).$$

Integrating,

$$E(A)+E(B)+E(C)-2E=\kappa^2\sin A\sin B\sin C,$$

since $\int_0 \cos\alpha dA=\int\sqrt{(1-\kappa^2\sin^2 A)}dA=E(A),$

and $v_2=0$ makes $B=0$, and $A+C=\pi$, or $E(A)+E(C)=2E$.

In Jacobi's notation

$$E\,\mathrm{am}\,v_1+E\,\mathrm{am}\,v_2+E\,\mathrm{am}\,v_3-2E=\kappa^2\mathrm{sn}\,v_1\mathrm{sn}\,v_2\mathrm{sn}\,v_3,$$

or
$$Zv_1+Zv_2+Zv_3=\kappa^2\mathrm{sn}\,v_1\mathrm{sn}\,v_2\mathrm{sn}\,v_3,$$

with
$$v_1+v_2+v_3=2K.$$

With
$$u+v+w=0,$$

$$Zu+Zv+Zw=-\kappa^2\mathrm{sn}\,u\,\mathrm{sn}\,v\,\mathrm{sn}\,w,$$

or
$$Zu+Zv-Z(u+v)=\kappa^2\mathrm{sn}\,u\,\mathrm{sn}\,v\,\mathrm{sn}(u+v).$$

Fagnano's Theorems.

177. The particular case of the Addition Theorem, obtained by putting $\gamma=\frac{1}{2}\pi$, or $u-v=K$, was discovered by Fagnano (1716), and leads to his theorems, namely, that if P, Q are two points on an ellipse of excentricity κ, whose excentric angles ϕ, ψ, measured from the minor axis, are such that

$$\Delta\phi\Delta\psi=\kappa', \text{ or } \tan\phi\tan\psi=1/\kappa'=a/b,$$

then the arc BP+arc BQ−arc $AB=a\kappa^2\sin\phi\sin\psi$,

or
$$\text{arc } BP-\text{arc } AQ=a\kappa^2\sin\phi\sin\psi=\kappa^2xx'/a;$$

and then
$$\tan^2\phi\tan^2\phi'=\frac{x^2x'^2}{(a^2-x^2)(a^2-x'^2)}=\frac{a^2}{b^2},$$

or
$$\kappa^2x^2x'^2-a^2(x^2+x'^2)+a^4=0.$$

On reference to fig. 23 it will be found that, if OY, OZ are the perpendiculars on the tangents at P and Q, then

(i.) $AOZ=\phi$, $AOY=\psi$,

(ii.) arc BP − arc $AQ=PY=QZ=VQ-PT$,
so that $VZ=PT$, and PY or $QZ=\kappa^2xx'/a$;
the tangents at P, Q meeting OA, OB in T, V;

(iii.) $OP^2-OQ^2=OY^2-OZ^2$; (iv.) $OY.OZ=ab$.

When P and Q coincide in F, then F is called Fagnano's point; and then

(i.) the arc BF − arc $AF=a-b$;

(ii.) the coordinates of F are $\sqrt{\frac{a^3}{a+b}}$, $\sqrt{\frac{b^3}{a+b}}$;

(iii.) $KF=a$, $FH=b$, $FG=a-b$, $OG=\sqrt{(ab)}$;

(iv.) the tangents at P, Q intersect in R on the confocal hyperbola FRD, through F, D, whose equation is

$$\frac{x^2}{a}-\frac{y^2}{b}=a-b;$$

(v.) the tangents at P and Q' intersect in R' on the confocal ellipse KDH, through K, D, H, whose equation is

$$\frac{x^2}{a}+\frac{y^2}{b}=a+b;$$

(vi.) $PR-\text{arc } PF=QR-\text{arc } QF$;

(vii.) the circle inscribed in the region bounded by AD, DB and the ellipse AB touches the ellipse at F; etc.

The proof of these theorems is left as an exercise.

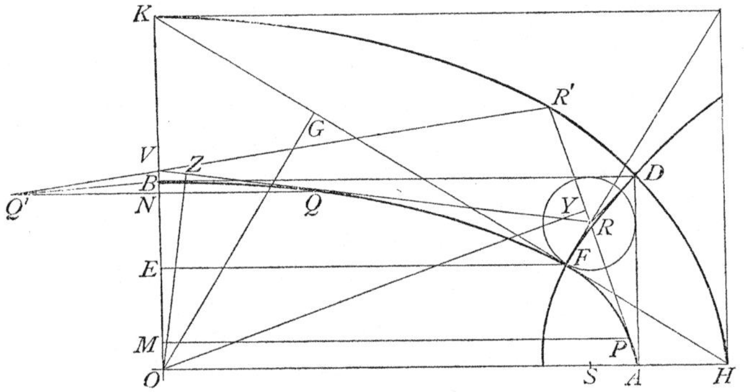

Fig. 23.

178. Denoting the arc AP by s, the perpendicular OY on the tangent at P by p, the angle AOY by ψ, then by Legendre's formula

$$\frac{ds}{d\psi}=\frac{d^2p}{d\psi^2}+p, \text{ while } PY=-\frac{dp}{d\psi},$$

so that $\quad s+PY=\int p\,d\psi$;

and in the ellipse

$$p=\sqrt{(a^2\cos^2\psi+b^2\sin^2\psi)}=a\Delta\psi,$$

while

$$PY=-dp/d\psi=a\kappa^2\sin\psi\cos\psi/\Delta\psi=a\kappa^2\sin\phi\sin\psi;$$

so that $\quad s+a\kappa^2\sin\phi\sin\psi=a\int_0\Delta\psi\,d\psi=aE\psi=\text{arc } BQ,$

or $\qquad \text{arc } BQ-\text{arc } AP=a\kappa^2\sin\phi\sin\psi,$

as at first, in Fagnano's Theorem.

Confocal Ellipses and Hyperbolas.

179. If we put

$$x+iy=c\sin(\phi+i\theta),$$

then $x=c\sin\phi\cosh\theta,\ y=c\cos\phi\sinh\theta$;

so that
$$\frac{x^2}{\cosh^2\theta}+\frac{y^2}{\sinh^2\theta}=c^2,$$

$$\frac{x^2}{\sin^2\phi}-\frac{y^2}{\cos^2\phi}=c^2,$$

the equations of a system of confocal ellipses and hyperbolas, since
$$\cosh^2\theta-\sinh^2\theta=\sin^2\phi+\cos^2\phi=1.$$

Then
$$\frac{dx^2}{d\phi^2}+\frac{dy^2}{d\phi^2}=\frac{dx^2}{d\theta^2}+\frac{dy^2}{d\theta^2}=c^2(\cosh^2\theta-\sin^2\phi);$$

so that, in an ellipse BP, along which θ is constant, the

$$\text{arc } BP=c\int\sqrt{(\cosh^2\theta-\sin^2\phi)}d\phi=aE\phi$$

as before, with $a=c\cosh\theta$, and the modulus equal to the excentricity $\operatorname{sech}\theta$.

For the confocal hyperbola, along which ϕ is constant, the arc is given by

$$c\int\sqrt{(\cosh^2\theta-\cos^2\phi)}d\theta,$$

which can be expressed by elliptic integrals of the first and second kind, of Legendre's form.

Putting

$$a=c\sin\phi,\ b=c\cos\phi,$$

the equation of the hyperbola is

$$(x/a)^2-(y/b)^2=1;$$

and now the coordinates of any point P on the hyperbola may be given by $a\operatorname{cosec}\chi$, $b\cot\chi$; and the tangent at P by

$$\frac{x}{a}\operatorname{cosec}\chi-\frac{y}{b}\cot\chi=1,$$

and then
$$\operatorname{amh}\theta=\tfrac{1}{2}\pi-\chi,$$

$$\cosh\theta=\operatorname{cosec}\chi,\ \sinh\theta=\cot\chi,\ \tanh\theta=\cos\chi, \text{ etc.}$$

The tangents at P, and at another point Q defined by χ', will therefore meet at a point R, where

$$\frac{x}{a}=\frac{\cot\chi'-\cot\chi}{\operatorname{cosec}\chi\cot\chi'-\operatorname{cosec}\chi'\cot\chi}=\frac{\sin(\chi-\chi')}{\cos\chi'-\cos\chi},\ \frac{y}{b}=\frac{\sin\chi-\sin\chi'}{\cos\chi'-\cos\chi}.$$

When we put

$$\chi=\operatorname{am}u,\ \chi'=\operatorname{am}v$$

the modular angle being ϕ, then as in § 173 for the ellipse,

$$\frac{x}{a}=\frac{s_2 c_2 d_1}{s_1 d_1 s_2 d_2}=\frac{c_2}{s_1 d_2}=\frac{\operatorname{cn}\frac{1}{2}(u-v)}{\operatorname{sn}\frac{1}{2}(u+v)\operatorname{dn}\frac{1}{2}(u-v)},$$

$$\frac{y}{b}=\frac{s_2 c_1 d_1}{s_1 d_1 s_2 d_2}=\frac{c_1}{s_1 d_2}=\frac{\operatorname{cn}\frac{1}{2}(u+v)}{\operatorname{sn}\frac{1}{2}(u+v)\operatorname{dn}\frac{1}{2}(u-v)},$$

and therefore, eliminating $\operatorname{cn}\frac{1}{2}(u-v)$ and $\operatorname{dn}\frac{1}{2}(u-v)$,

$$(x/\alpha^2)+(y/\beta)^2=1,$$

where $\alpha=\dfrac{a}{\kappa\operatorname{sn}\frac{1}{2}(u+v)}$, $\beta=\dfrac{b\operatorname{cn}\frac{1}{2}(u+v)}{\kappa'\operatorname{sn}\frac{1}{2}(u+v)}=\dfrac{a\operatorname{cn}\frac{1}{2}(u+v)}{\kappa\operatorname{sn}\frac{1}{2}(u+v)}$,

and $$\alpha^2-\beta^2=c^2=a^2+b^2,$$

so that R describes a confocal ellipse, when $u+v$ is constant.

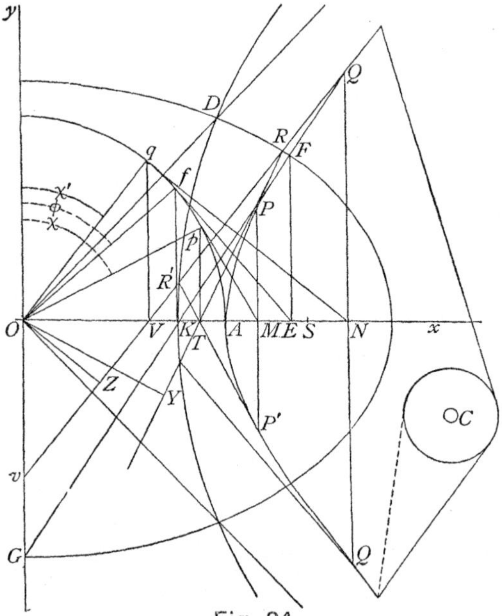

Fig. 24.

180. By putting $u+v=K$, we obtain theorems for the hyperbola (fig. 24) analogous to Fagnano's theorems for the ellipse.

Now (§ 123) $\alpha=c\sqrt{(1+\kappa')}$, $\beta=c\sqrt{\kappa'}$,

or $$\alpha^2=c(c+b),\quad \beta^2=cb\,;$$

and R describes the ellipse FD, whose equation is

$$\frac{x^2}{c+b}+\frac{y^2}{b}=c,$$

which will intersect the hyperbola in a point F, the analogue of Fagnano's point on the ellipse, the coordinates of which are

$$c\sin\phi\sqrt{(1+\cos\phi)},\quad c(\cos\phi)^{\frac{3}{2}}.$$

Now, as in § 57, with

$$\chi = \text{am } u,\ \chi' = \text{am } v, \text{ and } u+v=K,$$

$$\Delta\chi\Delta\chi' = \kappa' = \cos\phi,$$

and $$\cot\chi\cot\chi' = \kappa' = \cos\phi,$$

or $$\sinh\theta\sinh\theta' = \kappa',$$

and if x, y and x', y' denote the coordinates of P and Q,

$$x = a\,\text{cosec}\,\chi = a\Delta\chi'/\cos\chi',\quad x' = a\,\text{cosec}\,\chi' = a\Delta\chi/\cos\chi;$$

$$y = a \quad \cot\chi = a\kappa'\tan\chi',\qquad y' = a \quad \cot\chi' = a\kappa'\tan\chi;$$

and thus $$yy' = a^2\kappa' = c^2\cos^3\phi.$$

Drawing the perpendiculars OY, OZ from O on the tangents at P, Q, and denoting the angles AOY, AOZ by ω, ω'; then

$$\tan\omega = \frac{dx}{dy} = \frac{y/b^2}{x/a^2} = \tan\phi\cos\chi = \tan\phi\tanh\theta = \sin\phi\sin\chi'/\Delta\chi';$$

$$\sin\omega = \sin\phi\sin\chi',\ \cos\omega = \Delta\chi',\ \sin\omega' = \sin\phi\sin\chi,\ \cos\omega' = \Delta\chi.$$

Now denoting OY, OZ by p, p', then

$$p = \surd(a^2\cos^2\omega - b^2\sin^2\omega) = c\surd(\sin^2\phi - \sin^2\omega) = c\sin\phi\cos\chi';$$

$$pp' = c^2\sin^2\phi\cos\chi\cos\chi' = c^2\sin^2\phi\cos\phi\sin\chi\sin\chi' = c^2\cos\phi\sin\omega\sin\omega'.$$

Making use of the formulas

$$\frac{ds}{d\omega} = -\frac{d^2p}{d\omega^2} - p, \text{ and } PY = -\frac{dp}{d\omega},$$

then

$$PY - \text{arc } AP = \int p\,d\omega = c\int_0 \surd(\sin^2\phi - \sin^2\omega)\,d\omega$$

$$= c\int\sin^2\phi\cos^2\chi'\,d\chi'/\Delta\chi' = c\int(\Delta^2\chi' - \kappa'^2)\,d\chi'/\Delta\chi'$$

$$= c(E\chi' - \kappa'^2F\chi');$$

also $$PY = c\sin\omega\cos\omega/\surd(\sin^2\phi - \sin^2\omega)$$

$$= c\tan\chi'\Delta\chi' = c/\tan\chi\Delta\chi$$

$$= c\cosh\theta\sinh\theta/\surd(\cosh^2\theta - \sin^2\phi).$$

181. The arc AP of the hyperbola is now expressed in terms of an elliptic integral of the first and of the second kind; we can however express the arc by means of two elliptic integrals of the second kind, or by two elliptic arcs by means of Landen's transformation (§ 67).

We shall find that if we put

$$\omega + \chi' = 2\psi, \quad \text{or} \quad \sin(2\psi - \chi') = \sin\omega = \sin\phi\sin\chi',$$

then $$\tan\chi' = \frac{\sin 2\psi}{\sin\phi + \cos 2\psi}, \qquad \sec\chi' = \frac{(1+\sin\phi)\Delta(\psi, \gamma)}{\sin\phi + \cos 2\psi},$$

where $$\gamma^2 = \frac{4\sin\phi}{(1+\sin\phi)^2}, \qquad \gamma' = \frac{1-\sin\phi}{1+\sin\phi};$$

$$\sin \chi' = \frac{\sin 2\psi}{(1+\sin\phi)\Delta(\psi,\gamma)},\quad \sin\omega = \frac{\sin\phi\sin 2\psi}{(1+\sin\phi)\Delta(\psi,\gamma)},$$

$$\cos\omega = \Delta\chi' = \frac{1+\sin\phi\cos 2\psi}{(1+\sin\phi)\Delta(\psi,\gamma)},$$

and
$$\frac{d\omega}{\surd(\sin^2\phi-\sin^2\omega)} = \frac{d\chi'}{\Delta\chi'} = \frac{2d\psi}{(1+\sin\phi)\Delta(\psi,\gamma)},$$

$$\cos\omega + \surd(\sin^2\phi - \sin^2\omega) = \Delta\chi' + \kappa\cos\chi' = (1+\sin\phi)\Delta(\psi,\gamma);$$

so that

$$(1+\sin\phi)\Delta(\psi,\gamma)d\psi = \frac{(\Delta\chi'+\kappa\cos\chi')^2}{2\Delta\chi'}d\chi'$$

$$= \left(\Delta\chi' + 2\kappa\cos\chi' - \frac{1}{2}\frac{\kappa'^2}{\Delta\chi'}\right)d\chi',$$

Integrating,

$$(1+\kappa)E(\psi,\gamma) = E\chi' + \kappa\sin\chi' - \tfrac{1}{2}\kappa'^2F\chi';$$

and now the arc of the hyperbola

$$AP = PY + 2c\kappa\sin\chi' + cE\chi' - 2c(1+\kappa)E(\psi,\gamma).$$

182. If we put $\chi - \chi' = \tfrac{1}{2}\pi - \xi$,
then we find (§ 180)

$$\tan\xi = \frac{(1+\cos\phi)\tan\chi'}{1-\cos\phi\tan^2\chi'},\quad \sec\xi = \frac{\sec^2\chi'\Delta\chi'}{1-\cos\phi\tan^2\chi'},$$

$$\sin\xi = \frac{(1+\cos\phi)\sin\chi'\cos\chi'}{\Delta\chi'},$$

$$\Delta(\xi,\lambda) = \frac{1-(1-\cos\phi)\sin^2\chi'}{\Delta\chi'} = \frac{\Delta^2\chi'+\cos\phi}{(1+\cos\phi)\Delta\chi'},$$

and
$$\frac{d\xi}{\Delta(\xi,\lambda)} = (1+\cos\phi)\frac{d\chi'}{\Delta\chi'},\ \text{with } \lambda = \tan^2\tfrac{1}{2}\phi.$$

Now, $\sin(2\chi'-\xi) = \lambda\sin\xi$,
as in Landen's second transformation (§ 123); and

$$(1+\cos\phi)\Delta(\xi,\lambda)d\xi = (\Delta^2\chi'+\cos\phi)^2 d\chi'/\Delta^3\chi'$$

$$= \left(\Delta\chi' + 2\cos\phi\frac{1}{\Delta\chi'} + \frac{\cos^2\phi}{\Delta^3\chi'}\right)d\chi'$$

$$= 2\Delta\chi'd\chi' + 2\cos\phi\frac{d\chi'}{\Delta\chi'} - \sin^2\phi\, d\left(\frac{\sin\chi'\cos\chi'}{\Delta\chi'}\right).$$

Integrating,

$$(1+\cos\phi)E(\xi,\lambda) = 2E\chi' + 2\cos\phi F\chi' - \sin^2\phi\sin\chi'\cos\chi'/\Delta\chi';$$

and the arc AP can be expressed by means of $E\chi'$ and $E(\xi,\lambda)$.

When $\chi = \chi' = \text{am}\,\tfrac{1}{2}K$, then $\xi = \tfrac{1}{2}\pi$;
also (§ 175) $2E\chi' = E(\kappa) + 1 - \cos\phi$, while $2F\chi' = K$;
so that $(1+\kappa')E(\lambda) = E(\kappa) + \kappa'K.$

183. The following theorems, analogous to those of § 177, can easily be proved by the student:—

(i.) The difference between the infinite asymptote DT and the infinite arc FT is equal to $AD-\text{arc } AF$; so that the difference between the infinite asymptote OT and the infinite arc AT is equal to $OD+AD-2 \text{ arc } AF$;

(ii.) the coordinates of F are $(c+b)\surd\{(c-b)/c\}$, $\surd(b^3/c)$; and the tangent $FK=AD=b$, $KG=c$;

(iii.) the tangents at P, Q intersect in R on the confocal ellipse through F, whose equation is

$$\frac{x^2}{c+b}+\frac{y^2}{b}=c,$$

and the tangents at P', Q intersect in R' on the confocal hyperbola through D and K, whose equation is

$$\frac{x^2}{c-a}-\frac{y^2}{a}=c;$$

(iv.) $PR-\text{arc } PF=QR-\text{arc } QF$;

(v.) $P'R'+R'Q-\text{arc } P'Q$ is constant;

(vi.) the circle inscribed in the region bounded by the straight line AD, the asymptote DT and the hyperbola AQ touches the hyperbola at F;

(vii.) $PT=c\cot\chi\Delta\chi$, $QV=c\cot\chi'\Delta\chi'$, $Qv=c\Delta\chi'/\sin\chi'\cos\chi'$,

$$PT\,.\,QV=FK^2,\quad PY\,.\,QZ=c^2,$$

$$Qv-PT=QZ, \quad \text{or} \quad vZ=PT,$$

$$PR=c\frac{\Delta\chi}{\sin\chi}\frac{1-\cos(\chi-\chi')}{\cos\chi'-\cos\chi},\quad RQ=c\frac{\Delta\chi'}{\sin\chi'}\frac{1-\cos(\chi-\chi')}{\cos\chi'-\cos\chi},\ \text{etc.}$$

184. The geometrical theorems of § 173 for the ellipse hold with slight modification for the mechanical description of confocal ellipses and hyperbolas from a fixed hyperbola.

The threads from the reel must be led round distant points on the hyperbola APQ (fig. 24) and be wrapped on the curve; and now, starting from F, the confocal ellipse FRD will be described, if the threads are led off in the same direction.

At D, one thread DT must be supposed of infinite length; and, beyond D on the ellipse FD, the thread DT must be transferred to the other branch of the hyperbola.

By making the threads come off the reel in opposite directions, the confocal hyperbola DK can be described, starting from D or any other point R.

185. The integration of the functions of § 77 can now be expressed by means of the elliptic functions, and of the function E am u, defined by

$$E \text{ am } u = \int_0 \text{dn}^2 u du.$$

Then

$$\int_0 \kappa^2 \text{sn}^2 u du = u - E \text{ am } u$$

$$\int_0 \kappa^2 \text{cn}^2 u du = E \text{ am } u - \kappa'^2 u.$$

To integrate a reciprocal function, for instance $\text{nd}^2 u$, we notice that

$$\frac{d^2}{du^2} \log \text{dn } u = \kappa'^2 \text{nd}^2 u - \text{dn}^2 u,$$

so that

$$\int_0 \kappa'^2 \text{nd}^2 u du = E \text{ am } u - \kappa^2 \text{sn } u \text{ cn } u / \text{dn } u;$$

and so on.

Again, since

$$\text{cd}^2 u = \text{sn}^2(K - u),$$

$$\int_0 \kappa^2 \text{cd}^2 u du = u - \int_0 \text{dn}^2(K - u) du$$

$$= u - E + E \text{ am}(K - u)$$

$$= u - E \text{ am } u + \kappa^2 \text{sn } u \text{ cn } u / \text{dn } u;$$

and since

$$\kappa'^2 \text{nd}^2 u = \text{dn}^2(K - u),$$

$$\int \kappa'^2 \text{nd}^2 u du = E - E \text{ am}(K - u)$$

$$= E \text{ am } u - \kappa^2 \text{sn } u \text{ cn } u / \text{dn } u,$$

as before.

In Problem III., 86, we find

$$n \frac{dt}{d\theta} = \frac{\text{dn}^2 \theta}{\text{cn}^2 \theta} = \text{dc}^2 \theta,$$

and

$$nt = \int_0 \text{dc}^2 \theta d\theta = \theta - E \text{ am } \theta + \text{sn } \theta \text{ dn } \theta / \text{cn } \theta.$$

EXAMPLES.

1. Prove that the area of the Cassinian

$$r^4 - 2a^2 r^2 \cos 2\theta + a^4 = b^4$$

is

$$2 \int_0^{\frac{1}{2}\pi} (b^4 - a^4 \sin^2 \phi)^{\frac{1}{2}} d\phi, \text{ if } b > a;$$

or

$$2 \int_0^{\frac{1}{2}\pi} (a^4 - b^4 \sin^2 \phi)^{-\frac{1}{2}} b^4 \cos^2 \phi d\phi, \text{ if } a > b.$$

2. Rectify, by means of elliptic arcs (pointing out the geometrical connexion),

(i.) $y/b = \sin x/a$, $\cos x/a$, $\cosh x/a$, $\operatorname{dn} x/a$, $\operatorname{cn} x/a$, $\operatorname{sn} x/a, \ldots$;

(ii.) $r = b\cos(b\theta/a)$ or $a\cos(a\theta/b)$, the pedals of an epi- or hypo-cycloid;

(iii.) $r\cos(b\theta/a) = b$, or $r\cosh(b\theta/a) = b$, Cotes's spirals;

(iv.) the limaçon $r = a + b\cos\theta$, the trochoid, and the epi- and hypo-trochoids.

3. Express x as a function of s in the *Elastica* of § 97.

Prove that if the ordinate is made equal to p, the perpendicular on the tangent from the centre of an ellipse or hyperbola, and if the abscissa is made equal to the arc $AP \pm PY$, the curve will be an *Elastica* (Maclaurin, *Fluxions*, 1742.)

4. Prove that $$(1-\kappa^2)\frac{d^2K}{d\kappa^2} + \frac{1-3\kappa^2}{\kappa}\frac{dK}{d\kappa} - K = 0;$$

$$(1-\kappa^2)\frac{d^2E}{d\kappa^2} + \frac{1-\kappa^2}{\kappa}\frac{dE}{d\kappa} + E = 0.$$

Change the independent variable in these differential equations from κ to k, θ, or u, where

$$\kappa = \sqrt{k} = \sin\theta = \tanh u;$$

and reduce the resulting equations to the canonical form

$$\frac{1}{y}\frac{d^2y}{dx^2} + I = 0.$$

Solve the differential equations in which

$$I = \frac{1-kk'}{4k^2k'^2},\ \operatorname{cosec}^2 2\theta,\ -\operatorname{cosech}^2 2u,\ -\operatorname{sech}^2 2u, \ldots.$$

(Glaisher, *Q. J. M.*, XX., p. 313; Kleiber, *Messenger*, XVIII., p. 167.)

5. Prove that, if $u_1 + u_2 + u_3 + u_4 = 0$,

$$Zu_1 + Zu_2 + Zu_3 + Zu_4$$
$$= \frac{-\kappa^2 s_1 s_2 s_3 s_4}{1 + \kappa^2 s_1 s_2 s_3 s_4}\left(\frac{c_1 d_1}{s_1} + \frac{c_2 d_2}{s_2} + \frac{c_3 d_3}{s_3} + \frac{c_4 d_4}{s_4}\right)$$
$$= \frac{\kappa^2 c_1 c_2 c_3 c_4}{\kappa^2 c_1 c_2 c_3 c_4 - \kappa'^2}\left(\frac{s_1 d_1}{c_1} + \frac{s_2 d_2}{c_2} + \frac{s_3 d_3}{c_3} + \frac{s_4 d_4}{c_4}\right)$$
$$= \frac{\kappa^2 d_1 d_2 d_3 d_4}{\kappa'^2 + d_1 d_2 d_3 d_4}\left(\frac{s_1 c_1}{d_1} + \frac{s_2 c_2}{d_2} + \frac{s_3 c_3}{d_3} + \frac{s_4 c_4}{d_4}\right)$$
$$= \kappa\sqrt{(s_1^2 + s_2^2 + s_3^2 + s_4^2 - 2s_1 s_2 s_3 s_4 + 2c_1 c_2 c_3 c_4 - 2)}.$$

The Elliptic Integral of the Third Kind.

186. We can now make a fresh start, and prove the Addition Theorem for the Zeta Function independently; and then proceed to Jacobi's form of the Third Elliptic Integral. (*Fundamenta Nova*, 49; Glaisher, *Proc. L.M.S.* XVII. p. 153.)

Multiplying formulas (3) and (6), § 137,

$$\mathrm{dn}^2(u+v)-\mathrm{dn}^2(u-v)=\frac{-4\kappa^2\mathrm{sn}\,u\,\mathrm{cn}\,u\,\mathrm{dn}\,u\,\mathrm{sn}\,v\,\mathrm{cn}\,v\,\mathrm{dn}\,v}{(1-\kappa^2\mathrm{sn}^2u\,\mathrm{sn}^2v)^2}\ldots(1)$$

and, integrating with respect to v,

$$E\,\mathrm{am}(u+v)+E\,\mathrm{am}(u-v)=C-\frac{2\,\mathrm{cn}\,u\,\mathrm{dn}\,u/\mathrm{sn}\,u}{1-\kappa^2\mathrm{sn}^2u\,\mathrm{sn}^2v}$$

where C is the constant of integration, independent of v.

To determine C, first put $v=u$; then

$$C=E\,\mathrm{am}\,2u+\frac{2\,\mathrm{cn}\,u\,\mathrm{dn}\,u/\mathrm{sn}\,u}{1-\kappa^2\mathrm{sn}^4u};$$

so that, replacing $E\,\mathrm{am}\,u$ by $Eu/K+\mathrm{Z}u$,

$$\mathrm{Z}(u+v)+\mathrm{Z}(u-v)-\mathrm{Z}2u=\frac{2\,\mathrm{cn}\,u\,\mathrm{dn}\,u/\mathrm{sn}\,u}{1-\kappa^2\mathrm{sn}^4u}-\frac{2\,\mathrm{cn}\,u\,\mathrm{dn}\,u/\mathrm{sn}\,u}{1-\kappa^2\mathrm{sn}^2u\,\mathrm{sn}^2v}$$

$$=\frac{\mathrm{sn}\,2u}{\mathrm{sn}^2u}\left(1-\frac{1-\kappa^2\mathrm{sn}^4u}{1-\kappa^2\mathrm{sn}^2u\,\mathrm{sn}^2v}\right)=\kappa^2\mathrm{sn}\,2u\frac{\mathrm{sn}^2u-\mathrm{sn}^2v}{1-\kappa^2\mathrm{sn}^2u\,\mathrm{sn}^2v}$$

$$=\kappa^2\mathrm{sn}(u+v)\mathrm{sn}(u-v)\mathrm{sn}\,2u\ldots\ldots\ldots(2)$$

Replacing $u+v$, $u-v$, and $2u$ by u, v, and $u+v$, this becomes the formula given above, § 176,

$$\mathrm{Z}u+\mathrm{Z}v-\mathrm{Z}(u+v)=\kappa^2\mathrm{sn}\,u\,\mathrm{sn}\,v\,\mathrm{sn}(u+v).\ldots\ldots\ldots\ldots(2)^*$$

Again, put $u=0$ for the determination of C; then

$$C=2Eu+2\,\mathrm{cn}\,u\,\mathrm{dn}\,u/\mathrm{sn}\,u;$$

and now

$$\mathrm{Z}(u+v)+\mathrm{Z}(u-v)-2\mathrm{Z}u=\frac{-2\kappa^2\mathrm{sn}\,u\,\mathrm{cn}\,u\,\mathrm{dn}\,u\,\mathrm{sn}^2v}{1-\kappa^2\mathrm{sn}^2u\,\mathrm{sn}^2v}\ldots\ldots\ldots\ldots(3),$$

another form of the Addition Equation of the Zeta Function, leading immediately to Jacobi's form of the Third Elliptic Integral, as required in § 114.

187. Integrating this equation (3) again with respect to v, and employing Jacobi's notation of

$$\Pi(v,\,u)\ \text{for}\int_0\frac{\kappa^2\mathrm{sn}\,u\,\mathrm{cn}\,u\,\mathrm{dn}\,u\,\mathrm{sn}^2v\,dv}{1-\kappa^2\mathrm{sn}^2u\,\mathrm{sn}^2v},$$

where u is called the *parameter*, and v the *argument*, then

$$\Pi(v,\,u)=v\mathrm{Z}u-\tfrac{1}{2}\int_0\mathrm{Z}(u+v)dv-\tfrac{1}{2}\int_0\mathrm{Z}(u-v)dv.$$

Jacobi now introduces a new function Θu, called the *Theta Function*, defined by

$$\int_0 \mathrm{Z}u\,du = \log\frac{\Theta u}{\Theta 0},$$

or

$$\Theta u = \Theta 0 \exp\int_0 \mathrm{Z}u\,du; \qquad (4)$$

so that

$$\mathrm{Z}u = \frac{\Theta' u}{\Theta u}.$$

Now

$$\int_0 \mathrm{Z}(u+v)\,dv = \log\frac{\Theta(u+v)}{\Theta u},$$

$$\int_0 \mathrm{Z}(u-v)\,dv = \log\frac{\Theta u}{\Theta(u-v)};$$

and

$$\Pi(v, u) = v\mathrm{Z}u + \tfrac{1}{2}\log\frac{\Theta(u-v)}{\Theta(u+v)},$$

$$= \log e^{v\mathrm{Z}u}\sqrt{\frac{\Theta(u-v)}{\Theta(u+v)}}, \qquad (5),$$

so that the Third Elliptic Integral is expressed by Jacobi's Theta and Zeta Functions, the arguments being u and v, two in number only, and not three, n, κ, ϕ, as in Legendre's form.

188. Integrating equation (3) again with respect to u,

$$\int_0^u\int_0^v \{\mathrm{dn}^2(u+v) - \mathrm{dn}^2(u-v)\}\,dv\,du = \log(1-\kappa^2\mathrm{sn}^2 u\,\mathrm{sn}^2 v),$$

or

$$\log\frac{\Theta(u+v)}{\Theta v} + \log\frac{\Theta(u-v)}{\Theta v} - 2\log\frac{\Theta u}{\Theta 0} = \log(1-\kappa^2\mathrm{sn}^2 u\,\mathrm{sn}^2 v),$$

or

$$\frac{\Theta(u+v)\Theta(u-v)\Theta^2 0}{\Theta^2 u\,\Theta^2 v} = 1-\kappa^2\mathrm{sn}^2 u\,\mathrm{sn}^2 v, \qquad (6)$$

a formula which takes the place of the Addition Theorem for the Theta Functions.

For instance, putting $u=v$,

$$\Theta 2u = (1-\kappa^2\mathrm{sn}^4 u)\Theta^4 u/\Theta^3 0. \qquad (7)$$

Interchanging the *argument* and *parameter*, u and v, then

$$\Pi(u, v) = u\mathrm{Z}v + \tfrac{1}{2}\log\frac{\Theta(u-v)}{\Theta(u+v)},$$

so that $\Pi(u, v) - \Pi(v, u) = u\mathrm{Z}v - v\mathrm{Z}u, \qquad (8)$

and $\Pi(v, u)$ is thus made to depend upon $\Pi(u, v)$.

189. In Legendre's notation, $\Pi(n, \kappa, \phi)$ or simply $\Pi\phi$, is employed to denote his Elliptic Integral of the Third Kind

$$\int_0 \frac{d\phi}{(1+n\sin^2\phi)\Delta\phi},$$

n being called Legendre's *parameter* (§ 114); and with Jacobi's notation,

$$\Pi(n, \kappa, \operatorname{am} u) = \int_0 \frac{du}{1+n\operatorname{sn}^2 u}.$$

But Jacobi changes the notation, by putting $n = -\kappa^2\operatorname{sn}^2 a$, and by calling a the parameter; also by denoting the integral

$$\int_0 \frac{\kappa^2\operatorname{sn} a \operatorname{cn} a \operatorname{dn} a \operatorname{sn}^2 u\, du}{1-\kappa^2\operatorname{sn}^2 a \operatorname{sn}^2 u} \text{ by } \Pi(u, a),$$

and not the integral

$$\int_0 \frac{du}{1-\kappa^2\operatorname{sn}^2 a\operatorname{sn}^2 u}, \text{ which equals } u + \frac{\operatorname{sn} a\, \Pi(u, a)}{\operatorname{cn} a \operatorname{dn} a}.$$

In Legendre's notation, the Addition Equation of the elliptic integrals of the first kind

$$F\phi + F\psi = F\mu,$$

leads to

$$E\phi + E\psi - E\mu = \kappa^2 \sin\phi \sin\psi \sin\mu,$$

the Addition Theorem for the second elliptic integrals; and now for Legendre's elliptic integrals of the third kind, the Addition Theorem is (Legendre, *F. E. I.*, Chap. XVI.)

$$\Pi\phi + \Pi\psi - \Pi\mu = \frac{1}{\sqrt{a}} \tan^{-1} \frac{n\sqrt{a}\sin\phi\sin\psi\sin\mu}{1+n-n\cos\phi\cos\psi\cos\mu},$$

or

$$= \frac{1}{\sqrt{(-a)}} \tanh^{-1} \frac{n\sqrt{(-a)}\sin\phi\sin\psi\sin\mu}{1+n-n\cos\phi\cos\psi\cos\mu}, \quad (9)$$

according as a is positive or negative, where

$$a = (1+n)(1+\kappa^2/n);$$

this can be verified by differentiation.

This relation is very much simplified by the use of Jacobi's function $\Pi(u, a)$; and now with

$$\phi = \operatorname{am} u, \quad \psi = \operatorname{am} v, \quad \mu = \operatorname{am}(u+v),$$

it becomes

$$\Pi(u, a) + \Pi(v, a) - \Pi(u+v, a) = \tfrac{1}{2}\log\Omega,$$

where

$$\Omega = \frac{\Theta(u-a)\Theta(v-a)\Theta(u+v+a)}{\Theta(u+a)\Theta(v+a)\Theta(u+v-a)}, \quad \text{............(10)}$$

and Ω is capable of being expressed in a great variety of ways by means of the elliptic functions cn, sn, dn of combinations of u, v, a.

For $\left\{\frac{\Theta(u-a)\Theta(v-a)}{\Theta 0}\right\}^2 = \frac{\Theta(u-v)\Theta(u+v-2a)}{1-\kappa^2\text{sn}^2(u-a)\text{sn}(v-a)},$

$$\left\{\frac{\Theta(u+a)\Theta(v+a)}{\Theta 0}\right\}^2 = \frac{\Theta(u-v)\Theta(u+v+2a)}{1-\kappa^2\text{sn}^2(u+a)\text{sn}^2(v+a)},$$

$$\left\{\frac{\Theta a\Theta(u+v-a)}{\Theta 0}\right\}^2 = \frac{\Theta(u+v)\Theta(u+v-2a)}{1-\kappa^2\text{sn}^2 a\,\text{sn}^2(u+v-a)},$$

$$\left\{\frac{\Theta a\Theta(u+v+a)}{\Theta 0}\right\}^2 = \frac{\Theta(u+v)\Theta(u+v+2a)}{1-\kappa^2\text{sn}^2 a\,\text{sn}^2(u+v+a)},$$

(§ 188), so that (*Fundamenta Nova*, § 54)

$$\Omega^2 = \frac{1-\kappa^2\text{sn}^2(u+a)\text{sn}^2(v+a)}{1-\kappa^2\text{sn}^2(u-a)\text{sn}^2(v-a)}\,\frac{1-\kappa^2\text{sn}^2 a\,\text{sn}^2(u+v-a)}{1-\kappa^2\text{sn}^2 a\,\text{sn}^2(u+v+a)}. \quad\ldots(11)$$

One of the simplest expressions, equivalent to that given above in (9) in Legendre's notation, is

$$\Omega = \frac{1-\kappa^2\text{sn}\,u\,\text{sn}\,v\,\text{sn}\,a\,\text{sn}(u+v-a)}{1+\kappa^2\text{sn}\,u\,\text{sn}\,v\,\text{sn}\,a\,\text{sn}(u+v+a)}, \quad\ldots\ldots\ldots\ldots(12)$$

and a systematic collection of different forms of Ω is given by Glaisher (*Messenger of Mathematics*, X.).

190. According as Legendre's a or $(1+n)(1+\kappa^2/n)$ is positive or negative, so his Integral of the Third Kind $\Pi(n, \kappa, \phi)$ falls into one of two classes, the first called *circular*, the second *logarithmic*, or *hyperbolic*, as we shall call it.

In the corresponding classification of Jacobi's form, the parameter a is *imaginary* or *real*; and it is remarkable that in dynamical problems, it is the *circular* form, with *imaginary* Jacobian parameter a, which is of almost invariable occurrence.

When Legendre's

$$a \text{ or } (1+n)(1+\kappa^2/n)$$

is positive, and the corresponding Elliptic Integral of the Third Kind is *circular*, then Jacobi's parameter is imaginary; and

(i.) with n positive, we must put $n=-\kappa^2\text{sn}^2 ia$;

(ii.) $-\kappa^2 > n > -1$, we must, according to § 56, put

$$n = -\kappa^2\text{sn}^2(K+ib),$$

as in § 114; and now the integral is expressed by

$$\Pi(u, ia) \text{ or } \Pi(u, K+ib),$$

involving Theta and Zeta functions of the imaginary arguments ia or $K+ib$; for which there is no theorem, short of expansion, to express the result in a real form.

We shall find however, in the applications, that this imaginary form constitutes no real practical drawback.

Taking for example the result of § 114, then, by (6) § 188,

$$\rho = h\sqrt{\left(\frac{A-D\,.\,D-C}{AC}\right)}\sqrt{\{\Theta(u+a)\Theta(u-a)\}}\frac{\Theta 0}{\Theta u \Theta a},$$

with $u=nt$, and $a=K+t'iK'$; while

$$i(\phi-\mu t)=\frac{\operatorname{cn} a \operatorname{dn} a}{\operatorname{sn} a}u+\Pi(u,a),$$

$$\exp i(\phi-\mu t)=\exp\left(\frac{\operatorname{cn} a \operatorname{dn} a}{\operatorname{sn} a}u+u\mathrm{Z}a\right)\sqrt{\frac{\Theta(u-a)}{\Theta(u+a)}};$$

so that, by multiplication,

$$(x+iy)(\cos\mu t-i\sin\mu t),\text{ or }\rho\exp i(\phi-\mu t)$$

$$=h\sqrt{\left(\frac{A-D\,.\,D-C}{AC}\right)}\frac{\Theta(u-a)\Theta 0}{\Theta u\Theta a}\exp\left(\frac{\operatorname{cn} a \operatorname{dn} a}{\operatorname{sn} a}+\mathrm{Z}a\right)u;\;\ldots(13)$$

which, when resolved into its real and imaginary part, gives the vector of the herpolhode, or its coordinates with respect to axes resolving with constant angular velocity μ.

191. Take Jacobi's $\Pi(u, a)$, and split up the quantity under the sign of integration into a quotient and partial fractions; therefore

$$\frac{1}{2}\frac{\operatorname{cn} a \operatorname{dn} a}{\operatorname{sn} a}\left\{\int\frac{du}{1-\kappa\operatorname{sn} a\operatorname{sn} u}+\int\frac{du}{1+\kappa\operatorname{sn} a\operatorname{sn} u}\right\}$$

$$=u\operatorname{cn} a\operatorname{dn} a/\operatorname{sn} a+\Pi(u,a);$$

while

$$\frac{1}{2}\frac{\operatorname{cn} a \operatorname{dn} a}{\operatorname{sn} a}\left\{\int\frac{du}{1-\kappa\operatorname{sn} a\operatorname{sn} u}-\int\frac{du}{1+\kappa\operatorname{sn} a\operatorname{sn} u}\right\}$$

$$=\int\frac{\kappa\operatorname{cn} a\operatorname{dn} a\operatorname{sn} u}{1-\kappa^2\operatorname{sn}^2 a\operatorname{sn}^2 u}du$$

$$=\int\{\tfrac{1}{2}\kappa\operatorname{sn}(a+u)-\tfrac{1}{2}\kappa\operatorname{sn}(a-u)\}du$$

$$=\frac{1}{2}\log\frac{\operatorname{dn}(a+u)-\kappa\operatorname{cn}(a+u)}{\operatorname{dn}(a-u)+\kappa\operatorname{cn}(a-u)}\cdot\frac{\operatorname{dn} a+\kappa\operatorname{cn} a}{\operatorname{dn} a-\kappa\operatorname{cn} a}\;(\S 76).$$

Therefore, by addition and subtraction,

$$\frac{\operatorname{cn} a\operatorname{dn} a}{\operatorname{sn} a}\int_0\frac{du}{1-\kappa\operatorname{sn} a\operatorname{sn} u}=u\left(\mathrm{Z}a+\frac{\operatorname{cn} a\operatorname{dn} a}{\operatorname{sn} a}\right)$$

$$+\frac{1}{2}\log\frac{\Theta(a-u)}{\Theta(a+u)}\cdot\frac{\operatorname{dn}(a-u)-\kappa\operatorname{cn}(a-u)}{\operatorname{dn}(a+u)+\kappa\operatorname{cn}(a+u)}\cdot\frac{\operatorname{dn} a+\kappa\operatorname{cn} a}{\operatorname{dn} a-\kappa\operatorname{cn} a},$$

$$\frac{\operatorname{cn} a\operatorname{dn} a}{\operatorname{sn} a}\int\frac{du}{1+\kappa\operatorname{sn} a\operatorname{sn} u}=u\left(\mathrm{Z}a+\frac{\operatorname{cn} a\operatorname{dn} a}{\operatorname{sn} a}\right)$$

$$+\frac{1}{2}\log\frac{\Theta(a-u)}{\Theta(a+u)}\cdot\frac{\operatorname{dn}(a-u)+\kappa\operatorname{cn}(a-u)}{\operatorname{dn}(a+u)-\kappa\operatorname{cn}(a+u)}\cdot\frac{\operatorname{dn} a-\kappa\operatorname{cn} a}{\operatorname{dn} a+\kappa\operatorname{cn} a}.$$

192. Again, taking the formula (7), § 137,

$$\frac{\operatorname{sn}^2 a - \operatorname{sn}^2 u}{1 - \kappa^2 \operatorname{sn}^2 a \operatorname{sn}^2 u} = \operatorname{sn}(a+u)\operatorname{sn}(a-u),$$

and differentiating logarithmically with respect to a,

$$\frac{\operatorname{sn} a \operatorname{cn} a \operatorname{dn} a}{\operatorname{sn}^2 a - \operatorname{sn}^2 u} + \frac{\kappa^2 \operatorname{sn} a \operatorname{cn} a \operatorname{dn} a \operatorname{sn}^2 u}{1 - \kappa^2 \operatorname{sn}^2 a \operatorname{sn}^2 u}$$

$$= \frac{1}{2} \frac{\operatorname{cn}(a+u)\operatorname{dn}(a+u)}{\operatorname{sn}(a+u)} + \frac{1}{2} \frac{\operatorname{cn}(a-u)\operatorname{dn}(a-u)}{\operatorname{sn}(a-u)};$$

and then integrating with respect to u,

$$\int_0 \frac{\operatorname{sn} a \operatorname{cn} a \operatorname{dn} a\, du}{\operatorname{sn}^2 a - \operatorname{sn}^2 u} = \frac{1}{2} \log \frac{\operatorname{sn}(a+u)}{\operatorname{sn}(a-u)} - \Pi(u, a)$$

$$= -uZa + \frac{1}{2} \log \frac{\operatorname{sn}(a+u)}{\operatorname{sn}(a-u)} \frac{\Theta(a+u)}{\Theta(a-u)}$$

$$= -uZa + \frac{1}{2} \log \frac{\mathrm{H}(a+u)}{\mathrm{H}(a-u)}, \quad \ldots\ldots\ldots\ldots.(14)$$

introducing Jacobi's function $\mathrm{H}u$, called the *Eta Function*, defined by the equation (*Fundamenta Nova*, § 61),

$$\operatorname{sn} u = \frac{1}{\sqrt{\kappa}} \frac{\mathrm{H}u}{\Theta u}. \quad \ldots\ldots\ldots\ldots\ldots\ldots\ldots\ldots.(15)$$

This form (14) and Jacobi's $\Pi(u, a)$ are the two forms of the hyperbolic integral of the third kind to which Legendre's form can be reduced for negative values of a.

When $\quad 0 > n > -\kappa^2$, we put $n = -\kappa^2 \operatorname{sn}^2 a$,
and obtain Jacobi's form $\Pi(u, a)$ of (5).

When $\quad -1 > n > -\infty$, we put $n = -1/\operatorname{sn}^2 a$,
and obtain the above form (14).

This form again can be split up into partial fractions; and a similar procedure shows that, since

$$\int \frac{du}{\operatorname{sn} u} = \log \frac{\operatorname{sn} u}{\operatorname{dn} u + \operatorname{cn} u}, \text{ or } \log \frac{\operatorname{dn} u - \operatorname{cn} u}{\operatorname{sn} u},$$

therefore, by equations (4) and (7), § 137,

$$\int_0 \frac{\operatorname{cn} a \operatorname{dn} a \operatorname{sn} u\, du}{\operatorname{sn}^2 a - \operatorname{sn}^2 u}$$

$$= \tfrac{1}{2} \int \frac{\operatorname{sn}(a+u) - \operatorname{sn}(a-u)}{\operatorname{sn}(a+u)\operatorname{sn}(a-u)} du$$

$$= \tfrac{1}{2} \int \frac{du}{\operatorname{sn}(a-u)} - \tfrac{1}{2} \int \frac{du}{\operatorname{sn}(a+u)}$$

$$=\tfrac{1}{2}\log\frac{\operatorname{sn}(a+u)}{\operatorname{sn}(a-u)}\,\frac{\operatorname{dn}(a-u)+\operatorname{cn}(a-u)}{\operatorname{dn}(a+u)-\operatorname{cn}(a+u)}\,\frac{\operatorname{dn}a-\operatorname{cn}a}{\operatorname{dn}a+\operatorname{cn}a},$$

$$=\tfrac{1}{2}\log\frac{\operatorname{sn}(a-u)}{\operatorname{sn}(a+u)}\,\frac{\operatorname{dn}(a+u)+\operatorname{cn}(a+u)}{\operatorname{dn}(a-u)-\operatorname{cn}(a-u)}\,\frac{\operatorname{dn}a-\operatorname{cn}a}{\operatorname{dn}a+\operatorname{cn}a}. \tag{16}$$

Therefore, by addition and subtraction of (14) and (16),

$$\int_0 \frac{\operatorname{cn}a\operatorname{dn}a\,du}{\operatorname{sn}a-\operatorname{sn}u}$$

$$=-uZa+\tfrac{1}{2}\log\frac{\Theta(a+u)}{\Theta(a-u)}\,\frac{\operatorname{dn}(a+u)+\operatorname{cn}(a+u)}{\operatorname{dn}(a-u)-\operatorname{cn}(a-u)}\,\frac{\operatorname{dn}a-\operatorname{cn}a}{\operatorname{dn}a+\operatorname{cn}a},$$

$$\int_0 \frac{\operatorname{cn}a\operatorname{dn}a\,du}{\operatorname{sn}a+\operatorname{sn}u}$$

$$=-uZa+\tfrac{1}{2}\log\frac{\Theta(a+u)}{\Theta(a-u)}\,\frac{\operatorname{dn}(a+u)-\operatorname{cn}(a+u)}{\operatorname{dn}(a-u)+\operatorname{cn}(a-u)}\,\frac{\operatorname{dn}a+\operatorname{cn}a}{\operatorname{dn}a-\operatorname{cn}a}.$$

By means of equation (6), § 188, and the formulas of § 123, these relations may be written

$$\int_0 \frac{\operatorname{cn}a\operatorname{dn}a\,du}{\operatorname{sn}a-\operatorname{sn}u}$$

$$=-uZa+\log\frac{\Theta^2\frac{1}{2}(a+u)}{\Theta^2\frac{1}{2}(a-u)}\,\frac{\operatorname{sn}\frac{1}{2}a\operatorname{cn}\frac{1}{2}(a+u)\operatorname{dn}\frac{1}{2}(a+u)}{\operatorname{sn}\frac{1}{2}(a-u)\operatorname{cn}\frac{1}{2}a\operatorname{dn}\frac{1}{2}a},$$

$$\int_0 \frac{\operatorname{cn}a\operatorname{dn}a\,du}{\operatorname{sn}a+\operatorname{sn}u}$$

$$=-uZa+\log\frac{\Theta^2\frac{1}{2}(a+u)}{\Theta^2\frac{1}{2}(a-u)}\,\frac{\operatorname{sn}\frac{1}{2}(a+u)\operatorname{cn}\frac{1}{2}a\operatorname{dn}\frac{1}{2}a}{\operatorname{sn}\frac{1}{2}a\operatorname{cn}\frac{1}{2}(a-u)\operatorname{dn}\frac{1}{2}(a-u)}.$$

The student may prove, by a similar procedure, that

$$\int\frac{\operatorname{sn}a\operatorname{dn}a\,du}{\operatorname{cn}u-\operatorname{cn}a}=\tfrac{1}{2}\log\frac{1-\operatorname{cn}(a+u)}{1-\operatorname{cn}(a-u)}-\Pi(u,a),$$

$$\int\frac{\operatorname{sn}a\operatorname{dn}a\,du}{\operatorname{cn}u+\operatorname{cn}a}=\tfrac{1}{2}\log\frac{1+\operatorname{cn}(a-u)}{1+\operatorname{cn}(a+u)}+\Pi(u,a),$$

$$\int\frac{\kappa^2\operatorname{sn}a\operatorname{cn}a\,du}{\operatorname{dn}u-\operatorname{dn}a}=\tfrac{1}{2}\log\frac{1-\operatorname{dn}(a+u)}{1-\operatorname{dn}(a-u)}-\Pi(u,a),$$

$$\int\frac{\kappa^2\operatorname{sn}a\operatorname{cn}a\,du}{\operatorname{dn}u+\operatorname{dn}a}=\tfrac{1}{2}\log\frac{1+\operatorname{dn}(a-u)}{1+\operatorname{dn}(a+u)}+\Pi(u,a),$$

$$\int\frac{\operatorname{sn}a\operatorname{cn}a\operatorname{dn}a-\operatorname{sn}u\operatorname{cn}u\operatorname{dn}u}{\operatorname{sn}^2a-\operatorname{sn}^2u}du=\log\frac{\operatorname{sn}(a+u)\Theta(a+u)\Theta0}{\Theta a\Theta u}e^{-uZa},$$

$$\int\frac{\operatorname{sn}u\operatorname{dn}u-\operatorname{sn}a\operatorname{dn}a}{\operatorname{cn}u-\operatorname{cn}a}du=uZa-\log\frac{\Theta(a+u)\Theta0}{\Theta a\Theta u}\,\frac{1-\operatorname{cn}(a+u)}{1-\operatorname{cn}a}.$$

Euler's Pendulum.

193. Consider for instance the rolling oscillations on a horizontal plane of a body with a cylindrical base, such as a rocking stone, or a cradle.

Then the Principle of Energy, considering the line of contact as the instantaneous axis of rotation, leads to the equation

$$\tfrac{1}{2}(c^2-2ch\cos\theta+h^2+k^2)(d\theta/dt)^2=gh(\text{vers}\,\alpha-\text{vers}\,\theta),$$

where θ denotes the inclination to the vertical of the plane through the axis and the centre of gravity at any time t, α the extreme value of θ, c the radius of the cylindrical surface, h the distance of the C. G. from the axis of the cylinder, and k the radius of gyration about the parallel axis through the C. G.

When $c=0$, this equation reduces to ordinary pendulum oscillations, as in (3) § 3; but in the general case we have the oscillations of what is sometimes called Euler's Pendulum.

Then
$$\frac{dt^2}{d\theta^2}=\frac{\{(c-h)^2+k^2\}\cos^2\tfrac{1}{2}\theta+\{(c+h)^2+k^2\}\sin^2\tfrac{1}{2}\theta}{4gh(\sin^2\tfrac{1}{2}\alpha-\sin^2\tfrac{1}{2}\theta)}$$
$$=\frac{(c-h)^2+k^2+\{(c+h)^2+k^2\}\tan^2\tfrac{1}{2}\theta}{4gh\cos^2\tfrac{1}{2}\alpha(\tan^2\tfrac{1}{2}\alpha-\tan^2\tfrac{1}{2}\theta)};$$

and now, if we put

$$\tan\tfrac{1}{2}\theta=\tan\tfrac{1}{2}\alpha\cos\phi,$$

$$\frac{dt^2}{d\phi^2}=\frac{c^2-2ch\cos\alpha+h^2+k^2-\{(c+h)^2+k^2\}\sin^2\tfrac{1}{2}\alpha\sin^2\phi}{gh(1-\sin^2\tfrac{1}{2}\alpha\sin^2\phi)^2},$$

or
$$n\frac{dt}{d\phi}=\sqrt{\left(\frac{c^2-2ch\cos\alpha+h^2+k^2}{ch}\right)}\frac{\Delta\phi}{1-\sin^2\tfrac{1}{2}\alpha\sin^2\phi},$$

on putting $n^2=g/c$, and

$$\kappa^2=\frac{(c+h)^2+k^2}{c^2-2ch\cos\alpha+h^2+k^2}\sin^2\tfrac{1}{2}\alpha,\ \kappa'^2=\frac{(c-h)^2+k^2}{c^2-2ch\cos\alpha+h^2+k^2}\cos^2\tfrac{1}{2}\alpha.$$

To reduce this to Jacobi's canonical form, put $\phi=\text{am}\,u$, and $\sin^2\tfrac{1}{2}\alpha=\kappa^2\text{sn}^2a$; then $\text{dn}^2a=\cos^2\tfrac{1}{2}\alpha$,

and
$$\text{sn}^2a=\frac{c^2-2ch\cos\alpha+h^2+k^2}{(c+h)^2+k^2},\ \text{cn}^2a=\frac{4ch\cos^2\tfrac{1}{2}\alpha}{(c+h)^2+k^2};$$

so that
$$n\frac{dt}{du}=2\frac{\text{sn}\,a\,\text{dn}\,a}{\text{cn}\,a}\frac{\text{dn}^2u}{1-\kappa^2\text{sn}^2a\,\text{sn}^2u}$$
$$=2\frac{\text{sn}\,a\,\text{dn}\,a}{\text{cn}\,a}-\frac{2\kappa^2\text{sn}\,a\,\text{cn}\,a\,\text{dn}\,a\,\text{sn}^2u}{1-\kappa^2\text{sn}^2a\,\text{sn}^2u},$$

and
$$nt=2\frac{\text{sn}\,a\,\text{dn}\,a}{\text{cn}\,a}u-2\Pi(u,a)$$

while $\tan\tfrac{1}{2}\theta=\tan\tfrac{1}{2}\alpha\,\text{cn}\,u$.

In the ordinary pendulum, where $c=0$, this reduces, as in § 8, to

$$\tan\tfrac{1}{2}\theta=\tan\tfrac{1}{2}\alpha\ \mathrm{cn}(K-nt),$$

equivalent to

$$\sin\tfrac{1}{2}\theta=\sin\tfrac{1}{2}\alpha\ \mathrm{sn}\ nt\,;$$

where n now denotes $\surd\{gh/(h^2+k^2)\}$.

As another application of the Third Elliptic Integral the student may rectify the inverse (or pedal) of an ellipse or hyperbola, with respect to any point; examining the particular case when the point is the centre; also the case of the Lemniscate, the inverse or pedal of a rectangular hyperbola, with respect to the centre (R. A. Roberts, *Integral Calculus*, p. 310).

Examples.

1. Prove that, if $k+k'=1$,

$$\int_0^{\frac{1}{2}\pi}\int_0^{\frac{1}{2}\pi}\frac{(k\cos^2\phi+k'\cos^2\psi)d\phi d\psi}{\surd(1-k\sin^2\phi)\surd(1-k'\sin^2\psi)}=\tfrac{1}{2}\pi\,;$$

and deduce *Legendre's relation* of § 171.

2. $$\int_0^{1/k}\int_0^1\frac{k(y-x)dxdy}{\surd(x\,.\,1-x\,.\,1-kx)\surd(-y\,.\,1-y\,.\,1-ky)}=2\pi.$$

3. $$\int_1^{1/\kappa}\int_{-1}^1\frac{y-x}{(1+\kappa x)(1+\kappa y)}\ \frac{dxdy}{\surd(1-x^2.1-\kappa^2x^2)\surd(y^2-1.1-\kappa^2y^2)}=\frac{\pi}{\kappa\kappa'^2}$$ (§ 66).

4. $$\int_{e_2}^{e_1}\int_{e_3}^{e_2}\frac{(y-x)dxdy}{\surd(4.x-e_1.x-e_2.x-e_3)\surd(-4.y-e_1.y-e_2.y-e_3)}=\tfrac{1}{2}\pi$$ (§ 51).

5. $$\int_a^{a+H}\int_{a-H}^a\frac{(y-x)dxdy}{\surd\{(a-x).(x-m)^2+n^2\}\surd\{(y-a).(y-m)^2+n^2\}}=4\pi$$ (§ 47).

6. $$\int_\gamma^\delta\int_\beta^\gamma\frac{(\beta-a)(\gamma-a)(\delta-a)(y-x)dxdy}{(x-a)(y-a)\surd(-XY)}=2\pi$$ (§ 153).

7. Denoting $K-E$, $K'-E'$, $E-\kappa'^2K$, $E'-\kappa^2K'$ by J, J', G, G' respectively (Glaisher, *Q. J. M.*, XX.), prove that

$$\tfrac{1}{2}\pi=\kappa\kappa'^2\left(K'\frac{dK}{d\kappa}-K\frac{dK'}{d\kappa}\right)=\frac{\kappa'^2}{\kappa}\left(E'\frac{dJ}{d\kappa}-J\frac{dE'}{d\kappa}\right)$$

$$=\kappa\ \left(J'\frac{dE}{d\kappa}-E\frac{dJ'}{d\kappa}\right)=\frac{1}{\kappa}\ \left(G\frac{dG'}{d\kappa}-G'\frac{dG}{d\kappa}\right).$$

CHAPTER VII.

ELLIPTIC INTEGRALS IN GENERAL, AND THEIR APPLICATIONS.

194. The general algebraical function, the integral of which leads to elliptic integrals, is of the form

$$\frac{S+T\sqrt{X}}{U+V\sqrt{X}},$$

where S, T, U, V are rational integral algebraical functions of x, and X is of the third or fourth degree in x.

We first rationalize the denominator, so that

$$\frac{S+T\sqrt{X}}{U+V\sqrt{X}}=\frac{(S+T\sqrt{X})(U-V\sqrt{X})}{U^2-V^2X}=\frac{M}{D}+\frac{N}{D}\frac{1}{\sqrt{X}},$$

suppose; and now the integration of the rational part M/D is effected by elementary methods, when it is resolved into its quotient and partial fractions.

In the irrational part $N/D\sqrt{X}$, the rational fraction N/D is also resolved, into a quotient, having a typical term x^m, and into partial fractions, having typical terms

$$1/(x-a) \text{ or } 1/(x-a)^n.$$

By differentiation, we find that

$$\frac{d}{dx}(x^{m-3}\sqrt{X})=\{(m-1)ax^m+4(m-\tfrac{3}{2})bx^{m-1}+6(m-2)cx^{m-2}$$
$$+4(m-\tfrac{5}{2})dx^{m-3}+(m-3)ex^{m-4}\}/\sqrt{X};$$

so that, integrating, and denoting $\int x^m dx/\sqrt{X}$ by u_m,

$$x^{m-3}\sqrt{X}=(m-1)au_m+4(m-\tfrac{3}{2})bu_{m-1}+6(m-2)cu_{m-2}$$
$$+4(m-\tfrac{5}{2})du_{m-3}+(m-3)eu_{m-4},$$

a formula of reduction by means of which the integral u_m is made to depend ultimately on the integrals u_2, u_1, and u_0.

Similarly, by differentiation and integration, denoting

$$\int dx/(x-a)^n\sqrt{X} \text{ by } v_n,$$

we can determine another formula of reduction, of the form

$$\frac{\sqrt{X}}{(x-a)^{n-1}}=Av_n+Bv_{n-1}+Cv_{n-2}+Dv_{n-3}+Ev_{n-4},$$

by means of which the integral v_n is made to depend ultimately on the integrals v_1, v_0, v_{-1}, and v_{-2}; or rather, on v_1, u_0, u_1, u_2; since v_0 and u_0 are the same, and

$$v_{-1}=u_1-au_0, \quad v_{-2}=u_2-2au_1+a^2u_0.$$

By the various substitutions of Chapter II., u_0 is reduced to Legendre's First Elliptic Integral, while at the same time the integrals u_1, u_2, and v_1 are reduced to elliptic integrals of the Second and Third Kind.

When $x-a$ is a factor of X, the substitution $x-a=1/y$ shows that v_1 becomes $\int ydy/\sqrt{Y}$, where Y is a cubic function of y, and v_1 now reduces to the Second Elliptic Integral.

But without carrying out this work in detail, now only of antiquarian interest, we adopt instead the Weierstrassian notation; and by means of the substitutions of the previous chapter we express x and $\sqrt{X}$ rationally in terms of $\wp u$ and $\wp' u$; so that the integration is reduced ultimately to that of $A+B\wp' u$ with respect to u, A and B being rational functions of $\wp u$.

195. We must at this stage introduce the functions

$$\zeta u \text{ and } \sigma u,$$

the functions employed by Weierstrass, in conjunction with his function $\wp u$.

The function ζu, called the *zeta* function, is defined by

$$\zeta' u=-\wp u, \text{ or } \zeta u=-\int \wp u du;$$

while the function σu, called the *sigma* function, is defined by

$$\frac{d}{du}\log \sigma u=\zeta u,$$

or

$$\log \sigma u=\int \zeta u du, \quad \sigma u=\exp\int \zeta u du;$$

and thus

$$\frac{d^2 \log \sigma u}{du^2}=-\wp u.$$

Taking the definition of s or $\wp u$ in § 50,

$$u=\int_{\wp u}^{\infty}(4s^3-g_2{}^3-g_3)^{-\frac{1}{2}}ds,$$

expand in descending powers of s, and integrate; then

$$u=\int_{s}^{\infty}\tfrac{1}{2}s^{-\frac{3}{2}}(1-\tfrac{1}{4}g_2s^{-2}-\tfrac{1}{4}g_3s^{-3})^{-\frac{1}{2}}ds$$

$$=\int_{s}^{\infty}(\tfrac{1}{2}s^{-\frac{3}{2}}+\ * \ +\tfrac{1}{16}g_2s^{-\frac{7}{2}}+\tfrac{1}{16}g_3s^{-\frac{9}{2}}+\dots)ds$$

$$=\quad s^{-\frac{1}{2}}+\ * \ +\frac{g_2}{40}\ s^{-\frac{5}{2}}+\frac{g_3}{56}\ s^{-\frac{7}{2}}+\dots,$$

the * marking the place of a missing term in the expansion.

Therefore, by Reversion of Series, since u^2 is a rational function of s, we obtain, in the neighbourhood of $u=0$,

$$s \text{ or } \wp u=\frac{1}{u^2}+\ * \ +\frac{g_2u^2}{20}+\frac{g_3u^4}{28}+\dots.$$

To obtain further terms of the expansion, assume

$$\wp u=\frac{1}{u^2}+\ * \ +c_1u^2+c_2u^4+c_3u^6+\dots+c_nu^{2n}+\dots;$$

and since $\wp'^2u=4\wp^3u-g_2\wp u-g_3$,

$$\wp''u=6\wp^2u-\tfrac{1}{2}g_2,$$

$$\wp'''u=12\wp u\wp' u,$$

we can obtain from the last equation a recurring formula for the determination of the coefficients c; and as far as u^8,

$$\wp u=\frac{1}{u^2}+\ * \ +\frac{g_2u^2}{20}+\frac{g_3u^4}{28}+\frac{g_2{}^2u^6}{2^4.3.5^2}+\frac{3g_2g_3u^8}{2^4.5.7.11}+\dots.$$

The expansion of the zeta function is now

$$\zeta u=\frac{1}{u}+\ * \ -\frac{g_2u^3}{60}-\frac{g_3u^5}{140}-\frac{g_2{}^2u^7}{2^4.3.5^2.7}-\frac{g_2g_3u^9}{2^4.3.5.7.11}-\dots;$$

so that, defined more strictly,

$$\zeta u=\frac{1}{u}+\int_0\left(\frac{1}{u^2}-\wp u\right)du.$$

Similarly we shall find, for the sigma function,

$$\sigma u=u+\ * \ -\frac{g_2u^5}{2^4.3.5}-\frac{g_3u^7}{2^3.3.5.7}-\frac{g_2{}^2u^9}{2^9.3^2.5.7}-\frac{g_2g_3u^{11}}{2^7.3^2.5^2.7.11}-\dots,$$

so that, strictly defined,

$$\log\sigma u=\log u+\int_0\left(\zeta u-\frac{1}{u}\right)du, \text{ or } \sigma u=u\exp\int_0\left(\zeta u-\frac{1}{u}\right)du.$$

Homogeneity.

196. From considerations of *homogeneity* it follows, that if u is changed into u/m, and at the same time if g_2 and g_3 are changed into $m^4 g_2$ and $m^6 g_3$, then s or $\wp u$ is changed into $m^2 s$ or $m^2 \wp u$; so that

$$\wp(u;\ g_2, g_3) = \frac{1}{m^2}\wp\left(\frac{u}{m};\ m^4 g_2, m^6 g_3\right);$$

$$\wp'(u;\ g_2, g_3) = \frac{1}{m^3}\wp'\left(\frac{u}{m};\ m^4 g_2, m^6 g_3\right);$$

and similarly

$$\zeta(u;\ g_2, g_3) = \frac{1}{m}\zeta\left(\frac{u}{m};\ m^4 g_2, m^6 g_3\right),$$

$$\sigma(u;\ g_2, g_3) = m\,\sigma\left(\frac{u}{m};\ m^4 g_2, m^6 g_3\right).$$

At the same time the discriminant Δ becomes changed to $m^{12}\Delta$, but the absolute invariant J is left unchanged (§ 53); we may in this manner alter the argument u proportionally; for instance by taking $m = \sqrt[4]{(e_1 - e_3)}$ we can make the argument the same as in the corresponding elliptic functions (§ 51).

When m is chosen so that $m^{12}\Delta = 1$, or $m = \Delta^{-\frac{1}{12}}$, the elliptic integral is said to be *normalised* (Klein).

Suppose, for instance, that $g_2 = 0$, and m, m^2 are the imaginary cube roots of unity, $-\frac{1}{2} \pm \frac{1}{2} i\sqrt{3}$;

then $$m^3 = 1, \text{ and } u/m = m^2 u;$$

so that $$\wp(m^2 u;\ 0, g_3) = m^2 \wp(u;\ 0, g_3),$$
$$\wp(m\,u;\ 0, g_3) = m\,\wp(u;\ 0, g_3),$$

while $$\wp' u = \wp' m u = \wp' m^2 u.$$

Again $$\zeta(u;\ 0, g_3) = \frac{1}{m}\zeta\frac{u}{m} = \frac{1}{m^2}\zeta\frac{u}{m^2},$$

$$\sigma(u;\ 0, g_3) = m\sigma\frac{u}{m} = m^2\sigma\frac{u}{m^2}.$$

This is the simplest illustration of the theory of *Complex Multiplication of Elliptic Functions*, of which we shall make use hereafter; the general theory is required in the integration of the equation

$$\frac{M dy}{\sqrt{(4y^3 - g_2 y - g_3)}} = \frac{dx}{\sqrt{(4x^3 - g_2 x - g_3)}}$$

for particular numerical values of g_2 and g_3, when $1/M$ is a complex number of the form $a + ib\sqrt{n}$; in this instance $g_2 = 0$, and M is an imaginary cubic root of unity.

197. With the aid of these three functions of Weierstrass, $\wp u$, ζu, and σu, it is possible to express any elliptic integral, and we can thus complete the problem left unfinished in § 194.

The function ζu is analogous to Jacobi's Zeta function; and with $s = \wp u$, it may be defined by the relation

$$\zeta u = \int \frac{s\,ds}{\sqrt{S}} = \int (4s^3 - g_2 s - g_3)^{-\frac{1}{2}} s\,ds$$

$$= \int (\tfrac{1}{2}s^{-\frac{1}{2}} + * + \tfrac{1}{16}g_2 s^{-\frac{5}{2}} + \tfrac{1}{16}g_3 s^{-\frac{7}{2}} + \ldots)ds$$

$$= \quad s^{\frac{1}{2}} + * - \tfrac{1}{24}g_2 s^{-\frac{3}{2}} - \tfrac{1}{40}g_3 s^{-\frac{5}{2}} - \ldots.$$

Thus, for instance, from § 153, with appropriate limits,

$$\zeta u = \int \tfrac{1}{12}a \frac{a-\beta\,.\,a-\gamma\,.\,a-\delta}{x-a}\left(\frac{x-\beta}{a-\beta} + \frac{x-\gamma}{a-\gamma} + \frac{x-\delta}{a-\delta}\right)\frac{dx}{\sqrt{X}},$$

where $$u = \int \frac{dx}{\sqrt{X}}.$$

To obtain the Addition Equation of the zeta function analogous to (2) and (3) of § 186, take the formula (F) of § 144,

$$\wp u + \wp v + \wp(u+v) = \tfrac{1}{4}\left(\frac{\wp' u - \wp' v}{\wp u - \wp v}\right)^2;$$

implying also the formula, obtained by changing the sign of v,

$$\wp u + \wp v + \wp(u-v) = \tfrac{1}{4}\left(\frac{\wp' u + \wp' v}{\wp u - \wp v}\right)^2;$$

so that, by subtraction,

$$\wp(u-v) - \wp(u+v) = \frac{\wp' u\,\wp' v}{(\wp u - \wp v)^2}. \qquad \ldots\ldots\ldots\ldots\ldots\ldots(\alpha)$$

Integrating (α) with respect to v,

$$\zeta(u-v) + \zeta(u+v) + C = \frac{\wp' u}{\wp u - \wp v},$$

where C, the arbitrary constant of integration, may be obtained by putting $v=0$, when $\wp v = \infty$; so that $C = -2\zeta u$, and

$$\zeta(u-v) + \zeta(u+v) - 2\zeta u = \frac{\wp' u}{\wp u - \wp v}. \qquad \ldots\ldots\ldots\ldots\ldots\ldots(\beta)$$

An interchange of u and v gives

$$-\zeta(u-v) + \zeta(u+v) - 2\zeta v = \frac{-\wp' v}{\wp u - \wp v}; \qquad \ldots\ldots\ldots\ldots\ldots\ldots(\beta')$$

so that, by addition,

$$\zeta(u+v) - \zeta u - \zeta v = \tfrac{1}{2}\frac{\wp' u - \wp' v}{\wp u - \wp v} \qquad \ldots\ldots\ldots\ldots\ldots\ldots(\gamma)$$

$$= \sqrt{\{\wp u + \wp v + \wp(u+v)\}}$$

the Addition Equation, analogous to (2*) § 186.

With $$u+v+w=0,$$
this may be written, analogous to § 176,

$$\zeta u+\zeta v+\zeta w=-\sqrt{(\wp u+\wp v+\wp w)}.$$

198. We can now take the function $A+B\wp' u$ of § 194, and suppose that A and B are resolved into their quotient and partial fractions.

Writing $p, p', p'', \dots$ for $\wp u$ and its successive derivatives, then the relations

$$p'^2 = 4p^3-g_2p-g_3$$
$$p'' = 6p^2-\tfrac{1}{2}g_2,$$
$$p'''=12pp', \text{ etc.},$$

enable us to express the quotient or integral part of $A+B\wp' u$ in the form

$$C=c_0+c_1\wp u+c_2\wp' u+c_3\wp'' u+\dots.$$

Considering next a partial fraction of $A+B\wp' u$ of the form

$$\frac{P+Q\wp' u}{\wp u-a},$$

we replace a by $\wp v$, and write the partial fraction in the form

$$H\frac{\wp' u-\wp' v}{\wp u-\wp v}+K\frac{\wp' u+\wp' v}{\wp u-\wp v}$$
$$=2H\{\zeta(u+v)-\zeta u-\zeta v\}+2K\{\zeta(u-v)-\zeta u+\zeta v\}.$$

All such partial fractions can thus be expressed by a series of terms,

$$L=l_1\zeta(u-v_1)+l_2\zeta(u-v_2)+l_3\zeta(u-v_3)+\dots,$$

where the sum of the coefficients l is zero for each partial fraction, and therefore for the whole series; so that

$$l_1+l_2+l_3+\dots=0.$$

Again, by repeated differentiation of equations (β) and (β') (§ 197), with respect to u or v, we obtain equations, such as

$$\frac{\wp' u\,\wp' v}{(\wp u-\wp v)^2}=\wp(u+v)-\wp(u+v),$$

$$\frac{\wp'^2 v}{(\wp u-\wp v)^2}=\wp(u+v)+\wp(u-v)+2\wp v-\frac{\wp'' v}{\wp u-\wp v},$$

by means of which partial fractions of the form

$$\frac{P+Q\wp' u}{(\wp u-\wp v)^2}, \text{ or generally } \frac{P+Q\wp' u}{(\wp u-\wp v)^n},$$

can be expressed by terms of the form $\wp(u+v)$, $\wp(u-v)$, and by their derivatives; as well as by terms of the form L and C.

Thus, finally, $A+B\wp' u$, or any rational function of $\wp u$ and $\wp' u$, can always be expressed as the sum $L+P$ of two series of terms, $\quad L=l_1\zeta(u-v_1)+l_2\zeta(u-v_2)+l_3\zeta(u-v_3)+\dots,$

where $\quad l_1+l_2+l_3+\dots=0,$

and $\quad P=c+\Sigma m\,\wp^{(r)}(u-v);$

and now the integral can immediately be written down, involving, in general, the sigma, zeta, and $\wp$ function, as well as its derivatives.

When the sigma and zeta functions are absent, the integral is a function of $\wp u$ and $\wp' u$, and is not properly elliptic, but only algebraical.

This method of integration is taken from Halphen's *Fonctions Elliptiques*, I., chap. vii.

Halphen points out that to obtain the coefficients in the series of terms

$$l\zeta(u-v)+m_0\wp(u-v)+m_1\wp'(u-v)+m_2\wp''(u-v)+\dots,$$

corresponding to the same v, it is only necessary to take the coefficients of $(u-v)^{-1}, (u-v)^{-2}, (u-v)^{-3}, \dots$ in the expansion of $A+B\wp' u$ in *ascending* powers of $u-v$; the coefficient l being Cauchy's *residue*.

199. Integrating (β) with respect to v, then

$$\int_0 \frac{\wp' u\,dv}{\wp u-\wp v}=\log\frac{\sigma(u+v)}{\sigma(u-v)}-2v\zeta u, \ldots\ldots\ldots\ldots\ldots(\beta_1)$$

which may be considered a canonical form of the Third Elliptic Integral, in Weierstrass's notation.

Thus, for instance, in § 113,

$$i\phi'=\tfrac{1}{2}\int\frac{\wp' v\,du}{\wp u-\wp v}$$

$$=\tfrac{1}{2}\log\frac{\sigma(u+v)}{\sigma(u-v)}-u\zeta v,$$

or

$$e^{i\phi'}=e^{-u\zeta v}\sqrt{\frac{\sigma(u+v)}{\sigma(u-v)}}.$$

By integration of (γ), with respect to u and v,

$$\int\frac{1}{2}\frac{\wp' u-\wp' v}{\wp u-\wp v}du=\log\frac{\sigma(u+v)}{\sigma u\,\sigma v}-u\zeta v=\log\frac{\sigma(u+v)}{\sigma u\,\sigma v}e^{-u\zeta v}, \ldots(\gamma_1)$$

$$\int\frac{1}{2}\frac{\wp' u-\wp' v}{\wp u-\wp v}dv=\log\frac{\sigma(u+v)}{\sigma u\,\sigma v}-v\zeta u=\log\frac{\sigma(u+v)}{\sigma u\,\sigma v}e^{-v\zeta u}; \ldots(\gamma_2)$$

either of which may be taken as a canonical form of the Third Elliptic Integral; and also as illustrating the interchange of *amplitude* u and *parameter* v, as in the Jacobian Elliptic Integral of the Third Kind, $\Pi(u, v)$, in § 188.

Or otherwise, interchanging u and v in (β_1), or integrating (β'),

$$\int \frac{\wp' v\, du}{\wp u - \wp v} = \log \frac{\sigma(u-v)}{\sigma(u+v)} + 2u\zeta v, \ldots\ldots\ldots\ldots (\beta_2)$$

so that, by addition of (β_1) and (β_2),

$$\int \frac{\wp' u\, dv + \wp' v\, du}{\wp u - \wp v} = 2u\zeta v - 2v\zeta u, \ldots\ldots\ldots\ldots (\delta)$$

a form of the theorem of the interchange of *amplitude* and *parameter*, analogous to (8), § 188.

200. Integrating (β) with respect to u,

$$\log \frac{\sigma(v-u)}{\sigma v} + \log \frac{\sigma(v+u)}{\sigma v} - 2 \log \sigma u = \log(\wp u - \wp v),$$

or

$$\frac{\sigma(v+u)\sigma(v-u)}{\sigma^2 v\, \sigma^2 v} = \wp u - \wp v,$$

$$\frac{\sigma(u+v)\sigma(u-v)}{\sigma^2 u\, \sigma^2 u} = \wp v - \wp u,$$

$$= \frac{d^2}{du^2} \log \sigma u - \frac{d^2}{dv^2} \log \sigma v \ldots (\mathrm{K});$$

the fundamental formula is the use of Weierstrass's elliptic function, analogous to equation (6) of § 188.

As an application consider the herpolhode of 113; then

$$\rho = \frac{nh}{\mu} \sqrt{(\wp v - \wp u)} = \frac{nh}{\mu} \sqrt{\frac{\sigma(u+v)\sigma(u-v)}{\sigma^2 u \sigma^2 v}},$$

while

$$e^{i\phi'} = \sqrt{\frac{\sigma(u+v)}{\sigma(u-v)}} e^{-u\zeta v};$$

so that, in the curve described by H,

$$x + iy = \rho e^{i\phi'} = \frac{nh}{\mu} \frac{\sigma(u+v)}{\sigma u \sigma v} e^{-u\zeta v},$$

while in the herpolhode described by P we must multiply this function by $e^{i\mu t}$ or $\cos \mu t + i \sin \mu t$.

Putting $u = v$ in (K), we obtain

$$\frac{\sigma 2u}{\sigma^4 u} = -\mathrm{lt} \frac{\wp u - \wp v}{\sigma(u-v)} = -\wp' u.$$

This may be obtained by integration of the formula of § 149,

$$\wp 2u = \wp u - \frac{1}{4} \frac{d^2}{du^2} \log \wp u.$$

If u, v, w, x denote any four arguments,

$$\begin{aligned}&\sigma(u-v)\sigma(u+v)\sigma(w-x)\sigma(w+x)\\+&\sigma(v-w)\sigma(v+w)\sigma(u-x)\sigma(u+x)\\+&\sigma(w-u)\sigma(w+u)\sigma(v-x)\sigma(v+x)=0,\ldots\ldots\ldots\ldots(\mathrm{L})\end{aligned}$$

since it is of the form

$$(U-V)(W-X)+(V-W)(U-X)+(W-U)(V-X),$$

where $\qquad U-V=-\sigma^2u\sigma^2v(\wp u-\wp v)$, etc.

201. We notice that the Third Elliptic Integral can be expressed very simply as the logarithm of a function, so that we may write (γ_1) in the form

$$\int_0 \frac{1}{2}\frac{\wp' u-\wp' v}{\wp u-\wp v}du=\log\phi(u,v),$$

where $$\phi(u,v)=\frac{\sigma(u+v)}{\sigma u\,\sigma v}e^{-u\zeta v},$$

and $\phi(u,v)$ is called by Hermite a *doubly periodic function of the second kind.*

Changing the sign of u, or v,

$$\phi(u,-v)=\phi(-u,v)=-\frac{\sigma(u-v)}{\sigma u\,\sigma v}e^{u\zeta v};$$

so that $$\phi(u,v)\phi(u,-v)=\wp u-\wp v.$$

202. Suppose $\wp v=e_1$, e_2, or e_3; then, according to § 54, we can take $v=\omega_1$, $\omega_1+\omega_3$, or ω_3, to correspond; and now

$$\wp' v=0,\text{ and }\log\phi(u,v)=\tfrac{1}{2}\log(\wp u-\wp v);$$

so that

$$\phi(u,\omega_1)=\phi(u,-\omega_1)=\sqrt{(\wp u-e_1)},\text{ etc.};$$

and $\phi(u,v)$ is an elliptic function for these values of v.

We may thus put

$$\sqrt{(\wp u-e_1)}=\frac{\sigma(u+\omega_1)}{\sigma u\,\sigma\omega_1}e^{-u\zeta\omega_1},\text{ or }\frac{\sigma_1 u}{\sigma u},$$

where $$\sigma_1 u\text{ denotes }\frac{\sigma(u+\omega_1)}{\sigma\omega_1}e^{-u\zeta\omega_1}.$$

Similarly,

$$\sqrt{(\wp u-e_2)}=\frac{\sigma_2 u}{\sigma u},\quad \sqrt{(\wp u-e_3)}=\frac{\sigma_3 u}{\sigma u},$$

where $$\sigma_2 u=\frac{\sigma(u+\omega_1+\omega_3)}{\sigma(\omega_1+\omega_3)}e^{-u\zeta(\omega_1+\omega_3)},\quad \sigma_3 u=\frac{\sigma(u+\omega_3)}{\sigma\omega_3}e^{-u\zeta\omega_3}.$$

Also $\wp' u=-2\sqrt{(\wp u-e_1.\wp u-e_2.\wp u-e_3)}=-2\sigma_1 u\,\sigma_2 u\,\sigma_3 u/\sigma^3 u$,
and (§ 200) $$\sigma 2u=2\sigma u\,\sigma_1 u\,\sigma_2 u\,\sigma_3 u.$$

Denoting by α, β, γ the three numbers 1, 2, 3, taken in any order, then the relation

$$\sigma_\alpha u = \sigma u \surd(\wp u - e_\alpha)$$

gives, by a combination of the expansions of σu and $\wp u$ in § 195,

$$\sigma_\alpha u = 1 - \tfrac{1}{2}e_\alpha u^2 - \tfrac{1}{48}(6e_\alpha{}^2 - g_2)u^4 - \ldots.$$

so that $\sigma_\alpha u$ is an *even* function of u, and unaffected by Homogeneity (§ 196).

Thus, for instance, from ex. 9, p. 174,

$$\sigma_\alpha 2u + \sigma_\beta 2u = \{\surd(\wp 2u - e_\alpha) + \surd(\wp 2u - e_\beta)\}\sigma 2u$$

$$= -2\frac{(\wp u - e_\alpha)(\wp u - e_\beta)}{\wp' u}\sigma 2u = \sigma_\alpha{}^2 u\, \sigma_\beta{}^2 u.$$

The symbol η_α is employed to denote $\zeta\omega_\alpha$, so that η is the analogue of Legendre's E of § 77.

With positive discriminant Δ (§ 53), we find (exs. 4, 5, p. 199),

$$\eta_1\omega_3 - \eta_3\omega_1 = \tfrac{1}{2}i\pi;$$

and with negative Δ (§ 62),

$$\eta_2\omega_2' - \eta_2'\omega_2 = i\pi;$$

formulas analogous to *Legendre's relation* of § 171.

203. Denoting $\wp u$, $\wp v$, $\wp w$ by x, y, z, then (§ 165) if

$$u + v + w = 0,$$

$$(x+y+z)(4xyz - g_3) = (yz + zx + xy + \tfrac{1}{4}g_2)^2 \ldots\ldots\ldots\ldots\ldots\text{(I.)}$$

Denoting also $(x - e_\alpha)(y - e_\alpha)(z - e_\alpha)$ by $s_\alpha{}^2$, then since

$$e_\alpha{}^3 = \tfrac{1}{4}g_2 e_\alpha + \tfrac{1}{4}g_3,$$

$$s_\alpha{}^2 = xyz - \tfrac{1}{4}g_3 - (yz + zx + xy + \tfrac{1}{4}g_2)e_\alpha + (x+y+z)e_\alpha{}^2$$

$$s_\alpha = \frac{yz + zx + xy - 2(x+y+z)e_\alpha}{2\surd(x+y+z)},$$

by means of (I); and this is of the form $A + Be_\alpha$, so that

$$(e_2 - e_3)s_1 + (e_3 - e_1)s_2 + (e_1 - e_2)s_3 = 0;$$

or $(e_2 - e_3)\sigma_1 u\,\sigma_1 v\,\sigma_1 w + (e_3 - e_1)\sigma_2 u\,\sigma_2 v\,\sigma_2 w + (e_1 - e_2)\sigma_3 u\,\sigma_3 v\,\sigma_3 w = 0,$

since $$s_\alpha = \frac{\sigma_\alpha u\ \sigma_\alpha v\ \sigma_\alpha w}{\sigma u\ \sigma v\ \sigma w}.$$

(W. Burnside, *Messenger of Mathematics*, Oct. 1891.)

As an exercise the student may prove that, with

$$u + v + w + x = 0,$$

$$(e_2 - e_3)\sigma_1 u\,\sigma_1 v\,\sigma_1 w\,\sigma_1 x + (e_3 - e_1)\sigma_2 u\,\sigma_2 v\,\sigma_2 w\,\sigma_2 x$$

$$+ (e_1 - e_2)\sigma_3 u\,\sigma_3 v\,\sigma_3 w\,\sigma_3 x + (e_2 - e_3)(e_3 - e_1)(e_1 - e_2)\sigma u\,\sigma v\,\sigma w\,\sigma x = 0,$$

the analogue, in Weierstrass's notation, to Cayley's theorem, given in ex. 1, ii., p. 140.

204. The solution of Lamé's differential equation, which may be written in Weierstrass's notation

$$\frac{1}{y}\frac{d^2y}{du^2}=n(n+1)\wp u+h, \qquad (1)$$

is given, when $n=1$, by the function $\phi(u, v)$ of § 201.

For, differentiating ϕ logarithmically with respect to u,

$$\frac{1}{\phi}\frac{d\phi}{du}=\frac{1}{2}\frac{\wp' u-\wp' v}{\wp u-\wp v}=\zeta(u+v)-\zeta u-\zeta v,$$

and differentiating again,

$$\frac{1}{\phi}\frac{d^2\phi}{du^2}-\frac{1}{\phi^2}\frac{d\phi^2}{du^2}=-\wp(u+v)+\wp u,$$

so that

$$\frac{1}{\phi}\frac{d^2\phi}{du^2}=\frac{1}{4}\left(\frac{\wp' u-\wp' v}{\wp u-\wp v}\right)^2-\wp(u+v)+\wp u$$

$$=2\wp u+\wp v,$$

Lamé's differential equation, with $n=1$, and $h=\wp v$.

The general solution of

$$\frac{1}{y}\frac{d^2y}{du^2}=2\wp u+\wp v \qquad (2)$$

is therefore

$$y=C\phi(u, v)+C'\phi(u, -v), \text{ or } C\phi(u, v)+C'\phi(-u, v).$$

When h or $\wp v=e_1$, e_2, or e_3, the solution is one of Lamé's functions, as in § 202.

One solution is now $\sqrt{(\wp u-e_a)}$, where $a=1$, 2, or 3; the other being

$$\{\zeta(u+\omega_a)-e_a u\}\sqrt{(\wp u-e_a)},$$

as may be verified by differentiation, or determined independently from a knowledge of the particular solution $\sqrt{(\wp u-e_a)}$.

205. *The revolving chain, resumed.*

We are now able to complete the solution (§ 80) of the tortuous revolving chain, by obtaining an analytical expression for its projection on a plane perpendicular to the axis of revolution.

Putting $\quad y=r\cos\psi,\ z=r\sin\psi,$

then we have found in § 80, p. 70, that, when the notation of Legendre and Jacobi is employed,

$$\frac{d\psi}{dx}=\frac{H}{Tr^2}=\frac{H/T}{b^2\text{sn}^2(Kx/a)+c^2\text{cn}^2(Kx/a)},$$

which, on putting $u=Kx/a$, and

$$\kappa^2\text{sn}^2v=-(b^2-c^2)/c^2,\ \text{dn}^2v=b^2/c^2,$$

so that, with $\kappa^2=(b^2-c^2)/(d^2-c^2)$,

$$\text{sn}^2v=-(d^2-c^2)/c^2,\ \text{cn}^2v=d^2/c^2,$$

becomes $$\frac{di\psi}{du}=-\frac{\text{cn}\,v\,\text{dn}\,v/\text{sn}\,v}{1-\kappa^2\text{sn}^2u\,\text{sn}\,v},$$

so that $$i\psi=-u\frac{\text{cn}\,v\,\text{dn}\,v}{\text{sn}\,v}-\Pi(u,v).\ \dots\dots\dots\dots(1)$$

Since sn^2v is negative, we may, by (67) § 73, put $v=t'iK'$, where t' is a real proper fraction.

Now $$r=c\sqrt{(1-\kappa^2\text{sn}^2u\,\text{sn}^2v)}$$

$$=c\Theta 0\sqrt{\frac{\Theta(u+v)\Theta(u-v)}{\Theta^2u\,\Theta^2v}},\ \dots\dots\dots\dots(2)$$

while $$e^{i\psi}=\sqrt{\frac{\Theta(u+v)}{\Theta(u-v)}}\exp\left(-\frac{\text{cn}\,v\,\text{dn}\,v}{\text{sn}\,v}-Zv\right)u\,;$$

so that $$y+iz=c\Theta 0\frac{\Theta(u+v)}{\Theta u\,\Theta v}\exp\left(-\frac{\text{cn}\,v\,\text{dn}\,v}{\text{sn}\,v}-Zv\right)u\,;\ \dots(3)$$

which, when resolved into its real and imaginary part, will give y and z as functions of u or Kx/a, and thus represent the equation of the chain.

206. The procedure is more rapid with Weierstrass's notation.

Writing $y^2+z^2=r^2$, we have found that (§ 80)

$$\left(\frac{dr^2}{dx}\right)^2=\frac{n^4w^2}{T^2}(r^6-Ar^4+Br^2-C),$$

so that we may put

$$r^2=k^2(\wp u-\wp v),\ \dots\dots\dots\dots(1)$$

provided that $\dfrac{du}{dx}=\dfrac{\frac{1}{2}n^2wk}{T}$,

and g_2, g_3 are suitably chosen.

Since v is the value of u which makes r^2 vanish, therefore

$$k^4\wp'^2v\frac{du^2}{dx^2}=-\frac{4H^2}{T^2},$$

the value of $(dr^2/dx)^2$ when $r^2=0$ (§ 80); so that

$$\wp'^2v=-16H^2/n^4w^2k^6,\ \dots\dots\dots\dots(2)$$

and $\wp'v$ is therefore a pure imaginary, which we take to be negative imaginary, so that $v=t'\omega_3$ (§ 54).

Now $$\frac{d\psi}{du}=\frac{H}{Tr^2}\frac{dx}{du}=\frac{2H}{n^2wk^3}\frac{1}{\wp u-\wp v}=\frac{\frac{1}{2}i\wp'v}{\wp u-\wp v},$$

or
$$\frac{di\psi}{du}=\frac{-\frac{1}{2}\wp' v}{\wp u-\wp v}=\tfrac{1}{2}\zeta(v+u)+\tfrac{1}{2}\zeta(v-u)-\zeta v \qquad (3)$$

from (β') (§ 197); so that

$$i\psi=\tfrac{1}{2}\log\frac{\sigma(v+u)}{\sigma(v-u)}-u\zeta v,$$

$$e^{i\psi}=\sqrt{\frac{\sigma(v+u)}{\sigma(v-u)}}e^{-u\zeta v}, \qquad (4)$$

while
$$r=k\sqrt{\frac{\sigma(v+u)\sigma(v-u)}{\sigma^2 v\,\sigma^2 u}}, \qquad (5)$$

and
$$y+iz=k\frac{\sigma(v+u)}{\sigma v\,\sigma u}e^{-u\zeta v}$$
$$=k\phi(u,v),$$
$$y-iz=k\phi(u,-v), \qquad (6)$$

giving the form of the chain.

For a revolving chain fixed at two points, we must have r^2 restricted to lie between positive values, b^2 and c^2, and therefore $\wp u$ must be restricted to lie between e_2 and e_3; so that with du/dx constant, we must put $u=x\omega_1/a+\omega_3$.

For a chain *attracted* to the axis with intensity proportional to the distance, and thus taking up a form of *minimum* moment of inertia, we have $u=x\omega_1/a$; and now $\wp u$ can become infinite, and the chain reach to infinite distance.

In this and other mechanical problems, the parameter of the elliptic integral of the third kind is almost always imaginary; the apparent awkwardness of this imaginary parameter is removed when we proceed to express the vector $y+iz$ by a doubly periodic function of the second kind $\phi(u, v)$, whose logarithm is the elliptic integral of the third kind; and thence determine y and z theoretically by resolving $\phi(u, v)$ into its real and imaginary part.

Familiar instances of the same procedure are met with in Elementary Mathematics; thus

$$x+iy=c\cos(nt+ia), \text{ or } c\cosh(nt+i\beta),$$

will represent elliptic or hyperbolic motion about the centre.

Generally, with $x+iy=z$, $X+iY=Z=F'z$; then

$$\ddot{z}=F'z,\ \tfrac{1}{2}\dot{z}^2=Fz+h,\ t=\int\frac{dz}{\sqrt{(2Fz+2h)}}$$

will give the motion of a particle of unit mass under component forces (X, Y). (Lecornu, *Comptes Rendus*, t. 101, p. 1244.)

207. *The Tortuous Elastica.*

A procedure, similar to that just employed for the revolving chain, will show that the equation of the curve assumed by a round wire of uniform flexibility in all directions can be expressed by the equation

$$y+iz=k\phi(u,\ v)$$

and $$z=k\zeta u+\gamma u,$$

where $$u=s\omega_1/c+\omega_3,$$

s denoting the length of an arc of the wire, and $2c$ the length of a complete wave.

(*Proc. London Math. Society*, XVIII., p. 277.)

The elastic wire differs thus from the revolving chain in having $u=s\omega_1/c+\omega_3$, instead of $u=x\omega_1/a+\omega_3$ (§ 97).

To establish these equations, take the axis Ox as the axis of the applied wrench, consisting of a force X along Ox and a couple L in a plane perpendicular to Ox; denote the torsional couple about the tangent at any point by G, and the flexural rigidity of the wire by B.

Then the component couples of resilience about the axes Ox, Oy, Oz are taken to be

$$B(y'z''-y''z'),\ B(z'x''-z''x'),\ B(x'y''-x''y')$$

the accents denoting differentiation with respect to the arc s; the equations of equilibrium are therefore

$$B(y'z''-y''z')=Gx'+L \qquad (1)$$

$$B(z'x''-z''x')=Gy'+Xz \qquad (2)$$

$$B(x'y''-x''y')=Gz'-Xy \qquad (3)$$

(Binet and Wantzel, *Comptes Rendus*, 1844).

Differentiating each equation with respect to s, multiplying respectively by x', y', z', and adding, gives

$G'=0$; so that G is constant.

Multiply equations (1), (2), (3) by x', y', z', and add; then

$$G-X(yz'-y'z)=0,$$

so that $$yz'-y'z=r^2d\psi/ds=G/X, \text{ a constant};$$

and $$yz''-y''z=0.$$

Again, multiplying (2) by y, (3) by z, and adding, gives

$$Bx''(yz'-y'z)-Bx'(yz''-y''z)=G(yy'+zz'),$$

or $$Bx''=X(yy'+zz'),$$

so that, integrating, $$Bx'=\tfrac{1}{2}X(y^2+z^2)+H.$$

Then $$B^2x''^2 = X^2(yy'+zz')^2$$
$$= X^2\{(y^2+z^2)(y'^2+z'^2)-(yz'-y'z)^2\}$$
$$= 2X(Bx'-H)(1-x'^2)-G^2,$$

a cubic function of x'; so that, by inversion of the elliptic integral, x' or y^2+z^2 is an elliptic function of the arc s, which may be written

$$y^2+z^2=k^2(\wp\omega-\wp u), \qquad (4)$$

or $$Bx'=\tfrac{1}{2}Xk^2(\wp\omega-\wp u)+H,$$

provided $$\frac{du}{ds}=\frac{1}{2}\frac{Xk}{B};$$

and now $$\frac{dx}{du}=k(\wp\omega-\wp u)+\frac{2H}{Xk},$$

$$\frac{x}{k}=\zeta u+\left(\wp\omega+\frac{2H}{Xk^2}\right)u; \qquad (5)$$

also $$\frac{di\psi}{du}=\frac{iG}{Xr^2}\frac{ds}{du}=\frac{2iBG}{X^2k^3}\frac{1}{\wp\omega-\wp u}=\frac{\frac{1}{2}\wp'\omega}{\wp\omega-\wp u}. \qquad (6)$$

By Kirchhoff's *Kinetic Analogue*, it follows that the axis of a Spherical Pendulum, Gyrostat, or Top can be made to follow in direction the tangent of a certain Tortuous Elastica, when the point of contact of the tangent on the elastica moves with constant velocity; so that, if x, y, z are the coordinates of a point fixed in the axis of the Gyrostat, and Ox is vertical,

$$y+iz=k\frac{d}{du}\frac{\sigma(u+\omega)}{\sigma u\,\sigma\omega}\exp(\lambda-\zeta\omega)u,$$

$$x=k(\wp v-\wp u),$$

where now $$u=nt+\omega_3,$$

and $2\omega_1/n$ is the period of the oscillations of the Top, or Spherical Pendulum.

The Spherical Pendulum and the Top.

208. To prove these formulas independently for the spherical pendulum, let the weight of the bob be W lb., and let the tension of the thread be a force of NlW poundals; then the equations of motion are, with the axis of x drawn vertically *downwards*,

$$\frac{d^2x}{dt^2}+Nx=g,\quad \frac{d^2y}{dt^2}+Ny=0,\quad \frac{d^2z}{dt^2}+Nz=0; \qquad (1)$$

subject to the condition, l denoting the length of the thread,

$$x^2+y^2+z^2=l^2.$$

The equation of energy is

$$\tfrac{1}{2}(\dot{x}^2+\dot{y}^2+\dot{z}^2)=g(x+c); \quad\text{.....................(2)}$$

while $$y\dot{z}-\dot{y}z=h, \text{ a constant.} \quad\text{..............(3)}$$

Now, $$x\ddot{x}+y\ddot{y}+z\ddot{z}+Nl^2=gx,$$

so that $$Nl^2=gx+\dot{x}^2+\dot{y}^2+\dot{z}^2=g(3x+2c);$$

thus giving the tension of the thread.

Hermite writes (*Sur quelques applications des fonctions elliptiques*, 1885)

$$(y+iz)(\dot{y}-i\dot{z})=y\dot{y}+z\dot{z}-i(y\dot{z}-\dot{y}z)$$
$$=-x\dot{x}-ih,$$

so that the *norm* of each side is

$$(y^2+z^2)(\dot{y}^2+\dot{z}^2)=x^2\dot{x}^2+h^2.$$

Then

$$(l^2-x^2)\{2g(x+c)-\dot{x}^2\}=x^2\dot{x}^2+h^2,$$

or $$l^2\dot{x}^2=2g(x+c)(l^2-x^2)-h^2$$
$$=-2gx^3-2gcx^2+2gl^2x+2gcl^2-h^2;$$

so that x is a simple elliptic function of t, which we may write

$$x=k(\wp v-\wp u), \quad\text{......................(4)}$$

where $u=nt+\omega_3$, for $\wp u$ to lie between e_2 and e_3.

Then $$l^2k^2n^2\wp'^2u=2gk^3(\wp u-\wp v)^3-2gck^2(\wp u-\wp v)^2$$
$$-2gkl^2(\wp u-\wp v)+2gcl^2-h^2$$
$$=\tfrac{1}{2}gk^3(4\wp^3u-g_2\wp u-g_3),$$

provided $$n^2=\tfrac{1}{2}gk/l^2, \text{ and } \wp v=-\tfrac{1}{3}c/k;$$

while g_2 and g_3 are suitably chosen.

The value of $\wp' v$ is found by noticing that $x=0$ when $u=v$; and thus $$l^2k^2n^2\wp'^2v=2gcl^2-h^2.$$

Now Hermite writes

$$\frac{d^2}{dt^2}(y+iz)+N(y+iz)=0,$$

$$\frac{1}{y+iz}\frac{d^2}{du^2}(y+iz)=-N\frac{dt^2}{du^2}=-\frac{2Nl^2}{gk}=-2\frac{3x+2c}{k}=6\wp u+6\wp v,$$

Lamé's differential equation for $n=2$, with $h=6\wp v$.

The formal solution of this equation is reserved for the present; but it can be inferred for this case by taking the equation (3) and writing it

$$\frac{d\psi}{du}=\frac{h}{n(y^2+z^2)},$$

or $$\frac{di\psi}{du}=\frac{ih/n}{l^2-x^2}=\frac{\tfrac{1}{2}ih/nl}{l-x}+\frac{\tfrac{1}{2}ih/nl}{l+x}. \quad\text{..................(5)}$$

We now put

$$l-x=k(\wp u-\wp a), \quad l+x=k(\wp b-\wp u)\,; \quad\ldots\ldots\ldots(6)$$

and since $l^2\dot{x}^2=-h^2$, when $x=\pm l$, or when $u=a$, or b, therefore

$$\wp'^2 a=\wp'^2 b=-\frac{h^2}{k^2n^2l^2}.$$

With k positive, and $\wp b>\wp u>\wp a$, we take $\wp' a$ negative imaginary, and $\wp' b=-\wp' a$ positive imaginary, so that (§ 54), $a=p\omega_3$, $b=\omega_1+q\omega_3$, where p and q are real proper fractions.

Then
$$\frac{di\psi}{du}=\frac{-\frac{1}{2}\wp' a}{\wp u-\wp a}+\frac{\frac{1}{2}\wp' b}{\wp b-\wp u}, \quad\ldots\ldots\ldots\ldots\ldots\ldots\ldots\ldots(7)$$

and integrating, by equation (β), § 199,

$$i\psi=\tfrac{1}{2}\log\frac{\sigma(u+a)}{\sigma(u-a)}-u\zeta a+\tfrac{1}{2}\log\frac{\sigma(b+u)}{\sigma(b-u)}-u\zeta b. \quad (8)$$

Now
$$\frac{y+iz}{y-iz}=e^{2i\psi}=\frac{\sigma(u+a)\sigma(b+u)}{\sigma(u-a)\sigma(b-u)}\exp(-2\zeta a-2\zeta b)u,$$

while

$$(y+iz)(y-iz)=y^2+z^2=l^2-x^2=k^2(\wp u-\wp a)(\wp b-\wp u)$$
$$=k^2\frac{\sigma(u+a)\sigma(u-a)}{\sigma^2 a\,\sigma^2 u}\,\frac{\sigma(b+u)\sigma(b-u)}{\sigma^2 b\,\sigma^2 u},$$

so that
$$y+iz=k\frac{\sigma(u+a)\sigma(u+b)}{\sigma a\,\sigma b\,\sigma^2 u}\exp(-\zeta a-\zeta b)u,$$
$$y-iz=k\frac{\sigma(u-a)\sigma(b-u)}{\sigma a\,\sigma b\,\sigma^2 u}\exp(+\zeta a+\zeta b)u; \quad\ldots\ldots\ldots(9)$$

thus giving the solution of Lamé's differential equation for $n=2$.

209. It is interesting to verify that these values of $y+iz$ and $y-iz$ are solutions of Lamé's equation for $n=2$.

Denoting $y+iz$ by ϕ, and differentiating logarithmically,

$$\frac{1}{\phi}\frac{d\phi}{du}=\zeta(u+a)-\zeta u-\zeta a+\zeta(b+u)-\zeta u-\zeta b$$
$$=\frac{1}{2}\frac{\wp' u-\wp' a}{\wp u-\wp a}+\frac{1}{2}\frac{\wp' b-\wp' u}{\wp b-\wp u};$$

and differentiating again,

$$\frac{1}{\phi}\frac{d^2\phi}{du^2}=\left(\frac{1}{\phi}\frac{d\phi}{du}\right)^2-\wp(u+a)+\wp u-\wp(b+u)+\wp u$$
$$=\frac{1}{4}\left(\frac{\wp' u-\wp' a}{\wp u-\wp a}\right)^2+\frac{1}{2}\frac{\wp' u-\wp' a}{\wp u-\wp a}\frac{\wp' b-\wp' u}{\wp b-\wp u}+\frac{1}{4}\left(\frac{\wp' b-\wp' u}{\wp b-\wp u}\right)^2$$
$$+2\wp u-\wp(u+a)-\wp(b+u)$$

$$=4\wp u+\wp a+\wp b+\frac{1}{2}\frac{\wp' u-\wp' a}{\wp u-\wp a}\frac{\wp' b-\wp' u}{\wp b-\wp u}.$$

But with $\wp' a=-\wp' b$,

$$\frac{1}{2}\frac{\wp' u-\wp' a}{\wp u-\wp a}\frac{\wp' b-\wp' u}{\wp b-\wp u}=\frac{1}{2}\frac{\wp'^2 u-\wp'^2 b}{(\wp u-\wp a)(\wp b-\wp u)}=2(\wp u+\wp a+\wp b),$$

so that $$\frac{1}{\phi}\frac{d^2\phi}{du^2}=6\wp u+3\wp a+3\wp b,$$

Lamé's differential equation for $n=2$, with $h=3\wp a+3\wp b$, in place of the previous value of $h=6\wp v$.

From Kirchhoff's Kinetic Analogue in § 207 we may put

$$y+iz=k\frac{\sigma(u+a)\sigma(b+u)}{\sigma a\,\sigma b\,\sigma^2 u}\exp(-\zeta a-\zeta b)u$$
$$=k\frac{d}{du}\left\{\frac{\sigma(u+a+b)}{\sigma(a+b)\sigma u}\exp(-\zeta a-\zeta b)u\right\}$$
$$=k\frac{d}{du}\{\phi(u,\ a+b)e^{\lambda u}\},$$

where $$\lambda=\zeta(a+b)-\zeta a-\zeta b.$$

With $\wp'(a-b)=\wp' a=-\wp' b$,

therefore $\zeta(a-b)=\zeta a-\zeta b$;

and, changing the sign of a,

$$\frac{\sigma(u-a)\sigma(b+u)}{\sigma a\,\sigma b\,\sigma^2 u}\exp(\zeta a-\zeta b)u=\frac{d}{du}\phi(u,\ -a+b).$$

(Halphen, *F. E.*, I., p. 230.)

210. In the slightly more general case of the motion of the Top, we shall find it convenient to draw the axis Ox vertically *upwards*, and to call θ the angle which the axis OC of the top makes with the vertical Ox.

Then, from the principles of the Conservation of Energy and Momentum, we obtain the equations (Routh, *Rigid Dynamics*)

$$\tfrac{1}{2}A(d\theta/dt)^2+\tfrac{1}{2}A\sin^2\theta(d\psi/dt)^2=Wg(c-h\cos\theta), \ldots\ldots(1)$$
$$A\sin^2\theta(d\psi/dt)+Cr\cos\theta=G, \ldots\ldots\ldots\ldots(2)$$

where r denotes the constant angular velocity of the top about its axis of figure OC, $d\psi/dt$ the angular velocity of the vertical plane through Ox and OC, h the distance of the centre of gravity G from O, W lb. the weight of the top, and C, A its moments of inertia about the axis of figure OC, and about any axis through O at right angles to OC.

Putting $A/Wh = l = OP$, as in the simple pendulum, then P is the centre of oscillation for plane vibrations.

The elimination of $d\psi/dt$ between equations (1) and (2) gives

$$\tfrac{1}{2}l\sin^2\theta\left(\frac{d\theta}{dt}\right)^2 = g\left(\frac{c}{h} - \cos\theta\right)(1-\cos^2\theta) - \tfrac{1}{2}l\left(\frac{G - Cr\cos\theta}{A}\right)^2$$

$$= g(\cos\theta - \cos\alpha)(\cos\theta - \cos\beta)(\cos\theta - d), \ldots\ldots(3)$$

suppose; the inclination of the axis of the top to the vertical being supposed to oscillate between α and β,

$$\alpha > \theta > \beta, \text{ or } \cos\alpha < \cos\theta < \cos\beta < d.$$

Guided by equation (17), p. 37, we put

$$\cos\theta = \cos\alpha\cos^2\phi + \cos\beta\sin^2\phi,$$

$$\cos\theta - \cos\alpha = (\cos\beta - \cos\alpha)\sin^2\phi,$$

$$\cos\beta - \cos\theta = (\cos\beta - \cos\alpha)\cos^2\phi\ ; \ \ldots\ldots\ldots\ldots\ldots(4)$$

and therefore,

$$\left(\frac{d\phi}{dt}\right)^2 = \frac{1}{2}\frac{g}{l}(d - \cos\theta)$$

$$= \frac{1}{2}\frac{g}{l}\{d - \cos\alpha - (\cos\beta - \cos\alpha)\sin^2\phi\}$$

$$= n^2(1 - \kappa^2\sin^2\phi),$$

where $\kappa^2 = \dfrac{\cos\beta - \cos\alpha}{d - \cos\alpha}$, $\kappa'^2 = \dfrac{d - \cos\beta}{d - \cos\alpha}$,

and $ln^2 = \tfrac{1}{2}g(d - \cos\alpha)$.

Now we may put $\phi = \operatorname{am} nt$, and

$$\cos\theta = \cos\alpha\,\operatorname{cn}^2 nt + \cos\beta\,\operatorname{sn}^2 nt, \ldots\ldots\ldots\ldots\ldots(5)$$

so that the projection on the vertical Ox of the motion of a point on OC resembles ordinary plane pendulum motion.

When $d = 1$ and $\cos\alpha = -1$, then

$$n^2 = g/l,\quad \kappa^2 = \cos^2\tfrac{1}{2}\beta = \sin^2\tfrac{1}{2}(\pi - \beta)\ ;$$

G and Cr vanish, and the oscillations are in a vertical plane.

But, in the general state of motion,

$$A\frac{d\psi}{dt} = \frac{G - Cr\cos\theta}{\sin^2\theta}$$

$$= \frac{1}{2}\frac{G + Cr}{1 + \cos\theta} + \frac{1}{2}\frac{G - Cr}{1 - \cos\theta}$$

$$= \frac{1}{2}\frac{G + Cr}{1 + \cos\alpha + (\cos\beta - \cos\alpha)\operatorname{sn}^2 nt} + \frac{1}{2}\frac{G - Cr}{1 - \cos\alpha - (\cos\beta - \cos\alpha)\operatorname{sn}^2 nt},$$

so that ψ is expressed by two Third Elliptic Integrals.

Putting $\cos\theta = \pm 1$ in equation (3), show that

$$\left(\frac{G+Cr}{A}\right)^2 = 2\frac{g}{l}(1+\cos\alpha)(1+\cos\beta)(d+1)\,;$$

$$\left(\frac{G-Cr}{A}\right)^2 = 2\frac{g}{l}(1-\cos\alpha)(1-\cos\beta)(d-1),$$

while, in accordance with Jacobi's notation, we put

$$\kappa^2\text{sn}^2 v_1 = -\frac{\cos\beta-\cos\alpha}{1+\cos\alpha},\quad \kappa^2\text{sn}^2 v_2 = \frac{\cos\beta-\cos\alpha}{1-\cos\alpha}\,;$$

so that, finally, with $u=nt$, we find

$$\frac{di\psi}{du} = \frac{\text{cn}\,v_1\text{dn}\,v_1/\text{sn}\,v_1}{1-\kappa^2\text{sn}^2v_1\text{sn}^2u} + \frac{\text{cn}\,v_2\text{dn}\,v_2/\text{sn}\,v_2}{1-\kappa^2\text{sn}^2v_2\text{sn}^2u}\,; \quad\text{.........(6)}$$

and, as in the spherical pendulum (§ 208), we take

$$v_1 = ipK',\quad v_2 = K+iqK',$$

where p and q are real proper fractions.

In the Weierstrassian notation, we put, as in (6), § 208,

$$1+\cos\theta = k(\wp u - \wp a),\quad 1-\cos\theta = k(\wp b - \wp u)\,;$$

and thence (§ 224) $c-h\cos\theta = hk\{\wp(a+b)-\wp u\}$.

We thus obtain $$\frac{di\psi}{du} = \frac{-\frac{1}{2}\wp' a}{\wp u - \wp a} + \frac{\frac{1}{2}\wp' b}{\wp b - \wp u}\,; \quad\text{..................(7)}$$

but now the relation $\wp' a = -\wp' b$ holds only when $Cr=0$, or when the motion of the top is comparable with that of the spherical pendulum; on the other hand, the relation $\wp' a = \wp' b$ implies that $G=0$.

The Kinetic Analogue of the Top with the Tortuous Elastica (§ 207) is obtained by putting

$$a+b=\omega,\ \text{and}\ \lambda = \zeta(a+b)-\zeta a-\zeta b.$$

In the Steady Motion of the Top, $\alpha=\beta$, $\kappa=0$, $K=\frac{1}{2}\pi$; and the elliptic functions degenerate into circular functions.

We thus obtain the condition for the steady motion, and the period of the small oscillations, given in Routh's *Rigid Dynamics*.

211. A similar procedure will solve the general equations of motion of a solid figure of revolution, moving under no forces through an infinitely extended incompressible frictionless liquid; the work will be found in Appendix III. of Basset's *Hydrodynamics*, vol. I; also in Halphen's *Fonctions elliptiques*, II., chap. IV. The problem is of practical interest from its bearing upon the determination of the amount of spin requisite to secure the stability of an elongated projectile.

(*Proceedings, Royal Artillery Institution*, 1879.)

212. We again resume the consideration of the motion of a body under no forces, first mentioned in § 32, as affording a good practical illustration of the necessity for the introduction of various analytical theorems of Elliptic Functions.

Geometrical Representation of the Motion of a Body under No Forces, according to MacCullagh, Siacci, and Gebbia.

Quadrics concyclic with the momental ellipsoid, that is, having the same circular sections, are given by (Smith, *Solid Geometry*, § 170)

$$(A-H)x^2+(B-H)y^2+(C-H)z^2=Dh^2;$$

and now, if we produce the instantaneous axis of rotation OP to meet the concyclic quadric in P', and denote OP' by R',

$$(A-H)p^2+(B-H)q^2+(C-H)r^2=Dh^2\omega^2/R'^2,$$

while $$Ap^2+Bq^2+Cr^2=Dh^2\omega^2/R^2,$$

so that, by subtraction,

$$H(p^2+q^2+r^2)=Dh^2\omega^2\left(\frac{1}{R^2}-\frac{1}{R'^2}\right), \text{ or } \frac{h^2}{R^2}-\frac{h^2}{R'^2}=\frac{H}{D}.$$

Along the polhode, $R=h\sec\theta$, where θ denotes the angle between the instantaneous axis OP and the fixed axis of resultant angular momentum OC; and then

$$\frac{h^2}{R'^2}=\cos^2\theta-\frac{H}{D}, \qquad (1)$$

the polar equation of a quadric surface of revolution.

Since R^2 is less than $h^2\sec^2\theta$ for all points adjacent to P on the momental ellipsoid, therefore in the concyclic quadric

$$\frac{1}{R'^2} \text{ is greater than } \frac{\cos^2\theta}{h^2}-\frac{H}{Dh^2},$$

except at the point P', and therefore the concyclic quadric touches this quadric surface of revolution at P' and rolls upon it during the motion.

We may also take concyclic quadrics, given by

$$(H-A)x^2+(H-B)y^2+(H-C)z^2=Dh^2,$$

and now $$\frac{h^2}{R'^2}=\frac{H}{D}-\frac{h^2}{R^2}=\frac{H}{D}-\cos^2\theta, \qquad (2)$$

the polar equation of a quadric of revolution.

In particular, if $H=D$, then $R'\sin\theta=h$, the polar equation of a cylinder of revolution, outside which this concyclic hyperboloid rolls during the motion (Siacci, *In memoriam D. Chelini, Collectanea mathematica*, 1881.)

213. By reciprocation of these theorems, we prove MacCullagh's theorem, "that the ellipsoid of gyration,

$$\frac{x^2}{A}+\frac{y^2}{B}+\frac{z^2}{C}=\frac{1}{M},$$

always moves in contact with two fixed points on the axis of resultant angular momentum, equidistant from the centre"; and we also deduce Gebbia's extension of MacCullagh's theorem, that "confocals of the ellipsoid of gyration, the polar reciprocals of the concyclic ellipsoids of the momental ellipsoid, slide without rolling on fixed quadric surfaces of revolution."

In particular, the polar reciprocal of Siacci's cylinder of revolution is a circle, upon which a certain confocal to the ellipsoid of gyration slides without rolling.

Geometrical Representation of the Motion, according to Sylvester, Darboux, and Mannheim.

214. In Sylvester's splendid generalization of Poinsot's representation of the motion of the body, it is proved that a confocal to the momental ellipsoid rolls upon a plane perpendicular to the axis of resultant angular momentum OC at a constant distance from O, which plane rotates about OC with constant angular velocity, and therefore gives a geometrical representation of the time. (*Phil. Trans.*, 1866.)

The proof of this theorem depends upon two geometrical propositions, in connexion with confocal quadric surfaces—

(i.) "The locus of the pole of a fixed tangent plane to a quadric surface, with respect to any confocal, is the normal to the first surface;"

(ii.) "the difference of the squares of the perpendiculars from the centre on two parallel tangent planes of two confocals is constant and equal to the difference of the squares of the corresponding semi-axes."

Thus, in fig. 25, if OP' is a surface confocal with the momental ellipsoid OP, then Q, the pole of the invariable plane CP with respect to the surface OP', will lie in the normal PQ to the momental ellipsoid at P; while the surface OP' will touch a plane $C'P'$, parallel to the invariable plane CP, and such that $OC'^2=OC^2-\lambda^2$, λ^2 denoting the difference of the squares of corresponding semi-axes of the confocals.

Since C is a fixed point during the motion of the body, therefore C' is also fixed.

Drawing the plane QL through Q, parallel to the invariable plane, and denoting OC by h, as before; then since Q is the pole of CP,

$$OQ\,.\,OV = OP'^2, \text{ or } OL\,.\,OC = OC'^2 = h^2 - \lambda^2,$$

so that $$OL = h - \lambda^2/h, \qquad LC = \lambda^2/h.$$

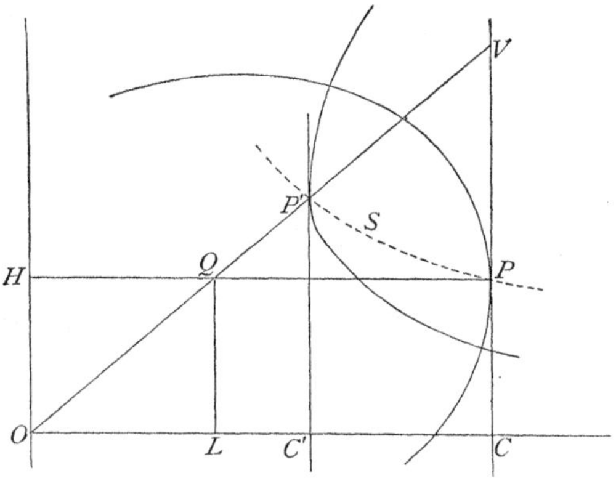

Fig. 25.

Again, denoting as before (§ 104) by μ the constant component of the angular velocity of the body about OC, so that the resultant angular velocity of the body about OP is $\mu \operatorname{cosec} OPC$, then the velocity of the point P' in the body is

$$\mu \operatorname{cosec} OPC\,.\,OP'\,.\,\sin POP' = \mu\,.\,P'V',$$

where V' is the point in which the line OP cuts the plane $C'P'$.

Therefore the angular velocity of P' about the invariable line OC is

$$\mu\frac{P'V'}{C'P'} = \mu\frac{PV}{CP} = \mu\frac{PQ}{OC} = \mu\frac{\lambda^2}{h^2},$$

a constant; so that if the surface OP' rolls without slipping on the plane $C'P'$, this plane must revolve about OC with constant angular velocity $\mu\lambda^2/h^2$.

The point P' lies in the plane $OQPC$; and since

$$\frac{C'P}{CP} = \frac{C'P'}{LQ} = \frac{OC'}{OL} = \frac{OC}{OC'},$$

therefore $$OC'\,.\,C'P' = OC\,.\,CP,$$

and P' lies on the rectangular hyperbola PP'; this is the geometrical property principally employed by Prof. Sylvester.

(*Solid Geometry*, Salmon, §§ 167, 180; Smith, §§ 163, 167.)

The angular velocity of the vector $C'P'$ with respect to the revolving plane $C'P'$ being $\frac{d\phi}{dt}-\mu\frac{\lambda^2}{h^2}$, it follows that, if ρ', ϕ' denote the polar coordinates of a point P' on the herpolhode described by P' on the revolving plane $C'P'$, then

$$\frac{\rho'^2}{\rho^2}=\frac{OC'^2}{OL^2}=\frac{h^2}{h^2-\lambda^2},$$

and $$\rho'^2\frac{d\phi'}{dt}=\mu\left(1-\frac{\lambda^2}{h^2}\right)\rho'^2+\frac{A-D\,.\,B-D\,.\,C-D}{ABC}\frac{\mu h^4}{h^2-\lambda^2},$$

equations similar to those required for the herpolhode of P.

In particular, if we take $\lambda^2=h^2$, then $OC'=0$, and the confocal OP' is a cone; and the plane through O rotates with constant angular velocity μ, while the cone, called by Poinsot the *rolling and slipping cone*, rolls on this revolving plane, the angular velocity about the line of contact OH being ν.

If we consider the curve described on this revolving plane by the point H, the foot of the perpendicular from P on the plane, then ρ, ϕ' being the polar coordinates of H (§ 113),

$$\frac{d\phi'}{dt}=\frac{d\phi}{dt}-\mu=\frac{A-D\,.\,B-D\,.\,C-D}{ABC}\frac{h^2}{\rho^2}\mu,$$

so that the point H describes on the revolving plane an orbit as if attracted to O; and, as in § 89, we shall find that the requisite central force is of the form $A\rho+B\rho^3$.

(Pinczon, *Comptes Rendus*, April, 1887.)

This is otherwise evident, by noticing that the vector $x+iy$ of this curve satisfies Lamé's equation (§ 204)

$$\frac{d^2}{dt^2}(x+iy)=(2\wp u+\wp v)(x+iy),$$

where $$\rho^2=k^2(\wp v-\wp u),$$

so that $$\frac{d^2x}{dt^2}=\left(3\wp v-2\frac{\rho^2}{k^2}\right)x,\quad \frac{d^2y}{dt^2}=\left(3\wp v-2\frac{\rho^2}{k^2}\right)y.$$

A value of λ may be found which makes the herpolhode of P' a closed curve; and this closed polhode is an algebraical curve, when v is an aliquot part of a period, the corresponding elliptic integrals of the third kind becoming *pseudo-elliptic*.

Abel has devoted great attention to the subject of *pseudo-elliptic* integrals (*Œuvres*, XI.), and the *algebraical herpolhode* affords an interesting application of his theorems (§ 218).

The Addition Theorem for the Third Elliptic Integral.

215. Theorems (9) and (10) of § 189 show that, employing the function $\phi(u, v)$ of § 201,

$$\log \phi(u_1, v) + \log \phi(u_2, v) = \log \phi(u_1+u_2, v) + \log \Omega,$$

or

$$\frac{\phi(u_1, v)\phi(u_2, v)}{\phi(u_1+u_2, v)} = \Omega,$$

or

$$\frac{\sigma(v+u_1)\sigma(v+u_2)\sigma(u_1+u_2)}{\sigma(v+u_1+u_2)\sigma v\, \sigma u_1 \sigma u_2} = \Omega, \quad\text{.......}(1)$$

where, expressed by elliptic functions of u_1, u_2, and v,

$$\Omega = \frac{\wp\tfrac{1}{2}(u_1-u_2) - \wp(v+\tfrac{1}{2}\,.\,u_1+u_2)}{\wp\tfrac{1}{2}(u_1+u_2) - \wp(v+\tfrac{1}{2}\,.\,u_1+u_2)} \cdot \frac{\wp'\tfrac{1}{2}(u_1+u_2)}{\wp\tfrac{1}{2}(u_1+u_2) - \wp\tfrac{1}{2}(u_1-u_2)} \text{....}(2)$$

Also, as in equation (8), § 188,

$$\log \phi(v, u) = \log \phi(u, v) + u\zeta v - v\zeta u;$$

so that

$$\log \phi(v, u_1) + \log \phi(v, u_2)$$
$$= \log \phi(v, u_1+u_2) - \{\zeta u_1 + \zeta u_2 - \zeta(u_1+u_2)\}v + \log \Omega, \text{...}(3)$$

the Addition Theorem for the *parameters* u_1, u_2.

These theorems have been generalized by Abel for the addition of any number of *amplitudes* or *parameters* in the Third Elliptic Integral, and the proof is a simple extension of his method, employed in § 162 (*Œuvres*, XXI.).

Denoting by a any arbitrary quantity, equation (7) of § 162 may be written

$$\frac{1}{a-x_r}\frac{dx_r}{\surd X_r} = \frac{\theta x_r}{(a-x_r)\psi' x_r}.$$

Now, since θa is of lower degree in a than ψa, and

$$\psi a = C\Pi(a-x_r),$$

it follows that, when resolved into partial fractions,

$$\frac{\theta a}{\psi a} = \Sigma \frac{\theta x_r}{(a-x_r)\psi' x_r};$$

and therefore, writing $\mathrm{f}x$ and ϕx for P and Q respectively, and A for the value of X when $x=a$,

$$\Sigma \frac{1}{a-x_r}\frac{dx_r}{\surd X_r} = \frac{\theta a}{\psi a} = 2\frac{\phi a \delta \mathrm{f}a - \mathrm{f}a \delta \phi a}{(\mathrm{f}a)^2 - (\phi a)^2 A}$$

$$= \frac{1}{\surd A}\frac{\delta \mathrm{f}a - \delta\phi a\,.\surd A}{\mathrm{f}a - \phi a\,.\surd A} - \frac{1}{\surd a}\frac{\delta \mathrm{f}a + \delta\phi a\,.\surd A}{\mathrm{f}a + \phi a\,.\surd A},$$

or

$$\Sigma \frac{\surd A}{a-x_r}\frac{dx_r}{\surd X_r} = \delta \log \frac{\mathrm{f}a - \phi a \surd A}{\mathrm{f}a + \phi a \surd A} = -2\delta \tanh^{-1}\left(\frac{\phi a}{\mathrm{f}a}\surd A\right) \text{...}(4)$$

Integrating, with the notation (§§ 197, 199),

$$\int_0 \frac{\frac{1}{2}\wp' v\, du}{\wp u - \wp v} = \tfrac{1}{2}\log(\wp u - \wp v) - \log \phi(u, v) = \tfrac{1}{2}\log \frac{\sigma(v-u)}{\sigma(v+u)} e^{2u\zeta v},$$

where $x = \wp u, \ \sqrt{X} = -\wp' u, \ a = \wp v, \ \sqrt{A} = -\wp' v;$

$$\Sigma \log \frac{\phi(u_r, v)}{\phi(u'_r, v)} = \tfrac{1}{2}\log \frac{\mathrm{f}a + \phi a . \sqrt{A}}{\mathrm{f}a - \phi a . \sqrt{A}} \frac{\mathrm{f}'a - \phi' a . \sqrt{A}}{\mathrm{f}'a + \phi' a . \sqrt{A}} \Pi \frac{\wp u_r - \wp v}{\wp u'_r - \wp v}, \qquad (5)$$

so that

$$\Pi \frac{\phi(u_r, v)}{\phi(u'_r, v)} \text{ or } \Pi \frac{\sigma(v+u_r)}{\sigma(v+u'_r)}$$

is expressible by elliptic functions, $\wp$ and $\wp'$, of v; provided that, as in (11), § 162,

$$\Sigma \int_{x'_r}^{x_r} dx_r / \sqrt{X_r} = 0, \text{ or } \Sigma u_r = \Sigma u'_r, \ \ldots\ldots\ldots\ldots\ldots\ldots\ldots (6)$$

the coefficients in $\mathrm{f}a$ and ϕa being determined as functions of $\wp u_r$ and $\wp' u_r$ by the *plexus* of equations (4) in § 162; $\mathrm{f}'a$ and $\phi' a$ being the same functions of u'_r.

Thus the function

$$\Pi \frac{\sigma(v+u_r)}{\sigma(v+u'_r)}, \ \ldots\ldots\ldots\ldots\ldots\ldots\ldots\ldots\ldots (7)$$

is an elliptic function of v provided that the sum of the values $-u_r$ of v which make the function vanish is equal to the sum of the values $-u'_r$ which make the function infinite; in other words, briefly expressed, provided the sum of the *zeroes* u is equal to the sum of the *infinities* u'.

In particular, with the u'_r's all zero, $\Sigma u_r = 0$; and in equation (6), § 162, we can put

$$\psi a = (\mathrm{f}a)^2 - (\phi a)^2 A = \Pi(\wp v - \wp u_r);$$

so that $\Sigma \log \phi(u_r, v) = \log(\mathrm{f}a + \phi a . \sqrt{A}) + \text{constant}.$

Thus $\Pi\phi(u_r, v)$, or $\dfrac{\sigma(v+u_1)\sigma(v+u_2) \ldots \sigma(v+u_\mu)}{(\sigma v)^\mu}, \ \ldots\ldots (8)$

when $u_1 + u_2 + u_3 + \ldots + u_\mu = 0, \ \ldots\ldots\ldots\ldots\ldots\ldots (9)$

is a rational *integral* function of $\wp v$ and $\wp' v$, which may be written, as in § 198,

$$C = c_0 + c_1 \wp v + c_2 \wp' v + \ldots + c_\mu \wp^{(\mu-1)} v. \ \ldots\ldots\ldots\ldots (10)$$

So also, since (§ 201)

$$\phi(-U, v)\phi(U, v) = \wp U - \wp v,$$

therefore, writing U for $-u_\mu$,

$$\sum_{r=1}^{r=\mu-1} \log \phi(u_r, v) = \log \phi(U, v) + \log \Omega + \text{a constant}, \quad\ldots\ldots(11)$$

where $$\Omega = C/(\wp U - \wp v).$$

In particular, when $U = e_\alpha$, $\phi(U, v) = \surd(\wp v - e_\alpha)$ (§ 202), and

$$\prod_{r=1}^{r=\mu-1} \phi(u_r, v) = C/\surd(\wp v - e_\alpha); \quad\ldots\ldots\ldots\ldots\ldots\ldots(12)$$

when $$u_1 + u_2 + u_3 + \ldots + u_{\mu-1} = \omega_\alpha.$$

By an interchange of amplitude and parameter,

$$\Sigma \log \phi(u, v_r) - \Sigma \log \phi(u, v'_r) = \log \Omega - \rho u, \quad\ldots\ldots\ldots(13)$$

provided that $$\Sigma v_r = \Sigma v'_r.$$

Ω being a function of $\wp u$, $\wp' u$, $\wp v$, $\wp' v$; and

$$\rho = \Sigma(\zeta v_r - \zeta v'_r).$$

216. A further application of Abel's Theorem of § 162 shows that ρ is expressible as a function of $\wp v$ and $\wp' v$; this is the generalization of the Addition Theorem for the Second Elliptic Integral, given in § 186.

For $$\Sigma \int \frac{x_r dx_r}{\surd X_r} = \Sigma \int \frac{x_r \theta x_r}{\psi' x_r},$$

and this case can be determined as a degenerate case of the preceding result; since, making $a = \infty$,

$$\Sigma \int \frac{x_r dx_r}{\surd X_r} = \text{lt}\, \Sigma \int \left(\frac{a^2}{a - x_r} - a\right) \frac{dx_r}{\surd X_r} = \text{lt}\, \Sigma \int \frac{a^2}{a - x_r} \frac{dx_r}{\surd X_r}$$

$=$ the coefficient of $1/a^2$ in the expansion in ascending powers of $1/a$ of

$$\frac{1}{\surd A} \log \frac{\text{f}a - \phi a \,.\, \surd A}{\text{f}a + \phi a \,.\, \surd A}. \quad\ldots\ldots\ldots\ldots\ldots\ldots\ldots(14)$$

Thus, with $X = 4x^3 - g_2 x - g_3$, and $x = \wp v$,

then $$\zeta v = \int x dx / \surd X;$$

and $$\rho \text{ or } \Sigma(\zeta v_r - \zeta v'_r) = -2\, \text{lt} \frac{a^2}{\surd A} \tanh^{-1} \frac{\phi a}{\text{f}a} \surd A, \; (a = \infty). \quad (15)$$

Jacobi calls $\surd A$ the *factor of the Third Elliptic Integral.* (*Werke*, II., p. 494.)

217. Similar results hold when, as in § 167, X is supposed resolved into two factors, X_1 and X_2.

Denoting $P^2X_1 - Q^2X_2$ by ψx,
and varying the arbitrary coefficients in P and Q, and consequently the roots of $\psi x = 0$, as in § 162, then

$$\psi' x_r dx_r + 2P\delta P \,.\, X_1 - 2Q\delta Q \,.\, X_2 = 0,$$

while
$$P\surd X_1 + Q\surd X^2 = 0;$$

so that
$$\psi' x_r dx_r - 2(Q\partial P - P\delta Q)\surd(X_1X_2) = 0,$$

or
$$\frac{dx_r}{\surd X_r} = 2\frac{Q\partial P - P\delta Q}{\psi' x_r} = \frac{\theta x_r}{\psi' x_r},$$

and
$$\Sigma dx_r/\surd X_r = 0, \text{ or } \Sigma u_r = \Sigma u'_r.$$

Again, as in § 215,

$$\Sigma\frac{1}{a - x_r}\frac{dx_r}{\surd X_r} = \frac{\theta a}{\psi a} = 2\frac{\phi a \delta\, \mathrm{f}a - \mathrm{f}a\, \delta\phi a}{(\mathrm{f}a)^2 A_1 - (\phi a)^2 A_2}$$
$$= \frac{1}{\surd A}\frac{\delta \mathrm{f}a \,.\, \surd A_1 - \delta\phi a \,.\, \surd A_2}{\mathrm{f}a \,.\, \surd A_1 - \phi a \,.\, \surd A_2} - \frac{1}{\surd A}\frac{\delta \mathrm{f}a \,.\, \surd A_1 + \delta\phi a \,.\, \surd A_2}{\mathrm{f}a \,.\, \surd A_1 + \phi a \,.\, \surd A_2}$$
$$= \frac{1}{\surd A}\,\delta \log \frac{\mathrm{f}a \,.\, \surd A_1 - \phi a \,.\, \surd A_2}{\mathrm{f}a \,.\, \surd A_1 + \phi a \,.\, \surd A_2}.$$

Thus, as an application to the formulas of §§ 174, 176, 186, and 189, take, as in § 38 (Durège, *Elliptische Functionen*, § 36),

$$X = X_1 X_2, \text{ where } X_1 = x,\ X_2 = (1-x)(1-kx).$$

Then, with $x = \mathrm{sn}^2 u$,

$$\int_0 \frac{dx}{\surd X} = 2u, \quad \int_0 \frac{xdx}{\surd X} = \frac{2}{k}(u - E \operatorname{am} u),$$

and
$$\int \frac{a}{a-x}\frac{dx}{\surd X} = 2\Pi(n, \kappa, \phi),$$

in Legendre's notation, with $\phi = \operatorname{am} u$, and $n = -1/a$.

Now, if, as in §§ 164, 165, we take

$$P \text{ or } \mathrm{f}x = p + x, \text{ and } Q \text{ or } \phi x = q,$$

and denote by x_1, x_2, x_3, the roots of the equation (7), § 167,

$$\psi x, \text{ or } P^2X_1 - Q^2X_2, \text{ or } (p+x)^2 x - q^2(1-x)(1-kx) = 0;$$

then
$$x_1 x_2 x_3 = q^2,$$
$$1 - x_1 \,.\, 1 - x_2 \,.\, 1 - x_3 = (1+p)^2,$$
$$x_1 + x_2 + x_3 - kx_1x_2x_3 = -2p;$$

so that, as in § 164,

$$(2 - x_1 - x_2 - x_3 + kx_1x_2x_3)^2 = 4(1 - x_1 \,.\, 1 - x_2 \,.\, 1 - x_3),$$

where
$$u_1 + u_2 + u_3 = 0.$$

Again,

$$\int_0^{} \frac{a}{a-x_1}\frac{dx_1}{\sqrt{X_1}}+\int_0^{} \frac{a}{a-x_2}\frac{dx_2}{\sqrt{X_2}}+\int_0^{} \frac{a}{a-x_3}\frac{dx_3}{\sqrt{X_3}}$$

$$=\frac{a}{\sqrt{A}}\log\frac{\mathfrak{f}a\,.\,\sqrt{A_1}-\phi a\,.\,\sqrt{A_2}}{\mathfrak{f}a\,.\,\sqrt{A_1}+\phi a\,.\,\sqrt{A_2}} \text{ or } -\frac{2a}{\sqrt{A}}\tanh^{-1}\frac{\phi a}{\mathfrak{f}a}\frac{\sqrt{A_2}}{\sqrt{A_1}}, \quad (16)$$

since x_1, x_2, x_3 vanish when p and q are made zero; and this is equivalent to the result of equation (9), § 189, with $a=-1/n$,

$$\frac{A}{a^2}=\frac{1-a\,.\,1-ka}{a}=-(1+n)\left(1+\frac{k}{n}\right)=-a;$$

and
$$\frac{\phi a}{\mathfrak{f}a}\frac{\sqrt{A_2}}{\sqrt{A_1}}=\frac{q\sqrt{(1-a\,.\,1-ka)}}{(p+a)\sqrt{a}}=-\frac{nq\sqrt{(-a)}}{1-np}$$

$$=-\frac{n\sqrt{(-a)}x_1x_2x_3}{1+n-n\sqrt{(1-x_1\,.\,1-x_2\,.\,1-x_3)}}.$$

Similarly, for the Second Elliptic Integral,

$$\int_0^{}\frac{x_1dx_1}{\sqrt{X_1}}+\int_0^{}\frac{x_2dx_2}{\sqrt{X_2}}+\int_0^{}\frac{x_3dx_3}{\sqrt{X_3}}$$

$$=-\text{lt}\,\frac{2a^2}{\sqrt{(a\,.\,1-a\,.\,1-ka)}}\tanh^{-1}\frac{q\sqrt{(1-a\,.\,1-ka)}}{(p+a)\sqrt{a}} \quad (a=\infty)$$

$$=-2\text{lt}\left\{\frac{aq}{p+a}+\tfrac{1}{3}(1-a\,.\,1-ka)\frac{q^3}{(p+a)^3}+\dots\right\}$$

$$=-2q=-2\sqrt{(x_1x_2x_3)};\quad\dots\dots\dots(17)$$

as before, in §§ 174, 176, and 186.

218. Abel's *pseudo-elliptic integrals* are derived by making the u's equal in equations (7), (12); or the v's equal in equation (13); also by making their sum equal to a period ω_a, or the sum of multiples of periods, such as $p\omega_1+q\omega_3$.

Now $\mu\log\phi(u, v)$ is of the form $\log\Omega-\rho u$,

or $\phi(u, v)^\mu$ is of the form $e^{-\rho u}\Omega$,

where Ω is a rational integral function of $\wp u$ and $\wp' u$ of the form of C in (8), sometimes qualified by a divisor $\sqrt{(\wp u-e_a)}$.

We begin with the simplest case of an *algebraical herpolhode* by taking $v=\omega_1+\frac{1}{2}\omega_3$; and then, from equations (39) and (40), § 54, we can infer that the value of s, between e_1 and e_2, which

makes
$$\frac{e_1-e_2\,.\,e_1-e_3}{e_1-s}=\frac{e_1-e_2\,.\,e_2-e_3}{s-e_2}$$

is
$$s \text{ or } \wp v=e_3+\sqrt{(e_1-e_3\,.\,e_2-e_3)}.$$

Denoting $\wp u$ by s, $\wp' u$ by $\surd S$, and $\wp v$ by a, we infer that

$$\int \frac{ds}{(s-a)\surd S}$$

is pseudo-elliptic, that is, can be expressed in terms of

$$\int ds/\surd S \text{ and of } \tan^{-1}(Q\surd S/P).$$

In fact, by differentiation of

$$\theta = \tfrac{1}{2}\cos^{-1}\frac{\surd(s-e_1 . s-e_2)}{a-s} = \tfrac{1}{2}\sin^{-1}\frac{\{\surd(e_1-e_3)-\surd(e_2-e_3)\}\surd(s-e_3)}{a-s},$$

$$\frac{d\theta}{ds} = \tfrac{1}{2}\{\surd(e_1-e_3)-\surd(e_2-e_3)\}\frac{s+a-2e_3}{(s-a)\surd S}$$

$$= \frac{\surd(e_1-e_3)-\surd(e_2-e_3)}{2\surd S} - \frac{\frac{1}{2}i\wp' v}{(s-a)\surd S},$$

since $\quad i\wp' v = -2\surd(e_1-e_3 . e_2-e_3)\{\surd(e_1-e_3)-\surd(e_2-e_3)\}.$

In the herpolhode, therefore, of § 113, p. 108

$$\phi - \mu t = \tfrac{1}{2}\int \frac{i\wp' v\, du}{\wp v - \wp u} = \theta - \tfrac{1}{2}\{\surd(e_1-e_3)-\surd(e_2-e_3)\}nt,$$

or $$\theta = \phi - \mu t + \tfrac{1}{2}\{\surd(e_1-e_3)-\surd(e_2-e_3)\}nt,$$

and therefore, relatively to axes revolving with constant angular velocity,

$$\mu - \tfrac{1}{2}\{\surd(e_1-e_3)-\surd(e_2-e_3)\}n,$$

the herpolhode will be the algebraical curve, given by

$$\theta = \tfrac{1}{2}\cos^{-1}\frac{\surd(s-e_1 . s-e_2)}{a-s},$$

$$(a-s)\cos 2\theta = \surd(s-e_1 . s-e_2),$$

$$(a-s)^2\cos^2 2\theta = (a-s)^2 - (e_3+2a)(a-s) + (a-e_1)(a-e_2),$$

$$(a-s)^2\sin^2 2\theta + \{\surd(e_1-e_3)+\surd(e_2-e_3)\}^2(a-s)$$
$$- \surd(e_1-e_3 . e_2-e_3)\{\surd(e_1-e_3)-\surd(e_2-e_3)\}^2 = 0;$$

where, as in § 113, $a-s$, or $\wp v - \wp u = \dfrac{\mu^2}{n^2}\dfrac{\rho^2}{h^2}$.

Referred to Cartesian coordinates, in which

$$\rho^2 = x^2+y^2, \quad \rho^2\sin 2\theta = 2xy,$$

this equation becomes

$$\left[4x^2 + \{\surd(e_1-e_3)-\surd(e_2-e_3)\}^2\frac{n^2}{\mu^2}h^2\right]$$
$$\times\left[4y^2 + \{\surd(e_1-e_3)-\surd(e_2-e_3)\}^2\frac{n^2}{\mu^2}h^2\right] = (e_1-e_2)^2\frac{n^4}{\mu^4}h^4;$$

of the form $$(x^2+b^2)(y^2+b^2) = a^4. \qquad \text{.....................(18)}$$

The relation $\wp v - e_3 = \sqrt{(e_1 - e_3 \,.\, e_2 - e_3)}$,
combined with the equations of §§ 110, 113, leads to the relation

$$\frac{A-D\,.\,D-C}{D^2} = \frac{A-B\,.\,B-C}{B^2};$$

and either $B=D$, which gives the *separating* polhode ; or

$$\frac{1}{D} = \frac{1}{A} - \frac{1}{B} + \frac{1}{C};$$

the relation for this algebraical herpolhode.

Now, from §§ 108-110,

$$e_c - e_b = \left(\frac{D}{C} - \frac{D}{B}\right)^2 \frac{\mu^2}{n^2}, \quad e_a - e_b = \left(\frac{D}{B} - \frac{D}{A}\right)^2 \frac{\mu^2}{n^2};$$

while, with $A > B > D > C$, and $e_c = e_1$, $e_a = e_2$, $e_b = e_3$,

$$\sqrt{(e_1 - e_3)} - \sqrt{(e_2 - e_3)} = \left(\frac{D}{C} - 2\frac{D}{B} + \frac{D}{A}\right)\frac{\mu}{n} = \left(1 - \frac{D}{B}\right)\frac{\mu}{n},$$

$$e_1 - e_2 = e_c - e_a = \left(\frac{D}{C} - \frac{D}{A}\right)\left(1 - \frac{D}{B}\right)\frac{\mu^2}{n^2}.$$

To determine the confocal surface which will describe this algebraical herpolhode by rolling on a fixed tangent plane, we must equate the angular velocity of the axes to $\mu\lambda^2/h^2$; and now

$$\frac{\lambda^2}{h^2} = \frac{1}{2}\left(1 + \frac{D}{B}\right).$$

The squares of the semi-axes of the confocal are therefore

$$\frac{Dh^2}{A} - \lambda^2 = \left(\frac{D}{A} - \frac{1}{2} - \frac{1}{2}\frac{D}{B}\right)h^2 = -\frac{1}{2}\left(\frac{D}{C} - \frac{D}{A}\right)h^2,$$

$$\frac{Dh^2}{B} - \lambda^2 = \left(\frac{D}{B} - \frac{1}{2} - \frac{1}{2}\frac{D}{B}\right)h^2 = -\frac{1}{2}\left(1 - \frac{D}{B}\right)h^2,$$

$$\frac{Dh^2}{C} - \lambda^2 = \left(\frac{D}{C} - \frac{1}{2} - \frac{1}{2}\frac{D}{B}\right)h^2 = \frac{1}{2}\left(\frac{D}{C} - \frac{D}{A}\right)h^2;$$

while the square of the distance from the centre of the tangent plane on which this confocal rolls is given by

$$h^2 - \lambda^2 = \frac{1}{2}\left(1 - \frac{D}{B}\right)h^2.$$

The confocal is therefore a hyperboloid of two sheets, of the form

$$-\frac{x^2}{a^2} - \frac{y^2}{b^2} + \frac{z^2}{a^2} = 1;$$

and in rolling on a fixed tangent plane at a distance b from the centre, it will trace out the algebraical herpolhode (18), being the preceding herpolhode, changed in scale in the ratio of h to b (Halphen, *F. E.*, II., p. 285).

219. A more complicated case can be constructed by taking $v=\omega_1+\frac{1}{3}\omega_3$; but now we must choose particular numerical values for g_2 and g_3.

If we select the modular angle of 15°, then $2\kappa\kappa'=\frac{1}{2}$, and in (C), § 53, $J=5^3\div 4$, $J-1=11^2\div 4$; so that, by choosing $\Delta=108$, then $g_2=15, \quad g_3=11$;

and $e_1=\frac{1}{2}+\sqrt{3}, \quad e_2=-1, \quad e_3=\frac{1}{2}-\sqrt{3}$.

It is easily verified that, with the above value of v, $\wp v=\frac{1}{2}$; for $\wp 2v=-\frac{3}{2}=\wp 4v$; also this value of $\wp v$ or s makes, in equations (39) and (40), § 54,

$$\wp^{-1}\left(\frac{e_1-e_2 \,.\, e_2-e_3}{s-e_2};\ g_2,\ -g_3\right)=2\wp^{-1}\left(\frac{e_1-e_2 \,.\, e_1-e_3}{e_1-s};\ g_2,\ -g_3\right).$$

The corresponding elliptic integral of the third kind in the herpolhode will now be pseudo-elliptic; we find, in fact, that,

if $$\theta=\tfrac{1}{3}\sin^{-1}\frac{3\sqrt{(4s^2-4s-11)}}{(2s-1)^{\frac{3}{2}}}=\tfrac{1}{3}\cos^{-1}\frac{(2s-7)\sqrt{(2s+2)}}{(2s-1)^{\frac{3}{2}}},$$

$$\frac{d\theta}{ds}=\frac{1}{\sqrt{2}}\frac{2s+5}{2s-1}\frac{1}{\sqrt{S}}=\tfrac{1}{2}\sqrt{2}\frac{du}{ds}-\frac{\frac{1}{2}i\wp'v}{\wp u-\wp v}\frac{du}{ds},$$

since $i\wp'v=-3\sqrt{2}$; so that, in the herpolhode,

$$\phi-\mu t=\int\frac{\frac{1}{2}i\wp'v\,du}{\wp v-\wp u}=-\tfrac{1}{2}\sqrt{2}\,nt+\theta;$$

and therefore, relatively to axes revolving with constant angular velocity $\mu-\frac{1}{2}\sqrt{2}n$, the herpolhode will be the algebraic curve

$$\theta=\tfrac{1}{3}\sin^{-1}\frac{3\sqrt{(4s^2-4s-11)}}{(2s-1)^{\frac{3}{2}}},$$

or $$(1-2s)^3\sin^2 3\theta+9(1-2s)^2-108=0,$$

in which $$1-2s=2(\wp v-\wp u)=2\frac{\mu^2}{n^2}\frac{\rho^2}{h^2}=3\frac{\rho^2}{c^2}, \text{ suppose};$$

and now $$\rho^6\sin^2 3\theta+3c^2\rho^4-4c^6=0, \quad\text{..................}(19)$$

a curve, consisting of six equal waves, arranged on a circle.

With (i.) $A>B>D>C$, and

$$e_c=\tfrac{1}{2}+\sqrt{3},\ e_a=-1,\ e_b=\tfrac{1}{2}-\sqrt{3},\ \wp v=\tfrac{1}{2};$$

then (§ 113) $$\wp v-e_b=\quad\sqrt{3}=\quad\frac{\mu^2}{n^2}\frac{A-D\,.\,D-C}{AC},$$

$$\wp v-e_c=-\sqrt{3}=-\frac{\mu^2}{n^2}\frac{A-D\,.\,B-D}{AB};$$

so that $$\frac{A-D\,.\,D-C}{AC}=\frac{A-D\,.\,B-D}{AB}.$$

Then, either $A-D=0$, which would give a stable rotation about the axis A; or

$$\frac{2}{D}=\frac{1}{B}+\frac{1}{C}; \quad \text{..........................(20)}$$

so that D is the harmonic mean between B and C.

Again, $$\wp v-e_a=\frac{3}{2}=\frac{\mu^2}{n^2}\frac{B-D\,.\,D-C}{BC},$$

so that $$\tfrac{1}{2}\sqrt{3}=\frac{A-D}{B-D}\frac{B}{A},$$

or $$\frac{1}{D}-\frac{1}{A}=\frac{\sqrt{3}}{2}\left(\frac{1}{D}-\frac{1}{B}\right),\quad \frac{1}{B}+\frac{1}{C}-\frac{2}{A}=\frac{\sqrt{3}}{2}\left(\frac{1}{C}-\frac{1}{B}\right),$$

or $$\frac{1}{C}-\frac{1}{A}=-(2+\sqrt{3})^2\left(\frac{1}{B}-\frac{1}{A}\right); \quad \text{........(21)}$$

which is impossible, with $A>B>C$.

But (ii.), with $A>D>B>C$, we find that D is the harmonic mean between A and B; also

$$\frac{1}{C}-\frac{1}{A}=(2+\sqrt{3})^2\left(\frac{1}{B}-\frac{1}{A}\right), \quad \text{..................(22)}$$

so that $2+\sqrt{3}$ is the ratio of the semi-axes of the focal ellipse of the momental ellipsoid, and $\sqrt[4]{3}(\sqrt{3}-1)$ is the excentricity.

Another algebraic herpolhode can be constructed by taking $v=\omega_1+\frac{2}{3}\omega_3$; and, with $g_2=15$, $g_3=11$, we find that

$$\wp v=-\tfrac{5}{2}+\sqrt{3},\quad i\wp' v=-3\sqrt{2(2-\sqrt{3})}.$$

Now, if

$$\theta=\tfrac{1}{3}\sin^{-1}\frac{6(\sqrt{3}-1)\sqrt{(s-e_2\,.\,s-e_3)}}{(2s-2\sqrt{3}+5)^{\frac{3}{2}}}=\tfrac{1}{3}\cos^{-1}\frac{(2s-10+7\sqrt{3})\sqrt{(2s-2e_1)}}{(2s-2\sqrt{3}+5)^{\frac{3}{2}}},$$

$$\frac{d\theta}{ds}=-\frac{\sqrt{2}(\sqrt{3}-1)}{2\sqrt{S}}-\frac{3\sqrt{2(2-\sqrt{3})}}{(2s-2\sqrt{3}+5)\sqrt{S}};$$

so that

$$\int\frac{\frac{1}{2}i\wp' v\,du}{\wp v-\wp u}=\int\frac{-3\sqrt{2(2-\sqrt{3})}\,ds}{(2s-2\sqrt{3}+5)\sqrt{S}}$$

$$=\tfrac{3}{2}\sqrt{2}(\sqrt{3}-1)\int\frac{ds}{\sqrt{S}}+\tfrac{1}{3}\sin^{-1}\frac{6(\sqrt{3}-1)\sqrt{(s-e_2\,.\,s-e_3)}}{(2s-2\sqrt{3}+5)^{\frac{3}{2}}};$$

and now the algebraic herpolhode, with respect to revolving axes, is given by

$$(2s-2\sqrt{3}+5)^{\frac{3}{2}}\sin 3\theta=6(\sqrt{3}-1)\sqrt{(s-e_2\,.\,s-e_3)},$$

reducing to an equation of the form

$$\rho^6\sin^2 3\theta+P\rho^4+Q\rho^2+R=0 \quad \text{..................(23)}$$

With (i.) $A > B > D > C$, and

$$e_c = \tfrac{1}{2} + \sqrt{3},\quad e_a = -1,\quad e_b = \tfrac{1}{2} - \sqrt{3};\quad \wp v = -\tfrac{5}{2} + \sqrt{3},$$

$$\wp v - e_a = -\tfrac{3}{2} + \sqrt{3} = \frac{\mu^2}{n^2}\frac{B-D\,.\,D-C}{BC},$$

$$\wp v - e_b = -3 + 2\sqrt{3} = \frac{\mu^2}{n^2}\frac{A-D\,.\,D-C}{AC},$$

$$\wp v - e_c = -3 \qquad = -\frac{\mu^2}{n^2}\frac{A-D\,.\,B-D}{AB}.$$

Therefore $$\frac{A-D\,.\,D-C}{AC} = 2\frac{B-D\,.\,D-C}{BC},$$

and rejecting the factor $D-C$,

$$1 - \frac{D}{A} = 2\left(1 - \frac{D}{B}\right),\ \text{or}\ \frac{1}{A} + \frac{1}{D} = \frac{2}{B} \quad\ldots\ldots\ldots\ldots(24)$$

Also $$\frac{D-C}{A-D}\cdot\frac{A}{C} = \frac{2\sqrt{3}-3}{6},\ \text{or}\ \frac{1}{C} - \frac{1}{D} = \frac{2\sqrt{3}-3}{6}\left(\frac{1}{D} - \frac{1}{A}\right),$$

$$\text{or}\ \frac{1}{C} - \frac{1}{B} = \frac{2}{\sqrt{3}}\left(\frac{1}{B} - \frac{1}{A}\right),\quad \frac{1}{C} - \frac{1}{B} = (\sqrt{3}-1)^2\left(\frac{1}{C} - \frac{1}{A}\right),\quad\ldots\ldots\ldots(25)$$

so that the excentricity of the focal ellipse of the momental ellipsoid is $\sqrt{3}-1$.

With (ii.) $A > D > B > C$, we are led to an impossible result.

Points of Inflexion on the Herpolhodes.

220. The original herpolhodes drawn by Poinsot (*Théorie nouvelle de la rotation des corps*) were represented with points of inflexion, as curves undulating between two concentric circles on the *invariable plane*.

But it was pointed out by Hess, in 1880, and de Sparre (*Comptes Rendus*, Nov., 1884), that such points of inflexion cannot exist on Poinsot's original herpolhodes, which are curves always concave to the centre, as drawn in Routh's *Rigid Dynamics*, Chap. IX.; like the horizontal projection of the path of the bob of a conical pendulum, or like the path of the Moon relative to the Sun, a good figure of which is given in the *English Mechanic*, p. 337, June, 1891, by Mr. H. P. Slade.

The herpolhodes described on planes parallel to the invariable plane in Sylvester's representation are capable, however, of possessing points of inflexion, when the confocal of the momental ellipsoid attains a certain shape. (Hess, *Das Rollen einer Fläche zweiten Grades auf einer invariabeln Ebene*, Munich, 1880; de Sparre, *Comptes Rendus*, Aug., 1885.)

Denoting by h the constant distance from the centre of the plane upon which a quadric surface rolls, de Sparre shows that the herpolhode on the plane has points of inflexion, when the quadric is

(i.) an ellipsoid

$$\frac{x^2}{a^2}+\frac{y^2}{b^2}+\frac{z^2}{c^2}=1,\quad a^2<b^2<c^2,\ \text{if } h^2>b^2, \text{ and } \frac{1}{a^2}>\frac{1}{b^2}+\frac{1}{c^2};$$

(in a momental ellipsoid, $A<B+C$, or $\frac{1}{a^2}<\frac{1}{b^2}+\frac{1}{c^2}$, so that points of inflexion cannot exist on the herpolhode);

(ii.) a hyperboloid of one sheet

$$\frac{x^2}{a^2}+\frac{y^2}{b^2}-\frac{z^2}{c^2}=1,\quad a^2<b^2,\ \text{if } h^2<a^2, \text{ and } \frac{1}{a^2}>\frac{1}{b^2}+\frac{1}{c^2};$$

(iii.) a hyperboloid of two sheets

$$\frac{x^2}{a^2}-\frac{y^2}{b^2}-\frac{z^2}{c^2}=1,\quad b^2<c^2,\ \text{if } \frac{1}{b^2}>\frac{1}{a^2}+\frac{1}{c^2}, \text{ whatever the value of } h.$$

These herpolhodes being similar to the original herpolhode of the momental ellipsoid, when referred to axes rotating with constant angular velocity $\mu\lambda^2/h^2$, can be considered as defined by the polar coordinates ρ, θ, given in terms of the time t, by the equations of § 113,

$$\rho^2=k^2(\wp v-\wp u), \quad\text{.........................}(1)$$

$$\frac{d\theta}{dt}=m+\frac{\frac{1}{2}i\wp' v}{\wp v-\wp u}n \quad\text{.........................}(2)$$

with $\quad u=nt+\omega_3,\ v=\omega_1+t'\omega_3,\ m/\mu=1-\lambda^2/h^2.$

Denoting the velocity in the curve by V, and its radius of curvature by R, then, resolving normally,

$$\frac{V^3}{R}=\frac{d\rho}{dt}\frac{1}{\rho}\frac{d}{dt}\left(\rho^2\frac{d\theta}{dt}\right)-\rho\frac{d\theta}{dt}\left(\frac{d^2\rho}{dt^2}-\rho\frac{d\theta^2}{dt^2}\right),$$

which will be found to reduce to an equation of the form

$$\frac{V^3}{R}=P\rho^2+Qk^2; \quad\text{.........................}(3)$$

where $\quad P=m^3+3mn^2\wp v+n^3 i\wp' v,$

$$Q=\tfrac{3}{2}m^2 ni\wp' v-mn^2\wp'' v-\tfrac{1}{8}n^3 i\wp''' v;$$

and the corresponding herpolhodes will have points of inflexion when λ is chosen so that $P\rho^2+Q$ can vanish.

Thus Halphen points out that the algebraical herpolhode of § 218 will have points of inflexion, if $b^2<\frac{1}{2}a^2$.

221. The polhode being given by the intersection of the two quadric surfaces
$$Ax^2+By^2+Cz^2=Dh^2,$$
$$A^2x^2+B^2y^2+C^2z^2=D^2h^2,$$
we may in consequence write
$$lx^2=\frac{B-C}{A}(a^2+\lambda),\quad ly^2=\frac{C-A}{B}(b^2+\lambda),\quad lz^2=\frac{A-B}{C}(c^2+\lambda),$$
where
$$(B-C)a^2+(C-A)b^2+(A-B)c^2=lDh^2,$$
$$A(B-C)a^2+B(C-A)b^2+C(A-B)c^2=lD^2h^2;$$
and then
$$\frac{x^2}{a^2+\lambda}+\frac{y^2}{b^2+\lambda}+\frac{z^2}{c^2+\lambda}=1,$$
the equation of a system of confocal quadrics, on choosing l such that
$$l=\frac{B-C}{A}+\frac{C-A}{B}+\frac{A-B}{C}.$$
Then
$$b^2-c^2=-\frac{D.A-D.B-C}{ABC}h^2,\quad c^2-a^2=-\frac{D.B-D.C-A}{ABC}h^2,$$
$$a^2-b^2=-\frac{D.C-D.A-B}{ABC}h^2.$$
By varying λ along the polhode, we find
$$\frac{2}{x}\frac{dx}{dt}=\frac{1}{a^2+\lambda}\frac{d\lambda}{dt},\quad\text{or}\quad\frac{dx}{dt}=\frac{1}{2}\frac{x}{a^2+\lambda}\frac{d\lambda}{dt},\ \ldots,$$
so that the polhode is an orthogonal trajectory of the confocal surfaces, for any one of which λ is constant; and two ellipsoids can be drawn on which the curve is a polhode, of which the generating lines of the confocal hyperboloid through the points are normals.

When these confocals are hyperboloids of one sheet, the generating lines may be made of material rods or wires, jointed at the points of crossing; and now any such a system of rods forming a hyperboloid is capable of deformation, and assumes in succession the shape of the confocal hyperboloids; the trajectory of any fixed point on a rod being orthogonal to the hyperboloids, and therefore capable of being a polhode, if the hyperboloids are coaxial with the momental ellipsoid of the body. (*Messenger of Mathematics*, 1878; *Senate House Solutions for* 1878; Larmor, *Proceedings Cam. Phil. Society*, 1884, *Jointed Wickerwork*; Darboux and Mannheim, *Comptes Rendus*, 1885 and 1886.)

Darboux has shown (Despeyrous, *Cours de mécanique*, t. II., Notes XVII., XVIII.) that if we hold a given generator fixed, then any point fixed in any other generator will describe a sphere; thus, if a rod moves with three points P, Q, R on it connected by means of bars to three fixed centres A, B, C in a straight line, any other point S of the rod will describe a sphere about a centre D in the line ABC, such that the A. R. $(ABCD)$ is equal to the A. R. $(PQRS)$.

The point where the line PQR meets the generator parallel to ABC will describe a plane, the corresponding centre being at an infinite distance; and generally, if one generator is held fixed, any point on the parallel generator will describe a plane.

The herpolhode can now be described by taking a jointed hyperboloid, similar and similarly situated, and of half the size of the former one used for describing the polhode, with one generator fixed along the invariable line OC, and with the parallel generator along the normal PQ at P; and now, if P is moved in a direction perpendicular to the hyperboloid at P, it will describe a plane curve, which is the herpolhode.

222. Any point fixed in a body moving under no forces, whose co-ordinates with respect to the principal axes are represented by a, b, c, will have component velocities

$$cq-br,\quad ar-cp,\quad bp-aq, \text{ parallel to the principal axes;}$$

and will describe a curve whose projection on the invariable plane will be given, in polar co-ordinates ρ and ϕ, by (§§ 104-113)

$$\rho^2 = a^2+b^2+c^2-\left(\frac{aAp+bBq+cCr}{D\mu}\right)^2$$

$$=\frac{(bCr-cBq)^2+(cAp-aCr)^2+(aBq-bAq)^2}{D^2\mu^2},$$

$$\rho^2\frac{d\phi}{dt} = \{(b^2+c^2)p-abq-car\}\frac{Ap}{D\mu}$$

$$+\{(c^2+a^2)q-bcr-abp\}\frac{Bq}{D\mu}$$

$$+\{(a^2+b^2)r-cap-bcq\}\frac{Cr}{D\mu},$$

the moment of the velocity about the invariable line OC; and p, q, r are given as functions of t in §§ 32, 106, and 108.

The equations are much simplified when the point is fixed on one of the principal axes, when two of the three quantities a, b, c vanish; and it will be a useful exercise for the student to prove that, in these cases, the curve of projection on the invariable plane with respect to axes rotating with angular velocity G/A, G/B, G/C respectively, is given by an equation of the form

$$x+iy=k\phi(u, \omega_a-v), \text{ or } k\phi(u, \omega_b-v), \text{ or } k\phi(u, \omega_c-v).$$

Another useful exercise is to deduce Poinsot's relations when the co-ordinate axes fixed in the body are not principal axes.

Now, if the equation of the momental ellipsoid is

$$Ax^2+By^2+Cz^2-2A'yz-2B'zx-2C'xy=Dh^2;$$

and if p, q, r denote as before the component angular velocities, and h_1, h_2, h_3 the components of angular momentum about the axes, the three equations of motion under no forces are

$$\frac{dh_1}{dt}-h_2r+h_3q=0, \quad \frac{dh_2}{dt}-h_3p+h_1r=0, \quad \frac{dh_3}{dt}-h_1q+h_2p=0,$$

where

$$h_1=Ap-C'q-B'r, \quad h_2=Bq-A'r-C'p, \quad h_3=Cr-B'p-A'q;$$

and these equations are solvable by elliptic functions.

(Dissertation *Ueber die Integration eines Differentialgleichungssystems*; Paul Hoyer, Berlin, 1879.)

223. The numerical results obtained in the preceding algebraical herpolhodes can be utilized in the corresponding problems of the revolving chain (§§ 205-206) and of the Tortuous Elastica (§ 207).

Putting $t'=\frac{1}{2}$, or $v=\frac{1}{2}\omega_3$ in § 206,

then $$\wp v=e_3-\sqrt{(e_1-e_3\,.\,e_2-e_3)},$$

$$i\wp'v=2\sqrt{(e_1-e_3\,.\,e_2-e_3)}\{\sqrt{(e_1-e_3)}+\sqrt{(e_2-e_3)}\};$$

and $$\psi=\int\frac{\frac{1}{2}i\wp'v\,du}{\wp u-\wp v}$$

$$=\tfrac{1}{2}\cos^{-1}\frac{\sqrt{(\wp u-e_1\,.\,\wp u-e_2)}}{\wp u-\wp v}-\tfrac{1}{2}\{\sqrt{(e_1-e_3)}+\sqrt{(e_2-e_3)}\}(u-\omega_3),$$

or $$(s-\wp v)\cos[2\psi+\{\sqrt{(e_1-e_3)}+\sqrt{(e_2-e_3)}\}x\omega_1/a]=\sqrt{(s-e_1.s-e_2)},$$

where $$s-\wp v=r^2/k^2.$$

In the corresponding problem of the Tortuous Elastica of § 207, it is merely requisite to replace x by the arc s.

The working out of the analogies for the other algebraical herpolhodes is left as an exercise; merely mentioning that

$$\wp(\tfrac{2}{3}\omega_3;\ 15,\ 11) = -\tfrac{3}{2},$$

and that, if

$$\theta = \tfrac{1}{3}\sin^{-1}\frac{\sqrt{2}\sqrt{(4s^3-15s-11)}}{(2s+3)^{\frac{3}{2}}} = \tfrac{1}{3}\cos^{-1}\frac{6s+7}{(2s+3)^{\frac{3}{2}}},$$

$$\frac{d\theta}{ds} = \frac{1}{\sqrt{2}}\frac{2s+1}{2s+3}\frac{1}{\sqrt{S}} = \frac{1}{\sqrt{2}}\frac{1}{\sqrt{S}} - \frac{i\wp' v}{2s+3}\frac{1}{\sqrt{S}},$$

or

$$\int\frac{\frac{1}{2}i\wp' v\,du}{\wp u - \wp v} = \frac{u}{\sqrt{2}} - \frac{1}{3}\sin^{-1}\frac{\frac{1}{2}\wp' u}{(\wp u - \wp v)^{\frac{3}{2}}}.$$

224. The analytical expressions in §§ 208, 210 for the motion of the Spherical Pendulum and of the Top or Gyrostat show, by comparison with the equations of the herpolhode in § 200, that this motion may be considered as compounded of two Poinsot representations of the motion of a body under no forces, as given in §§ 104, 214 (Jacobi, *Werke*, II., p. 477).

The relations connecting these two component Poinsot motions have engaged the attention of Darboux (Despeyrous, *Cours de mécanique*, II., Note XIX.), of Halphen (*F. E.*, II., Chap. III.), and of Routh (*Q. J. M.*, XXIII.).

We may put the conclusions arrived at by these mathematicians in the following condensed form, depending on fundamental dynamical and geometrical considerations.

(i.) If the vector OH represents the axis of resultant angular momentum, then H lies in a horizontal plane through the point G, where the vertical vector OG represents G, the constant component of angular momentum about the vertical.

(ii.) If the plane drawn through H, perpendicular to the axis of the Top, cuts this axis in C, then $OC = Cr$, the constant component of angular momentum about OC, the axis of the Top.

(iii.) These two planes, one horizontal and through G, which we shall call *the invariable plane of* G, and the other through C and perpendicular to OC, which we shall call *the invariable plane of* C, intersect in a line HK perpendicular to the vertical plane GOC; and if HK meets the plane GOC in K, then

$$CH^2 - GH^2 = CK^2 - GK^2 = OG^2 - OC^2 = G^2 - C^2r^2.$$

(iv.) The instantaneous axis of rotation OI lies in the plane HOC; and if OI meets CH in I, the resultant angular velocity

about OI is OI/C; also $CI/CH=C/A$,
and the velocity of C is $r \,.\, CI$.

(v.) By equation (i.) of § 210, the square of the velocity of C is

$$(2C^2r^2Wg/A)(c-h\cos\theta);$$

so that

$$CI^2=(2C^2Wg/A)(c-h\cos\theta),$$
$$CH^2=2AWg(c-h\cos\theta)$$
$$=2AWghk(\wp w-\wp u), \text{ suppose.}$$

Then, by equation (3) of § 210, with $u=nt+\omega_3$,

$$\tfrac{1}{2}ln^2k^2\wp'^2u=gk^3(\wp u-\wp a)(\wp u-\wp b)(\wp u-\wp w)-(\alpha+\beta\wp u)^2;$$

and therefore, when $u=a, b, w$, we have three equations of the form $i\wp'a=\alpha+\beta\wp a, -i\wp'b=\alpha+\beta\wp b, i\wp'w=\alpha+\beta\wp w$;
so that, according to § 165, we may put $w=b-a$.

(vi.) Now $GH^2=2AWghk\{\wp(b-a)-\wp u\}-G^2+C^2r^2$

$$=2AWghk(\wp w'-\wp u), \text{ suppose,}$$

where

$$\wp w'-\wp(a+b)=-(G^2-C^2r^2)/2AWghk;$$

and since

$$i\frac{G+Cr}{\sqrt{(2AWghk)}}=-\tfrac{1}{2}k\wp'a,\quad i\frac{G-Cr}{\sqrt{(2AWghk)}}=\tfrac{1}{2}k\wp'b,$$

and

$$2=k(\wp b-\wp a),$$

therefore

$$\wp w'-\wp(b-a)=-\frac{\wp'a\wp'b}{(\wp b-\wp a)^2},$$

and therefore (§ 151) we may put $w'=b+a$.

(vii.) The point H moves in the invariable plane of G with velocity equal to the impressed couple of gravity, and parallel to the axis of the couple; so that the velocity of H is in the direction HK, and equal to $Wgh\sin\theta$; and the moment of this velocity about G is $Wgh\sin\theta \,.\, GK$.

But

$$GK\sin\theta=OC-OG\cos\theta,$$

so that

$$\rho^2(d\phi/dt)=Wgh(Cr-G\cos\theta),$$

if ρ, ϕ denote the polar coordinates of H in the invariable plane of G.

Now

$$\rho^2=2AWghk\{\wp(b+a)-\wp u\},$$

and

$$\cos\theta=k(\wp u-\tfrac{1}{2}\wp a-\tfrac{1}{2}\wp b);$$

so that finally we shall find, after reduction,

$$\frac{d\phi}{dt}=\frac{G}{2A}+\frac{\tfrac{1}{2}i\wp'(b+a)}{\wp(b+a)-\wp u}n;$$

and therefore H describes in the invariable plane of G a herpolhode with parameter $b+a$.

(viii.) Similar considerations will show that the curve described by H in the invariable plane of C is also a herpolhode, with parameter $b-a$.

If in equation (2) of § 210 we replace Cr by Ar', the motion of OC is unaltered, but now the momental ellipsoid at O becomes a sphere, and OH is the instantaneous axis of rotation; so that the motion of OC is produced by rolling the cone, whose base is the herpolhode described by H in the invariable plane of C, on the cone whose base is the herpolhode in the invariable plane of G, the angular velocity being proportional to OH.

(ix.) But in the general case, where OI is the instantaneous axis, the curve described by I in the invariable plane of C is similar to the curve described by H, and is therefore a herpolhode.

Now from (v.), drawing CM, IN perpendicular to OG,

$$\begin{aligned} OI^2 &= OC^2 + CI^2 \\ &= C^2r^2 + (2C^2Wg/A)(c - OG + GM) \\ &= C^2r^2 + \frac{2C^2Wg}{A}\left(c - OG + \frac{A}{C-A}\cdot GN\right), \end{aligned}$$

so that OI^2 varies as the height of I above a certain horizontal plane; and the locus of I is therefore a sphere, to which the point O and this plane are related as limiting point and radical plane.

The motion of the Top can therefore be produced by rolling the herpolhode described by I in the invariable plane of C on this sphere, with angular velocity proportional to OI.

(x.) It still remains to be shown that the cone described by OI in space round OG is a herpolhode cone; this is left as an exercise.

Darboux shows that two such hyperboloids as those described in § 221, with a pair of generating lines, PQ, PQ' in coincidence, and the opposite generators OG, OC of the same system intersecting in a fixed point O, may be used to represent the motion of OC, the axis of a Top, when OG is held vertical; the point P of intersection of the coincident generators being made to describe herpolhodes in the invariable planes of G and C, by being moved in the direction of the common normal of the hyperboloids.

225. The numerical results of the pseudo-elliptic integrals of §§ 218, 219, and 223 can be utilised for the construction of similar degenerate cases of the motion of the Top.

Thus, if $a=\frac{1}{2}\omega_3,\ b=\omega_1+\frac{1}{2}\omega_3,$

then $b+a=\omega_1+\omega_3,\ b-a=\omega_1;$

and we shall find $\cos\alpha=0$, $\cos\beta=\kappa$, $d=\sec\beta$, and

$$C^2r^2=2AWgh\sec\beta,\quad G^2=2AWgh\cos\beta.$$

The spherical curve described by C is now given by

$$\sin\theta\sin(nt\cos\beta-\psi)=\sqrt{\{\cos\theta(\cos\beta-\cos\theta)\}},$$
$$\sin\theta\cos(nt\cos\beta-\psi)=\sqrt{(1-\cos\beta\cos\theta)}.$$

With $a=\frac{1}{2}\omega_3,\ b=\omega_1-\frac{1}{2}\omega_3,$ and $b+a=\omega_1,$

we find that $\cos\alpha$, $\cos\beta$, and d are unaltered, but Cr and G are interchanged; and C now describes the spherical curve

$$\sin\theta\sin(nt-\psi)=\sqrt{\{\cos\theta(\sec\beta-\cos\theta)\}},$$
$$\sin\theta\cos(nt-\psi)=\sqrt{(1-\sec\beta\cos\theta)}.$$

Again, with $a=\frac{2}{3}\omega_3,\ b=\omega_1-\frac{1}{3}\omega_3,\ g_2=15,\ g_3=11;$

so that $\wp a=-\frac{3}{2}$, $\wp b=\frac{1}{2}$, we find that

$k=1$, $\cos\alpha=-\sqrt{3}+1$, $\cos\beta=-\frac{1}{2}$, $d=\sqrt{3}+1$, $C^2r^2=4AWgh;$

and the spherical curve described by C is given by

$$\sin^3\theta\sin 3\psi=(-1-2\cos\theta)^{\frac{3}{2}},$$
$$\sin^3\theta\cos 3\psi=(1+\cos\theta+\cos^2\theta)\sqrt{(2+2\cos\theta-\cos^2\theta)}.$$

To realise this motion practically, place a homogeneous sphere, of radius c, inside a fixed spherical bowl of radius a, in contact at an angular distance of 60° from the lowest point, and spin the sphere about the common normal with angular velocity

$$\sqrt{\left\{35\frac{g}{c}\left(\frac{a}{c}-1\right)\right\}}.$$

The sphere if released will roll on the interior in this curve.

As another numerical illustration we may take

$$g_2=48,\quad g_3=44;$$

when $\wp(\omega_1+\frac{1}{3}\omega_3)=2,\ \wp\frac{2}{3}\omega_3=-4;$

$$\wp'(\omega_1+\tfrac{1}{3}\omega_3)=-\wp'\tfrac{2}{3}\omega_3=6i\sqrt{3}.$$

Also, with $g_2=30,\quad g_3=28,\quad \omega_3/\omega_1=i\sqrt{2},$

$$\wp\tfrac{1}{3}\omega_3=-5-\tfrac{3}{2}\sqrt{6},\quad \wp\tfrac{2}{3}\omega_3=1-\tfrac{3}{2}\sqrt{6},\ \text{etc.}$$

226. It is convenient to represent the two parts of ψ by ψ_1 and ψ_2, such that

$$\frac{d\psi_1}{dt}=\frac{1}{2}\,\frac{G+Cr}{A}\,\frac{1}{1+\cos\theta},\quad \frac{d\psi_1}{du}=\frac{\frac{1}{2}i\wp'a}{\wp u-\wp a};$$

$$\frac{d\psi_2}{dt}=\frac{1}{2}\,\frac{G-Cr}{A}\,\frac{1}{1-\cos\theta},\quad \frac{d\psi_2}{du}=\frac{\frac{1}{2}i\wp'b}{\wp b-\wp u};$$

also to put $\chi=\psi_1-\psi_2$, whence Euler's angle $\phi=\chi+(A-C)rt/A$, and

$$\frac{d\chi}{dt}=\frac{Cr-G\cos\theta}{A\sin^2\theta},$$

an expression obtained by interchanging G and Cr in ψ.

With $a=p\omega_3$, $b=\omega_1+q\omega_3$, a change of q into $-q$ interchanges G and Cr, while a change of p into $-p$ interchanges G and $-Cr$: both changes of sign change G and $-G$ and Cr into $-Cr$, and thus reverse the motion.

The following degenerate cases of the motion of the Top will afford an exercise on the preceding results of §§ 210, 224:—

A. With $b-a=\omega_1$, or $q-p=0$,

$$d=\frac{G}{Cr}=\frac{c}{h}=\frac{1+\cos\alpha\cos\beta}{\cos\alpha+\cos\beta},$$

$$C^2r^2/2AWgh=\cos\alpha+\cos\beta;$$

and by § 215, χ is now *pseudo-elliptic;* and

$$\chi=\surd(\cos\alpha+\cos\beta)\surd(\tfrac{1}{2}g/l)t-\xi,$$

where
$$\xi=\tan^{-1}\sqrt{\frac{(\cos\beta-\cos\theta)(\cos\theta-\cos\alpha)}{1+\cos\alpha\cos\beta-(\cos\alpha+\cos\beta)\cos\theta}}$$

$$=\sin^{-1}\frac{\surd\{(\cos\beta-\cos\theta)(\cos\theta-\cos\alpha)\}}{\sin\theta}$$

$$=\cos^{-1}\frac{\surd\{1+\cos\alpha\cos\beta-(\cos\alpha+\cos\beta)\cos\theta\}}{\sin\theta}.$$

The angular velocity of H round G in the invariable plane or G is now constant and equal to $\frac{1}{2}G/A$.

B. With $b-a=\omega_1+\omega_3$, or $q-p=1$,

$$\cos\beta=\frac{G}{Cr}=\frac{c}{h}=\frac{1+d\cos\alpha}{\cos\alpha+d},$$

$$C^2r^2/2AWgh=\cos\alpha+d,$$

and the spherical curve described by C has cusps on the circle given by $\theta=\beta$; and now

$$\chi=\surd(\cos\alpha+d)\surd(\tfrac{1}{2}g/l)t-\xi',$$

where
$$\xi'=\tan^{-1}\sqrt{\frac{(d-\cos\theta)(\cos\theta-\cos\alpha)}{1+d\cos\alpha-(\cos\alpha+d)\cos\theta}},\ \text{etc.}$$

The angular velocity of H round G is again equal to $\frac{1}{2}G/A$.

C. With $b+a=\omega_1$, or $q+p=0$,

$$d=\frac{Cr}{G}=\frac{1+\cos a\cos\beta}{\cos a+\cos\beta};$$

and now ψ is pseudo-elliptic, and given by

$$\psi=\surd(\cos a+\cos\beta)\surd(\tfrac{1}{2}g/l)t-\xi;$$

while the angular velocity of H round C in the invariable plane of C is constant and equal to $\frac{1}{2}Cr/A$.

D. With $b+a=\omega_1+\omega_3$, or $q+p=1$,

$$\cos\beta=\frac{Cr}{G}=\frac{1+d\cos a}{\cos a+d},$$

$$\psi=\surd(\cos a+d)\surd(\tfrac{1}{2}g/l)t-\xi',$$

and the angular velocity of H round C in the invariable plane of C is again $\frac{1}{2}Cr/A$.

E. With $q=1$, $b=\omega_1+\omega_3$, $G-Cr=0$, and ψ_2 disappears; and now $\cos\beta=c/h=1$, the Top being spun originally in the upright position.

Now if the Top falls ultimately to the extreme inclination a, we find that $\quad C^2r^2/2AWgh=1+\cos a;$

and subsequently, after a time t,

$$\sin\tfrac{1}{2}\theta=\sin\tfrac{1}{2}a\operatorname{sech}\{\sin\tfrac{1}{2}a\surd(g/l)t\},$$

$$\psi=\frac{Crt}{2A}-\sin^{-1}\sqrt{\frac{\cos\theta-\cos a}{1+\cos\theta}};$$

so that the integrals for t and ψ are pseudo-elliptic.

F. With $q=0$, $b=\omega_1$, $G-Cr=0$, and ψ_2 again disappears; but now $d=1$, and the Top does not rise to the vertical position.

For numerical illustrations of this motion, take

$$a=\tfrac{2}{3}\omega_3, \text{ and } g_2=15,\ g_3=11, \text{ when } \wp a=-\tfrac{3}{2};$$

or $$g_2=48,\ g_3=44, \text{ when } \wp a=-4.$$

G. With $p=1$, $a=\omega_3$, $G+Cr=0$, and ψ_1 disappears; now $\cos a=-1$, and the Top passes through its lowest position.

For numerical examples of pseudo-elliptic cases, employ the results $\wp(\omega_1+\frac{1}{3}\omega_3;\ 15, 11)=\frac{1}{2}$, and $\wp(\omega_1+\frac{1}{3}\omega_3;\ 48, 44)=2$.

H. With $p=1$ and $q=1$, $G=0$ and $Cr=0$; and the motion reduces to plane revolutions, as in § 18.

I. With $p=1$ and $q=0$, $G=0$ and $Cr=0$; and the motion reduces to plane oscillations, as in § 3.

K. With $p=1, q=0$, $d=1$, $\cos\beta=-1$, $\cos a=-1$, the pendulum is at rest in its lowest position.

The Trajectory of a Projectile, for the Cubic Law of Resistance.

227. An immediate application of the function $\phi(u, v)$ of § 201 occurs in the solution of the motion of a body under gravity in a resisting medium, in which it is assumed that the resistance of the medium is in the direction opposite to motion, and that it varies as the *cube* of the velocity.

Refer the motion to oblique coordinate axes, one Ox in the direction of projection at the point of infinite velocity, and the other Oy drawn vertically downwards.

Denote by w the *terminal* velocity of the projectile in the medium; so that if W denotes the weight in pounds, the resistance of the air at a velocity v is a force of $W(v/w)^3$ pounds, and the retardation produced is $g(v/w)^3$.

The equations of motion are then

$$\frac{d^2x}{dt^2}=-\frac{g}{w^3}\left(\frac{ds}{dt}\right)^3\frac{dx}{ds}, \quad\ldots\ldots(1)$$

$$\frac{d^2y}{dt^2}=-\frac{g}{w^3}\left(\frac{ds}{dt}\right)^3\frac{dy}{ds}+g. \quad\ldots\ldots(2)$$

Eliminating the term due to the resistance,

$$\frac{dx}{dt}\frac{d^2y}{dt^2}-\frac{d^2x}{dt^2}\frac{dy}{dt}=g\frac{dx}{dt},$$

or, writing p for dy/dx,

$$\frac{dp}{dt}=g\frac{dt}{dx}, \text{ or } \frac{dp}{dt}\frac{dx}{dt}=g. \quad\ldots\ldots(3)$$

If Ox makes an angle α with the horizon, then

$$\frac{ds^2}{dt^2}=\frac{dy^2}{dt^2}-2\frac{dy}{dt}\frac{dx}{dt}\sin\alpha+\frac{dx^2}{dt^2}$$

$$=\frac{dx^2}{dt^2}(p^2-2p\sin\alpha+1),$$

and now equation (1) becomes

$$\frac{d^2x}{dt^2}=-\frac{g}{w^3}\left(\frac{ds}{dt}\right)^2\frac{dx}{dt}$$

$$=-\frac{g}{w^3}\left(\frac{dx}{dt}\right)^3(p^2-2p\sin\alpha+1),$$

or

$$w^3\left(\frac{dx}{dt}\right)^{-4}\frac{d^2x}{dt^2}=-\frac{dp}{dt}(p^2-2p\sin\alpha+1), \quad\ldots\ldots(4)$$

Integrating, noticing that $dx/dt=\infty$, when $p=0$,

$$\tfrac{1}{3}w^3\left(\frac{dx}{dt}\right)^{-3}=\tfrac{1}{3}p^3-p^2\sin\alpha+p=\tfrac{1}{3}P,$$

suppose, where $p^3-3p^2\sin\alpha+3p$ is denoted by P;

or $$\frac{dx}{dt}=wP^{-\frac{1}{3}}. \quad\text{.......................(5)}$$

Then, from (3), $$\frac{dp}{dx}=g\left(\frac{dx}{dt}\right)^{-2}=\frac{g}{w^2}P^{\frac{2}{3}},$$

so that $$\frac{g}{w^2}\frac{dx}{dp}=P^{-\frac{2}{3}},$$

$$\frac{g}{w^2}\frac{dy}{dp}=pP^{-\frac{2}{3}},$$

and $$\frac{gx}{w^2}=\int_0 P^{-\frac{2}{3}}dp, \quad\text{.......................(6)}$$

$$\frac{gy}{w^2}=\int_0 pP^{-\frac{2}{3}}dp; \quad\text{.......................(7)}$$

while $$\frac{dp}{dt}=\frac{g}{w}P^{\frac{1}{3}},$$

$$\frac{gt}{w}=\int_0 P^{-\frac{1}{3}}dp. \quad\text{.......................(8)}$$

228. The integration required in (6) is similar to that of ex. 8, p. 65, discussed also in § 157; we substitute

$$z=m^2P^{\frac{1}{3}}/p,$$

where m is some arbitrary constant factor; and then

$$4z^3-g_3=\{(4m^6-g_3)p^2-12m^6p\sin\alpha+12m^6\}/p^2,$$

which is a perfect square, when

$$4m^6-g^3=3m^6\sin^2\alpha,\ \text{ or }\ g_3=m^6(4-3\sin^2\alpha);$$

so that $$\sqrt{(4z^3-g_3)}=m^3\sqrt{3}(2-p\sin\alpha)/p,$$

and $$\frac{6z^2dz}{\sqrt{(4z^3-g_3)}}=-\frac{2m^2\sqrt{3}dp}{p^2},$$

or $$\frac{dz}{\sqrt{(4z^3-g_3)}}=-\frac{m^3\sqrt{3}dp}{3p^2z^2}=-\frac{dp}{m\sqrt{3}P^{\frac{2}{3}}}=-\frac{gdx}{w^2},$$

on choosing $m^2=\frac{1}{3}$; so that

$$\frac{gx}{w^2}=\int^{\infty}\frac{dz}{\sqrt{(4z^3-g_3)}},$$

$$z=\wp\left(\frac{gx}{w^2};\ 0,\ g_3\right). \quad\text{.......................(9)}$$

Then $$\wp'\frac{gx}{w^2}=\frac{p\sin\alpha-2}{3p};$$

and supposing $x=a$ at the vertical asymptote, where $p=\infty$,

$$\wp'\frac{ga}{w^2}=\frac{\sin\alpha}{3},\quad \wp'\frac{ga}{w^2}=\frac{1}{3},$$

so that $$\wp'\frac{ga}{w^2}-\wp'\frac{gx}{w^2}=\frac{2}{3p},$$

or $$p=\frac{dy}{dx}=\frac{\frac{2}{3}}{\wp'\frac{ga}{w^2}-\wp'\frac{gx}{w^2}}=\frac{6\wp^2\frac{ga}{w^2}}{\wp'\frac{ga}{w^2}-\wp'\frac{gx}{w^2}};\ \ldots\ldots\ldots(10)$$

and, integrating, $$y=\int_0\frac{6\wp^2\frac{ga}{w^2}}{\wp'\frac{ga}{w^2}-\wp'\frac{gx}{w^2}}dx,$$

the equation of the trajectory.

It is convenient to write u and v for gx/w^2 and ga/w^2;

and now $$\frac{gy}{w^2}=\int_0\frac{6\wp^2v\,du}{\wp'v-\wp'u},\ \ldots\ldots\ldots\ldots\ldots\ldots\ldots(11)$$

to be integrated by the preceding rules of § 198.

Rationalizing the denominator $\wp'v-\wp'u$, it becomes

$$\wp'^2v-\wp'^2u \text{ or } 4(\wp^3v-\wp^3u),$$

since $g_2=0$; and resolved into linear factors, it becomes

$$4(\wp v-\wp u)(\omega\wp v-\wp u)(\omega^2\wp v-\wp u),$$

where ω, ω^2 denote the imaginary cube roots of unity, viz.,

$$\omega=-\tfrac{1}{2}+\tfrac{1}{2}\sqrt{3}i,\ \omega^2=-\tfrac{1}{2}-\tfrac{1}{2}\sqrt{3}i.$$

Now, resolved into partial fractions,

$$\frac{6\wp^2v}{\wp'v-\wp'u}=\frac{6\wp^2v(\wp'v+\wp'u)}{4(\wp^3v-\wp^3u)}$$

$$=\frac{1}{2}\frac{\wp'v+\wp'u}{\wp v-\wp u}+\frac{1}{2}\omega\frac{\wp'v+\wp'u}{\omega\wp v-\wp u}+\frac{1}{2}\omega^2\frac{\wp'v+\wp'u}{\omega^2\wp v-\wp u}$$

$$=\frac{1}{2}\frac{\wp'v+\wp'u}{\wp v-\wp u}+\frac{1}{2}\omega\frac{\wp'\omega v+\wp'u}{\wp\omega v-\wp u}+\frac{1}{2}\omega^2\frac{\wp'\omega^2v+\wp'u}{\wp\omega^2v-\wp u}\ldots(12)$$

on making use of the results of § 196, when $g_2=0$.

Then

$$\frac{gy}{w^2}=\int_0\frac{1}{2}\frac{\wp'v+\wp'u}{\wp v-\wp u}du+\omega\int_0\frac{1}{2}\frac{\wp'\omega v+\wp'u}{\wp\omega v-\wp u}du+\omega^2\int_0\frac{1}{2}\frac{\wp'\omega^2v+\wp'u}{\wp\omega^2v-\wp u}du,$$

which is prepared for integration as required in § 198; and since

$$\int \frac{1}{2}\frac{\wp' v+\wp' u}{\wp v-\wp u}du=\int\{\zeta(v-u)+\zeta u-\zeta v\}du$$

$$=-\log\sigma(v-u)+\log\sigma u-u\zeta v+\text{constant}$$

$$=-\log\frac{\sigma(v-u)}{\sigma v\,\sigma u}e^{u\zeta v}=-\log\phi(-u,\ v);$$

therefore the result of the integration may be expressed by

$$\frac{gy}{w^2}=-\log\phi(-u,\ v)-\omega\log\phi(-u,\ \omega v)-\omega^2\log\phi(-u,\ \omega^2 v)\text{....}(13)$$

The conditions of Homogeneity of § 196 also show that the last equation (13) may be written

$$\frac{gy}{w^2}=-3u\zeta v-\log\frac{\sigma(v-u)}{\sigma v}-\omega\log\frac{\sigma(\omega v-u)}{\sigma\omega v}-\omega^2\log\frac{\sigma(\omega^2 v-u)}{\sigma\omega^2 v},$$

or simply

$$\frac{gy}{w^2}=-3u\zeta v-\log\sigma(v-u)-\omega\log\sigma(\omega v-u)-\omega^2\log\sigma(\omega^2 v-u),\ (14)$$

subject to the condition that $y=0$, when u or $x=0$.

The equation is left in the complex imaginary form, as there exists no theorem for the expression of

$$\log\sigma(\omega v-u)\text{ in the form }P+iQ;$$

unless we introduce a new function $\Phi(a,\ a)$, defined by (Halphen, *F. E.*, I., p. 151)

$$\Phi(a,\ a)=\int_0^a\{\zeta(a+ia)+\zeta(a-ia)\}da.$$

229. For the expression of the time t in the trajectory, equation (8) leads to

$$\frac{gt}{w}=\int\frac{6\wp v\,\wp u}{\wp' v-\wp' u}du$$

$$=\int\frac{1}{2}\frac{\wp' v+\wp' u}{\wp v-\wp u}du+\omega^2\int\frac{1}{2}\frac{\wp'\omega v+\wp' u}{\wp\omega v-\wp u}du+\omega\int\frac{1}{2}\frac{\wp'\omega^2 v+\wp' u}{\wp\omega^2 v-\wp u}du,\ (15)$$

when resolved, as before for y, into partial fractions; so that

$$\frac{gt}{w}=-\log\phi(-u,\ v)-\omega^2\log\phi(-u,\ \omega v)-\omega\log\phi(-u,\ \omega^2 v),$$

or $$=-\log\frac{\sigma(v-u)}{\sigma v}-\omega^2\log\frac{\sigma(\omega v-u)}{\sigma\omega v}-\omega\log\frac{\sigma(\omega^2 v-u)}{\sigma\omega^2 v},$$

or simply

$$=-\log\sigma(v-u)-\omega^2\log\sigma(\omega v-u)-\omega\log\sigma(\omega^2 v-u),\ \ (16)$$

subject to the condition that $t=0$, when x or $u=0$.

By addition,

$$\frac{gy}{w^2}+\frac{gt}{w}=-3\log\phi(-u,\,v)+\log\phi(-u,\,v)\phi(-u,\,\omega v)\phi(-u,\,\omega^2 v);$$

$$=-\log\left\{\frac{\sigma(v-u)}{\sigma v\,\sigma u}e^{u\zeta v}\right\}^3+\log\frac{\sigma(v-u)\sigma(\omega v-u)\sigma(\omega^2 v-u)}{\sigma v\,\sigma\omega v\,\sigma\omega^2 v\,\sigma^3 u},$$

and this last term, when expressed in a real form, is equal to

$$\log\tfrac{1}{2}(\wp' v-\wp' u).$$

(Halphen, *F. E.*, I., p. 232.)

This can be proved independently; for

$$\frac{gy}{w^2}+\frac{gt}{w}+3\log\phi(-u,\,v)$$

$$=\int\frac{6\wp v(\wp v+\wp u)}{\wp' v-\wp' u}du+3\int\frac{1}{2}\frac{\wp' v+\wp' u}{\wp v-\wp u}du$$

$$=\int\frac{-6\wp^2 u\,du}{\wp' v-\wp' u}=\log(\wp' v-\wp' u)+\text{a constant.} \quad\text{..........(17)}$$

230. For the purpose of the expression of y and t in ascending powers of x or u, it is useful to employ the function $\frac{\sigma(v-u)}{\sigma v}e^{u\zeta v}$, which we may denote by $\psi(-u,\,v)$ or ψ; so that $\psi(-u,\,v)=\sigma u\,\phi(-u,\,v)$, and $\psi=1$, when $u=0$.

We may now write

$$gy/w^2=-\log\psi(-u,\,v)-\omega\log\psi(-u,\,\omega v)-\omega^2\log\psi(-u,\,\omega^2 v),$$
$$gt/w\;=-\log\psi(-u,\,v)-\omega^2\log\psi(-u,\,\omega v)-\omega\log\psi(-u,\,\omega^2 v).$$

Differentiating logarithmically,

$$\frac{1}{\psi}\frac{\partial\psi}{\partial u}=-\zeta(v-u)+\zeta v$$
$$=-u\wp v+\frac{u^2}{2!}\wp' v-\frac{u^3}{3!}\wp'' v+\dots$$

on expanding the second side by Taylor's Theorem; so that, integrating again,

$$\log\psi(-u,\,v)=-\frac{u^2}{2!}\wp v+\frac{u^3}{3!}\wp' v-\frac{u^4}{4!}\wp'' v+\dots, \quad\text{..............(18)}$$

Then, with $g_2=0$, and $\wp\omega v=\omega\wp v$, etc.,

$$\log\psi(-u,\,\omega\,v)=-\frac{u^2}{2!}\omega\wp v+\frac{u^3}{3!}\wp' v-\frac{u^4}{4!}\omega^2\wp'' v+\dots, \quad\text{.........(19)}$$

$$\log\psi(-u,\,\omega^2 v)=-\frac{u^2}{2!}\omega^2\wp v+\frac{u^3}{3!}\wp' v-\frac{u^4}{4!}\omega\wp'' v+\dots; \quad\text{..........(20)}$$

so that $$\frac{gy}{w^2}=3\left(\frac{u^4}{4!}\wp''v-\frac{u^7}{7!}\wp^{(v)}v+\frac{u^{10}}{10!}\wp^{(viii)}v-\dots\right),\dots\dots\dots(21)$$

$$\frac{gt}{w}=3\left(\frac{u}{2!}\wp v-\frac{u^5}{5!}\wp'''v+\frac{u^8}{8!}\wp^{(vi)}v-\dots\right),\dots\dots\dots(22)$$

and here $u=gx/w^2$, $g_2=0$, $g_3=\frac{1}{27}(4-3\sin^2\alpha)$, $\wp v=\frac{1}{3}$,
$\wp'v=\frac{1}{3}\sin\alpha$, $\wp v''=\frac{2}{3}$, $\wp v'''=\frac{4}{3}\sin\alpha$, $\wp^{(iv)}v=4-\frac{4}{3}\sin\alpha$, $\wp^{(v)}v=\frac{40}{3}\sin\alpha,\dots$

231. When p_1, p_2, p_3 denote the values of p corresponding to three points defined by the values x_1, x_2, x_3 of x, or u_1, u_2, u_3 of u, such that

$$x_1+x_2+x_3=0, \text{ or } u_1+u_2+u_3=0,$$

then, according to § 145,

$$(P_1P_2P_3)^{\frac{1}{3}}=p_1p_2p_3-(p_2p_3+p_3p_1+p_1p_2)\sin\alpha+p_1+p_2+p_3. \quad (23)$$

This Theorem follows also as a corollary of Abel's Theorem, as applied in § 166; and it is interesting to proceed to the determination, in a similar manner, of the corresponding values of

$$y_1+y_2+y_3, \text{ and } t_1+t_2+t_3.$$

Changing, in § 166, x into p and y into $P^{\frac{1}{3}}$, then from (7) § 166,

$$\frac{g}{w^2}(dy_1+dy_2+dy_3)=p_1P_1^{-\frac{2}{3}}dp_1+p_2P_2^{-\frac{2}{3}}dp_2+p_3P_3^{-\frac{2}{3}}dp_3$$

$$=\frac{3}{a^3-1}\left(\frac{p_1^2\delta a+p_1\delta\beta}{p_3-p_1\cdot p_1-p_2}+\frac{p_2^2\delta a+p_2\delta\beta}{p_1-p_2\cdot p_2-p_3}+\frac{p_3^2\delta a+p_3\delta\beta}{p_2-p_3\cdot p_3-p_1}\right)=-\frac{3\delta a}{a^3-1},$$

$$\frac{g}{w}(dt_1+dt_2+dt_3)=P_1^{-\frac{1}{3}}dp_1+P_2^{-\frac{1}{3}}dp_2+P_3^{-\frac{1}{3}}dp_3$$

$$=\frac{3}{a^3-1}\left\{\frac{(ap_1+\beta)(p_1\delta a+\delta\beta)}{(p_3-p_1)(p_1-p_2)}+\dots\right\}=-\frac{3a\delta a}{a^3-1}.$$

Therefore

$$\frac{g}{w^2}(y_1+y_2+y_3)=\int^{\infty}\frac{3da}{a^3-1}$$

$$=-\log(a-1)-\omega\log(a-\omega)-\omega^2\log(a-\omega^2),\dots(24)$$

$$\frac{g}{w}(t_1+t_2+t_3)=\int^{\infty}\frac{3ada}{a^3-1}$$

$$=-\log(a-1)-\omega^2\log(a-\omega)-\omega\log(a-\omega^2);\,..(25)$$

where $$a=\frac{P_2^{\frac{1}{3}}-P_3^{\frac{1}{3}}}{p_2-p_3}=\frac{P_3^{\frac{1}{3}}-P_1^{\frac{1}{3}}}{p_3-p_1}=\frac{P_1^{\frac{1}{3}}-P_2^{\frac{1}{3}}}{p_1-p_2};\dots\dots\dots\dots(26)$$

and $a=\infty$, when $p_1=p_2=p_3=0$.

As a corollary from the preceding expressions for y and t in terms of x or u, it follows that

$$\frac{\sigma(v-u_1)\sigma(v-u_2)\sigma(v-u_3)}{\sigma^3v\,\sigma u_1\sigma u_2\sigma u_3}=\frac{1}{a-1}.$$

232. By taking $x_3=0$ and $p_3=0$, then

$$p_1+p_2-p_1 p_2 \sin\alpha=0, \text{ or } 1/p_1+1/p_2=\sin\alpha,$$

when $$x_1+x_2=0, \text{ or } u_1+u_2=0.$$

Now, from equations (13) and (16),

$$\frac{g}{w^2}(y_1+y_2)=-\log(\wp u-\wp v)-\omega\log(\wp u-\omega\wp v)-\omega^2\log(\wp u-\omega^2\wp v)$$

$$=-\frac{1}{2}\log\frac{(\wp u-\wp v)^3}{\wp^3 u-\wp^3 v}-\surd 3\tan^{-1}\frac{\surd 3\wp v}{2\wp u+\wp v},$$

$$\frac{g}{w}(t_1+t_2)=-\log(\wp u-\wp v)-\omega^2\log(\wp u-\omega\wp v)-\omega\log(\wp u-\omega^2\wp v)$$

$$=-\frac{1}{2}\log\frac{(\wp u-\wp v)^3}{\wp^3 u-\wp^3 v)}+\surd 3\tan^{-1}\frac{\surd 3\wp v}{2\wp u+\wp v}.$$

In particular, when $u=\omega_2$, then

$$\phi(-u,\omega_2)=\surd(\wp u-e_2),$$

and $$\frac{gy}{w^2}=-3\omega_2\zeta v-\frac{1}{4}\log\frac{(\wp v-e_2)^3}{\frac{1}{4}\wp'^2 v}-\frac{1}{2}\surd 3\tan^{-1}\frac{\surd 3\wp v}{2e_2+\wp v},$$

$$\frac{gt}{w}=\qquad -\frac{1}{4}\log\frac{(\wp v-e_2)^3}{\frac{1}{4}\wp'^2 v}+\frac{1}{2}\surd 3\tan^{-1}\frac{\surd 3\wp v}{2e_2+\wp v},$$

so that the expressions for y and t are pseudo-elliptic; and, at this point, $$p=2\sin\alpha.$$

233. We may now investigate the properties of certain points on the trajectory.

When $$u=2\omega_2-v,$$

then $$\wp u=\tfrac{1}{3},\ \wp' u=-\tfrac{1}{3}\sin\alpha, \text{ and } p=\operatorname{cosec}\alpha,$$

so that the tangent is perpendicular to Ox.

The velocity in the trajectory is given by

$$w(p^2-2p\sin\alpha+1)^{\frac{1}{2}}(p^3-3p^2\sin\alpha+3p)^{-\frac{1}{3}},$$

and this is a minimum, by logarithmic differentiation, when

$$\frac{p-\sin\alpha}{p^2-2p\sin\alpha+1}-\frac{p^2-2p\sin\alpha+1}{p^3-3p^2\sin\alpha+3p}=0,$$

or $$p^2\cos^2\alpha+p\sin\alpha-1=0. \quad\text{.................(27)}$$

If the tangent AB makes an angle β with Ox at the point A,

then $$p=\frac{\sin\beta}{\cos(\alpha-\beta)},$$

so that the relation becomes

$$\tan\alpha=-2\cot 2\beta=\tan\beta-\cot\beta. \quad\text{.........(28)}$$

Then $$\surd(4+\tan^2\alpha)=\tan\beta+\cot\beta=2\operatorname{cosec}2\beta,$$

or $$\surd(3g_3)=\tfrac{1}{3}\surd(4-3\sin^2\alpha)=\tfrac{2}{3}\cos\alpha\operatorname{cosec}2\beta.$$

The relation (28) is equivalent to a number of other relations, such as

$$\tan(2\beta-\alpha)=\tan\alpha-\tan 2\beta=\tan\alpha+2\cot\alpha,$$

$$\tan(\alpha-\beta)=\cot^3\beta,$$

$$\tan\alpha=\{\cot(\alpha-\beta)\}^{\frac{1}{3}}-\{\tan(\alpha-\beta)\}^{\frac{1}{3}},$$

$$3\tan\alpha+\tan^3\alpha=2\cot 2(\alpha-\beta)=\cot(\alpha-\beta)-\tan(\alpha-\beta),$$

$$\tan\alpha=\{\cot(\alpha-\beta)\}^{\frac{1}{3}}-\{\tan(\alpha-\beta)\}^{\frac{1}{3}}, \text{ etc.}$$

Also, since $$p=\frac{2}{\sin\alpha-3\wp' u},$$

therefore, at these points of minimum velocity,

$$\wp'^2 u=\tfrac{1}{9}(4-3\sin^2\alpha)=3g_3, \text{ and } \wp^3 u=g_3,$$

and therefore $\wp 2u=\wp u$, or $u=\frac{2}{3}\omega_2$, as in § 166.

The integrals for y and t at these points of minimum velocity are therefore pseudo-elliptic, and depend on

$$\int\frac{ds}{(s^3-1)\sqrt{(4s^3-1)}} \text{ and } \int\frac{sds}{(s^3-1)\sqrt{(4s^3-1)}},$$

integrals first considered by Euler (Legendre, *F. E.*, I., Chap. XXVI.).

We find, by differentiation, that

$$\frac{d}{ds}\tanh^{-1}\frac{\sqrt{3}(2s-1)}{\sqrt{(4s^3-1)}}=-\tfrac{1}{2}\sqrt{3}\frac{2s+1}{s-1}\frac{1}{\sqrt{(4s^3-1)}}, \ldots\ldots(29)$$

$$\frac{1}{2}\frac{d}{ds}\log\frac{\sqrt{(4s^3-1)}+\sqrt{3}}{\sqrt{(4s^3-1)}+\sqrt{3}(2s-1)}$$

$$=\frac{\frac{1}{2}\sqrt{3}}{\sqrt{(4s^3-1)}}-\frac{3}{\sqrt{(4s^3-1)}+\sqrt{3}}\frac{s-1}{\sqrt{(4s^3-1)}}, \ldots(30)$$

$$\tfrac{1}{3}\sqrt{3}\frac{d}{ds}\tan^{-1}\sqrt{3}\frac{\sqrt{(4s^3-1)}-\sqrt{3}(2s+1)}{\sqrt{(4s^3-1)}-\sqrt{3}(2s-1)}$$

$$=\frac{\frac{1}{2}\sqrt{3}}{\sqrt{(4s^3-1)}}+\frac{3}{\sqrt{(4s^3-1)}+\sqrt{3}}\frac{s-1}{\sqrt{(4s^3-1)}}, \ldots(31)$$

by means of which the results can be constructed; and noticing that, if $s=\wp v$, $\sqrt{(4s^3-1)}=\wp' v$, $g_2=0$, $g_3=1$, then

$$\frac{\sqrt{(4s^3-1)}+\sqrt{3}}{\sqrt{(4s^3-1)}+\sqrt{3}(2s-1)}=\wp(v-\tfrac{2}{3}\omega_2),$$

$$\sqrt{3}\frac{\sqrt{(4s^3-1)}-\sqrt{3}(2s+1)}{\sqrt{(4s^3-1)}+\sqrt{3}(2s-1)}=\wp'(v-\tfrac{2}{3}\omega_2),$$

we find finally, when $u=\frac{2}{3}\omega_2$,

$$gy/w^2=\tfrac{2}{3}\omega_2\zeta\omega_2-2\omega_2\zeta v+\tfrac{1}{2}\log\wp(v-\tfrac{2}{3}\omega_2)-\tfrac{1}{3}\sqrt{3}\tan^{-1}\wp'(v-\tfrac{2}{3}\omega_2), \ (32)$$

$$gt/w=2v\zeta\omega_2-\tfrac{2}{3}\omega_2\zeta\omega_2+\tfrac{1}{2}\log\wp(v-\tfrac{2}{3}\omega_2)+\tfrac{1}{3}\sqrt{3}\tan^{-1}\wp'(v-\tfrac{2}{3}\omega_2). \ (33)$$

234. Denoting by θ the angle which the tangent at any point makes with Ox, the tangent at O, the point of infinite velocity, and by ϕ the angle which it makes with the tangent at A, the point of minimum velocity, then $\theta=\beta-\phi$, and

$$p=\frac{\sin\theta}{\cos(\alpha-\theta)}=\frac{\sin(\beta-\phi)}{\cos(\alpha-\beta+\phi)};$$

so that $$\frac{\sin\alpha-3\wp' u}{2}=\frac{1}{p}=\frac{\cos(\alpha-\beta+\phi)}{\sin(\beta-\phi)},$$

and $$3\wp' u=\frac{\sin\alpha\sin(\beta-\phi)-2\cos(\alpha-\beta+\phi)}{\sin(\beta-\phi)}$$

$$=-2\cos\alpha\frac{\cos(\beta-\phi)+\frac{1}{2}\tan\alpha\sin(\beta-\phi)}{\sin(\beta-\phi)}$$

$$=-2\cos\alpha\frac{\cos(\beta-\phi)-\cot 2\beta\sin(\beta-\phi)}{\sin(\beta-\phi)}$$

$$=-2\cos\alpha\operatorname{cosec}2\beta\frac{\sin(\beta+\phi)}{\sin(\beta-\phi)};$$

and since

$$\wp'\tfrac{2}{3}\omega_2=-\surd(3g_3)=-\tfrac{1}{3}\surd(4-3\sin^2\alpha)=-\tfrac{2}{3}\cos\alpha\operatorname{cosec}2\beta,$$

therefore $$\frac{\sin(\beta+\phi)}{\sin(\beta-\phi)}=\frac{\wp' u}{\wp'\frac{2}{3}\omega_2},$$

or $$\frac{\tan\phi}{\tan\beta}=\frac{\wp' u-\wp'\frac{2}{3}\omega_2}{\wp' u+\wp'\frac{2}{3}\omega_2}.\qquad\qquad(34)$$

Therefore, at points defined by u_1, u_2, where the tangents make equal angles with the tangent at A,

$$\wp' u_1\,.\,\wp' u_2=\wp'^2\tfrac{2}{3}\omega_2.$$

Thus, if $u_1=0$, then $u_2=\omega_2$; and the tangent where $u=\omega_2$ makes an angle 2β with Ox.

By the principle of *Homogeneity* of § 196, we can select any arbitrary value of g_3, and it is convenient to take $g_3=1$; and

now, if $\dfrac{gx}{w^2}=\dfrac{u}{m}$, then $\wp\dfrac{gx}{w^2}=m^2\wp u$, $\wp'\dfrac{gx}{w^2}=m^3\wp' u$,

where $m^6=g_3$, $m=(4-3\sin^2\alpha)^{\frac{1}{6}}/\surd 3$.

With $g_2=0$, $g_3=1$, we have found, in § 166,

$$\wp\tfrac{2}{3}\omega_2=1,\ \wp'\tfrac{2}{3}\omega_2=-\surd 3,\ \wp'\tfrac{4}{3}\omega_2=\surd 3.$$

Again, if $\dfrac{v}{m}=\dfrac{ga}{w^2}$, then

$\wp v=(4-3\sin^2\alpha)^{-\frac{1}{3}}$, $\wp' v=\surd 3\sin\alpha(4-3\sin^2\alpha)^{-\frac{1}{2}}=-\surd 3\cos 2\beta$; so that, as α increases from 0 to $\frac{1}{2}\pi$, $\wp' v$ increases from 0 to $\surd 3$, and v increases from ω_2 to $\frac{4}{3}\omega_2$.

Denoting the analytical expression for $\tan\phi/\tan\beta$ in (34) by X, then X is independent of α or β, and therefore a Table of numerical values of X, with u or mgx/w^2 for argument, will serve for all trajectories.

It will be a useful numerical exercise for the student to prove that corresponding values of u and X are

$$\tfrac{1}{6}\omega_2, \quad \frac{\sqrt[4]{3}(\sqrt{3}+1)-\sqrt{2}}{2\sqrt[6]{2}};$$

$$\tfrac{1}{3}\omega_2, \quad \frac{1}{\sqrt[3]{2}};$$

$$\tfrac{1}{2}\omega_2, \quad \frac{\sqrt{3}+1-\sqrt{2}\sqrt[4]{3}}{2};$$

$$\tfrac{2}{3}\omega_2, \quad 0;$$

$$\tfrac{5}{6}\omega_2, \quad \frac{\sqrt{3}-1-\sqrt{2}\sqrt[4]{3}}{2};$$

$$\omega_2, \quad -1;$$

$$\tfrac{7}{6}\omega_2, \quad -\frac{\sqrt[4]{3}(\sqrt{3}+1)+\sqrt{2}}{2\sqrt[6]{2}};$$

$$\tfrac{4}{3}\omega_2, \quad \infty.$$

Examples.

Prove that, with $g_2=0$, $g_3=1$,

1. $\wp(u-\tfrac{2}{3}\omega_2)=\dfrac{\wp'u+\sqrt{3}}{\wp'u+\sqrt{3}(2\wp u-1)}.$

2. $\wp'(u-\tfrac{2}{3}\omega_2)=\sqrt{3}\dfrac{\wp'u-\sqrt{3}(2\wp u+1)}{\wp'u+\sqrt{3}(2\wp u-1)}.$

3. $\wp(u-\tfrac{2}{3}\omega_2)\wp(u+\tfrac{2}{3}\omega_2)=\dfrac{\wp^3u-1}{(\wp u-1)^3}.$

4. $\displaystyle\int\frac{du}{\wp u-1}=-\tfrac{2}{3}u-\tfrac{2}{9}\sqrt{3}\tanh^{-1}\sqrt{3}\frac{2\wp u-1}{\wp' u}.$

5. $\displaystyle\int\frac{du}{\wp'u+\sqrt{3}}=\tfrac{1}{6}\sqrt{3}u-\tfrac{1}{12}\log\wp(u-\tfrac{2}{3}\omega_2)-\tfrac{1}{18}\sqrt{3}\tan^{-1}\wp'(u-\tfrac{2}{3}\omega_2).$

6. $\displaystyle\int\frac{\wp u\,du}{\wp'u+\sqrt{3}}=\quad-\tfrac{1}{12}\log\wp(u-\tfrac{2}{3}\omega_2)+\tfrac{1}{18}\sqrt{3}\tan^{-1}\wp'(u-\tfrac{2}{3}\omega_2).$

7. Integrate $(\wp u)^{-1}$, $(\wp u)^{-2}$, $(\wp u)^{-3}$.

CHAPTER VIII.

THE DOUBLE PERIODICITY OF THE ELLIPTIC FUNCTIONS.

235. Besides pointing out the advantage of the direct Elliptic Functions obtained by the inversion of the Elliptic Integrals (§ 5), Abel made an equally important step (Crelle, II., 1827) in showing that the Elliptic Functions are *doubly-periodic* functions, having a real period, $4K$ or $2K$, as already defined in § 11, and an *imaginary period*, $4K'i$ or $2K'i$, where, as before in § 11,

$$K' = \int_0^{\frac{1}{2}\pi} d\psi / \sqrt{(1-\kappa'^2 \sin^2\psi)} = F\kappa'.$$

Doubly-periodic functions make their appearance when we consider functions of a complex argument $w = u + vi$.

Denoting $x + yi$ by z, we have already discussed in § 179 the system of confocal conics given by

$$z = c \sin w, \text{ or } c \cos w, \text{ when } u \text{ or } v \text{ is constant.}$$

In this case $$w = \int \frac{dz}{\sqrt{(c^2 - z^2)}},$$

and the *poles* of this integral, as defined in § 54, are given by $z = \pm c$, the foci of the confocal system of conics.

Changing the origin to a focus, then

$$w = \int \frac{dz}{\sqrt{(z \,.\, 2c - z)}},$$

and
$$z = 2c \sin^2 \tfrac{1}{2} w,$$
$$2c - z = 2c \cos^2 \tfrac{1}{2} w,$$
$$dz/dw = c \sin w.$$

Denoting by r, r' the focal distances of a point, then

$$r^2 = (x + yi)(x - yi) = 4c^2 \sin^2 \tfrac{1}{2}(u + vi) \sin^2 \tfrac{1}{2}(u - vi),$$

or
$$r = 2c \sin \tfrac{1}{2}(u+vi) \sin \tfrac{1}{2}(u-vi),$$
$$r' = 2c \cos \tfrac{1}{2}(u+vi) \cos \tfrac{1}{2}(u-vi);$$
so that
$$r'+r = 2c \cos vi = 2c \cosh v,$$
$$r'-r = 2c \cos u,$$
giving the confocal ellipses and hyperbolas, for which v and u are constants.

It is convenient to denote $x-yi$ by z' and $u-vi$ by w'; and now the Jacobian
$$J \text{ or } \frac{\partial(x, y)}{\partial(u, v)} = c^2 \sin w \sin w' = \tfrac{1}{4} rr'.$$

236. Now, if we consider the integral (11) of § 38,
$$w = \int \frac{dz}{\sqrt{(z \,.\, 1-z \,.\, 1-kz)}},$$
then
$$z = \text{sn}^2 \tfrac{1}{2} w,$$
$$1-z = \text{cn}^2 \tfrac{1}{2} w,$$
$$1-kz = \text{dn}^2 \tfrac{1}{2} w,$$
$$dz/dw = \text{sn} \tfrac{1}{2} w \text{ cn} \tfrac{1}{2} w \text{ dn} \tfrac{1}{2} w;$$
and the *poles* of the integral are given by $z=0$, 1, and $1/k$.

Denoting by r, r', r'' the distances of a point from these poles or foci O, O', O'' in fig. 26, then
$$r' = \text{sn} \tfrac{1}{2} w \text{ sn} \tfrac{1}{2} w', \quad r = \text{cn} \tfrac{1}{2} w \text{ cn} \tfrac{1}{2} w', \quad kr'' = \text{dn} \tfrac{1}{2} w \text{ dn} \tfrac{1}{2} w';$$
or by means of formulas (2), (3), (5), (28), (29) of § 137, with $\tfrac{1}{2}w$ and $\tfrac{1}{2}w'$ for u and v, and therefore u and iv for $u+v$ and $u-v$,
$$r' = \frac{\text{cn } vi - \text{cn } u}{\text{dn } vi + \text{dn } u} = \frac{1}{\kappa^2} \frac{\text{dn } vi - \text{dn } u}{\text{cn } vi + \text{cn } u},$$
$$r = \frac{\text{cn } vi \text{ dn } u + \text{cn } u \text{ dn } vi}{\text{dn } vi + \text{dn } u} = \frac{\kappa'^2}{\kappa^2} \frac{\text{dn } vi - \text{dn } u}{\text{cn } vi \text{ dn } u - \text{cn } u \text{ dn } vi},$$
$$kr'' = \frac{\text{cn } vi \text{ dn } u + \text{cn } u \text{ dn } vi}{\text{cn } vi + \text{cn } u} = \kappa'^2 \frac{\text{cn } vi - \text{cn } u}{\text{cn } vi \text{ dn } u - \text{cn } u \text{ dn } vi}.$$

From these relations, by the alternate elimination of u and v,
$$\left.\begin{aligned} r + r' \text{dn } vi &= \text{cn } vi \\ r - r' \text{dn } u &= \text{cn } u \end{aligned}\right\},$$
or
$$\left.\begin{aligned} kr'' + kr' \text{cn } vi &= \text{dn } vi \\ kr'' - kr' \text{cn } u &= \text{dn } u \end{aligned}\right\},$$
or
$$\left.\begin{aligned} kr'' \text{dn } vi - kr \text{ cn } vi &= 1-k \\ kr'' \text{dn } u - kr \text{ cn } u &= 1-k \end{aligned}\right\},$$
the vectorial equations of one and the same system of confocal orthogonal Cartesian Ovals (fig. 26); also $J = krr'r''$. (Darboux, *Annales scientifiques de l'école normale supérieure*, IV., 1867.)

As we travel round one of these curves and make complete circuits, each enclosing a pair of poles of the integral w, defined either by 0 and 1, or 1 and $1/k$, the integral increases by constant quantities $4K$ or $4K'i$, the corresponding periods of the elliptic function $\text{sn}^2\frac{1}{2}w$, as in § 55.

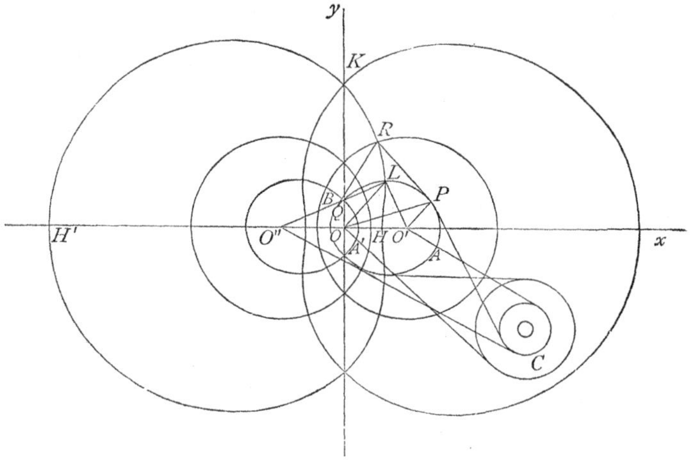

Fig. 26.

By making $k=0$, we obtain the degenerate case of the confocal conics, and now $K=\frac{1}{2}\pi$, while $K'=\infty$; so that the circular functions have a real period 2π and an infinite imaginary period; on the other hand, the hyperbolic functions, as illustrated by the confocal ellipses, have an infinite real period and an imaginary period $2\pi i$.

Mr. J. Hammond has shown, in the *American Journal of Mathematics*, vol. I., how these Cartesian Ovals may be described mechanically, by means of reels of thread, as in the case of the confocal conics of § 173.

He takes two reels of thread, of different diameters, fastened together, and pivoted on the same axis at C. Now, if the threads are led through a pair of the foci, O and O', the curves

$$r \pm lr' = c$$

will be described, if the diameters are in the ratio of l to 1.

By leading the threads round an oval, as in fig. 26, theorems can be obtained, connecting arcs of confocal Cartesian Ovals, analogous to those of Graves and Chasles for elliptic arcs.

237. By inversion of this system of confocal Cartesian Ovals, we shall obtain another system of orthogonal quartic curves, with four concyclic foci A, B, C, D, defined by the vectors $z=\alpha, \beta, \gamma, \delta$, suppose; and now

$$w=\int dz/\sqrt{(z-\alpha \,.\, z-\beta \,.\, z-\gamma \,.\, z-\delta)};$$

or, writing w for $w/\sqrt{(\alpha-\gamma \,.\, \beta-\delta)}$, then, from § 66,

$$\frac{\beta-\delta \,.\, z-\alpha}{\alpha-\delta \,.\, z-\beta}=\operatorname{sn}^2\tfrac{1}{2}w,\quad \frac{\alpha-\beta \,.\, z-\delta}{\alpha-\delta \,.\, z-\beta}=\operatorname{cn}^2\tfrac{1}{2}w,\quad \frac{\alpha-\beta \,.\, z-\gamma}{\alpha-\gamma \,.\, z-\beta}=\operatorname{dn}^2\tfrac{1}{2}w.$$

Denoting by r_1, r_2, r_3, r_4 the distances of a point from the foci A, B, C, D, then, from these equations,

$$\text{mod.}\ \frac{\beta-\delta}{\alpha-\delta}\frac{r_1}{r_2}=\operatorname{sn}\tfrac{1}{2}w \operatorname{sn}\tfrac{1}{2}w',\quad \text{mod.}\ \frac{\alpha-\beta}{\alpha-\delta}\frac{r_4}{r_2}=\operatorname{cn}\tfrac{1}{2}w \operatorname{cn}\tfrac{1}{2}w',$$

$$\text{mod.}\ \frac{\alpha-\beta}{\alpha-\gamma}\frac{r_3}{r_2}=\operatorname{dn}\tfrac{1}{2}w \operatorname{dn}\tfrac{1}{2}w';$$

so that we obtain the vectorial equations of these orthogonal quartic curves on replacing r', r, r'' in the equations of the Cartesian Ovals by these expressions.

(*Proc. Cam. Phil. Society*, vol. IV.; Holzmuller, *Einführung in die Theorie der isogonalen Verwandtschaften*, 1882.)

238. We now proceed to express the elliptic functions of the imaginary argument vi by functions of a real argument v.

We know that $\cos vi=\cosh v$, $\sin vi=i\sinh v$, $\tan vi=i\tanh v$; and that the function ϕ or $\operatorname{amh} u$, and its inverse function

$$u \text{ or } \operatorname{amh}^{-1}\phi=\log(\sec\phi+\tan\phi)=\cosh^{-1}\sec\phi, \text{ etc.},$$

connects the circular functions of ϕ, for which $\kappa=0$, with the hyperbolic functions of u in § 16, for which $\kappa=1$; and then

$$\cosh u=\sec\phi,\ \sinh u=\tan\phi,\ \tanh u=\sin\phi,\ \tanh\tfrac{1}{2}u=\tan\tfrac{1}{2}\phi.$$

Now, if $\qquad \phi=\operatorname{amh}\psi i,$

then $\qquad \cos\phi\cosh\psi i=1$, or $\cos\phi\cos\psi=1$,

a symmetrical relation, so that

$$\psi=\operatorname{amh}\phi/i;$$

and
$$\sin\phi=\tanh\psi i=i\tan\psi,$$
$$\cos\phi=\operatorname{sech}\psi i=\sec\psi,$$
$$\tan\phi=\sinh\psi i=i\sin\psi, \text{ etc.}$$

Also
$$d\phi=i\operatorname{sech}\psi i d\psi=i\sec\psi d\psi,$$
$$\Delta(\phi,\kappa)=\sqrt{(1+\kappa^2\tan^2\psi)}=\sec\psi\Delta(\psi,\kappa'),$$

so that
$$\frac{d\phi}{\Delta(\phi,\kappa)}=\frac{id\psi}{\Delta(\psi,\kappa')}.$$

If $$\psi = \text{am}(v, \kappa'),$$
then $$\phi = \text{am}(vi, \kappa);$$
and
$$\text{sn}(vi, \kappa) = i\frac{\text{sn}(v, \kappa')}{\text{cn}(v, \kappa')}, \text{ or } i\,\text{sc}(v, \kappa'), \text{ or } i\,\text{tn}(v, \kappa');$$
$$\text{cn}(vi, \kappa) = \frac{1}{\text{cn}(v, \kappa')}, \text{ or } \text{nc}(v, \kappa');$$
$$\text{dn}(vi, \kappa) = \frac{\text{dn}(v, \kappa')}{\text{cn}(v, \kappa')}, \text{ or } \text{dc}(v, \kappa'),$$
connecting the elliptic functions of imaginary argument vi and modulus κ with the elliptic functions of real argument v and complementary modulus κ'.

Putting $v = K'$, we notice that $\text{sn}\,K'i$, $\text{cn}\,K'i$, and $\text{dn}\,K'i$ are infinite; and putting $v = 2K'$, then
$$\text{sn}\,2K'i = 0,\ \text{cn}\,2K'i = -1,\ \text{dn}\,2K'i = -1;$$
also
$$\text{sn}\,4K'i = 0,\ \text{cn}\,4K'i = 1,\ \text{dn}\,4K'i = 1.$$

239. The Addition Theorems of § 116 may now be written
$$\text{cn}(u + vi) = (\text{cn}\,u\ \text{cn}\,v - i\,\text{sn}\,u\,\text{dn}\,u\,\text{sn}\,v\,\text{dn}\,v) \div D,$$
$$\text{sn}(u + vi) = (\text{sn}\,u\,\text{dn}\,v + i\,\text{cn}\,u\,\text{dn}\,u\ \text{sn}\,v\ \text{cn}\,v) \div D,$$
$$\text{dn}(u + vi) = (\text{dn}\,u\,\text{cn}\,v\,\text{dn}\,v - i\kappa^2\text{sn}\,u\ \text{cn}\,u\ \text{sn}\,v) \div D,$$
$$D = \text{cn}^2 v + \kappa^2 \text{sn}^2 u\ \text{sn}^2 v;$$
remembering that the modulus of the elliptic functions of v is κ', while that of the functions of u is κ.

Thus, putting $v = K'$,
$$\text{cn}(u + K'i) = -i\frac{\text{dn}\,u}{\kappa\,\text{sn}\,u},\ \text{sn}(u + K'i) = \frac{1}{\kappa\,\text{sn}\,u},\ \text{dn}(u + K'i) = -i\frac{\text{cn}\,u}{\text{sn}\,u};$$
so that, putting $u = K$,
$$\text{cn}(K + K'i) = -i\kappa'/\kappa,\quad \text{sn}(K + K'i) = 1/\kappa,\quad \text{dn}(K + K'i) = 0.$$

Writing C, S, D for $\text{cn}\,2u$, $\text{sn}\,2u$, $\text{dn}\,2u$, then (§ 123)
$$\text{sn}^2(u + \tfrac{1}{2}K'i) = \frac{1 - \text{cn}(2u + K'i)}{1 + \text{dn}(2u + K'i)} = \frac{1}{\kappa}\frac{\kappa S + Di}{S - Ci}, \text{ etc.}$$

Generally, when m and n denote any integers, we find that
$$\text{cn}(u + 2mK + 2nK'i) = (-1)^{m+n}\text{cn}\,u,$$
$$\text{sn}(u + 2mK + 2nK'i) = (-1)^{m}\ \text{sn}\,u,$$
$$\text{dn}(u + 2mK + 2nK'i) = (-1)^{n}\ \text{dn}\,u;$$
so that $4K$ and $2K'i$ are the periods of $\text{sn}\,u$,

$2K$ and $4K'i$ are the periods of $\text{dn}\,u$;

the periods of $\text{cn}\,u$ being $2(K + K'i)$ and $2(K - K'i)$.

In § 164, we may now write

$$u_1+u_2+u_3=4mK+4nK'i;$$

or in the notation of the Theory of Numbers,

$$u_1+u_2+u_3\equiv 0 \text{ (mod. } 4K,\ 4K'i).$$

240. A combination of the transformations of §§ 29 and 238, to the reciprocal and to the complementary modulus, gives

$$\operatorname{cn}(vi,\kappa)=\frac{1}{\operatorname{cn}(v,\kappa')}=\frac{1}{\operatorname{dn}(\kappa'v,1/\kappa')}=\frac{\operatorname{cn}(\kappa'vi,i\kappa/\kappa')}{\operatorname{dn}(\kappa'vi,i\kappa/\kappa')},$$

$$\operatorname{sn}(vi,\kappa)=\frac{i\operatorname{sn}(v,\kappa')}{\operatorname{cn}(v,\kappa')}=\frac{i\operatorname{sn}(\kappa'v,1/\kappa')}{\kappa'\operatorname{dn}(\kappa'v,1/\kappa')}=\frac{\operatorname{sn}(\kappa'vi,i\kappa/\kappa')}{\kappa'\operatorname{dn}(\kappa'vi,i\kappa/\kappa')},$$

$$\operatorname{dn}(vi,\kappa)=\frac{\operatorname{dn}(v,\kappa')}{\operatorname{cn}(v,\kappa')}=\frac{\operatorname{cn}(\kappa'v,1/\kappa')}{\operatorname{dn}(\kappa'v,1/\kappa')}=\frac{1}{\operatorname{dn}(\kappa'vi,i\kappa/\kappa')}.$$

Thus $\quad \operatorname{cn}(\kappa'u,i\kappa/\kappa')=\operatorname{cd}(u,\kappa)=\operatorname{sn}(K-u,\kappa),$

or $\quad \operatorname{am}(\kappa'u,i\kappa/\kappa')=\tfrac{1}{2}\pi-\operatorname{am}(K-u,\kappa);$

as is otherwise evident, when we notice that, if

$$u=\int_0^\psi(1-\kappa^2\cos^2\phi)^{-\frac{1}{2}}d\phi=\frac{1}{\kappa'}\int_0^\psi\left(1+\frac{\kappa^2}{\kappa'^2}\sin^2\psi\right)^{-\frac{1}{2}}d\psi,$$

so that $\quad \psi=\operatorname{am}(\kappa'u,i\kappa/\kappa'),$

then $\quad K-u=\int_\psi^{\frac{1}{2}\pi}(1-\kappa^2\cos^2\psi)^{-\frac{1}{2}}d\psi=\int_0^\phi(1-\kappa^2\sin^2\phi)^{-\frac{1}{2}}d\phi,$

or $\quad \phi=\operatorname{am}(K-u,\kappa),$

provided $\quad \psi=\tfrac{1}{2}\pi-\phi.$

241. As an application, take the values of v_1 and v_2 in § 210;

$$\operatorname{dn}^2v_1=\frac{1+\cos\beta}{1+\cos\alpha},\quad \operatorname{sn}^2v_1=-\frac{d-\cos\alpha}{1+\cos\alpha},\quad \operatorname{cn}^2v_1=\frac{d+1}{1+\cos\alpha},$$

$$\operatorname{dn}^2v_2=\frac{1-\cos\beta}{1-\cos\alpha},\quad \operatorname{sn}^2v_2=\frac{d-\cos\alpha}{1-\cos\alpha},\quad \operatorname{cn}^2v_2=-\frac{d-1}{1-\cos\alpha};$$

so that, with $v_1=pK'i$, $v_2=K+qK'i$, where p and q are real proper fractions (§ 56), then

$$\frac{1-\cos\alpha}{1+\cos\alpha}=-\frac{\operatorname{sn}^2v_1}{\operatorname{sn}^2v_2}=-\frac{\operatorname{sn}^2pK'i\operatorname{dn}^2qK'i}{\operatorname{cn}^2qK'i},$$

$$\frac{1-\cos\beta}{1+\cos\beta}=-\frac{\operatorname{sn}^2v_1}{\operatorname{sn}^2v_2}\frac{\operatorname{dn}^2v_2}{\operatorname{dn}^2v_1}=-\frac{\kappa'^2\operatorname{sn}^2pK'i}{\operatorname{dn}^2pK'i\operatorname{cn}^2qK'i},$$

$$\frac{d-1}{d+1}=\frac{\operatorname{sn}^2v_1}{\operatorname{sn}^2v_2}\frac{\operatorname{cn}^2v_2}{\operatorname{cn}^2v_1}=\frac{\kappa'^2\operatorname{sn}^2pK'i\operatorname{sn}^2qK'i}{\operatorname{cn}^2pK'i\operatorname{cn}^2qK'i}.$$

Thence, expressed in a real form,

$$\frac{1-\cos\alpha}{1+\cos\alpha}=\frac{\mathrm{sn}^2 pK'\,\mathrm{dn}^2 qK'}{\mathrm{cn}^2 pK'},$$

or (§ 135) $\tan\frac{1}{2}\alpha=\tan\frac{1}{2}[\mathrm{am}\{(p+q)K',\kappa'\}+\mathrm{am}\{(p-q)K',\kappa'\}]$,

$$\alpha=\mathrm{am}\{(p+q)K',\kappa'\}+\mathrm{am}\{(p-q)K',\kappa'\}.$$

Also (§ 29) $\dfrac{1-\cos\beta}{1+\cos\beta}=\dfrac{\kappa'^2\mathrm{sn}^2 pK'\,\mathrm{cn}^2 qK'}{\mathrm{dn}^2 pK'}$

$$=\frac{\mathrm{sn}^2(p\kappa'K',\,1/\kappa')\,\mathrm{dn}^2(q\kappa'K',\,1/\kappa')}{\mathrm{cn}^2(p\kappa'K',\,1/\kappa')},$$

so that $\beta=\mathrm{am}\{(p+q)\kappa'K',\,1/\kappa'\}+\mathrm{am}\{(p-q)\kappa'K',\,1/\kappa'\}$.

And $\dfrac{d-1}{d+1}=\kappa'^2\mathrm{sn}^2 pK'\,\mathrm{sn}^2 qK'$

$$=-\frac{\mathrm{sn}^2(ipK',\,\kappa)\,\mathrm{dn}^2\{(1-q)iK'-K,\,\kappa\}}{\mathrm{cn}^2(ipK',\,\kappa)},$$

or $d=\cos[\mathrm{am}\{(p+q-1)iK'+K,\,\kappa\}+\mathrm{am}\{(p-q+1)iK'-K,\,\kappa\}]$.

In the Spherical Pendulum, $Cr=0$; and therefore (§ 210)

$$\frac{1-\cos\alpha}{1+\cos\alpha}\,\frac{1-\cos\beta}{1+\cos\beta}\,\frac{d-1}{d+1}=1;$$

and $\dfrac{d-1}{d+1}=\kappa'^2\,\mathrm{sn}^2 pK'\,\mathrm{sn}^2 qK'=\dfrac{\mathrm{sn}\,qK'\,\mathrm{cn}\,pK'\,\mathrm{dn}\,pK'}{\mathrm{sn}\,pK'\,\mathrm{cn}\,qK'\,\mathrm{dn}\,qK'},$

or $\mathrm{sn}(p-q)K'=\mathrm{sn}\,pK'\,\mathrm{cn}\,qK'\,\mathrm{dn}\,qK'$.

Thence

$$d=-\frac{\mathrm{sn}(q+p)K'}{\mathrm{sn}(q-p)K'},\quad \cos\beta=-\frac{\mathrm{cn}(q+p)K'}{\mathrm{cn}(q-p)K'},\quad \cos\alpha=-\frac{\mathrm{dn}(q+p)K'}{\mathrm{dn}(q-p)K'}.$$

242. With Jacobi's notation of § 189, the expression for $i\psi$ in § 210 becomes

$$i\psi=\left(\frac{\mathrm{cn}\,v_1\mathrm{dn}\,v_1}{\mathrm{sn}\,v_1}+\frac{\mathrm{cn}\,v_2\mathrm{dn}\,v_2}{\mathrm{sn}\,v_2}\right)u+\Pi(u,\,v_1)+\Pi(u,\,v_2)$$

$$=\left(\frac{\mathrm{cn}\,v_1\mathrm{dn}\,v_1}{\mathrm{sn}\,v_1}+\mathrm{Z}v_1+\frac{\mathrm{cn}\,v_2\mathrm{dn}\,v_2}{\mathrm{sn}\,v_2}+\mathrm{Z}v_2\right)u+\tfrac{1}{2}\log\frac{\Theta(u-v_1)\Theta(u-v_2)}{\Theta(u+v_1)\Theta(u+v_2)};$$

and now, if we divide ψ into its *secular* and *periodic* part, in the form $\psi=\Psi u/K+\psi'$,
then Ψ is called the *apsidal angle*, in the motion of the Top or of the Spherical Pendulum, as seen illustrated for instance in a Giant Stride; and

$$i\Psi=\left(\frac{\mathrm{cn}\,v_1\mathrm{dn}\,v_1}{\mathrm{sn}\,v_1}+\mathrm{Z}v_1+\frac{\mathrm{cn}\,v_2\mathrm{dn}\,v_2}{\mathrm{sn}\,v_2}+\mathrm{Z}v_2\right)K+\tfrac{1}{2}\log\frac{\Theta(K-v_1)\Theta(K-v_2)}{\Theta(K+v_1)\Theta(K+v_2)},$$

which must now be expressed in a real form.

From § 172,

$$iZ(vi,\kappa)=i\int_0^{vi}(\mathrm{dn}^2vi-E/K)dvi$$

$$=\frac{E}{K}v-\int\frac{\mathrm{dn}^2(v,\kappa')}{\mathrm{cn}^2(v,\kappa')}dv$$

$$=\frac{E}{K}v-v+E\,\mathrm{am}(v,\kappa')-\frac{\mathrm{sn}\,v\,\mathrm{dn}\,v}{\mathrm{cn}\,v}\quad(\S\ 185)$$

$$=\left(\frac{E}{K}+\frac{E'}{K'}-1\right)v+\mathrm{Z}(v,\kappa')-\frac{\mathrm{sn}\,v\,\mathrm{dn}\,v}{\mathrm{cn}\,v}$$

$$=\frac{\pi v}{2KK'}+\mathrm{Z}(v,\kappa')-\frac{\mathrm{sn}\,v\,\mathrm{dn}\,v}{\mathrm{cn}\,v},$$

by means of Legendre's relation of § 171.

Thus, with $v_1=pK'i$,

$$i\left(\frac{\mathrm{cn}\,v_1\mathrm{dn}\,v_1}{\mathrm{sn}\,v_1}+\mathrm{Z}v_1\right)=\frac{\pi p}{2K}+\mathrm{Z}(pK',\kappa')+\frac{\mathrm{cn}\,pK'\mathrm{dn}\,pK'}{\mathrm{sn}\,pK'}.$$

Again, by (2)*, § 186, since $\mathrm{Z}K=0$,

$$\mathrm{Z}(K+u)=\mathrm{Z}u-\kappa^2\mathrm{sn}\,u\,\mathrm{sn}(K+u);$$

therefore, with $v_2=K+qK'i$,

$$i\left(\frac{\mathrm{cn}\,v_2\mathrm{dn}\,v_2}{\mathrm{sn}\,v_2}+\mathrm{Z}v_2\right)=\frac{\pi q}{2K}+\mathrm{Z}(qK',\kappa').$$

Also, if p and q are proper fractions, the logarithmic term of $i\Psi$ vanishes (§ 264); so that, finally,

$$\frac{\Psi}{K}=\frac{\pi}{2K}(p+q)+\mathrm{Z}(pK',\kappa')+\mathrm{Z}(qK',\kappa')+\frac{\mathrm{cn}\,pK'\mathrm{dn}\,pK'}{\mathrm{sn}\,pK'}.$$

In the Spherical Pendulum,

$$\mathrm{cn}\,pK'\mathrm{dn}\,pK'/\mathrm{sn}\,pK'=\kappa'^2\mathrm{sn}\,pK'\mathrm{sn}\,qK'\mathrm{sn}(p-q)K'$$
$$=\mathrm{Z}qK'+\mathrm{Z}(p-q)K'-\mathrm{Z}pK';$$

so that $$\frac{\Psi}{K}=\frac{\pi}{2K}(p+q)+2\mathrm{Z}(qK',\kappa')+\mathrm{Z}\{(p-q)K',\kappa'\}.$$

With the Weierstrass notation, taking u in equation (8) of § 208 between the limits ω_3 and $\omega_1+\omega_3$, we find (§ 278)

$$i\Psi=(a+b)\zeta\omega_1-(\zeta a+\zeta b)\omega_1,$$

where $$a=p\omega_3,\quad b=\omega_1+q\omega_3.$$

In small oscillations near the lowest position, p and κ' are very nearly unity, while q and κ are small.

The Geometry of the Cartesian Oval.

243. Denote the angles POO', $PO'O$, $PO''O$ in fig. 26 by θ, θ', θ'' respectively; then with O as origin,

$$x+yi=\text{cn}^2\tfrac{1}{2}w, \quad x-yi=\text{cn}^2\tfrac{1}{2}w';$$

$$i\tan\tfrac{1}{2}\theta=\frac{\sqrt{(x+yi)}-\sqrt{(x-yi)}}{\sqrt{(x+yi)}+\sqrt{(x-yi)}}$$

$$=\frac{\text{cn}\,\frac{1}{2}w-\text{cn}\,\frac{1}{2}w'}{\text{cn}\,\frac{1}{2}w+\text{cn}\,\frac{1}{2}w'}=\sqrt{\left(\frac{1-\text{cn}\,u}{1+\text{cn}\,u}\cdot\frac{1-\text{cn}\,vi}{1+\text{cn}\,vi}\right)},$$

or, in a real form, with modulus κ' for the functions of v,

$$\tan\tfrac{1}{2}\theta=\sqrt{\left(\frac{1-\text{cn}\,u}{1+\text{cn}\,u}\cdot\frac{1-\text{cn}\,v}{1+\text{cn}\,v}\right)}=\frac{\text{sn}\,\frac{1}{2}u\,\text{dn}\,\frac{1}{2}u}{\text{cn}\,\frac{1}{2}u}\,\frac{\text{sn}\,\frac{1}{2}v\,\text{dn}\,\frac{1}{2}v}{\text{cn}\,\frac{1}{2}v};$$

$$\cos\theta=\frac{\text{cn}\,u+\text{cn}\,v}{1+\text{cn}\,u\,\text{cn}\,v}, \qquad \sin\theta=\frac{\text{sn}\,u\,\text{sn}\,v}{1+\text{cn}\,u\,\text{cn}\,v}.$$

With O'' as origin,

$$\kappa^2(x+yi)=\text{dn}^2\tfrac{1}{2}w;$$

and, similarly,

$$i\tan\tfrac{1}{2}\theta''=\frac{\text{dn}\,\frac{1}{2}w-\text{dn}\,\frac{1}{2}w'}{\text{dn}\,\frac{1}{2}w+\text{dn}\,\frac{1}{2}w'}=\sqrt{\left(\frac{1-\text{dn}\,u}{1+\text{dn}\,u}\cdot\frac{1-\text{dn}\,vi}{1+\text{dn}\,vi}\right)},$$

$$\tan\tfrac{1}{2}\theta''=\sqrt{\left(\frac{1-\text{dn}\,u}{1+\text{dn}\,u}\cdot\frac{\text{dn}\,v-\text{cn}\,v}{\text{dn}\,v+\text{cn}\,v}\right)}=\frac{\kappa^2\text{sn}\,\frac{1}{2}u\,\text{cn}\,\frac{1}{2}u}{\text{dn}\,\frac{1}{2}u}\,\frac{\text{sn}\,\frac{1}{2}v}{\text{cn}\,\frac{1}{2}v\,\text{dn}\,\frac{1}{2}v};$$

$$\cos\theta''=\frac{\text{cn}\,v+\text{dn}\,u\,\text{dn}\,v}{\text{dn}\,v+\text{dn}\,u\,\text{cn}\,v}, \quad \sin\theta''=\frac{\kappa^2\text{sn}\,u\,\text{sn}\,v}{\text{dn}\,v+\text{dn}\,u\,\text{cn}\,v}.$$

With O' as origin, and

$$x+yi=\text{sn}^2\tfrac{1}{2}w,$$

then

$$i\tan\tfrac{1}{2}\theta'=\frac{\text{sn}\,\frac{1}{2}w-\text{sn}\,\frac{1}{2}w'}{\text{sn}\,\frac{1}{2}w+\text{sn}\,\frac{1}{2}w'}.$$

To reduce this to a real form, similar to the above, we require two new formulas, not included in Jacobi's list (§ 137), but easily derivable from it, namely,

$$\{\text{dn}(u+v)\pm\text{cn}(u+v)\}\{\text{dn}(u-v)\pm\text{cn}(u-v)\}=(c_1d_2\pm c_2d_1)^2/D,$$

$$\{\text{dn}(u+v)\pm\text{cn}(u+v)\}\{\text{dn}(u-v)\mp\text{cn}(u-v)\}=\kappa'^2(s_1\mp s_2)^2/D.$$

Now, with $\frac{1}{2}w$ and $\frac{1}{2}w'$ for u and v, and u and vi for $u+v$ and $u-v$,

$$i\tan\tfrac{1}{2}\theta'=\sqrt{\left(\frac{\text{dn}\,u+\text{cn}\,u}{\text{dn}\,u-\text{cn}\,u}\cdot\frac{\text{dn}\,vi-\text{cn}\,vi}{\text{dn}\,vi+\text{cn}\,vi}\right)},$$

$$\tan\tfrac{1}{2}\theta'=\sqrt{\left(\frac{\text{dn}\,u+\text{cn}\,u}{\text{dn}\,u-\text{cn}\,u}\cdot\frac{1-\text{dn}\,v}{1+\text{dn}\,v}\right)}=\frac{\text{cn}\,\frac{1}{2}u\,\text{dn}\,\frac{1}{2}u}{\text{sn}\,\frac{1}{2}u}\,\frac{\text{sn}\,\frac{1}{2}v\,\text{cn}\,\frac{1}{2}v}{\text{dn}\,\frac{1}{2}v};$$

$$\cos\theta'=\frac{-\text{cn}\,u+\text{dn}\,u\,\text{dn}\,v}{\text{dn}\,u-\text{cn}\,u\,\text{dn}\,v}, \quad \sin\theta'=\frac{\kappa'^2\text{sn}\,u\,\text{sn}\,v}{\text{dn}\,u-\text{cn}\,u\,\text{dn}\,v}.$$

244. Again, denoting the angles which P subtends at $O'O''$, $O''O$, OO' by ϕ, ϕ', ϕ'' respectively, so that

$$\phi=\pi-\theta'-\theta'',\ \phi'=\theta-\theta'',\ \phi''=\pi-\theta-\theta';$$

then we shall find

$$\tan\tfrac{1}{2}\phi=\frac{\operatorname{sn}\tfrac{1}{2}u\operatorname{dn}\tfrac{1}{2}u}{\operatorname{cn}\tfrac{1}{2}u}\frac{\operatorname{cn}\tfrac{1}{2}v}{\operatorname{sn}\tfrac{1}{2}v\operatorname{dn}\tfrac{1}{2}v}=\sqrt{\left(\frac{1-\operatorname{cn}u}{1+\operatorname{cn}u}\frac{1+\operatorname{cn}v}{1-\operatorname{cn}v}\right)},$$

$$\tan\tfrac{1}{2}\phi'=\frac{\kappa'\operatorname{sn}\tfrac{1}{2}u}{\operatorname{cn}\tfrac{1}{2}u\operatorname{dn}\tfrac{1}{2}u}\frac{\kappa'\operatorname{sn}\tfrac{1}{2}v\operatorname{cn}\tfrac{1}{2}v}{\operatorname{dn}\tfrac{1}{2}v}=\sqrt{\left(\frac{\operatorname{dn}u-\operatorname{cn}u}{\operatorname{dn}u+\operatorname{cn}u}\cdot\frac{1-\operatorname{dn}v}{1+\operatorname{dn}v}\right)},$$

$$\tan\tfrac{1}{2}\phi''=\frac{\operatorname{sn}\tfrac{1}{2}u\operatorname{cn}\tfrac{1}{2}u}{\operatorname{dn}\tfrac{1}{2}u}\frac{\operatorname{cn}\tfrac{1}{2}v\operatorname{dn}\tfrac{1}{2}v}{\operatorname{sn}\tfrac{1}{2}v}=\sqrt{\left(\frac{1-\operatorname{dn}u}{1+\operatorname{dn}u}\cdot\frac{\operatorname{dn}v+\operatorname{cn}v}{\operatorname{dn}v-\operatorname{cn}v}\right)};$$

$$\cos\phi=\frac{\operatorname{cn}u-\operatorname{cn}v}{1-\operatorname{cn}u\operatorname{cn}v},\qquad \sin\phi=\frac{\operatorname{sn}u\operatorname{sn}v}{1-\operatorname{cn}u\operatorname{cn}v},$$

$$\cos\phi'=\frac{\operatorname{cn}u+\operatorname{dn}u\operatorname{dn}v}{\operatorname{dn}u+\operatorname{cn}u\operatorname{dn}v},\quad \sin\phi'=\frac{\kappa'^2\operatorname{sn}u\operatorname{sn}v}{\operatorname{dn}u+\operatorname{cn}u\operatorname{dn}v}.$$

$$\cos\phi''=\frac{-\operatorname{cn}v+\operatorname{dn}u\operatorname{dn}v}{\operatorname{dn}v-\operatorname{dn}u\operatorname{cn}v},\ \sin\phi''=\frac{\kappa^2\operatorname{sn}u\operatorname{sn}v}{\operatorname{dn}v-\operatorname{dn}u\operatorname{cn}v}.$$

Similarly, denoting by ω, ω', ω'' the angles which the normal at P to the oval along which v is constant makes with PO, PO', PO'', we shall find

$$\tan\omega'=\frac{\operatorname{sn}u\operatorname{cn}v}{\operatorname{sn}v},\ \tan\omega=\frac{\operatorname{sn}u\operatorname{dn}v}{\operatorname{dn}u\operatorname{sn}v},\ \tan\omega''=\frac{\operatorname{sn}u}{\operatorname{cn}u\operatorname{sn}v}.$$

Drawing the three circles through $O'PO''$, $O''PO$, OPO', and denoting the points in which the normal at P meets them again by Q, Q', Q'', we shall obtain similar simple expressions for PQ, OQ, ... (Williamson, *Diff. and Int. Calculus*).

245. The two ovals defined by v and $2K'-v$ form a complete curve; and so also the ovals defined by u and $2K-u$.

Denoting by P, P', Q, Q' the four corresponding points defined by (u, v), $(u, 2K'-v)$, $(2K-u, v)$, $(2K-u, 2K'-v)$; and denoting by p, p', q, q' their consecutive positions when u receives a small increment du, then

$$Pp=\sqrt{J}du=\kappa\sqrt{(rr'r'')}du$$

$$=\frac{\operatorname{cn}vi\operatorname{dn}u+\operatorname{cn}u\operatorname{dn}vi}{\operatorname{dn}vi+\operatorname{dn}u}\sqrt{\left(\frac{\operatorname{cn}vi-\operatorname{cn}u}{\operatorname{cn}vi+\operatorname{cn}u}\right)}du$$

$$=\frac{\operatorname{dn}u+\operatorname{cn}u\operatorname{dn}v}{\operatorname{dn}v+\operatorname{dn}u\operatorname{cn}v}\sqrt{\left(\frac{1-\operatorname{cn}u\operatorname{cn}v}{1+\operatorname{cn}u\operatorname{cn}v}\right)}du;$$

and changing u into $2K-u$, v into $2K'-v$,

$$Q'q'=\frac{\operatorname{dn}u-\operatorname{cn}u\operatorname{dn}v}{\operatorname{dn}v-\operatorname{dn}u\operatorname{cn}v}\sqrt{\left(\frac{1-\operatorname{cn}u\operatorname{cn}v}{1+\operatorname{cn}u\operatorname{cn}v}\right)}du.$$

Then $$Pp+Q'q'=\frac{2}{\kappa^2}\frac{\operatorname{dn}u\operatorname{dn}v}{1+\operatorname{cn}u\operatorname{cn}v}\sqrt{\left(\frac{1-\operatorname{cn}u\operatorname{cn}v}{1+\operatorname{cn}u\operatorname{cn}v}\right)}du$$
$$=\frac{2\operatorname{dn}v}{\kappa^2\operatorname{sn}^2v}\sqrt{(1-2\operatorname{cn}v\cos\theta+\operatorname{cn}^2v)}d\theta;$$
so that the sum of the arcs described by P and Q' is expressible as an elliptic arc.

Again $$Pp-Q'q'=\frac{2}{\kappa^2}\frac{\kappa^2\operatorname{cn}u-\kappa'^2\operatorname{cn}v}{1+\operatorname{cn}u\operatorname{cn}v}\sqrt{\left(\frac{1-\operatorname{cn}u\operatorname{cn}v}{1+\operatorname{cn}u\operatorname{cn}v}\right)}du,$$
which is expressible in the form
$$\frac{2\operatorname{cn}v}{\kappa^2\operatorname{sn}^2v}\sqrt{(1-2\operatorname{dn}v\cos\theta'+\operatorname{dn}^2v)}d\theta'$$
$$+\frac{2}{\kappa^2\operatorname{sn}^2v}\sqrt{(\operatorname{dn}^2v+2\operatorname{cn}v\operatorname{dn}v\cos\phi''+\operatorname{cn}^2v)}d\phi'';$$
so that the difference of the arcs described by P and Q' is expressible by the sum of two elliptic arcs; and thus the arc of the Cartesian Oval described by P is given by means of three elliptic arcs, which is Genocchi's Theorem (*Annali di Matematica*, VI., 1864; Mr. S. Roberts, *Proc. L. M. S.*, III., V.).

246. Let us examine the analytical properties and physical applications of the functions
$$\log\operatorname{cn}\tfrac{1}{2}w,\quad \log\operatorname{sn}\tfrac{1}{2}w,\quad \log\operatorname{dn}\tfrac{1}{2}w.$$
Denoting $\log\operatorname{cn}\frac{1}{2}w$ by $\phi_1+i\psi_1$, when resolved into its real and imaginary part, then
$$\phi_1+i\psi_1=\tfrac{1}{2}\log\operatorname{cn}\tfrac{1}{2}w\operatorname{cn}\tfrac{1}{2}w'+\tfrac{1}{2}\log\operatorname{cn}\tfrac{1}{2}w/\operatorname{cn}\tfrac{1}{2}w'$$
$$=\tfrac{1}{2}\log\frac{\operatorname{cn}\frac{1}{2}w\operatorname{dn}\frac{1}{2}w\operatorname{cn}\frac{1}{2}w'\operatorname{dn}\frac{1}{2}w'}{\operatorname{dn}\frac{1}{2}w\operatorname{dn}\frac{1}{2}w'}+i\tan^{-1}i\frac{\operatorname{cn}\frac{1}{2}w'-\operatorname{cn}\frac{1}{2}w}{\operatorname{cn}\frac{1}{2}w'+\operatorname{cn}\frac{1}{2}w}$$
$$=\tfrac{1}{2}\log\frac{\operatorname{cn}iv\operatorname{dn}u+\operatorname{dn}vi\operatorname{cn}u}{\operatorname{dn}vi+\operatorname{dn}u}+i\tan^{-1}i\sqrt{\left(\frac{1-\operatorname{cn}u}{1+\operatorname{cn}u}\cdot\frac{1-\operatorname{cn}vi}{1+\operatorname{cn}vi}\right)},$$
as in § 236, by means of formulas (3), (20), (28) of § 137; and now expressing the elliptic functions of vi, to modulus κ, in terms of functions of v, to modulus κ' understood; then
$$\phi_1=\tfrac{1}{2}\log\frac{\operatorname{dn}u+\operatorname{cn}u\operatorname{dn}v}{\operatorname{dn}v+\operatorname{dn}u\operatorname{cn}v},\quad \psi_1=\tan^{-1}\sqrt{\left(\frac{1-\operatorname{cn}u}{1+\operatorname{cn}u}\cdot\frac{1-\operatorname{cn}v}{1+\operatorname{cn}v}\right)}.$$
Denoting $\log\operatorname{sn}\frac{1}{2}w$ by $\phi_2+i\psi_2$, then
$$\phi_2+i\psi_2=\tfrac{1}{2}\log\operatorname{sn}\tfrac{1}{2}w\operatorname{sn}\tfrac{1}{2}w'+\tfrac{1}{2}\log\operatorname{sn}\tfrac{1}{2}w/\operatorname{sn}\tfrac{1}{2}w'$$
$$=\tfrac{1}{2}\log\frac{\operatorname{sn}\frac{1}{2}w\operatorname{dn}\frac{1}{2}w\operatorname{sn}\frac{1}{2}w'\operatorname{dn}\frac{1}{2}w'}{\operatorname{dn}\frac{1}{2}w\operatorname{dn}\frac{1}{2}w'}+i\tan^{-1}i\frac{\operatorname{sn}\frac{1}{2}w'-\operatorname{sn}\frac{1}{2}w}{\operatorname{sn}\frac{1}{2}w'+\operatorname{sn}\frac{1}{2}w}$$
$$=\tfrac{1}{2}\log\frac{\operatorname{cn}vi-\operatorname{cn}u}{\operatorname{dn}vi+\operatorname{dn}u}+i\tan^{-1}i\sqrt{\left(\frac{\operatorname{dn}u+\operatorname{cn}u}{\operatorname{dn}u-\operatorname{cn}u}\cdot\frac{\operatorname{dn}vi-\operatorname{cn}vi}{\operatorname{dn}vi+\operatorname{cn}vi}\right)}$$

$$=\tfrac{1}{2}\log\frac{1-\operatorname{cn}u\operatorname{cn}v}{\operatorname{dn}v+\operatorname{dn}u\operatorname{cn}v}+i\tan^{-1}\sqrt{\left(\frac{\operatorname{dn}u+\operatorname{cn}u}{\operatorname{dn}u-\operatorname{cn}u}\cdot\frac{1-\operatorname{dn}v}{1+\operatorname{dn}v}\right)}.$$

Similarly, denoting $\log\operatorname{dn}\tfrac{1}{2}w$ by $\phi_3+i\psi_3$, it

$$=\tfrac{1}{2}\log\frac{\operatorname{cn}vi\operatorname{dn}u+\operatorname{cn}u\operatorname{dn}vi}{\operatorname{cn}vi+\operatorname{cn}u}+i\tan^{-1}i\sqrt{\left(\frac{1-\operatorname{dn}u}{1+\operatorname{dn}u}\cdot\frac{1-\operatorname{dn}vi}{1+\operatorname{dn}vi}\right)}$$

$$=\tfrac{1}{2}\log\frac{\operatorname{dn}u+\operatorname{cn}u\operatorname{dn}v}{1+\operatorname{cn}u\operatorname{cn}v}+i\tan^{-1}\sqrt{\left(\frac{1-\operatorname{dn}u}{1+\operatorname{dn}u}\cdot\frac{\operatorname{dn}v-\operatorname{cn}v}{\operatorname{dn}v+\operatorname{cn}v}\right)}.$$

By (20), (21), (22), (23) of § 137, we prove, in a similar manner,

$$\log\sqrt{\frac{1+\operatorname{cn}w}{1-\operatorname{cn}w}}=\tfrac{1}{2}\log\frac{\operatorname{cn}vi+\operatorname{cn}u}{\operatorname{cn}vi-\operatorname{cn}u}+i\tan^{-1}\frac{-i\operatorname{sn}vi\operatorname{dn}u}{\operatorname{dn}vi\operatorname{sn}u}$$

$$=\tanh^{-1}(\operatorname{cn}u\operatorname{cn}v)+i\tan^{-1}(\operatorname{dn}u\operatorname{sn}v/\operatorname{sn}u\operatorname{dn}v),$$

$$\log\sqrt{\frac{1+\operatorname{dn}w}{1-\operatorname{dn}w}}=\tanh^{-1}(\operatorname{dn}u\operatorname{cn}v/\operatorname{dn}v)-i\tan^{-1}(\operatorname{cn}u\operatorname{sn}v/\operatorname{sn}u),$$

$$\log\sqrt{\left(\frac{\operatorname{dn}w+\operatorname{cn}w}{\operatorname{dn}w-\operatorname{cn}w}\right)}=\text{etc.}$$

247. These conjugate functions ϕ and ψ of the complex $u+vi$ are capable of representing the solution of various physical problems concerning a plane in which u and v are taken as rectangular co-ordinates, since they satisfy the conditions

$$\frac{\partial\phi}{\partial u}=\frac{\partial\psi}{\partial v},\qquad \frac{\partial\phi}{\partial v}=-\frac{\partial\psi}{\partial u},$$

$$\frac{\partial^2\phi}{\partial u^2}+\frac{\partial^2\phi}{\partial v^2}=0,\quad \frac{\partial^2\psi}{\partial u^2}+\frac{\partial^2\psi}{\partial v^2}=0.$$

Here u and v are not restricted to be rectangular co-ordinates, but they may represent the conjugate functions of confocal conics or Cartesian Ovals, as in §§ 179, 236, or of any orthogonal system, which divides up a plane into elementary squares or rectangles, as on a map or chart.

As in § 54, we take a *period rectangle* $OABC$, bounded by $u=0$, $u=2K$, $v=0$, $v=2K'$; and now, as the end of the vector w or $u+vi$, drawn from O, travels round the boundary $OABC$ of this period rectangle, the vector w assumes the values

$$2tK(0<t<1);\qquad 2K+2t'K'i(0<t'<1);$$

$$2tK+2K'i(1>t>0);\qquad 2t'K'i(1>t'>0).$$

When the sides of the period rectangle are a and b, we replace u and v by $2Kx/a$ and $2K'y/b$, where $K'/K=b/a$.

Taking the function $\log \operatorname{cn} \frac{1}{2}w$ or $\phi_1+i\psi_1$, then from O to A, $\psi_1=0$; from A to B, $\psi_1=\frac{1}{2}\pi$; from B to C, $\psi_1=\frac{1}{2}\pi$; and from C to O, $\psi_1=0$.

At A, where $u=2K$, $v=0$, then $\phi_1=-\infty$; and at C, where $u=0$, $v=2K$, $\phi_1=\infty$.

The functions ϕ_1 and ψ_1 therefore satisfy the conditions required of the potential and stream function, due to electrodes at A and C, of the plane motion of electricity or fluid, when bounded by the rectangle $OABC$.

The function ψ_1 will also represent the stationary temperature at any point of the rectangle, when the sides OA, OC are maintained at temperature zero, and the sides AB, BC at temperature $\frac{1}{2}\pi$.

When the period rectangle is a square, or $K=K'$, then $\psi_1=\frac{1}{4}\pi$ when $u+v=2K$, or along the diagonal AC; we thus obtain the permanent temperature inside an isosceles rectangular prism, when the base is maintained at one constant temperature, and the sides at another.

Similar considerations will show that the function $\log \operatorname{sn} \frac{1}{2}w$ or $\phi_2+i\psi_2$ will give the streaming motion in the same period rectangle, due to a source at O, and an equal sink at C.

The function ψ_2 is now zero along OA, AB, BC, and $\frac{1}{2}\pi$ along OC; and ψ_2 will therefore represent the stationary temperature when OC is maintained at temperature $\frac{1}{2}\pi$, while the other sides are maintained at zero temperature.

A superposition of four such cases will give the permanent temperature when the sides of the period rectangle are maintained at any four arbitrary constant temperatures. (F. Purser, *Messenger of Mathematics*, VI., p. 137.)

EXAMPLES.

1. Solve the equation

$$\kappa^2 \operatorname{sn}^4 u - 2\kappa^2 \operatorname{sn}^2 u + 1 = 0.$$

2. Investigate the curves given by

$$dz/dw=(1-z^3)^{\frac{2}{3}}.$$

3. Prove that the system of orthogonal curves given by

$$\xi+i\eta=\operatorname{sn}(u+vi)$$

are the stereographic projections of a system of confocal spheroconics (W. Burnside, *Messenger of Mathematics*, XX.).

Prove that the stereographic projection of the points

$$x = R\,\mathrm{sn}\,u\,\mathrm{dn}\,v, \quad y = R\,\mathrm{dn}\,u\,\mathrm{sn}\,v, \quad z = R\,\mathrm{cn}\,u\,\mathrm{cn}\,v,$$

on the sphere $x^2+y^2+z^2=R^2$,

whose latitude and longitude are θ, ϕ, are given by

$$\xi+\eta i = 2R\tan(\tfrac{1}{4}\pi-\tfrac{1}{2}\theta)(\cos\phi+i\sin\phi) = R\sqrt{\frac{1-\mathrm{cn}(u+vi)}{1+\mathrm{cn}(u+vi)}}.$$

Prove also that

$$\left(\frac{\partial x}{\partial u}\right)^2+\left(\frac{\partial y}{\partial u}\right)^2+\left(\frac{\partial z}{\partial u}\right)^2=\left(\frac{\partial x}{\partial v}\right)^2+\left(\frac{\partial y}{\partial v}\right)^2+\left(\frac{\partial z}{\partial v}\right)^2$$

$$= R^2(1-\kappa^2\mathrm{sn}^2u-\kappa'^2\mathrm{sn}^2v).$$

4. Discuss the physical interpretation of

$$\phi+i\psi = \tan^{-1}\frac{\kappa\kappa'\mathrm{sn}\,u\,\mathrm{sn}\,v}{\mathrm{dn}\,u\,\mathrm{dn}\,v}+i\tan^{-1}\frac{\kappa'\mathrm{cn}\,v}{\kappa\,\mathrm{cn}\,u};$$

and determine the single function from which it is derived;

also of $\phi+i\psi=\tanh^{-1}\dfrac{\kappa\,\mathrm{cn}\,u}{\mathrm{dn}\,n\,\mathrm{dn}\,v}+i\tan^{-1}\dfrac{\kappa\,\mathrm{sn}\,u\,\mathrm{sn}\,v}{\mathrm{cn}\,v}.$

Interpret these expressions when

$$x+yi=c\sin(u+vi).$$

5. Prove that, if $x+yi=\mathrm{sn}\,w$,

then

$$\phi+i\psi=\frac{i}{\kappa}\left(Zw+\frac{\pi w}{2KK'}\right)$$

gives the plane motion of liquid streaming past two obstacles given by $x=1$ and $1/\kappa$, $x=-1$ and $-1/\kappa$ (W. Burnside, *Messenger*, XX.).

The Double Periodicity of Weierstrass's Functions.

248. A procedure similar to that of § 236 will show that the Cartesian Ovals of fig. 26 are also the representation of the conjugate functions of the system $z=\wp w$, obtained from the definition of § 50,

$$w=\int_z^\infty \frac{dz}{\sqrt{(4z^3-g_2z-g_3)}},$$

or

$$dz/dw=\wp' w=-\sqrt{(4z^3-g_2z-g_3)},$$

where

$$4z^3-g_2z-g_3=4(z-e_1)(z-e_2)(z-e_3);$$

and $z=e_1$, e_2, e_3 define the three foci.

According to § 51,

$$\wp w-e_3=(e_1-e_3)\mathrm{ns}^2\sqrt{(e_1-e_3)}w=\ (e_2-e_3)\mathrm{sn}^2\{\sqrt{(e_1-e_3)}w+K'i\},$$

$$\wp w-e_2=(e_1-e_3)\mathrm{ds}^2\sqrt{(e_1-e_3)}w=\ (e_2-e_3)\mathrm{cn}^2\{\sqrt{(e_1-e_3)}w+K'i\},$$

$$\wp w-e_1=(e_1-e_3)\,\mathrm{cs}^2\sqrt{(e_1-e_3)}w=-(e_1-e_3)\mathrm{dn}^2\{\sqrt{(e_1-e_3)}w+K'i\},$$

by § 239; thus identifying these results with those of § 236.

With the notation of § 202,

$$\wp w - e_1 = \left(\frac{\sigma_1 w}{\sigma w}\right)^2, \qquad \wp w - e_2 = \left(\frac{\sigma_2 w}{\sigma w}\right)^2, \qquad \wp w - e_3 = \left(\frac{\sigma_3 w}{\sigma w}\right)^2;$$

and denoting the focal distances by r_1, r_2, r_3, and $u - vi$ by w',

$$r_1 = \frac{\sigma_1 w\, \sigma_1 w'}{\sigma w\, \sigma w'}, \qquad r_2 = \frac{\sigma_2 w\, \sigma_2 w'}{\sigma w\, \sigma w'}, \qquad r_3 = \frac{\sigma_3 w\, \sigma_3 w'}{\sigma w\, \sigma w'}.$$

249. To express these focal distances in a real form, as in § 236, we employ the Addition Theorem (K) of § 200, written

$$\sigma(u+v)\sigma(u-v) = \sigma^2 u\, \sigma^2 v\{(\wp v - e_\alpha) - (\wp u - e_\alpha)\}$$

$$= \sigma^2 u\, \sigma_\alpha{}^2 v - \sigma_\alpha{}^2 u\, \sigma^2 v. \quad \text{.................}(M)$$

Again, from § 154, $\wp(u+v) - e_\alpha$ is a perfect square; and we may write $x = \wp u, \quad y = \wp v, \quad s = \wp(u+v),$

$$N = \wp u - e_\alpha, \qquad D = \wp u - e_\beta \,.\, \wp u - e_\gamma;$$

$$\surd\{\wp(u+v) - e_\alpha\}$$

$$= \frac{\surd(\wp u - e_\alpha \,.\, \wp v - e_\beta \,.\, \wp v - e_\gamma) - \surd(\wp u - e_\beta \,.\, \wp u - e_\gamma \,.\, \wp v - e_\alpha)}{\wp v - \wp u}, \quad (N)$$

and now

$$\sigma_\alpha(u+v)\sigma(u-v) = \surd\{\wp(u+v) - e_\alpha\}\sigma^2 u\, \sigma^2 v(\wp v - \wp u)$$

$$= \sigma u\, \sigma_\alpha u\, \sigma_\beta v\, \sigma_\gamma v - \sigma_\beta u\, \sigma_\gamma u\, \sigma_\alpha v\, \sigma v, \ldots (O)$$

and changing the sign of v,

$$\sigma(u+v)\sigma_\alpha(u-v) = \sigma u\, \sigma_\alpha u\, \sigma_\beta v\, \sigma_\gamma v + \sigma_\beta u\, \sigma_\gamma u\, \sigma_\alpha v\, \sigma v. \ldots (P)$$

Again, by multiplication with (N) and reduction,

$$\frac{\sigma_\alpha(u+v)\sigma_\beta(u-v)}{\sigma(u+v)\ \sigma(u-v)}$$

$$= \frac{\surd(\wp u - e_\alpha \,.\, \wp u - e_\beta \,.\, \wp v - e_\alpha \,.\, \wp v - e_\beta) - (e_\alpha - e_\beta)\surd(\wp u - e_\gamma \,.\, \wp v - e_\gamma)}{\wp v - \wp u},$$

or

$$\sigma_\alpha(u+v)\sigma_\beta(u-v) = \sigma_\alpha u\, \sigma_\beta u\, \sigma_\alpha v\, \sigma_\beta v - (e_\alpha - e_\beta)\sigma u\, \sigma_\gamma u\, \sigma v\, \sigma_\gamma v, \ (Q)$$

$$\sigma_\alpha(u-v)\sigma_\beta(u+v) = \sigma_\alpha u\, \sigma_\beta u\, \sigma_\alpha v\, \sigma_\beta v + (e_\alpha - e_\beta)\sigma u\, \sigma_\gamma u\, \sigma v\, \sigma_\gamma v. \ (R)$$

Similarly,

$$\frac{\sigma_\alpha(u+v)\sigma_\alpha(u-v)}{\sigma(u+v)\ \sigma(u-v)} = \frac{(\wp u - e_\alpha)(\wp v - e_\alpha) - (e_\alpha - e_\beta)(e_\alpha - e_\gamma)}{\wp v - \wp u},$$

or

$$\sigma_\alpha(u+v)\sigma_\alpha(u-v) = \sigma_\alpha{}^2 u\, \sigma_\alpha{}^2 v - (e_\alpha - e_\beta)(e_\alpha - e_\gamma)\sigma^2 u\, \sigma^2 v. \quad \ldots\ldots (S)$$

(Schwarz, *Elliptische Functionen*, p. 51.)

Now, from these equations (O), (P), (Q), (R), with

w or $\frac{1}{2}(u+vi)$ for u, and w' or $\frac{1}{2}(u-vi)$ for v,

$$r_1=\frac{\sigma_1 w\, \sigma_2 w\, \sigma_1 w' \sigma_2 w'}{\sigma w\, \sigma_2 w\, \sigma w' \sigma_2 w'}=-(e_3-e_1)\frac{\sigma_1 u\, \sigma_2 vi+\sigma_2 u\, \sigma_1 vi}{\sigma_3 u\, \sigma_1 vi-\sigma_1 u\, \sigma_3 vi},$$

or $$r_1=\frac{\sigma_1 w\, \sigma_3 w\, \sigma_1 w' \sigma_3 w'}{\sigma w\, \sigma_3 w\, \sigma w' \sigma_3 w'}=-(e_1-e_2)\frac{\sigma_1 u\, \sigma_3 vi+\sigma_3 u\, \sigma_1 vi}{\sigma_1 u\, \sigma_2 vi-\sigma_2 u\, \sigma_1 vi},$$

with similar equations for r_2 and r_3; and thence the vectorial equations of the Cartesian Ovals analogous to those of § 236

$$\left.\begin{aligned} r_2\sigma_3 u-r_3\sigma_2 u &= (e_2-e_3)\sigma_1 u \\ r_2\sigma_3 vi-r_3\sigma_2 vi &= -(e_2-e_3)\sigma_1 vi \end{aligned}\right\}, \text{ etc.}$$

These vectorial equations again are the geometrical interpretation of the formula, immediately deducible from (N),

$$\sigma_\beta w\, \sigma_\beta w' \sigma_\gamma(w+w')-\sigma_\gamma w\, \sigma_\gamma \sigma_\beta(w+w')$$
$$=(e_\beta-e_\gamma)\sigma w\, \sigma w' \sigma_\alpha(w+w'), \quad \text{.........(T)}$$

Making $m^2=-1$ in the *homogeneity equations* of § 196, gives

$$\wp(vi\,;\; g_2, g_3)=-\;\wp(v\,;\; g_2, -g_3),$$

the equivalent of the equations of § 238, by which a change is made to a real argument and complementary modulus; while

$$\zeta(vi\,;\; g_2, g_3)=-\;i\zeta(v\,;\; g_2, -g_3),$$
$$\sigma(vi\,;\; g_2, g_3)=\;i\sigma(v\,;\; g_2, -g_3),$$
$$\sigma_\alpha(vi\,;\; g_2, g_3)=\;\sigma_\alpha(v\,;\; g_2, -g_3).$$

250. When a point has made a complete circuit of one of the ovals, enclosing a pair of foci, defined by e_2 and e_3, or e_1 and e_2, z will have regained its original value, but w will have increased or diminished by $2\omega_1$ or $2\omega_3$, defined as in §§ 51, 52 by the *rectilinear* integrals

$$\omega_1=\int_{e_1}^{\infty} ds/\sqrt{S}=\int_{e_3}^{e_2} ds/\sqrt{S},$$

$$\omega_3=\int_{e_2}^{e_1} ds/\sqrt{S}=\int_{-\infty}^{e_3} ds/\sqrt{S};$$

so that $2\omega_1$, $2\omega_3$ are the *periods* of the function $\wp u$, and

$$\wp(u+2m\omega_1+2n\omega_3)=\wp u.$$

To fix the ideas we have supposed the circuit of two poles of the integral made on the enclosing branch of a Cartesian Oval, but the result will be the same whatever be the curve, provided it makes the same number and nature of circuits.

Now, in § 165, we can have

$$u+v+w=2m\omega_1+2n\omega_3\equiv 0 \text{ (mod. } 2\omega_1, 2\omega_3).$$

251. In § 54 it has been shown how, as the vector of the argument w traces out the contour of the *period* rectangle, $\wp w$ assumes all real values; and $\wp w$ may be made to assume any arbitrary complex value at a point in the interior of the rectangle, given by a determinate vector $t\omega_1 + t'\omega_3$.

It is convenient to put $\omega_1 + \omega_3 = -\omega_2$, so that

$$\omega_1 + \omega_2 + \omega_3 = 0, \text{ with } e_1 + e_2 + e_3 = 0;$$

and now $$\wp\omega_1 = e_1, \quad \wp\omega_2 = e_2, \quad \wp\omega_3 = e_3;$$

while $$\wp'\omega_1 = \wp'\omega_2 = \wp'\omega_3 = 0.$$

The equations of § 54 show that

$$\wp(u \pm \omega_1) - e_1 = \frac{e_1 - e_2 \,.\, e_1 - e_3}{\wp u - e_1},$$

$$\wp(u \pm \omega_2) - e_2 = \frac{e_2 - e_3 \,.\, e_2 - e_1}{\wp u - e_2},$$

$$\wp(u \pm \omega_3) - e_3 = \frac{e_3 - e_1 \,.\, e_3 - e_1}{\wp u - e_3};$$

equations analogous to those of § 57, in Jacobi's notation.

Thus, from ex. 9, p. 174,

$$4\wp\, 2u = \wp u + \wp(u + \omega_1) + \wp(u + \omega_2) + \wp(u + \omega_3).$$

With negative discriminant, as in § 62, we take e_2 as real, and e_1, e_2 imaginary; also $\omega_1 = \frac{1}{2}(\omega_2 + \omega'_2)$, $\omega_3 = \frac{1}{2}(\omega_2 - \omega'_2)$; and

$$\wp\omega_1 = e_1, \quad \wp\omega_3 = e_3, \quad \wp\omega_2 = \wp\omega'_2 = e_2.$$

252. A great advantage of the Weierstrassian notation (at first rather baffling to one accustomed to the methods of Legendre and Jacobi) is that the dimensions of the elliptic integral are left arbitrary, and can be changed by an application of the *Principle of Homogeneity* of § 196.

When the canonical elliptic integral of § 50 is *normalized* in Klein's manner (§ 196) by multiplying by $\Delta^{\frac{1}{12}}$, then

$$\int \frac{\Delta^{\frac{1}{12}} ds}{\sqrt{(4s^3 - g_2 s - g_3)}} = \int \frac{d\sigma}{\sqrt{(4\sigma^3 - \gamma_2\sigma - \gamma_3)}},$$

where $$s = \Delta^{\frac{1}{6}}\sigma, \qquad g_2 = \Delta^{\frac{1}{3}}\gamma_2, \quad g_3 = \Delta^{\frac{1}{2}}\gamma_3;$$

and now $$\gamma_2^3 - 27\gamma_3^2 = 1,$$

so that the new discriminant is unity, and

$$J = \gamma_2^3, \quad J - 1 = 27\gamma_3^2.$$

If ϖ_1, ϖ_3 denote the real and imaginary half periods of the normalized integral, then

$$\varpi_1 = \omega_1 \Delta^{\frac{1}{12}}, \quad \varpi_3 = \omega_3 \Delta^{\frac{1}{12}}.$$

The general elliptic integral, written with homogeneous variables as in § 155, is also normalized by Klein by multiplying by the twelfth root of the discriminant of the corresponding quartic, and its half periods are now ϖ_1 and ϖ_3.

If we normalize, for instance, the canonical integral (11) of § 38, written with homogeneous variables x_1, x_2, in the form

$$\int (x_1 x_2 . x_2 - x_1 . x_2 - k x_1)^{-\frac{1}{2}} (x_2 dx_1 - x_1 dx_2),$$

then the invariants g_2, g_3, and the discriminant Δ of the quartic

$$x_1 x_2 . x_2 - x_1 . x_2 - k x_1,$$

being the expressions given in § 68, therefore

$$\Delta^{\frac{1}{12}} = \{\tfrac{1}{16} k(1-k)\}^{\frac{1}{6}} = \sqrt[3]{(\tfrac{1}{4}\kappa\kappa')}.$$

Now the half periods of integral (11), § 38, being $2K$, $2K'i$,

$$\varpi_1 = 2K\sqrt[3]{(\tfrac{1}{4}\kappa\kappa')}, \quad \varpi_3 = 2K'i\sqrt[3]{(\tfrac{1}{4}\kappa\kappa')}.$$

We are thereby enabled to change from Weierstrass's ω_1 and ω_3 to Jacobi's K and K', and to utilize the numerical results of Legendre's Tables. (Klein, *Math. Ann.*, XIV., p. 118.)

When the discriminant Δ is negative, we normalize by multiplying by $(-\Delta)^{\frac{1}{12}}$, and replace ω_1 and ω_3 by ω_2 and ω_2' (§ 62); but now the new discriminant $\gamma_2^3 - 27\gamma_3^2 = -1$, and

$$\omega_2(-\Delta)^{\frac{1}{12}} = 2K\sqrt[6]{(\tfrac{1}{4}\kappa\kappa')}, \quad \omega'_2(-\Delta^{\frac{1}{12}}) = 2K'i\sqrt[6]{(\tfrac{1}{4}\kappa\kappa')} \text{ (§§ 47, 58).}$$

For instance, if $g_2 = 0$ in § 50, $(-\Delta)^{\frac{1}{12}} = \sqrt[4]{3}\sqrt[6]{g_3}$; and in § 58, $J = 0$, or $2\kappa\kappa' = \frac{1}{2}$, $2\sqrt[6]{(\frac{1}{4}\kappa\kappa')} = \sqrt[3]{2}$; and now

$$\omega_2\sqrt[4]{3}\sqrt[6]{g_3} = K\sqrt[3]{2}, \quad \omega_2'\sqrt[4]{3}\sqrt[6]{g_3} = iK'\sqrt[3]{2};$$

while (§ 47) $\qquad \omega_2'/\omega_2 = K'i/K = i\sqrt{3}.$

Confocal Quadric Surfaces.

253. The symmetry and elegance of the Weierstrass notation is well exhibited in the physical applications relating to confocal surfaces of the second degree.

The equation of any one of a system of confocal quadrics being

$$\frac{x^2}{a^2+\lambda} + \frac{y^2}{b^2+\lambda} + \frac{z^2}{c^2+\lambda} = 1,$$

we put

$$a^2+\lambda = m^2(\wp u - e_1), \quad b^2+\lambda = m^2(\wp u - e_2), \quad c^2+\lambda = m^2(\wp u - e_3);$$

and now the integral

$$\int_\lambda^\infty \frac{d\lambda}{\sqrt{(a^2+\lambda \,.\, b^2+\lambda \,.\, c^2+\lambda)}} = \frac{2u}{m}.$$

With $e_1 > e_2 > e_3$, we must take $a^2 < b^2 < c^2$.

Three confocals can be drawn through any point x, y, z, an ellipsoid, a hyperboloid of one sheet, and a hyperboloid of two sheets.

Supposing the ellipsoid to be defined by λ or u, and the hyperboloid of one sheet in a similar manner by μ or v, and the hyperboloid of two sheets by ν or w; then in going round the period rectangle of § 54,

(i.) $u=p\omega$, $\infty>\wp u>e_1$, for the ellipsoids; starting with $p=0$ for the infinite sphere, and ending with $p=1$ for the inside of focal ellipse;

(ii.) $v=\omega_1+q\omega_3$, $e_1>\wp v>e_2$, for the hyperboloids of one sheet; starting with $q=0$ from the focal ellipse, and ending with $q=1$ for the focal hyperbola;

(iii.) $w=r\omega_1+\omega_3$, $e_2>\wp w>e_3$, for the hyperboloids of two sheets; starting with $q=1$ from the focal hyperbola, and ending with $q=0$ for the outside of the focal ellipse;

(iv.) the fourth side of the period rectangle gives imaginary surfaces.

254. Replacing b^2-a^2 and c^2-a^2 by β^2 and γ^2, so that

$$(y/\beta)^2+(z/\gamma)^2=1, \quad x=0,$$

are the equations of the focal ellipse of the confocal system, we should have to put, with Jacobi's notation,

$$a^2+\lambda=\gamma^2\mathrm{cs}^2(u,\kappa),\ b^2+\lambda=\gamma^2\mathrm{ds}^2(u,\kappa),\ c^2+\lambda=\gamma^2\mathrm{ns}^2(u,\kappa);$$
$$a^2+\mu=-\beta^2\mathrm{sn}^2(v,\kappa'),\ b^2+\mu=\beta^2\mathrm{cn}^2(v,\kappa'),\ c^2+\mu=\gamma^2\mathrm{dn}^2(v,\kappa');$$
$$a^2+\nu=-\gamma^2\mathrm{dn}^2(w,\kappa),\ b^2+\nu=-\gamma^2\mathrm{cn}^2(w,\kappa),\ c^2+\nu=\kappa^2\gamma^2\mathrm{sn}^2(w,\kappa);$$

where
$$\kappa^2=\frac{c^2-b^2}{c^2-a^2},\quad \kappa'^2=\frac{b^2-a^2}{c^2-a^2};$$

and now u, v, w will be Lamé's *parameters*, as given in Maxwell's *Electricity and Magnetism*, I., chap. X.

By solution of the three equations of the confocal quadrics,

$$x^2=\frac{a^2+\lambda\,.\,a^2+\mu\,.\,a^2+\nu}{a^2-b^2\,.\,a^2-c^2},\quad y^2=\frac{b^2+\lambda\,.\,b^2+\mu\,.\,b^2+\nu}{b^2-c^2\,.\,b^2-a^2},$$
$$z^2=\frac{c^2+\lambda\,.\,c^2+\mu\,.\,c^2+\nu}{c^2-a^2\,.\,c^2-b^2},$$

and thus x, y, z can be expressed as functions of u, v, w.

Employing the function s_a of § 203,

$$x^2=\frac{m^2s_1{}^2}{e_1-e_2\,.\,e_1-e_3},\quad y^2=\frac{m^2s_2{}^2}{e_2-e_3\,.\,e_2-e_1},\quad z^2=\frac{m^2s_3{}^2}{e_3-e_1\,.\,e_3-e_2}.$$

When $b^2=c^2$, the ellipsoids are oblate spheroids, and the hyperboloids of two sheets degenerate into planes through Ox; and now the orthogonal system is given by

$$\frac{x^2}{\cot^2 u}+\frac{y^2+z^2}{\operatorname{cec}^2 u}=\gamma^2, \quad\text{.....................(i.)}$$

$$-\frac{x^2}{\tanh^2 v}+\frac{y^2+z^2}{\operatorname{sech}^2 v}=\gamma^2, \quad\text{.....................(ii.)}$$

$$\frac{y^2}{\cos^2 w}-\frac{z^2}{\sin^2 w}=0; \quad\text{.....................(iii.)}$$

intersecting in the point

$$x=\gamma \cot u \tanh v,$$
$$y=\gamma \operatorname{cec} u \operatorname{sech} v \cos w,$$
$$z=\gamma \operatorname{cec} u \operatorname{sech} v \sin w.$$

When $b^2=a^2$, the ellipsoids are prolate spheroids, and the hyperboloids of one sheet are planes through Oz; now the orthogonal system is given by

$$\frac{x^2+y^2}{\operatorname{cech}^2 u}+\frac{z^2}{\coth^2 u}=\gamma^2, \quad\text{.....................(iv.)}$$

$$-\frac{x^2}{\sin^2 v}+\frac{y^2}{\cos^2 v}=0, \quad\text{.....................(v.)}$$

$$-\frac{x^2+y^2}{\operatorname{sech}^2 w}+\frac{z^2}{\tanh^2 w}=\gamma^2; \quad\text{.....................(vi.)}$$

intersecting in the point

$$x=\gamma \operatorname{cech} u \sin v \operatorname{sech} w,$$
$$y=\gamma \operatorname{cech} u \cos v \operatorname{sech} w,$$
$$z=\gamma \coth u \tanh w.$$

The degenerate case of confocal paraboloids, where the centre is at an infinite distance, may be written

$$\frac{y^2}{\cosh^2 \frac{1}{2}u}+\frac{z^2}{\sinh^2 \frac{1}{2}u}=8a(a\cosh u - x), \quad\text{...........(vii.)}$$

$$\frac{y^2}{\cos^2 \frac{1}{2}v}-\frac{z^2}{\sin^2 \frac{1}{2}v}=8a(a \quad \cos v - x), \quad\text{...........(viii.)}$$

$$\frac{y^2}{\sinh^2 \frac{1}{2}w}+\frac{z^2}{\cosh^2 \frac{1}{2}w}=8a(a\cosh w + x), \quad\text{...........(ix.)}$$

intersecting in the point

$$x=a(\cosh u+\cos v-\cosh w),$$
$$y=4a\cosh \tfrac{1}{2}u \cos \tfrac{1}{2}v \sinh \tfrac{1}{2}w,$$
$$z=4a\sinh \tfrac{1}{2}u \sin \tfrac{1}{2}v \cosh \tfrac{1}{2}w.$$

(*Proc. Lond. Math. Society*, XIX.)

255. We may take u, v, w as Lamé's *thermometric parameters*, and now Laplace's equation becomes (Maxwell, *Electricity*, I., chap. X.)

$$(\mu-\nu)\frac{\partial^2\phi}{\partial u^2}+(\nu-\lambda)\frac{\partial^2\phi}{\partial v^2}+(\lambda-\mu)\frac{\partial^2\phi}{\partial w^2}=0.$$

Thus $\phi=Au+Bv+Cw+D(u^2+v^2+w^2)$
$+2Evw+2Fwu+2Guv+Huvw$
is a particular solution of this equation; for instance, the electric potential between two confocal ellipsoids, defined by u_1 and u_2, maintained at potentials U_1 and U_2, is given by

$$U=\{U_1(u-u_2)+U_2(u_1-u)\}/(u_1-u_2).$$

When the solution ϕ is equal to UVW, the product of three functions, U a function of u only, V of v, and W of w only, then Laplace's equation becomes

$$(\mu-\nu)\frac{1}{U}\frac{d^2U}{du^2}+(\nu-\lambda)\frac{1}{V}\frac{d^2V}{dv^2}+(\lambda-\mu)\frac{1}{W}\frac{d^2W}{dw^2}=0\,;$$

so that we may put

$$\frac{1}{U}\frac{d^2U}{du^2}=g\lambda+h,\quad \frac{1}{V}\frac{d^2V}{dv^2}=g\mu+h,\quad \frac{1}{W}\frac{d^2W}{dw^2}=g\nu+h;$$

three equations of Lamé's form (§ 204), when $g=n(n+1)$.

256. The complete solution of Lamé's equation was first obtained by Hermite, in the form

$$U=CF(u)+C'F(-u).$$

Denoting by Y the product U_1U_2 of U_1 and U_2, or $F(u)$ and $F(-u)$, two particular solutions of the general linear differential equation of the second order, in its canonical form

$$\frac{1}{U}\frac{d^2U}{du^2}=I,$$

where I is some function of u, and denoting differentiation with respect to u by accents, then

$$Y'=U_1'U_2+U_1U_2',$$
$$Y''=U_1''U_2+2U_1'U_2'+U_1U_2''$$
$$=2IU_1U_2+2U_1'U_2',$$

or $$Y''-2IY=2U_1'U_2';$$

and $$Y'''-2IY'-2I'Y=2U_1''U_2'+2U_1'U_2''$$
$$=2I(U_1U_2'+U_1'U_2)=2IY',$$

or $$Y'''-4IY'-2I'Y=0,$$

the general solution of which linear differential equation is

$$AU_1^2+2BU_1U_2+CU_2^2.$$

A first integral of this differential equation is

$$2YY'' - Y'^2 - 4IY^2 + C^2 = 0,$$

where C is a constant, given by

$$U_1 U_2' - U_1' U_2 = C,$$

the integral of

$$U_1 U_2'' - U_1'' U_2 = 0.$$

In Lamé's differential equation

$$I = n(n+1)\wp u + h;$$

and now, changing to $x = \wp u$ as independent variable,

$$(4x^3 - g_2 x - g_3)\frac{d^3 Y}{dx^3} + 3(6x^2 - \tfrac{1}{2}g_2)\frac{d^2 Y}{dx^2}$$
$$- 4\{(n^2+n-3)x + h\}\frac{dY}{dx} - 2n(n+1)Y = 0,$$

and this equation for Y has, as a particular solution, a rational integral function of x or $\wp u$, of the nth order, which we may write

$$Y = \Pi(\wp u - \wp a),$$

and

$$h = (2n-1)\Sigma\wp a.$$

Now, by logarithmic differentiation,

$$\frac{U_2'}{U_2} + \frac{U_1'}{U_1} = \frac{Y'}{Y} = \sum\frac{\wp' u}{\wp u - \wp a},$$

while

$$\frac{U_2'}{U_2} - \frac{U_1'}{U_1} = \frac{C}{Y} = \Pi\frac{C}{(\wp u - \wp a)}.$$

Brioschi shows (*Comptes Rendus*, XCII.) that, when resolved into partial fractions, we may put

$$\frac{C}{\Pi(\wp u - \wp a)} = \sum\frac{\wp' a}{\wp u - \wp a},$$

provided that

$$\Sigma\wp' a = 0,\ \Sigma\wp a\wp' a = 0,\ \Sigma(\wp a)^2\wp' a = 0,\ \ldots,\ \Sigma(\wp a)^{n-2}\wp' a = 0,$$

and

$$\Sigma(\wp a)^{n-1}\wp' a = C.$$

Then

$$\frac{U_1'}{U_1} = \sum\frac{1}{2}\frac{\wp' u - \wp' a}{\wp u - \wp a},\quad \frac{U_2'}{U_2} = \sum\frac{1}{2}\frac{\wp' u + \wp' a}{\wp u - \wp a};$$

and, integrating,

$$Fu, \text{ or } U_1 = \Pi\frac{\sigma(u+a)}{\sigma a\,\sigma u}\exp(-u\zeta a) = \Pi\phi(u, a);$$

while U_2 or $F(-u)$ is obtained by changing the sign of u or a.

257. Hermite shows (*Comptes Rendus*, 1877) that the function $F(u)$ may be otherwise expressed by

$$F(u)=\left(\frac{d}{du}\right)^{n-1}\phi u-A_1\left(\frac{d}{du}\right)^{n-3}\phi u+A_2\left(\frac{d}{du}\right)^{n-5}\phi u-\dots$$

and ϕu, called the *simple element*, is of the form $e^{\lambda u}\phi(u, \omega)$, $\phi(u, \omega)$ being a solution for $n=1$ and $h=\wp\omega$ (§ 204).

To obtain the coefficients A_1, A_2, ... in $F(u)$, we suppose ϕu or $e^{\lambda u}\phi(u, \omega)$, Fu, $\wp u$ expanded in the neighbourhood of $u=0$ (§ 195), in the form (Halphen, *F. E. I.*, chap. VII.)

$$e^{\lambda u}\phi(u, \omega)=\frac{1}{u}+\lambda+(\lambda^2-\wp\omega)\frac{u}{2!}+(\lambda^3-3\lambda\wp\omega-\wp'\omega)\frac{u^3}{3!}+\dots,$$

$$(-1)^{n-1}Fu=\frac{(n-1)!}{u^n}-A_1\frac{(n-3)!}{u^{n-2}}+A_2\frac{(n-5)!}{u^{n-4}}-\dots.$$

Substituting in Lamé's differential equation

$$F''u=\{n(n+1)\wp u+h\}Fu,$$

we obtain, by equating coefficients,

$$A_1=\frac{(n-1)(n-2)}{2(2n-1)}h,$$

$$A_2=\frac{(n-1)(n-2)(n-3)(n-4)}{8(2n-1)(2n-3)}\left\{h^2-\frac{n(n+1)(2n-1)}{10}g_2\right\}, \dots.$$

On comparing the two forms of the solution Fu, we find that

$$\omega=\Sigma a, \text{ and } \lambda=\zeta\omega-\Sigma\zeta a.$$

Thus, for instance, when $n=2$, we find, as in § 209,

$$F(u)=\frac{\sigma(u+a)\sigma(u+b)}{\sigma a\,\sigma b\,\sigma^2 u}\exp(-\zeta a-\zeta b)u$$

$$=\frac{d}{du}\frac{\sigma(u+a+b)}{\sigma(a+b)\sigma u}\exp(-\zeta a-\zeta b)u.$$

When $n=3$,

$$Fu=\phi(u, a_1)\phi(u, a_2)\phi(u, a_3)$$

$$=\frac{d^2}{du^2}\phi(u, \omega)e^{\lambda u}-(\wp a_1+\wp a_2+\wp a_3)\phi(u, \omega)e^{\lambda u},$$

where

$$a_1+a_2+a_3=\omega,$$

$$\wp' a_1+\wp' a_2+\wp' a_3=0,$$

$$\wp a_1\wp' a_1+\wp a_2\wp' a_2+\wp a_3\wp' a_3=0,$$

$$\zeta\omega-\zeta a_1-\zeta a_2-\zeta a_3=\lambda.$$

This fails when $g_2=0$, and $a_1=v$, $a_2=\omega v$, $a_3=\omega^2 v$; but now (§ 229)

$$Fu=\tfrac{1}{2}(\wp' v-\wp' u).$$

CHAPTER IX.

THE RESOLUTION OF THE ELLIPTIC FUNCTIONS INTO FACTORS AND SERIES.

258. The well-known expressions for the circular and hyperbolic functions in the form of finite and infinite products (Chrystal, *Algebra*, II., p. 322; Hobson, *Trigonometry*, chap. XVII.) have their analogues for the Elliptic Functions, as laid down by Abel in *Crelle*, 2 and 3.

Granting the possibility of the resolution into linear factors, the individual factors are readily inferred from a consideration of the *zeroes* and *infinities* of the function.

Denote $2mK+2nK'i$ by Ω,

where m and n denote any integers, positive or negative,

denote also $\Omega+K$ or $(2m+1)K+2nK'i$ by Ω_1,

$\Omega+K+K'i$ or $(2m+1)K+(2n+1)K'i$ by Ω_2,

and $\Omega+K'i$ or $2mK+(2n+1)K'i$ by Ω_3.

Then considering the function

$$\operatorname{sn} u,$$

the zeroes are given by $u=\Omega$, and the infinities by $u=\Omega_3$ (§ 239); and thus we infer that, if $\operatorname{sn} u$ can be resolved into a convergent product of an infinite number of linear factors, the form is

$$\operatorname{sn} u = A\frac{u \prod\limits_{m=-\infty}^{m=\infty}{}' \prod\limits_{n=-\infty}^{n=\infty}{}' \left(1-\dfrac{u}{\Omega}\right)}{\prod\limits_{m=-\infty}^{m=\infty} \prod\limits_{n=-\infty}^{n=\infty} \left(1-\dfrac{u}{\Omega_3}\right)}, \quad\text{.................(1)}$$

the accents in the numerator denoting that the simultaneous zero values of m and n are excluded.

Similarly, $$\operatorname{cn} u = B\,\Pi\,\Pi\left(1-\frac{u}{\Omega_1}\right)\Big/D, \quad\text{.....................(2)}$$

$$\operatorname{dn} u = C\,\Pi\,\Pi\left(1-\frac{u}{\Omega_2}\right)\Big/D. \quad\text{.....................(3)}$$

the zeroes of $\operatorname{cn} u$ being given by $u=\Omega_1$, and the zeroes of $\operatorname{dn} u$ by $u=\Omega_2$, while the infinities are given as before by $u=\Omega_3$; D denoting the denominator in (1).

259. But now, in demonstrating the analytical equivalence of the expressions on the two sides of equations (1), (2), (3), it will fix the ideas if we employ a physical interpretation, such as that given in § 247.

It was shown there that the real and imaginary part (*norm* and *amplitude*) of

$$\log \operatorname{sn} w,$$

where $w=u+vi$, will represent in the rectangle $OABC$ the potential and current function of the flow of electricity (or of liquid, following the laws of electrical flow) from a positive electrode at O to a negative electrode at C, $\frac{1}{2}\pi$ ampères being the strength of the current; but here we take $OA=K$, $OC=K'$; and u, v are the coordinates of any point in the rectangle.

The infinite series of electrodes, which are the optical images by reflexion of these two electrodes at O and C, will form a system on an infinite conducting plane, such that, if the strength of the current at each electrode is 2π ampères, the resultant effect in the rectangle $OABC$ will be the same as before.

(Jochmann, *Zeitschrift für Mathematik*, 1865;
O. J. Lodge, *Phil. Mag.* 1876; *Q. J. M.*, XVII.)

Starting with a single electrode at O, of current 2π ampères, the potential and current function at any point whose vector is w or $u+vi$ are the *norm* and *amplitude* of $\log w$; and $\log w$ may be called the *vector function* of the electrode at O.

For an electrode at a point whose vector is $c=a+bi$, the vector function at $z=x+yi$ is $\log(z-c)$,

which may be written

$$\log(1-z/c),$$

disregarding the complex constant $\log(-c)$.

The vector of any optical image of O in the sides of the rectangle $OABC$ being given by Ω, the vector potential of the corresponding electrode is $\log(1-w/\Omega)$; and the vector function of the system of images of the positive electrode at O will be

$$\log w\Pi'\Pi'\left(1-\frac{w}{\Omega}\right).$$

Similarly the vector function of the system of images of the negative electrode at C will be

$$\log \Pi\,\Pi\left(1-\frac{w}{\Omega_3}\right).$$

But these functions, considered separately, represent a physical impossibility, and are analytically meaningless; their difference, however,

$$\log w\Pi'\Pi'\left(1-\frac{w}{\Omega}\right)\Big/\Pi\,\Pi\left(1-\frac{w}{\Omega_3}\right),$$

will represent the vector function of the whole system of positive and negative electrodes; and since this function satisfies the requisite conditions inside the rectangle $OABC$ as the function $\log \operatorname{sn} w$, we are led to infer equation (1), with suitable restrictions explained hereafter.

For $\log \operatorname{cn} w$, the positive electrode is placed at A, the negative electrode being still at C; the vectors of the positive electrode images are given by Ω_1; and now equation (2) is inferred; while for $\log \operatorname{dn} w$, the positive electrode is placed at B, and the vectors of its images are given by Ω_2, the negative electrode being at C; and we infer equation (3).

When in the rectangle $OABC$ we have $OA=a$, $OC=b$, we take $K'/K=b/a$, and write $K(x/a)+K'i(y/b)$ for $u+vi$, x, y now denoting the coordinates of a point.

260. We now proceed to express these doubly infinite products of factors, corresponding to the different integral values of m and n, by means of singly infinite factors for different values of n; that is, we combine all the factors for one value of n and the infinite series of values of m into a single expression; and here we employ the formulas for the trigonometrical functions expressed as infinite products.

Interpreted physically, we determine the vector function of an infinite series of electrodes, equispaced on a straight line parallel to OA.

Denoting the vectors of such a series of positive electrodes by $2ma+nbi$, the vector function is

$$\log \prod_{m=-\infty}^{m=\infty} (z-2ma-nbi), \text{ or } \log(z-nbi)\Pi'\left(1-\frac{z-nbi}{2ma}\right);$$

and provided that $(z-nbi)/2ma$ is ultimately zero when m is infinite, or that z/ma and n/m tend to the limit zero, we can write this vector function (Cayley, *Elliptic Functions*, p. 300)

$$\log \sin \tfrac{1}{2}\pi(z-nbi)/a, \quad \text{......................(4)}$$

Resolved into its *norm* and *amplitude*, this vector function is

$$\tfrac{1}{2}\log \tfrac{1}{2}[\cosh\{\pi(y-nb)/a\}-\cos \pi x/a]$$
$$+i\tan^{-1}[\tanh\{\tfrac{1}{2}\pi(y-nb)/a\}\cot(\tfrac{1}{2}\pi x/a)]. \quad \text{...(5)}$$

The amplitude or current function is therefore constant when $x=(2m+1)a$; and there is no flow across these lines, provided however, as is physically evident, we do not recede to such a large distance from the origin that we are not justified in taking $\operatorname{lt} z/2ma$ as zero.

261. We suppose that Oy passes through the centre of this infinite series of electrodes, or that m reaches to equal infinite positive and negative values; but now, at a very large distance from O, the electrodes on one side of a line, given by $x=(2m+1)a$, where m is a large number, will preponderate over the electrodes on the other side, and the resultant effect will be a uniform normal flow α across this line, to counteract which a term of the form $-\alpha z$ or $\log e^{-\alpha z}$ must be added to the vector function.

The analytical equivalent of this physical effect is illustrated by the theorem proved in Hobson's *Trigonometry*, p. 328, that, when the integers p and q are made infinite in any given ratio, then ϕz, the limit of the product

$$\left(1+\frac{z}{pa}\right)\cdots\left(1+\frac{z}{2a}\right)\left(1+\frac{z}{a}\right)z\left(1-\frac{z}{a}\right)\left(1-\frac{z}{2a}\right)\cdots\left(1-\frac{z}{qa}\right)$$
$$=\left(\frac{p}{q}\right)^{\frac{z}{a}}\sin \pi\frac{z}{a}. \quad \text{........................(6)}$$

The infinite product $\Pi(1+c_n x)$ is convergent for all finite values of x, if the series Σc_n is convergent; as is evident on expanding the logarithm of the product.

But Weierstrass shows (*Berlin Sitz.*, 1876) that the *divergent* product

$$z\left(1-\frac{z}{a}\right)\left(1-\frac{z}{2a}\right)\left(1-\frac{z}{3a}\right)\dots$$

can be made convergent if the exponential factor $e^{z/ma}$ is attached to the linear factor $1-z/ma$; or, interpreted electrically, if to the motion due to the electrode at ma, whose vector function is $\log(1-z/ma)$, we add a uniform streaming motion parallel to the vector ma, given by $\log e^{z/ma}$ or z/ma.

Now, denoting the harmonic series

$$1^{-1}+2^{-1}+3^{-1}+\dots+p^{-1} \text{ by } s_p,$$

$$\phi z = e^{(s_p-s_q)z/a}\sin(\pi z/a)=(p/q)^{z/a}\sin(\pi z/a),$$

since the limit of $s_p-\log p$ or $s_q-\log q$ is *Euler's constant*.

262. In a similar manner it is inferred that the vector function of an infinite series of positive electrodes, whose vectors are

$$(2m+1)a+nbi,$$

m reaching to equal positive and negative infinite values, is

$$\log\cos\tfrac{1}{2}\pi(z-nbi)/a=\tfrac{1}{2}\log\tfrac{1}{2}[\cosh\{\pi(y-nb)/a\}+\cos(\pi x/a)]$$
$$+i\tan^{-1}[\tanh\{\tfrac{1}{2}\pi(y-nb)/a\}\tan(\tfrac{1}{2}\pi x/a)], \quad (7)$$

having lines of equal amplitude given by $x=2ma$.

Therefore the vector function of a pair of lines of electrodes, whose vectors are $2ma\pm nbi$, is

$$\log\sin\{\tfrac{1}{2}\pi(z-nbi)/a\}\sin\{\tfrac{1}{2}\pi(z+nbi)/a\}$$
$$=\log\tfrac{1}{2}\{\cosh(n\pi b/a)-\cos(\pi z/a)\};$$

or, corrected by the addition of a constant, which makes the function vanish when $z=0$, the vector function is

$$\log\frac{\cosh(n\pi b/a)-\cos(\pi z/a)}{\cosh(n\pi b/a)-1}=\log\frac{1-2q^n\cos(\pi z/a)+q^{2n}}{(1-q^n)^2}, \quad (8)$$

where

$$q=e^{-\pi b/a}.$$

For a pair of lines of electrodes whose vectors are $(2m+1)a\pm nbi$, the vector function is

$$\log\cos\{\tfrac{1}{2}\pi(z-nbi)/a\}\cos\{\tfrac{1}{2}\pi(z+nbi)/a\},$$

which may be replaced by

$$\log\frac{\cosh(n\pi b/a)+\cos(\pi z/a)}{\cosh(n\pi b/a)+1}=\log\frac{1+2q^n\cos(\pi z/a)+q^{2n}}{(1+q^n)^2}. \quad (9)$$

For the line of electrodes along OA, whose vectors are $2ma$ or $(2m+1)a$, the vector function will be

$$\log\sin(\tfrac{1}{2}\pi z/a) \text{ or } \log\cos(\tfrac{1}{2}\pi z/a). \quad \dots\dots\dots\dots(10)$$

263. Under Cayley's restrictions, that m reaches to equal positive and negative infinite values, and n also; but that the infinite values of n are infinitely small compared with the infinite values of m (equivalent to taking the infinite array of the images of the electrodes as contained in an infinite rectangle, of which the length in the direction OA is infinitely greater than the breadth in the direction OB), we can now replace the doubly infinite products in (1), (2), (3) by singly infinite products, in the form

$$\operatorname{sn} u = A \sin(\tfrac{1}{2}\pi u/K) \prod_{n=1}^{n=\infty} \frac{1-2q^{2n}\cos(\pi u/K)+q^{4n}}{(1-q^{2n})^2} \div D, \quad (11)$$

$$\operatorname{cn} u = B \cos(\tfrac{1}{2}\pi u/K) \prod \frac{1+2q^{2n}\cos(\pi u/K)+q^{4n}}{(1+q^{2n})^2} \div D, \quad (12)$$

$$\operatorname{dn} u = C \prod \frac{1+2q^{2n-1}\cos(\pi u/K)+q^{4n-2}}{(1+q^{2n-1})^2} \div D, \quad (13)$$

where

$$D = \prod \frac{1-2q^{2n-1}\cos(\pi u/K)+q^{4n-2}}{(1-q^{2n-1})^2}. \quad \ldots\ldots(14)$$

By putting $u=0$, the values of A, B, C are seen to be $K/\frac{1}{2}\pi$, 1, 1; while $q=\exp(-\pi K'/K)$.

The common denominator D of the three elliptic functions, which represents physically a function whose logarithm is the vector function of the negative electrodes at points whose vectors are of the form Ω_3, is the equivalent of Jacobi's Theta Function of § 187; and we write

$$\Theta u = \Theta 0 \prod \frac{1-2q^{2n-1}\cos(\pi u/K)+q^{4n-2}}{(1-q^{2n-1})^2}$$

$$= \Theta 0 \prod \left\{1+\frac{\sin^2(\frac{1}{2}\pi u/K)}{\sinh^2(2n-1)\frac{1}{2}\pi K'/K}\right\}. \quad \ldots\ldots\ldots(15)$$

The numerator of $\operatorname{sn} u$ will now be the equivalent of the Eta Function, defined in § 192; and thus

$$\mathrm{H}u = \sqrt{\kappa}\, \operatorname{sn} u\, \Theta u$$

$$= \sqrt{\kappa} \frac{K}{\frac{1}{2}\pi} \Theta 0 \sin(\tfrac{1}{2}\pi u/K) \prod \frac{1-2q^{2n}\cos(\pi u/K)+q^{4n}}{(1-q^{2n})^2}$$

$$= \sqrt{\kappa} \frac{K}{\frac{1}{2}\pi} \Theta 0 \sin(\tfrac{1}{2}\pi u/K) \prod \left\{1+\frac{\sin^2(\frac{1}{2}\pi u/K)}{\sinh^2(n\pi K'/K)}\right\}. \quad \ldots(16)$$

The numerator of $\operatorname{cn} u$ is represented by the Eta Function of $u+K$, and the numerator of $\operatorname{dn} u$ by the Theta Function of $u+K$; and the factors are so chosen that

$$\operatorname{sn} u = \frac{1}{\sqrt{\kappa}} \frac{\mathrm{H}u}{\Theta u}, \quad \operatorname{cn} u = \frac{\sqrt{\kappa'}}{\sqrt{\kappa}} \frac{\mathrm{H}(u+K)}{\Theta u}, \quad \operatorname{dn} u = \sqrt{\kappa'} \frac{\Theta(u+K)}{\Theta u}. \quad (17)$$

Equation (6) of § 188 may now be written

$$\Theta(u+v)\Theta(u-v)\Theta^2 0 = \Theta^2 u\, \Theta^2 v - \mathrm{H}^2 u\, \mathrm{H}^2 v; \quad \ldots\ldots\ldots(18)$$

while, by means of (7), § 137,

$$\mathrm{H}(u+v)\mathrm{H}(u-v)\Theta^2 0 = \mathrm{H}^2 u\, \Theta^2 v - \Theta^2 u\, \mathrm{H}^2 v. \quad \ldots\ldots\ldots(19)$$

264. It is convenient to replace $\frac{1}{2}\pi u/K$ by a single letter x; and we shall now find that the constant factors are so adjusted as to give the expansions in a Fourier series in the form

$$\Theta u = 1 - 2q\cos 2x + 2q^4\cos 4x - 2q^9\cos 6x + \ldots, \quad \ldots\ldots(20)$$

$$\mathrm{H}u = 2q^{\frac{1}{4}}\sin x - 2q^{\frac{9}{4}}\sin 3x + 2q^{\frac{25}{4}}\sin 5x - \ldots. \quad \ldots\ldots\ldots\ldots(21)$$

It is easily shown algebraically that

$$\prod_{n=1}^{n=\infty} (1-q^{2n-1}z)(1-q^{2n-1}z^{-1})$$

$$= Q\{1 - q(z+z^{-1}) + q^4(z^2+z^{-2}) - q^9(z^3+z^{-3}) + \ldots\} \quad (20)^*$$

by changing z into q^2z and multiplying by qz, when the product on the left hand side merely changes sign; whence equation (20) is inferred from (15) by putting $z = e^{2xi}$; and equation (21) is obtained from (20)* by writing qz for z, and multiplying by $q^{\frac{1}{4}}z^{\frac{1}{2}}$.

Written in the exponential form,

$$\Theta u = \sum_{n=-\infty}^{n=\infty} i^{2n}q^{n^2}e^{2nxi}, \quad \mathrm{H}u = -\Sigma i^{2n-1}q^{(n-\frac{1}{2})^2}e^{(2n-1)xi}, \quad \ldots\ldots(22)$$

or with $q = e^{-a}$, $a = \pi K'/K$, and $b = xi$,

$$\Theta u = \Sigma i^{2n}e^{-n^2a+2nb}, \quad \mathrm{H}u = -\Sigma i^{2n-1}e^{-(n-\frac{1}{2})^2a+(2n-1)b}. \ldots(23)$$

Then $\Theta(u+K) = \Sigma q^{n^2}e^{2nxi} = \Sigma e^{-n^2a+2nb}$,

$$\mathrm{H}(u+K) = \Sigma q^{(n-\frac{1}{2})^2}e^{(2n-1)xi} = \Sigma e^{-(n-\frac{1}{2})^2a+(2n-1)b}; \quad \ldots\ldots(24)$$

and

$$\Theta(u+2K) = \Theta u,$$

$$\mathrm{H}(u+2K) = -\mathrm{H}u, \quad \ldots\ldots\ldots\ldots\ldots\ldots\ldots(25)$$

Changing u into $u+K'i$, or x into $x+\frac{1}{2}i\log q$, we find

$$\Theta(u+K'i) = iq^{-\frac{1}{4}}e^{-xi}\mathrm{H}u,$$

$$\mathrm{H}(u+K'i) = iq^{-\frac{1}{4}}e^{-xi}\Theta u, \quad \ldots\ldots\ldots\ldots\ldots(26)$$

agreeing in giving $\kappa \operatorname{sn} u \operatorname{sn}(u+K'i) = 1$, $\ldots\ldots\ldots\ldots\ldots\ldots$ (27)

and leading by differentiation to the formula

$$\mathrm{Z}(u+K'i) = \mathrm{Z}u + (\operatorname{cn} u \operatorname{dn} u/\operatorname{sn} u) - (\tfrac{1}{2}\pi i/K), \quad \ldots\ldots\ldots(28)$$

which, with (§ 176),

$$\mathrm{Z}(u+K) = \mathrm{Z}u - (\kappa^2\operatorname{sn} u \operatorname{cn} u/\operatorname{dn} u), \quad \ldots\ldots\ldots\ldots\ldots\ldots\ldots(29)$$

leads to

$$\mathrm{Z}(u+K+K'i) = \mathrm{Z}u - (\operatorname{sn} u \operatorname{dn} u/\operatorname{cn} u) - (\tfrac{1}{2}\pi i/K) \ldots\ldots\ldots(30)$$

265. Jacobi writes (*Werke*, I., p. 499) x for $\frac{1}{2}\pi u/K$, and θx for Θu, $\theta_1 x$ for $\mathrm{H}u$, $\theta_2 x$ for $\mathrm{H}(u+K)$, and $\theta_3 x$ for $\Theta(u+K)$; and now

$$\theta x = \Sigma i^{2n} q^{n^2} e^{2nxi}$$
$$= 1 - 2q\cos 2x + 2q^4\cos 4x - 2q^9\cos 6x + \dots \quad \dots\dots(31)$$
$$\theta_1 x = \Sigma i^{2n-1} q^{(n-\frac{1}{2})^2} e^{(2n-1)xi}$$
$$= 2q^{\frac{1}{4}}\sin x - 2q^{\frac{9}{4}}\sin 3x + 2q^{\frac{25}{4}}\sin 5x \quad - \dots \quad \dots\dots(32)$$
$$\theta_2 x = \Sigma q^{(n-\frac{1}{2})^2} e^{(2n-1)xi}$$
$$= 2q^{\frac{1}{4}}\cos x + 2q^{\frac{9}{4}}\cos 3x + 2q^{\frac{25}{4}}\cos 5x \quad + \dots \quad \dots\dots(33)$$
$$\theta_3 x = \Sigma q^{n^2} e^{2nxi}$$
$$= 1 + 2q\cos 2x + 2q^4\cos 4x + 2q^9\cos 6x + \dots \quad \dots\dots(34)$$

or, with $q = e^{-a}$, $b = xi$,

$$\theta x = \Sigma i^{2n} \exp(-n^2 a + 2nb),$$
$$\theta_3 x = \Sigma \exp(-n^2 a + 2nb),$$
$$\theta_1 x = \Sigma i^{2n-1} \exp\{-(n-\tfrac{1}{2})^2 a + (2n-1)b\},$$
$$\theta_2 x = \Sigma \exp\{-(n-\tfrac{1}{2})^2 a + (2n-1)b\}. \quad \dots\dots\dots\dots(35)$$

Conversely, starting with these θ functions as defined by these exponential series, it is possible to rewrite the whole theory of Elliptic Functions *ab initio* in the reverse order, and to deduce all the preceding results.

(Jacobi, *Werke*, I., p. 499; Clifford, *Math. Papers*, p. 443.)

For instance, we find that

$$\theta(x+\tfrac{1}{2}\pi) = \theta_3 x, \quad \theta(x+\tfrac{1}{2}i\log q) = -iq^{-\frac{1}{4}}e^{xi}\theta_1 x,$$
$$\theta_1(x+\tfrac{1}{2}\pi) = \theta_2 x, \quad \theta_1(x+\tfrac{1}{2}i\log q) = -iq^{-\frac{1}{4}}e^{xi}\theta x,$$
$$\theta_2(x+\tfrac{1}{2}\pi) = -\theta_1 x, \quad \theta_2(x+\tfrac{1}{2}i\log q) = q^{-\frac{1}{4}}e^{xi}\theta_3 x,$$
$$\theta_3(x+\tfrac{1}{2}\pi) = \theta x, \quad \theta_3(x+\tfrac{1}{2}i\log q) = q^{-\frac{1}{4}}e^{xi}\theta_2 x. \quad \dots\dots(36)$$

The quotient of two θ functions is thus a *doubly periodic function*, of *real period* 2π or π, and *imaginary period* $i\log q$.

The form of the θ and Θ function series shows that they satisfy partial differential equations of the form

$$\frac{d^2\theta}{dx^2} = -4\frac{d\theta}{d\log q}, \quad \dots\dots\dots\dots\dots\dots\dots\dots(37)$$

and the θ functions are therefore suitable for the solution of problems in the Conduction of Heat.

Thus, if $\theta(x\cos a + y\sin a, q)$ represents at any instant, $t=0$, the temperature at the point (x, y) of an infinite plane, of which γ denotes the *thermometric conductivity*, then at any subsequent time t, the temperature will be given by

$$\theta(x \cos \alpha + y \sin \alpha, \; qe^{-4\gamma^2 t}). \quad \text{.................(38)}$$

266. Similar considerations to those of § 258 enable us to resolve other expressions into factors; for instance,

$$\frac{\text{dn}\, u - \kappa \,\text{cn}\, u}{\kappa'}, \text{ or its reciprocal } \frac{\text{dn}\, u + \kappa \,\text{cn}\, u}{\kappa'},$$

so that $$\frac{\text{dn}\, u - \kappa \,\text{cn}\, u}{\kappa'} = \frac{\kappa'}{\text{dn}\, u + \kappa \,\text{cn}\, u} = \sqrt{\frac{\text{dn}\, u - \kappa \,\text{cn}\, u}{\text{dn}\, u + \kappa \,\text{cn}\, u}},$$

Now $\text{dc}\, u$, or $\text{sn}(K-u) = 1/\kappa$, when

$$u = (4m+1)K + (2n+1)K'i,$$

or $$\cos \tfrac{1}{2}\pi u/K = \cosh(2n-1)\tfrac{1}{2}\pi K'/K;$$

while $$\text{dc}\, u = -1/\kappa,$$

when $$\cos \tfrac{1}{2}\pi u/K = -\cosh(2n-1)\tfrac{1}{2}\pi K'/K;$$

and therefore we may put

$$\frac{\text{dn}\, u - \kappa \,\text{cn}\, u}{\kappa'} = C\Pi \frac{\cosh(2n-1)\tfrac{1}{2}\pi K'/K - \cos \tfrac{1}{2}\pi u/K}{\cosh(2n-1)\tfrac{1}{2}\pi K'/K + \cos \tfrac{1}{2}\pi u/K}$$

$$= C\Pi \frac{1 - 2q^{n-\frac{1}{2}}\cos(\tfrac{1}{2}\pi u/K) + q^{2n-1}}{1 + 2q^{n-\frac{1}{2}}\cos(\tfrac{1}{2}\pi u/K) + q^{2n-1}}, \quad \text{......(39)}$$

where the letter C is used to denote some constant factor.

Now, writing x for $\tfrac{1}{2}\pi u/K$, and supposing x and u real,

$$\log(1 - 2c\cos x + c^2) = \log(1 - ce^{xi}) + \log(1 - ce^{-xi})$$

$$= -2(c\cos x + \tfrac{1}{2}c^2 \cos 2x + \tfrac{1}{3}c^3 \cos 3x + \ldots),$$

$$\log(1 + 2c\cos x + c^2) = 2(c\cos x - \tfrac{1}{2}c^2 \cos 2x + \tfrac{1}{3}c^3 \cos 3x - \ldots),$$

$$\log \frac{1 - 2c\cos x + c^2}{1 + 2c\cos x + c^2} = -4(c\cos x + \tfrac{1}{3}c^3 \cos 3x + \tfrac{1}{5}c^5 \cos 5x + \ldots).$$

Therefore, expanding the logarithm of (39),

$$\log \frac{\text{dn}\, u - \kappa \,\text{cn}\, u}{\kappa'}$$

$$= \log C - 4\Sigma(q^{n-\frac{1}{2}}\cos x + \tfrac{1}{3}q^{3n-\frac{3}{2}}\cos 3x + \tfrac{1}{5}q^{5n-\frac{5}{2}}\cos 5x + \ldots)$$

$$= \log C - 4\left(\frac{q^{\frac{1}{2}}}{1-q}\cos x + \frac{1}{3}\frac{q^{\frac{3}{2}}}{1-q^3}\cos 3x + \frac{1}{5}\frac{q^{\frac{5}{2}}}{1-q^5}\cos 5x + \ldots\right)$$

$$= \log C - 2\Sigma \frac{1}{2m-1} \frac{\cos(2m-1)\tfrac{1}{2}\pi u/K}{\sinh(2m-1)\tfrac{1}{2}\pi K'/K}, \quad \text{................(40)}$$

and, differentiating,

$$\kappa \,\text{sn}\, u = \frac{\pi}{K}\Sigma \frac{\sin(2m-1)\tfrac{1}{2}\pi u/K}{\sinh(2m-1)\tfrac{1}{2}\pi K'/K}, \quad \text{..............(41)}$$

the expression of $\text{sn}\, u$ in a Fourier Series.

267. By forming the similar factorial expressions for

$$\kappa \operatorname{sn} u + i \operatorname{dn} u \quad \text{and} \quad \operatorname{sn} u + i \operatorname{cn} u,$$

and taking logarithms, we shall find

$$\log(\kappa \operatorname{sn} u + i \operatorname{dn} u)$$

$$= \text{constant} - 2i\Sigma \frac{1}{2m-1} \frac{\sin(2m-1)\frac{1}{2}\pi u/K}{\cosh(2m-1)\frac{1}{2}\pi K'/K}, \ldots (42)$$

$$\log(\operatorname{sn} u + i \operatorname{cn} u) = \text{constant} - i\Sigma \frac{1}{m} \frac{\sin m\pi u/K}{\cosh m\pi K'/K}, \ldots\ldots (43)$$

and, differentiating,

$$\kappa \operatorname{cn} u = \frac{\pi}{K} \Sigma \frac{\cos(2m-1)\frac{1}{2}\pi u/K}{\cosh(2m-1)\frac{1}{2}\pi K'/K}, \ldots\ldots\ldots\ldots (44)$$

$$\operatorname{dn} u = \frac{\pi}{2K} + \frac{\pi}{K} \Sigma \frac{\cos m\pi u/K}{\cosh m\pi K'/K}, \ldots\ldots\ldots\ldots (45)$$

and therefore, integrating,

$$\operatorname{am} u = \frac{\pi x}{2K} + \Sigma \frac{\sin m\pi u/K}{m \cosh m\pi K'/K}. \ldots\ldots\ldots\ldots (46)$$

We have now found that, in § 78,

$$B_n = \frac{1}{n \cosh n\pi K'/K}.$$

268. From § 263, we find, in a similar manner, that

$$\log \Theta u = \text{constant} + \log \Pi\{1 - 2q^{2n-1}\cos(\pi u/K) + q^{4n-2}$$

$$= \text{constant} - \Sigma \frac{1}{m} \frac{\cos(m\pi u/K)}{\sinh(m\pi K'/K)}; \ldots\ldots\ldots\ldots (47)$$

and, differentiating,

$$\mathrm{Z} u = \frac{\pi}{K} \Sigma \frac{\sin(m\pi u/K)}{\sinh(m\pi K'/K)}, \ldots\ldots\ldots\ldots (48)$$

$$\operatorname{dn}^2 u = \frac{E}{K} + \frac{\pi^2}{K^2} \Sigma \frac{m \cos(m\pi u/K)}{\sinh(m\pi K'/K)}, \ldots\ldots (49)$$

or

$$\kappa^2 \operatorname{sn}^2 u = 1 - \frac{E}{K} - \frac{\pi^2}{K^2} \Sigma \frac{m \cos(m\pi u/K)}{\sinh(m\pi K'/K)}. \ldots\ldots (50)$$

Now, referring back to § 78, we can put

$$C_n = \frac{\pi}{K} \frac{1}{\sinh n\pi K'/K} = \frac{\pi}{K} \frac{2q^n}{1-q^{2n}}.$$

Putting $u=0$ in (49) or (50) gives what is called "a q series,"

$$\Sigma \frac{m}{\sinh(m\pi K'/K)} = \Sigma \frac{2mq^m}{1-q^{2m}} = \frac{K(K-E)}{\pi^2}. \ldots\ldots (51)$$

As an exercise, the student may form the similar factorial expressions for

$$\frac{1-\operatorname{cn} u}{\operatorname{sn} u},\quad \frac{1-\operatorname{sn} u}{\operatorname{cn} u},\quad \frac{1-\operatorname{dn} u}{\kappa \operatorname{sn} u},\quad \frac{\operatorname{dn} u-\operatorname{cn} u}{\kappa' \operatorname{sn} u},\ \text{etc.,}$$

and their reciprocals

$$\frac{1+\operatorname{cn} u}{\operatorname{sn} u},\quad \frac{1+\operatorname{sn} u}{\operatorname{cn} u},\quad \frac{1+\operatorname{dn} u}{\kappa \operatorname{sn} u},\quad \frac{\operatorname{dn} u+\operatorname{sn} u}{\kappa' \operatorname{sn} u},\ \text{etc.;}$$

and thence determine, by logarithmic differentiation, the Fourier Series for $\operatorname{ns} u$, $\operatorname{cs} u$, $\operatorname{ds} u$, etc. (Glaisher, *Q. J. M.*, XVII.).

The applications of these expansions will be found in papers in the *Q. J. M.*, XVIII., XIX., XX.

269. As an application of these q series, consider the problem of the electrification of two insulated spheres, in presence of each other, of radii a and b, and at a distance c from centre to centre, when maintained at potentials V_a and V_b, with charges of E_a and E_b (Maxwell, *Electricity and Magnetism*, I., chap. XI.).

Then $$E_a=q_{aa}V_a+q_{ab}V_b,\quad E_b=q_{ab}V_a+q_{bb}V_b, \qquad (52)$$
where q_{aa}, q_{bb} are called the *coefficients of capacity*, and q_{ab} the *coefficient of induction.*

We take u and v as coordinates, given by the dipolar system

$$x+yi=k\tan\tfrac{1}{2}(u+vi), \qquad (53)$$

so that $u=$constant represents a circle through the poles $(0, \pm k)$, and $v=$constant represents an orthogonal circle, with the poles as limiting points.

Now, if we revolve this system about the axis Oy, which may be supposed vertical, the two spheres, if outside each other, may be supposed defined by

$$v=\alpha \text{ and } v=-\beta,$$

so that $a=k\operatorname{cosech}\alpha$, $b=k\operatorname{cosech}\beta$, $c=k(\coth\alpha+\coth\beta)$; and putting $\alpha+\beta=\varpi$, Maxwell shows, by Sir W. Thomson's method of successive images, that

$$q_{aa}=k\Sigma\operatorname{cosech}(n\varpi-\beta),\quad q_{ab}=-k\Sigma\operatorname{cosech} n\varpi,$$
$$q_{bb}=k\Sigma\operatorname{cosech}(n\varpi-\alpha), \qquad (54)$$

the summations extending for all positive integral values of n from 1 to ∞.

Here q_{ab} is called *Lambert's Series*; it is considered in the *Fundamenta Nova*, § 66.

Again, with $\alpha-\beta=x$,

$$q_{aa}=k\Sigma \text{ cosech } \tfrac{1}{2}\{(2n-1)\varpi+x\},$$

$$q_{bb}=k\Sigma \text{ cosech } \tfrac{1}{2}\{(2n-1)\varpi-x\};$$

and by the preceding formulas it can be shown that

$$q_{bb}-q_{aa}=k\kappa\frac{K'}{\varpi} \tan \text{am}\left(K'\frac{x}{\varpi}, \kappa'\right). \quad \ldots\ldots\ldots\ldots(55)$$

When the two spheres are equal, $x=0$, and

$$q_{aa}=q_{bb}=k\Sigma \text{ cosech } \tfrac{1}{2}(2n-1)\varpi=k\Sigma\frac{2q^{n-\frac{1}{2}}}{1-q^{2n-1}}.$$

When $\beta=0$, the sphere β becomes a plane; and now

$$q_{aa}=-q_{ab}=k\Sigma \text{ cosech } n\alpha=a \sinh \alpha\Sigma \text{ cosech } n\alpha;$$

which shows that the capacity of a sphere of radius a is raised from a to $a \sinh \alpha\Sigma$ cosech $n\alpha$ by the presence of an uninsulated plane at a distance $a \cosh \alpha$ from its centre.

Similar functions occur in the determination of the motion of two cylinders or spheres, defined by $v=\alpha$ and $-\beta$, when the interspace is filled with homogeneous frictionless liquid.

(W. M. Hicks, *Phil. Trans.*, 1880; *Q. J. M.*, XVII., XVIII.; Basset, *Hydrodynamics*, I., Chaps. X., XI.; C. Neumann, *Hydrodynamische Untersuchungen.*)

270. To illustrate geometrically the singly infinite product forms in § 263 of the elliptic functions, consider the analogous problems of electrodes at the corners of curvilinear rectangular plates, bounded by arcs of concentric circles and their radii.

The vectors from the centre as origin of a series of p electrodes, equally spaced round a circle of radius a, will be

$$a \exp 2r\pi i/p, \text{ where } r=1, 2, 3, \ldots, p;$$

and with polar coordinates r, θ, the vector of the point will be $r \exp i\theta$; so that for the p electrodes, each conducting a current of 2π ampères, the vector function is

$$\log \prod_{r=1}^{r=p}\{r \exp(i\theta)-a \exp(2r\pi i/p)\}=\log(r^p e^{ip\theta}-a^p), \quad \ldots\ldots(56)$$

by De Moivre's Theorem (Hobson, *Trigonometry*, Chap. XIII.).

Interpreted geometrically, the *norm* is the logarithm of the product of the distances of any point P from the electrodes, while the *amplitude* is the sum of the angles the lines joining the electrodes to P make with the vector $\theta=0$.

We thus prove incidentally one of Cotes's theorems, namely, that the square of the product of these distances is

$$(r^pe^{ip\theta}-a^p)(r^pe^{-ip\theta}-a^p)=r^{2p}-2a^pr^p\cos p\theta+a^{2p}, \ldots(57)$$

and, in addition, the theorem that the sum of the angles the vectors from the electrodes to P make with the vector $\theta=0$ is

$$\tan^{-1}\frac{r^p\sin p\theta}{r^p\cos p\theta-a^p}; \ldots\ldots\ldots\ldots(58)$$

and when the sum of these angles is constant, the locus of P is an oblique trajectory of the curves

$$r^p\cos p\theta \text{ or } r^p\sin p\theta=\text{constant}.$$

With a single negative electrode at the centre, of current $n\pi$ ampères, half the total current from the n electrodes on the circle will flow to O, the other half flowing off to infinity.

Now the vector potential is, on writing e^ρ for r/a,

$$\log(r^ne^{in\theta}-a^n)-\tfrac{1}{2}\log r^ne^{in\theta}$$
$$=\tfrac{1}{2}\log(\cosh n\rho-\cos n\theta)+i\tan^{-1}\frac{r^n\sin n\theta}{r^n\cos n\theta-a^n}-\tfrac{1}{2}in\theta. \ldots(59)$$

We can isolate a sector, bounded by $\theta=0$, $\theta=\pi/n$, and $r=a$; and the preceding expression will represent the vector function of the electrical flow of $\frac{1}{2}\pi$ ampères, with electrodes at the end of the vectors $r=a$, and at $r=0$.

The *amplitude* of this expression will also represent the temperature in this sector, if the radius $\theta=0$ is maintained at temperature 0, while the radius $\theta=\pi/n$ and the arc $r=a$ are maintained at temperature $\frac{1}{2}\pi$.

271. Now suppose that on the same circle $r=a$, an equal number p of negative electrodes are placed, equally spaced between the positive electrodes; the vectors of these electrodes being $a\exp(2r-1)\pi i/p$, the vector function is

$$-\log(r^pe^{ip\theta}+a^p);$$

or, if moved out radially on to a circle of radius b,

$$-\log(r^pe^{ip\theta}+b^p). \ldots\ldots\ldots\ldots(60)$$

The vector function of p equal electrodes at $a\exp 2r\pi i/p$, and of p equal negative electrodes at $a\exp(2r-1)\pi i/p$ will therefore be $\qquad \log(r^pe^{ip\theta}-a^p)/(r^pe^{ip\theta}+a^p)$;

which, when resolved into its *norm* and *amplitude*, is

$$\tfrac{1}{2}\log\frac{r^{2p}-2a^pr^p\cos p\theta+a^{2p}}{r^{2p}+2a^pr^p\cos p\theta+a^{2p}}+i\tan^{-1}\frac{2a^pr^p\sin p\theta}{r^{2p}-a^{2p}}$$

$$=-\tanh^{-1}\frac{\cos p\theta}{\cosh p\rho}+i\tan^{-1}\frac{\sin p\theta}{\sinh p\rho}, \qquad \text{..........(61)}$$

with $\rho=\log(r/a)$; this function will represent the state of electrical motion in a wedge bounded by $\theta=0$ and $\theta=\pi/p$.

272. The substitution in the preceding expressions in § 247 of the conjugate functions $p\theta$ and $\log(r/a)^p$ or $p\rho$ for u and v, leads to the solution of corresponding problems for curvilinear rectangles bounded by arcs of concentric circles and their radii; and now $q=(b/a)^p$, where a and b are the radii of the curved sides, while π/p is the angle between the straight radial sides; so that in the rectangle $OABC$,

$$OA=a\pi/p, \quad BC=b\pi/p, \quad OC=AB=a-b.$$

The vectors of the images of an electrode at O are now

$$aq^{2n/p}\exp 2r\pi i/p,$$

where n denotes any integer, positive or negative, and

$$r=1, 2, 3, \ldots, n.$$

For electrodes at A, B, C, the vectors of the images are

$$aq^{2n/p}\exp(2r-1)i\pi/p,$$
$$aq^{(2n-1)/p}\exp 2ri\pi/p,$$
$$aq^{(2n-1)/p}\exp(2r-1)i\pi/p.$$

For a given value of n, the vector potential of the electrodes, whose vectors on a circle of radius $aq^{n/p}$ are

$$aq^{n/p}\exp 2ri\pi/p \quad \text{or} \quad aq^{n/p}\exp(2r-1)\pi i/p$$

will be $\quad \log\Pi(r^pe^{ip\theta}-a^pq^n)$ or $\log\Pi(r^pe^{ip\theta}+a^pq^n)$. ....(62)

Now, suppose a positive electrode is placed at O and a negative electrode at C, with the corresponding system of images; the vector function is

$$\log\prod_{n=-\infty}^{n=\infty}(r^pe^{ip\theta}-a^pq^{2n})/(r^pe^{ip\theta}-a^pq^{2n-1})$$

$$=\log\frac{\left(\frac{re^{i\theta}}{a}\right)^p-1}{\left(\frac{re^{i\theta}}{a}\right)^{\frac{1}{2}p}}\prod_{n=1}^{n=\infty}\frac{\left\{1-q^{2n}\left(\frac{re^{i\theta}}{a}\right)^p\right\}\left\{1-q^{2n}\left(\frac{a}{re^{i\theta}}\right)^p\right\}}{\left\{1-q^{2n-1}\left(\frac{re^{i\theta}}{a}\right)^p\right\}\left\{1-q^{2n-1}\left(\frac{a}{re^{i\theta}}\right)^p\right\}}$$

on introducing a negative electrode, of current π ampères, at the origin; and, writing $\pi w/K$ for $p\theta+i\log(a/r)^p$, this becomes

$$\log\sin(\tfrac{1}{2}\pi w/K)\Pi\frac{1-2q^{2n}\cos(\pi w/K)+q^{4n}}{1-2q^{2n-1}\cos(\pi w/K)+q^{4n-2}}, \qquad \text{......(63)}$$

equivalent, as in § 263, on omitting constant terms, to $\log \operatorname{sn} w$.

A similar procedure with electrodes at A, C, and B, C, will lead to the singly infinite factorial expressions for $\operatorname{cn} u$ and $\operatorname{dn} u$.

Projecting these equipotential and stream lines stereographically on a sphere which touches the plane, we shall obtain the corresponding solutions for the flow of electricity on the surface of the sphere.

(Robertson Smith, *Proc. R. S. of Edinburgh*, vol. VII.; M. J. M. Hill and A. J. C. Allen, *Q. J. M.*, XVI., XVII.)

273. When these electrodes are replaced by straight parallel vortices, perpendicular to the plane, which is taken as horizontal, the potential and stream functions are interchanged.

Suppose a vortex is placed at a point P in the rectangle $OABC$; to introduce the restriction that there is no flow across the sides of the rectangle, we must suppose the motion due to vortices which are the optical reflexions of the point P in the sides of the rectangle; the sign of the vortex being positive or negative according as the corresponding image has been formed by an even or odd number of reflexions.

The vectors of the positive images will therefore be

$$2ma+2nbi \pm z,$$

and of the negative images

$$2ma+2nbi \pm z';$$

where $$z=x+yi, \quad z'=x-yi.$$

The resultant current and velocity function at $\zeta=\xi+\eta i$ will therefore be the norm and amplitude of

$$\log \Pi\Pi \frac{(2ma+2nbi+\zeta-z)(2ma+2nbi+\zeta+z)}{(2ma+2nbi+\zeta-z')(2ma+2nbi+\zeta+z')}. \qquad \text{....(64)}$$

At the point P, this vector function, due to all the other images, is therefore

$$\log \Pi\Pi \frac{(2ma+2nbi)(2ma+2nbi+2z)}{(2ma+2nbi+z-z')(2ma+2nbi+z+z')};$$

and writing $\dfrac{K'}{K}=\dfrac{b}{a}$, and $2K\dfrac{x}{a}+2K'i\dfrac{y}{b}=u+vi=w$,

this may, according to § 263, be replaced by

$$\log \frac{\mathrm{H}(u+vi)}{\mathrm{H}u\,\mathrm{H}vi}. \qquad \text{..........................(65)}$$

The stream function at P is therefore, disregarding constants,

$$\tfrac{1}{2}\log\frac{\mathrm{H}(u+vi)\mathrm{H}(u-vi)}{\mathrm{H}^2u\,\mathrm{H}^2vi}=\tfrac{1}{2}\log\frac{\Theta^2u\,\mathrm{H}^2vi-\mathrm{H}^2u\,\Theta^2vi}{\mathrm{H}^2u\,\mathrm{H}^2vi}\quad(\S\ 263)$$

$$=\tfrac{1}{2}\log\left(\frac{\Theta^2u}{\mathrm{H}^2u}-\frac{\Theta^2vi}{\mathrm{H}^2vi}\right)$$

$$=\tfrac{1}{2}\log(\mathrm{ns}^2u-\mathrm{ns}^2vi)$$

$$=\tfrac{1}{2}\log\{\mathrm{ns}^2(u,\kappa)+\mathrm{ns}^2(v,\kappa')-1\}\,;\ \ldots(66)$$

so that the curve described by the vortex is given by

$$\mathrm{ns}^2(2Kx/a,\kappa)+\mathrm{ns}^2(2K'y/b,\kappa')=\text{constant},\ \ldots\ldots\ldots(67)$$

and all the other image vortices keep up a symmetrical dance, by describing similar curves.

274. The vortex is stationary when at the centre of the rectangle; and now, changing to the centre as origin, the vectors of the images are $ma+nbi$, where $m+n$ is even for the positive, and odd for the negative images; so that the vector function of the motion is given by

$$\log\Pi\Pi\frac{(2ma+2nbi-z)\{(2m+1)a+(2n+1)bi-z\}}{\{2ma+(2n+1)bi-z\}\{(2m+1)a+2nbi-z\}}$$

$$=\ \log\frac{\mathrm{sn}\,\tfrac{1}{2}w\,\mathrm{dn}\,\tfrac{1}{2}w}{\mathrm{cn}\,\tfrac{1}{2}w}=\tfrac{1}{2}\log\frac{1-\mathrm{cn}\,w}{1+\mathrm{cn}\,w},\ \ldots\ldots\ldots\ldots\ldots\ldots(68)$$

Expressed as norm and amplitude, as in § 247, this function

$$=\tfrac{1}{4}\log\frac{1-\mathrm{cn}\,w}{1+\mathrm{cn}\,w}\cdot\frac{1-\mathrm{cn}\,w'}{1+\mathrm{cn}\,w'}+\tfrac{1}{4}\log\frac{1-\mathrm{cn}\,w}{1+\mathrm{cn}\,w}\cdot\frac{1+\mathrm{cn}\,w'}{1-\mathrm{cn}\,w'}$$

$$=\tfrac{1}{2}\log\frac{\mathrm{cn}\,vi-\mathrm{cn}\,u}{\mathrm{cn}\,vi+\mathrm{cn}\,u}+\tfrac{1}{2}\log\frac{\mathrm{sn}\,u\,\mathrm{dn}\,vi-\mathrm{dn}\,u\,\mathrm{sn}\,vi}{\mathrm{sn}\,u\,\mathrm{dn}\,vi+\mathrm{dn}\,u\,\mathrm{sn}\,vi}$$

$$=-\tanh^{-1}\frac{\mathrm{cn}\,u}{\mathrm{cn}\,vi}-\tanh^{-1}\frac{\mathrm{sn}\,u\,\mathrm{dn}\,vi}{\mathrm{dn}\,u\,\mathrm{sn}\,vi}$$

$$=-\tanh^{-1}(\mathrm{cn}\,u\,\mathrm{cn}\,v)+i\tan^{-1}\frac{\mathrm{sn}\,u\,\mathrm{dn}\,v}{\mathrm{dn}\,u\,\mathrm{sn}\,v},\ \ldots\ldots\ldots\ldots(69)$$

with $u=2Kx/a$, $v=2K'y/b$; the modulus of the elliptic functions of v being κ'.

The equation of a stream line of liquid is therefore given by

$$\mathrm{cn}\,u\,\mathrm{cn}\,v=\text{constant, or}$$

$$\mathrm{cn}(2Kx/a,\kappa)\mathrm{cn}(2K'y/b,\kappa')=\text{constant.}\ \ldots\ldots\ldots\ldots(70)$$

Close up to a vortex the velocity according to these expressions would become infinitely great, which is physically impossible; but a solid core may be substituted for this central portion, and the shape of this core has been investigated by J. H. Michell, *Phil. Trans.*, 1890.

275. When a point is placed inside an equilateral triangle, the Kaleidoscopic series of positive images is given by the vectors z, ωz, $\omega^2 z$, where $z = x + yi$, and ω is an imaginary cube root of unity; the negative images being given by z', $\omega z'$, $\omega^2 z'$, where $z' = -x + yi$; the origin being at a corner of the triangle, and the axis of x perpendicular to the opposite side (Fig. 27, i.).

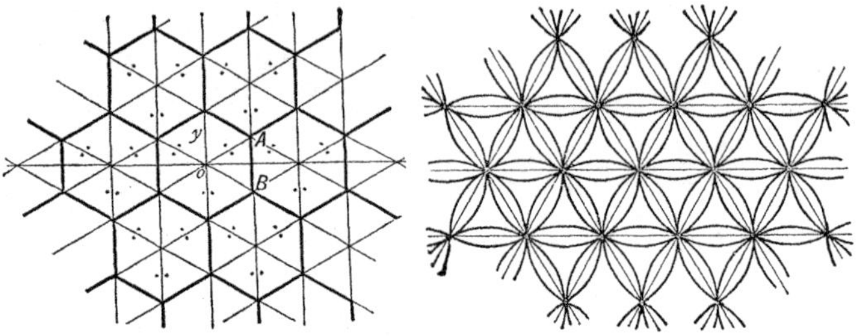

(i.) Fig. 27. (ii.)

In addition, similar groups of six images must be added, ranged round the centre of hexagons forming a tesselated pavement, the vectors of the centres of the hexagons being

$$2mh + 2nhi\sqrt{3} \text{ and } (2m+1)h + (2n+1)hi\sqrt{3},$$

where h denotes the altitude of the equilateral triangle.

In the corresponding doubly infinite products, the elliptic functions will have $K'/K = \sqrt{3}$, so that (§ 47), $\kappa = \sin 15°$, $2\kappa\kappa' = \frac{1}{2}$.

Then, in Weierstrass's notation, the vector potential at

$$\zeta = \xi + \eta i$$

for a single source or electrode inside the triangle will, neglecting constant terms and factors, be expressed by (§ 278)

$$\begin{aligned} \log \sigma(\zeta - z)\sigma(\zeta - \omega z)\sigma(\zeta - \omega^2 z) \\ \sigma_1(\zeta - z)\sigma_1(\zeta - \omega z)\sigma_1(\zeta - \omega^2 z) \\ \sigma(\zeta - z')\sigma(\zeta - \omega z')\sigma(\zeta - \omega^2 z') \\ \sigma_1(\zeta - z')\sigma_1(\zeta - \omega z')\sigma_1(\zeta - \omega^2 z'); \end{aligned} \quad \ldots\ldots\ldots\ldots(71)$$

while for a vortex or electrified wire, the vector potential is

$$\log \frac{\sigma(\zeta - z)\sigma(\zeta - \omega z)\sigma(\zeta - \omega^2 z)\sigma_1(\zeta - z)\sigma_1(\zeta - \omega z)\sigma_1(\zeta - \omega^2 z)}{\sigma(\zeta - z')\sigma(\zeta - \omega z')\sigma(\zeta - \omega^2 z')\sigma_1(\zeta - z')\sigma_1(\zeta - \omega z')\sigma_1(\zeta - \omega^2 z')}. \quad (72)$$

The nature of the resolution of these functions into their norm and amplitude is illustrated in §§ 227 to 231.

(O. J. Lodge, *Phil. Mag.*, 1876; O. Zimmermann, *Das logarithmische Potential einer gleichseitig dreieckigen Platte,* Diss. Jena, 1880; A. E. H. Love, *Vortex Motion in Certain Triangles,* Am. J. M., XI.)

So also for a rectangular boundary $OACB$, if we write

$$\alpha \text{ for } \xi - x + (\eta - y)i, \text{ or } \zeta - z,$$
$$\beta \text{ for } \xi + x + (\eta - y)i, \text{ or } \zeta + z',$$
$$\gamma \text{ for } \xi + x + (\eta + y)i, \text{ or } \zeta + z,$$
$$\delta \text{ for } \xi - x + (\eta + y)i, \text{ or } \zeta - z';$$

z, $-z'$, $-z$, z' being the vectors of the point P and its images by reflexion in the coordinate axes Ox, Oy, taken in order in the four quadrants; then the vectors of all the other images by reflexion in the sides of the rectangle $OABC$ being ranged in a similar manner round points whose vectors are $2ma+2nbi$, it follows from what has gone before that we may express the vector function at ζ of all their images, taken as positive, by

$$\log \sigma\alpha\, \sigma\beta\, \sigma\gamma\, \sigma\delta, \quad \text{......(73)}$$

with $\omega_1 = a, \quad \omega_3 = bi;$

disregarding constant factors, and exponential factors of the form $\exp(Au+Bu^2)$.

But when we represent the vector potential of a vortex or electrified wire at P, the vector potential becomes

$$\log \frac{\sigma\alpha\ \sigma\gamma}{\sigma\beta\ \sigma\delta}. \quad \text{......(74)}$$

276. As another illustration of the connexion of a regular Kaleidoscopic figure with Elliptic Functions, consider the solution of the *reciprocant*

$$(t^2+1)c - 10abt + 15a^3 = 0, \quad \text{......(75)}$$

where $t = \dfrac{dy}{dx}, \quad a = \dfrac{d^2y}{dx^2}, \quad b = \dfrac{d^3y}{dx^3}, \quad c = \dfrac{d^4y}{dx^4}.$

(Sylvester, *Lectures on the Theory of Reciprocants*, VI., 1888.)

Mr. J. Hammond has shown (*Nature*, Jan. 7, 1886, p. 231; *Proc. L. M. S.*, XVII., p. 128) that the integral of this equation (75) may be written

$$x + yi = \int \frac{(1+ti)dt}{\sqrt{\{\frac{1}{2}(\kappa - \lambda i)(1+ti)^6 + \frac{1}{2}(\kappa+\lambda i)(1-ti)^6\}}}. \quad \text{...(76)}$$

By turning the axes through an angle $\frac{1}{6}\tan^{-1}(\lambda/\kappa)$, we can make λ vanish; and now, replacing $\frac{1}{2}\kappa$ by unity,

$$x + yi = \int \frac{(1+ti)dt}{\sqrt{\{(1+ti)^6 + (1-ti)^6\}}}, \quad \text{......(77)}$$

$$\left(\frac{1+ti}{1-ti}\right)^2 = -\wp(x+yi;\ 0, 4), \quad \left(\frac{1-ti}{1+ti}\right)^2 = -\wp(x-yi;\ 0, 4), \quad \text{...(78)}$$

and $$\wp(x+yi)\wp(x-yi) = 1. \quad \text{......(79)}$$

Since (§ 196) $\wp\omega z=\omega\wp z,\quad \wp\omega^2 z=\omega^2\wp z,$
where ω is an imaginary cube root of unity, therefore

$$\wp\omega(x+yi)\,\wp\omega^2(x-yi)=1, \quad\ldots\ldots(80)$$

which shows that the curve is unchanged if turned through an angle of 60° about the origin (Fig. 27, ii.).

Captain MacMahon has shown that the intrinsic equation of this curve may be written

$$\cos 3\psi=\mathrm{dn}(s/c),\ \text{with}\ \kappa=\tfrac{1}{2}\sqrt{2}. \quad\ldots\ldots(81)$$

The student may also show that the equation of the curve may be written in one of the forms

$$\mathrm{am}(x\pm K,\kappa)=\mathrm{am}(y\pm K',\kappa'),$$
$$\kappa'^2\mathrm{tn}^2(x,\kappa)=\kappa^2\,\mathrm{tn}^2(y,\kappa'),$$
$$\kappa^2\mathrm{sn}^2(x,\kappa)=\kappa'^2\mathrm{sn}^2(y,\kappa'),$$
$$\mathrm{dn}(x,\kappa)\mathrm{dn}(y,\kappa')=\kappa, \quad\ldots\ldots(82)$$

with $\qquad \kappa=\sin 15°,\ \kappa'=\sin 75°.$

As a similar exercise, the student may solve the reciprocant

$$tc-5ab=0 \quad\ldots\ldots(83)$$

in the form $\qquad \wp x\,\wp y=\pm 1, \quad\ldots\ldots(84)$

and determine its intrinsic equation, drawing the corresponding curves (*Proc. London Math. Soc.*, XVII., p. 360).

277. When we expand, in ascending powers of u, the logarithm of a doubly infinite product, such as that in the numerator of sn u in equation (1), § 258, we find

$$\log u\Pi'\Pi'\left(1-\frac{u}{\Omega}\right)=\log u-u\Sigma\Omega^{-1}-\tfrac{1}{2}u^2\Sigma\Omega^{-2}-\tfrac{1}{3}u^3\Sigma\Omega^{-3}-\ldots(85)$$

Now, when the origin is taken at the centre of all the points whose vectors are Ω, the coefficients of u, u^3, u^5, ... vanish; but the value of the series is still indeterminate, until the infinite curve containing all these points has been defined.

For if P denotes this infinite product, and P' its value when the boundary has changed into a similar curve, then

$$\log P'-\log P=\tfrac{1}{2}u^2\Sigma\Omega^{-2}+\tfrac{1}{4}u^4\Sigma\Omega^{-4}+\ldots,$$

where the summation now extends over the region lying between the two boundaries; and now the limit of $\Sigma\Omega^{-2}$ is a definite number, A suppose, while the limit of $\Sigma\Omega^{-4}$, ... is zero.

Therefore

$$\log P'-\log P=\tfrac{1}{2}Au^2,\ \text{or}\ P'=Pe^{\frac{1}{2}Au^2}\ldots; \quad\ldots\ldots(86)$$

so that the value of the infinite product depends on the shape of the infinite boundary (Clifford, *Math. Papers*, p. 463).

But, as in § 261, Weierstrass removes this ambiguity by attaching to each linear factor of the product, such as

$$1-\frac{u}{\Omega},$$

an exponential factor $\exp\left(\frac{u}{\Omega}+\frac{1}{2}\frac{u^2}{\Omega^2}\right)$;

and, in the physical analogue, the corresponding electrode at Ω, whose vector function is $\log(1-u/\Omega)$, must have associated with it a uniform flow in the direction of the vector Ω, represented by u/Ω; and a streaming motion in rectangular hyperbolas, whose asymptotes are parallel and perpendicular to the vector Ω, represented by $\frac{1}{2}(u/\Omega)^2$.

Now in the expansion of the logarithm of the doubly infinite product P, when these exponential factors are introduced,

$$\log P=\log u-\tfrac{1}{4}u^4\Sigma\Omega^{-4}-\tfrac{1}{6}u^6\Sigma\Omega^{-6}-\ldots, \qquad \ldots\ldots\ldots\ldots(87)$$

an *absolutely convergent* series; that is, a series the value of which is independent of the order of the terms.

278. Making a new start *ab initio* with the *sigma function* (§ 195), as defined now by the equation

$$\sigma u=u\prod_{m=-\infty}^{m=\infty}{}'\prod_{n=-\infty}^{n=\infty}{}'\left(1-\frac{u}{\Omega}\right)\exp\left(\frac{u}{\Omega}+\frac{1}{2}\frac{u^2}{\Omega^2}\right), \qquad \ldots\ldots\ldots(\mathrm{U})$$

where $\Omega=2m\omega+2n\omega'$, and $\omega'/\omega i$ is a real positive quantity, so that ω, ω' correspond to ω_1, ω_3 or ω_2, ω_2' according as Δ is positive or negative, then σu is the analogue of Jacobi's Eta Function; in fact,

$$\sigma u=Ce^{Au^2}\mathrm{H}\sqrt{(e_1-e_3)}u=Ce^{Au^2}\theta_1(\tfrac{1}{2}\pi u/\omega), \qquad \ldots\ldots(88)$$

(§ 263), where C, A are certain constants; also $\log\sigma u$ is the same as $\log P$ in equation (87).

Now denoting, as in § 195,

$$\frac{d\log\sigma u}{du} \text{ by } \zeta u, \text{ and } \frac{d^2\log\sigma u}{du^2} \text{ or } \frac{d\zeta u}{du} \text{ by } -\wp u,$$

$$\zeta u=\frac{1}{u}+\Sigma'\left(\frac{1}{u-\Omega}+\frac{1}{\Omega}+\frac{u}{\Omega^2}\right)$$

$$=\frac{1}{u}-u^3\Sigma\Omega^{-4}-u^5\Sigma\Omega^{-6}-\ldots, \qquad \ldots\ldots\ldots\ldots(\mathrm{V})$$

by differentiation of (U) and (58); so that, on reference to § 195, we may put

$$g_2=60\Sigma\Omega^{-4}, \quad g_3=140\Sigma\Omega^{-6}, \qquad \ldots\ldots\ldots\ldots\ldots(\mathrm{W})$$

also $g_2^{\,2}=2^4.3.5^2.7\,\Sigma\Omega^{-8}$, $g_2g_3=2^4.3.5.7.11\,\Sigma\Omega^{-10}$, etc.

Differentiating (60) again,

$$\wp u = \frac{1}{u^2} + \Sigma' \left\{ \frac{1}{(u-\Omega)^2} - \frac{1}{\Omega^2} \right\}, \dots\dots\dots\dots\dots (X)$$

$$\wp' u = -\frac{2}{u^3} - \Sigma' \frac{2}{(u-\Omega)^3}. \dots\dots\dots\dots\dots\dots (Y)$$

Then $(\sigma u)/u$, $u\zeta u$, $u^2\wp u$, $u^3\wp' u$, $u^4\wp'' u, \dots$, are unaffected by the considerations of *homogeneity* of § 196; as for instance in the expansions in equations (21) and (22) on p. 249.

A change in (X) and (Y) of u into $u+2p\omega+2q\omega'$, where p and q are integers, merely leads to a rearrangement of terms; so that, as in § 250,

$$\wp(u+2p\omega+2q\omega') = \wp u.$$

Also, since in $\Omega = 2m\omega + 2n\omega'$, the arrangements (m, n) and $(-m, -n)$ exist in pairs, therefore

$$\wp'\omega = 0, \quad \wp'(\omega+\omega') = 0, \quad \wp'\omega' = 0;$$

and $$\wp'^2 u = 4 \, . \, \wp u - \wp\omega \, . \, \wp u - \wp(\omega+\omega') \, . \, \wp u - \wp\omega'$$
$$= 4\wp^3 u - g_2 \wp u - g_3, \dots\dots\dots\dots\dots\dots\dots (AA)$$

as originally defined otherwise in § 50.

A change of u into $u+2\omega$ in (V) shows that, by a rearrangement of terms,

$$\zeta(u+2\omega) = \zeta u + 2\eta, \dots\dots\dots\dots\dots\dots (89)$$

where η is a certain constant, determined by putting $u = -\omega$, so that

$$\eta = \zeta\omega. \dots\dots\dots\dots\dots\dots\dots (90)$$

Similarly

$$\zeta(u+2\omega') = \zeta u + 2\eta', \dots\dots\dots\dots\dots\dots (91)$$

where

$$\eta' = \zeta\omega'; \dots\dots\dots\dots\dots\dots\dots (92)$$

and, generally,

$$\zeta(u+2p\omega+2q\omega') = \zeta u + 2p\eta + 2q\eta'. \dots\dots\dots (BB)$$

Integrating (89) and (90),

$$\sigma(u+2\omega) = Ce^{2\eta u}\sigma u, \qquad \sigma(u+2\omega') = C'e^{2\eta' u}\sigma u;$$

where C and C' are determined by putting $u = -\omega$ and $-\omega'$; so that

$$\sigma(u+2\omega) = -e^{2\eta(u+\omega)}\sigma u, \ \sigma(u+2\omega') = -e^{2\eta(u+\omega')}\sigma u, \ (93)$$

and therefore

$$\sigma(u+2p\omega) = -(-1)^{p+1}e^{2p\eta(u+p\omega)}\sigma u, \dots\dots\dots\dots\dots (94)$$

$$\sigma(u+2q\omega') = -(-1)^{q+1}e^{2q\eta'(u+q\omega')}\sigma u, \dots\dots\dots\dots\dots (95)$$

and, generally,

$$\sigma(u+2p\omega+2q\omega') = -(-1)^{(p+1)(q+1)}e^{(2p\eta+2q\eta')(u+p\omega+q\omega')}\sigma u, \dots (CC)$$

obtained also by integration of (BB).

The doubly infinite products in (U) may be converted into singly infinite products; and now

$$\sigma u = \frac{2\omega}{\pi} e^{\frac{1}{2}\eta u^2/\omega} \sin\frac{\pi u}{2\omega} \Pi \frac{1-2q^{2n}\cos(\pi u/\omega)+q^{4n}}{(1-q^{2n})^2}, \ldots\text{(BB)}$$

where $q = e^{\pi i\omega'/\omega}$, and

$$2\eta\omega = \tfrac{1}{6}\pi^2 - \pi^2\Sigma\frac{4q^{2n}}{(1-q^{2n})^2} = \tfrac{1}{6}\pi^2 - \pi^2\Sigma\,\text{cosech}^2(n\omega'/\omega i), \ldots (97)$$

etc.; for the proof of these and other similar formulas merely stated here, the reader is referred to Schwarz and Halphen.

Also, denoting $\Omega+\omega$, $\Omega+\omega+\omega'$, $\Omega+\omega'$ by Ω_1, Ω_2, Ω_3, then the function $\sigma_a u$ of § 202 may be otherwise defined *ab initio* by the relation

$$\sigma_a u = e^{\frac{1}{2}e_a u^2}\Pi\Pi\left(1-\frac{u}{\Omega_a}\right)\exp\left(\frac{u}{\Omega_a}+\frac{1}{2}\frac{u^2}{\Omega_a{}^2}\right), \ldots\ldots\ldots\text{(EE)}$$

which will be found to lead to the preceding results.

Denoting $\dfrac{d^2}{du^2}\log\sigma_a u$ by $-\wp_a u$, we shall find that

$$\wp_a u = \wp(u+\omega_a), \quad a=1, 2, 3. \ldots\ldots\ldots\ldots\ldots\ldots (98)$$

(A. R. Forsyth, *Q. J. M.*, XXII.)

279. Returning to the function C of equations (8) and (10), § 215, and changing the sign of the u's, we may also write it

$$C = \frac{\sigma(v+u_1+u_2+\ldots+u_\mu)\sigma(v-u_1)\sigma(v-u_2)\ \ldots\ \sigma(v-u_\mu)}{(\sigma v)^{+1}}$$

$$= c_0 + c_1\wp v + c_2\wp' v + \ldots + c_\mu \wp^{(\mu-1)} v\,; \ldots\ldots\ldots\ldots\ldots\ldots\ldots\ldots (99)$$

and since we may suppose the u's and v to be all increased by equal amounts, the condition (9) of § 215 is no longer required.

Now, since C vanishes when $v=u_r$, where $r=1, 2, 3, \ldots, \mu$; therefore the coefficients c_0, c_1, c_2, ..., c_μ are determined by a series of equations of the form

$$0 = c_0 + c_1\wp u_r + c_2\wp' u_r + \ldots + c_\mu\wp^{(\mu-1)}u_r\,; \ldots\ldots\ldots (100)$$

and therefore the determinant

$$\begin{vmatrix} 1, & \wp v, & \wp' v, & \ldots, & \wp^{(\mu-1)} v \\ \ldots & \ldots & \ldots & \ldots & \ldots \\ 1, & \wp u_r, & \wp' u_r, & \ldots, & \wp^{(\mu-1)} w_r \\ \ldots & \ldots & \ldots & \ldots & \ldots \end{vmatrix} = MC, \ldots\ldots\ldots (101)$$

where M is a factor independent of v; and now this theorem, as a corollary of Abel's theorem, shows that the determinant also vanishes when $v = -u_1 - u_2 - \ldots - u_\mu$.

The symmetry of the determinant shows that M must be a symmetric function of the u's; or writing u_0 for v, and denoting the determinant by $\phi(u_0, u_1, u_2, \dots, u_\mu)$, then ϕ is a symmetric function of the u's, such that

$$\phi(u_0, u_1, \dots, u_\mu) = A\frac{\sigma(u_0+u_1+\dots+u_\mu)\Pi_{p,q}\sigma(u_p-u_q)}{(\sigma u_0)^{\mu+1}(\sigma u_1)^{\mu+1}\dots(\sigma u_\mu)^{\mu+1}}, \dots \text{(FF)}$$

$$(p<q,\ p,\ q=0,\ 1,\ 2,\ \dots,\ \mu),$$

and it will be found (Schwarz, § 14) that

$$A = (-1)^{\frac{1}{2}\mu(\mu-1)}\, 1!\, 2!\, 3! \dots \mu!.$$

Thus, for instance, with $\mu=2$,

$$\begin{vmatrix} 1, & \wp u, & \wp' u \\ 1, & \wp v, & \wp' v \\ 1, & \wp w, & \wp' w \end{vmatrix} = 2\frac{\sigma(u+v+w)\sigma(v-w)\sigma(w-u)\sigma(u-v)}{\sigma^3 u\, \sigma^3 v\, \sigma^3 w}.$$

By forming a similar function C' of the u''s, subject to the condition (6) of § 215, we see that (7) is an elliptic function of v, which can be expressed by C/C', where C and C' are given by determinants, as above.

Equation (CC) is also sufficient to prove that the function in (7) § 215 is doubly periodic.

As an application of the principles of this article and of §§ 209, 215, 216, 257, the student may prove that Ω of § 215 is, writing a for u_1, b for u_2, and u for v, given by the equations

$$\begin{aligned}\Omega &= \frac{\sigma(u+a)\sigma(u+b)\sigma(a+b)}{\sigma(u+a+b)\sigma u\, \sigma a\, \sigma b} \\ &= \begin{vmatrix} 1, & \wp u, & \wp^2 u \\ 1, & \wp a, & \wp^2 a \\ 1, & \wp b, & \wp^2 b \end{vmatrix} \div \begin{vmatrix} 1, & \wp u, & \wp' u \\ 1, & \wp a, & \wp' a \\ 1, & \wp b, & \wp' b \end{vmatrix} \\ &= \zeta(u+a+b) - \zeta u - \zeta a - \zeta b.\end{aligned}$$

We thus verify the equations of §§ 209, 257,

$$\begin{aligned}\frac{d}{du}\frac{\sigma(u+a+b)}{\sigma u\, \sigma(a+b)}e^{-u(\zeta a+\zeta b)} &= \frac{\sigma(u+a)\sigma(u+b)}{\sigma^2 u\, \sigma a\, \sigma b}e^{-u(\zeta a+\zeta b)} \\ &= \phi(u, a)\phi(u, b).\end{aligned}$$

When condition (6) of § 215 is not satisfied, then (7) reappears qualified by an exponential factor of the form $e^{\rho v}$ when v is increased by $2p\omega + 2q\omega'$; the function is then called by Hermite *a doubly periodic function of the second kind;* the function $\phi(u, v)$ defined in § 201 being the simplest instance of this kind of function.

280. Making the u's all equal, as in § 218, and interchanging u and v, the function

$$\chi u = \frac{\sigma(u+\mu v)\{\sigma(u-v)\}^{\mu}}{(\sigma u)^{\mu+1}(\sigma v)^{\mu(\mu+1)}}$$

is a doubly periodic function which can be expressed in the form of C; but now the coefficients c must be determined by a series of equations of the form

$$\begin{aligned} 0 &= c_0 + c_1\wp v + c_2\wp' v + \dots, \\ 0 &= \quad c_1\wp' v + c_2\wp'' v + \dots, \\ 0 &= \quad c_1\wp'' v + c_2\wp''' v + \dots, \\ &\dots\dots\dots\dots\dots\dots \end{aligned}$$

Expressed as a determinant we may now put

$$\chi u = \frac{1}{n!}\begin{vmatrix} \wp u - \wp v, & \wp' u - \wp' v, & \dots, & \wp^{(\mu-1)}u - \wp^{(\mu-1)}v \\ \wp' v, & \wp'' v, & \dots, & \wp^{(\mu)}u - \wp^{(\mu)}v \\ \wp'' v, & \wp''' v, & \dots & \\ \dots & \dots & \dots & \\ \wp^{(\mu-1)}v, & \wp^{(\mu)}v, & \dots & \end{vmatrix}.$$

Finally, making $u=v$, and dividing both sides by $(u-v)^{\mu}$, we find, in the limit,

$$\frac{\sigma(\mu+1)u}{\sigma(u)^{(\mu+1)^2}} = M\begin{vmatrix} \wp' u, & \wp'' u, & \dots, & \wp u^{(\mu)}u \\ \wp'' u, & \wp''' u, & \dots, & \wp^{(\mu+1)}u \\ \dots & \dots & \dots & \dots \\ \wp^{\mu}u, & \wp^{(\mu+1)}u, & \dots, & \wp^{(2\mu-1)}u \end{vmatrix}, \quad \dots\text{(GG)}$$

where $$M = \frac{(-1)^{\mu}}{(1!\,2!\,3!\dots\mu!)^2} \quad \text{(Schwarz, § 15)};$$

Halphen denotes this function of u by $\psi_{(\mu+1)}u$.

Thus for instance, as in § 200, with $\mu=1$,

$$\psi_2 u = \frac{\sigma 2u}{(\sigma u)^4} = -\wp' u.$$

Again, with $\mu=2$,

$$\psi_3 u = \frac{\sigma 3u}{(\sigma u)^9} = \tfrac{1}{4}(\wp' u\,\wp''' u - \wp''^2 u) = \wp'^2 u(\wp u - \wp 2u).$$

By logarithmic differentiation,

$$\frac{d^2}{du^2}\log\psi_n u = \frac{d^2}{du^2}\log\frac{\sigma nu}{(\sigma u)^{n^2}} = n^2(\wp u - \wp nu), \quad \dots\text{(HH)}$$

whence $\wp nu$ can be expressed rationally in terms of $\wp u, \wp' u, \dots$.

When $u=v$,

$$\text{lt}\frac{\chi u}{\{\sigma(u-v)\}^{\mu}} = \frac{\sigma(\mu+1)v}{(\sigma v)^{(\mu+1)^2}} = \psi_{(\mu+1)}v.$$

Also, when $u=0$,

$$\text{lt}(\sigma u)^{\mu+1}\chi u=(-1)^{\mu}\frac{\sigma\mu v}{(\sigma v)^{\mu^2}}=(-1)^{\mu}\psi_{\mu}v$$

$$=\text{lt}(\sigma u)^{\mu+1}\{a_0+a_1\wp u+\ldots+a_{\mu}\wp^{(\mu-1)}u)\}$$

$$=a_{\mu}(-1)^{\mu+1}\mu!\,; \quad\ldots\ldots\ldots\ldots\ldots\ldots\ldots(102)$$

and therefore $a_{\mu}=0$, when $\mu v=2p\omega_1+2q\omega_3$.

281. In the pseudo-elliptic integrals (§ 218)

$$\mu v=0 \text{ (mod. } \omega_1,\ \omega_3);$$

and now, knowing the number μ, the coefficients c_0, c_1, c_2, ... in C or χu are readily calculated from a knowledge of the values of $\wp v$, $\wp' v$, $\wp'' v$, ... ; in this way the results employed in §§ 218, 219, 223, 225, 233 were inferred.

Thus, for instance, in § 219, we know that

$$\mu=3,\quad \mu v=3\omega_1+\omega_3\,;$$

$$\wp v=\tfrac{1}{2},\quad \wp' v=3i\sqrt{2},\quad \wp'' v=-6,\quad \wp''' v=18i\sqrt{2},\quad \wp'''' v=-252,\ \ldots;$$

so that the ratios of c_0, c_1, c_2, ... can be calculated from the equations

$$0=c_0+\tfrac{1}{2}c_1+3i\sqrt{2}c_2-6c_3,$$
$$0=3i\sqrt{2}c_1-6c_2+18i\sqrt{2}c_3,$$
$$0=-6c_1+18i\sqrt{2}c_2-252c_3.$$

Taking an arbitrary value of c_3, say $\tfrac{2}{3}$, we find, by solution,

$$c_0=-9,\quad c_1=-10,\quad c_2=-3i\sqrt{2}\,;$$

$$\chi u=\tfrac{3}{2}c_3(\tfrac{2}{3}\wp'' u-3i\sqrt{2}\,\wp' u-10\,\wp u-9)$$

$$=\tfrac{3}{2}c_3\{(2\,\wp u+2)(2\,\wp u-7)-3i\sqrt{2}\,\wp' u\}.$$

Now
$$\chi u=\frac{\sigma(u+3\omega_1+\omega_3)\sigma^3(u-v)}{\sigma^4 u\,\sigma^{12}v}$$

$$=C\sqrt{(\wp u-e_2)}\left\{\frac{\sigma(u-v)}{\sigma u\,\sigma v}e^{\rho u}\right\}^3;$$

so that, in the algebraical herpolhode referred to axes rotating with a certain angular velocity, we may put

$$(x+iy)^3=A\chi u(\wp u-e_2)^{-\frac{1}{2}},$$

thus leading to the results of § 219.

As other numerical examples the student may investigate the results of §§ 218, 223, 225, 233; also the example due to Abel (*Œuvres*, I., p. 142), where $\mu=5$, $g_2=12$, $g_3=19$, and $v=\tfrac{2}{5}\omega_2'$ or $\tfrac{4}{5}\omega_2'$, when $\wp v=-2$ or 1; we then find that the values of c_0, c_1, c_2, c_3, c_4, c_5 are proportional to

$$-288,\quad -36,\quad -48i\sqrt{3},\quad 12,\quad i\sqrt{3},\quad 0\,;$$

or
$$-396,\quad -252,\quad -12i\sqrt{3},\quad -24,\quad i\sqrt{3},\quad 0.$$

Writing s for $\wp u$, then we may put

$$\chi u = -288 - 36\wp u - 48i\sqrt{3}\wp' u + 12\wp'' u + i\sqrt{3}\wp''' u$$
$$= \quad 36(2s^2 - s - 10) + 12i\sqrt{3}(s-4)\sqrt{(4s^3 - 12s - 19)},$$
$$\chi u = -396 - 252\wp u - 12i\sqrt{3}\wp' u - 24\wp'' u + i\sqrt{3}\wp''' u$$
$$= - \quad 36(4s^2 + 7s + 7) + 12i\sqrt{3}(s-1)\sqrt{(4s^3 - 12s - 19)}.$$

We thence infer that the corresponding pseudo-elliptic integrals involve

$$\tan^{-1}\frac{(s-4)\sqrt{(4s^3-12s-19)}}{\sqrt{3}(2s^2-s-10)} = \cos^{-1}\frac{\sqrt{3}(2s^2-s-10)}{2(s-1)^{\frac{5}{2}}} = \ldots,$$

$$\text{or } \tan^{-1}\frac{(s-1)\sqrt{(4s^3-12s-19)}}{\sqrt{3}(4s^2+7s+7)} = \cos^{-1}\frac{\sqrt{3}(4s^2+7s+7)}{2(s+2)^{\frac{5}{2}}} = \ldots;$$

and now by differentiation we infer that

$$\int \frac{2s+13}{s-1} \frac{ds}{\sqrt{(4s^3-12s-19)}} = \frac{2}{\sqrt{3}} \tan^{-1}\frac{(s-4)\sqrt{(4s^3-12s-19)}}{\sqrt{3}(2s^2-s-10)},$$

$$\int \frac{4s-7}{s+2} \frac{ds}{\sqrt{(4s^3-12s-19)}} = \frac{2}{\sqrt{3}} \tan^{-1}\frac{(s-1)\sqrt{(4s^3-12s-19)}}{\sqrt{3}(4s^2+7s+7)}.$$

Thus, in the Weierstrassian notation,

$$\int \frac{\frac{1}{2}i\wp' v du}{\wp u - \wp v} = \quad \tfrac{1}{5}\tan^{-1}\frac{(\wp u - 4)\wp' u}{\sqrt{3}(2\wp^2 u - \wp u - 10)} - \tfrac{1}{5}\sqrt{3}u,$$

$$\text{or} \qquad = -\tfrac{1}{5}\tan^{-1}\frac{(\wp u - 1)\wp' u}{\sqrt{3}(4\wp^2 u + 7\wp u + 7)} + \tfrac{2}{5}\sqrt{3}u,$$

with $g_2 = 12$, $g_3 = 19$, according as $\wp v = 1$ or -2.

These results may be employed in the construction of degenerate cases of the catenaries discussed in §§ 80, 205, 206.

Thus, for instance, the curve given by

$$r^2 = k^2(\wp u + 2),$$
$$r^5 \cos(2\sqrt{3}u - 5\theta) = \sqrt{3}k(4r^4 - 9k^2 r^2 + 9k^4),$$

is a plane catenary for a central attraction $n^2 wr$ per unit of length, in which (§ 80)

$$t = \tfrac{1}{2}n^2 w(r^2 - 3k^2), \quad tp = \tfrac{3}{4}\sqrt{3}n^2 wk^3.$$

So also a tortuous catenary is given by the equations

$$r^2 = k^2\{\wp(\tfrac{2}{3}x/k) - 1\},$$
$$r^5 \cos(5\theta + \tfrac{2}{3}\sqrt{3}x/k) = \sqrt{3}k(2r^4 + 3k^2 r^2 - 9k^4),$$

under an attraction $n^2 wr$ to the axis Ox.

282. Other pseudo-elliptic integrals are formed by the sum of two or more elliptic integrals of the third kind, when the sum of the parameters is of the form $p\omega+q\omega'$, as in § 226, for the expressions of ξ and ξ'.

We shall denote the integral of the third kind in the form (β_1), § 199, by $\Phi(u, v)$, as this we have found is the form of most frequent occurrence in the dynamical applications; and now (β_1) shows that

$$\Phi(u, a)+\Phi(u, b)-\Phi(u, a+b)$$
$$=\{\zeta a+\zeta b-\zeta(a+b)\}u+\tfrac{1}{2}\log\frac{\sigma(a-u)\sigma(b-u)\sigma(a+b+u)}{\sigma(a+u)\sigma(b+u)\sigma(a+b-u)}$$
$$=-\tfrac{1}{2}\frac{\wp'a-\wp'b}{\wp a-\wp b}u+\tfrac{1}{2}\log\frac{\wp(a+u)-\wp(b+u)}{\wp(a-u)-\wp(b-u)}\cdot\frac{\wp u-\wp(a+b-u)}{\wp u-\wp(a+b+u)},$$

by reason of (γ), § 197, and (K), § 200.

When $a+b=\omega_a$, $\wp'(a+b)=0$, $\Phi(u, a+b)=0$; and now

$$\Phi(u, a)+\Phi(u, b)=-\frac{\frac{1}{2}\wp'a}{\wp a-e_a}u+\tfrac{1}{4}\log\frac{\wp(a+u)-e_a}{\wp(a-u)-e_a}.$$

By equation (N), § 249, we may write

$$\tfrac{1}{4}\log\frac{\wp(a+u)-e_a}{\wp(a-u)-e_a}=\tanh^{-1}\sqrt{\left(\frac{\wp u-e_a\,.\,\wp a-e_\beta\,.\,\wp a-e_\gamma}{\wp a-e_a\,.\,\wp a-e_\beta\,.\,\wp u-e_\gamma}\right)}$$
$$=\tanh^{-1}\frac{\wp'a}{\wp a-e_a}\frac{\wp u-e_a}{\wp'u},\quad\text{or}\quad i\tan^{-1}\frac{i\wp'a}{\wp a-e_a}\frac{\wp u-e_a}{\wp'u},$$

the latter form to be employed in dynamical problems, where $\wp'a$ is always imaginary; thence the expressions given for ξ and ξ' in § 226 can be inferred.

As an application we can put $a+b=\omega_1+\omega_3$ or ω_3 in § 209, and thence deduce a degenerate case of the Spherical Pendulum.

Examples.

1. Prove the following q series:—

(i.) $1+2q+2q^4+2q^9+\ldots=\Theta K=\sqrt{(K/\frac{1}{2}\pi)}$;

(ii.) $\dfrac{2q^{\frac{1}{4}}+2q^{\frac{9}{4}}+2q^{\frac{25}{4}}+\ldots}{1+2q+2q^4+\ldots}=\dfrac{\mathrm{H}K}{\Theta K}=\sqrt{\kappa}$;

(iii.) $\dfrac{1-2q+2q^4-\ldots}{1+2q+2q^4+\ldots}=\dfrac{\Theta 0}{\Theta K}=\sqrt{\kappa'}$;

(iv.) $(1-2q+2q^4-\ldots)^4+(2q^{\frac{1}{4}}+2q^{\frac{9}{4}}+\ldots)^4=(1+2q+2q^4+\ldots)^4$;

(v.) $\sqrt{(\kappa\kappa')}\risingdotseq 2q^{\frac{1}{4}}$, $q\risingdotseq\frac{1}{16}\kappa^2\kappa'^2$, $J\risingdotseq 1/1728q^2$, or $-1/1728q$, according as Δ is positive or negative, when q and κ or κ' is small.

2. With the notation of § 265, prove the theorem

$$\begin{aligned}\theta_3(w)\theta_3(x)\theta_3(y)\theta_3(z)-\theta_2(w)\theta_2(x)\theta_2(y)\theta_2(z)\\ -\theta(w)\theta(x)\theta(y)\theta(z)+\theta_1(w)\theta_1(x)\theta_1(y)\theta_1(z)\\ =2\theta_1(s)\theta_1(s-y-z)\theta_1(s-z-x)\theta_1(s-x-y),\end{aligned}$$

where $2s=w+x+y+z.$

Deduce the formulas

(i.) $\kappa^2\kappa'^2\text{sn}\, u\ \text{sn}\, v\ \text{sn}\, r\ \text{sn}\, s$

$$-\kappa^2\text{cn}\, u\ \text{cn}\, v\ \text{cn}\, r\ \text{cn}\, s+\text{dn}\, u\ \text{dn}\, v\ \text{dn}\, r\ \text{dn}\, s-\kappa'^2=0,$$

provided $u+v+r+s=0.$

(ii.) $\kappa^2\text{sn}\,\tfrac{1}{2}(u+v+r+s)\text{sn}\,\tfrac{1}{2}(u+v-r-s)$

$$\times\,\text{sn}\,\tfrac{1}{2}(u-v+r-s)\text{sn}\,\tfrac{1}{2}(u-v-r+s)$$

$$=\frac{(\text{dn}\,u\,\text{dn}\,v\,\text{dn}\,r\,\text{dn}\,s-\kappa^2\text{cn}\,u\,\text{cn}\,v\,\text{cn}\,r\,\text{cn}\,s+\kappa^2\kappa'^2\text{sn}\,u\,\text{sn}\,v\,\text{sn}\,r\,\text{sn}\,s-\kappa'^2)}{(\text{dn}\,u\,\text{dn}\,v\,\text{dn}\,r\,\text{dn}\,s-\kappa^2\text{cn}\,u\,\text{cn}\,v\,\text{cn}\,r\,\text{cn}\,s-\kappa^2\kappa'^2\text{sn}\,u\,\text{sn}\,v\,\text{sn}\,r\,\text{sn}\,s+\kappa'^2)}.$$

3. Show that

$$\begin{aligned}(e_2-e_3)\sigma_1(u)\sigma_1(3u)+(e_3-e_1)\sigma_2(u)\sigma_2(3u)+(e_1-e_2)\sigma_3(u)\sigma_3(3u)\\ =2(e_2-e_3)(e_3-e_1)(e_1-e_2)\sigma^2(u)\sigma^2(2u).\end{aligned}$$

4. Show that Weierstrass' function $\sigma(u)$ satisfies the partial differential equations

$$4g_2\frac{\partial\sigma}{\partial g_2}+6g_3\frac{\partial\sigma}{\partial g_3}+\sigma-u\frac{\partial\sigma}{\partial u}=0,$$

$$\frac{\partial^2\sigma}{\partial u^2}-12g_3\frac{\partial\sigma}{\partial g_2}-\tfrac{2}{3}g_2{}^2\frac{\partial\sigma}{\partial g_3}+\tfrac{1}{12}g_2u^2\sigma=0.$$

Show that the second of these equations is also satisfied by the function

$$\sigma_a(u)/\{(e_a-e_\beta)(e_a-e_\gamma)\}^{\frac{1}{4}};$$

and write down the differential equation satisfied by $\sigma_a u$.

5. Prove that the projection of a geodesic on a quadric of revolution on a plane perpendicular to the axis is analytically similar to a herpolhode (Halphen, II., Chap. VI.).

6. Evaluate the surface of an ellipsoid.

7. Construct some degenerate cases of trajectories or catenaries on a sphere, or on a vertical paraboloid or cone, employing the numerical results of the pseudo elliptic integrals.

CHAPTER X.

THE TRANSFORMATION OF ELLIPTIC FUNCTIONS.

283. By the *Theory of Transformation* is meant the expression, in terms of the elliptic functions of modulus κ and argument u, of an elliptic function with respect to a new modulus λ and of a proportional argument u/M; and then M is called the *multiplier*, and the relation connecting the moduli λ and κ is called the *modular equation*.

A particular case of Transformation has already been introduced in *Landen's Transformation* (§§ 28, 67, 71, 123, 181, 182) in its application to Pendulum Motion, and to the Rectification of the Hyperbola.

In accordance with the plan of this treatise, we begin with a physical application of the Theory of Transformation, before proceeding to the analytical treatment of the subject.

Suppose then in § 259 that an odd number, n, of such rectangles as $OABC$ are placed in contact, side by side, so as to form a single rectangle OA_nB_nC, of length $OA_n = na$, and height $OC = b$; and now put

$$OA_n/OC = na/b = K/K',$$

$$OA\ /OC = \ a/b = \Lambda/\Lambda',$$

so that $$\Lambda'/\Lambda = nK'/K; \qquad (1)$$

where K, K' denote the quarter periods with respect to the modulus κ (§ 11), and Λ, Λ' with respect to the modulus λ.

Let us begin by placing a positive electrode at O, and an equal negative electrode at C; then, inside the rectangle OB, the vector function will be

$$\log \operatorname{sn} \Lambda z/a = \log \operatorname{sn}(\Lambda x/a + \Lambda' iy/b),$$

with $$z = x + yi.$$

But, inside the rectangle OB_n, the vector function of these electrodes and their images will be that due to positive electrodes at $2sa$ and negative electrodes at $2sa+bi$, where s assumes all integral values from 0 to $n-1$; and the vector function of this system is (§§ 259, 275)

$$\log \prod_{s=0}^{s=n-1} \operatorname{sn} K(z-2sa)/na = \log \Pi \operatorname{sn}(Kx/na + K'iy/b - 2sK/n).$$

The physical equivalence of these two forms of the vector function, as seen from two different points of view, shows that

$$\operatorname{sn}(\Lambda z/a) = A \prod_{s=0}^{s=n-1} \operatorname{sn}(Kz/na - 2sK/n),$$

or $$\operatorname{sn}(u/M, \lambda) = A\, \Pi \operatorname{sn}(u - 2sK/n), \quad \text{.....(2)}$$

where $$u/M = \Lambda z/a, \quad u = Kz/na;$$

so that $$M = K/n\Lambda = K'/\Lambda'; \quad \text{.....(3)}$$

this is the formula for the *first real transformation* of the sn function, of the nth order.

Similar considerations will show that

$$\operatorname{cn}(u/M, \lambda) = B\, \Pi \operatorname{cn}(u - 2sK/n), \quad \text{.....(4)}$$

$$\operatorname{dn}(u/M, \lambda) = C\, \Pi \operatorname{dn}(u - 2sK/n). \quad \text{.....(5)}$$

If, as in § 263, we put

$$q = \exp(-\pi K'/K), \text{ and } r = \exp(-\pi\Lambda'/\Lambda);$$

then $$r = q^n, \quad \text{.....(6)}$$

and λ is less than κ.

It simplifies matters to place the rectangle OB in the middle of n such rectangles placed side by side, and now s ranges from $-\frac{1}{2}(n-1)$ to $\frac{1}{2}(n+1)$; and combining equal positive and negative values of s, we find, according to (7) § 137,

$$\operatorname{sn}(u/M, \lambda) = A \operatorname{sn} u \prod_{s=1}^{s=\frac{1}{2}(n-1)} \frac{\operatorname{sn}^2 u - \operatorname{sn}^2 2s\omega}{1 - \kappa^2 \operatorname{sn}^2 2s\omega \operatorname{sn}^2 u}, \quad \text{......(7)}$$

where $$\omega = K/n;$$

or $$y = \frac{x}{M} \Pi \frac{1 - x^2/a^2}{1 - \kappa^2 a^2 x^2}, \quad \text{.....(8)}$$

connecting $y = \operatorname{sn}(u/M, \lambda)$ and $x = \operatorname{sn}(u, \kappa)$, $a = \operatorname{sn}(2sK/n)$.

284. Next suppose that n equal rectangles, such as $OABC$, are piled on each other, so as to form a single rectangle OAB_nC_n, where $OA = a$, $OC_n = nb$; and now put

$$OA/OC_n = a/nb = K/K',$$
$$OA/OC = a/b = \Lambda/\Lambda';$$

so that $$K'/K = n\Lambda'/\Lambda. \quad\text{.........(9)}$$

The physical equivalence of a positive electrode at O and an equal negative electrode at C, and of their images in the rectangle $OABC$, with the positive electrodes at $2sK'iy/b$ and the negative electrodes at $(2s+1)K'iy/b$ in the rectangle OAB_nC_n and their images, shows in a similar manner that

$$\text{sn}(\Lambda z/a, \lambda) = A\,\Pi\,\text{sn}(Kx/a + K'iy/nb - 2sK'i/n),$$

where s may assume all integral values from 0 to $n-1$, but preferably, from $-\frac{1}{2}(n-1)$ to $\frac{1}{2}(n+1)$; or

$$\text{sn}(u/M, \lambda) = A\,\Pi\,\text{sn}(u - 2sK'i/n, \kappa), \quad\text{.........(10)}$$

where $$u/M = \Lambda z/a, \quad u = Kz/a;$$

so that $$M = K/\Lambda = K'/n\Lambda'; \quad\text{.........(11)}$$

and now, with

$$q = \exp(-\pi K'/K), \quad r = \exp(-\pi\Lambda'/\Lambda),$$

we have $$r = q^{1/n}, \quad\text{.........(12)}$$

and now λ is greater than κ.

Similar considerations show that, by placing positive and negative electrodes at A and C, or B and C, we shall obtain the formulas

$$\text{cn}(u/M, \lambda) = B\,\Pi\,\text{cn}(u - 2sK'i/n); \quad\text{.........(13)}$$
$$\text{dn}(u/M, \lambda) = C\,\Pi\,\text{dn}(u - 2sK'i/n); \quad\text{.........(14)}$$

these are the formulas for the *second real transformation* of the elliptic functions, of the nth order.

A similar physical interpretation of Transformation may be given in connexion with the curvilinear rectangles bounded by concentric circular arcs and their radii, as discussed in § 270.

285. Besides the first and second real transformations in which q is changed into q^n and $q^{1/n}$, now denoted by r_∞ and r_0, there are in addition $n-1$ imaginary transformations, when n is a prime number, in which q is changed into $\omega^p q^{1/n}$, denoted by r_p, where $p = 1, 2, 3, \ldots, n-1$, and ω is an imaginary nth root of unity; so that, corresponding to a given value of κ, the modular equation of the nth order, if prime will be of the $(n+1)$th degree in λ, having the roots

$$\lambda_\infty, \lambda_0, \lambda_1, \lambda_2, \ldots, \lambda_{n-1},$$

of which two only, λ_∞ and λ_0, will be real; $\lambda_\infty < \kappa < \lambda_0$.

We need only consider the Transformations of prime order, as a Transformation of composite order, mn, can be made to depend on the transformations of the mth and nth order.

The different transformations of the mnth order are formed by changing q into $q^{m/n}$; so that the number of transformations for any number in general is the number of divisors of mn; reducing to $n+1$, as before, for a prime number n.

For a transformation of order n^2 there is one real transformation for which q remains unaltered, and we thus obtain the formulas for Multiplication of the argument u by n.

286. After this physical introduction, we can proceed to the general algebraical theory of Transformation, as developed by Jacobi in his *Fundamenta nova theoriæ functionum ellipticarum*, 1829.

The theory in its generality consists in the determination of y as a rational algebraical function of x, of the form

$$y = U/V, \qquad \text{.........................(15)}$$

where U and V are rational integral functions of x,

$$\left.\begin{aligned} U &= a_n x^n + a_{n-1} x^{n-1} + \ldots + a_1 x + a_0, \\ V &= b_n x^n + b_{n-1} x^{n-1} + \ldots + b_1 x + b_0, \end{aligned}\right\} \qquad \text{............(16)}$$

so as to satisfy a differential relation of the form

$$\frac{M\,dy}{\sqrt{Y}} = \frac{dx}{\sqrt{X}}, \qquad \text{.........................(17)}$$

where

$$\left.\begin{aligned} X &= ax^4 + 4bx^3 + 6cx^2 + 4dx + e, \\ Y &= Ay^4 + 4By^3 + 6Cy^2 + 4Dy + E, \end{aligned}\right\} \qquad \text{............(18)}$$

Making the substitution of (15), we find that we must have

$$\frac{M\left(\dfrac{dU}{dx}V - U\dfrac{dV}{dx}\right)}{\sqrt{(AU^4 + 4BU^3V + 6CU^2V^2 + 4DUV^3 + EV^4)}} = \frac{dx}{\sqrt{X}},$$

and the first condition requisite is that

$$AU^4 + 4BU^3V + 6CU^2V^2 + 4DUV^3 + EV^4 = XT^2, \ldots(19)$$

where T is a rational integral function of x, of the $(2n-2)$th degree; and now, if we can make

$$T = M\left(\frac{dU}{dx}V - U\frac{dV}{dx}\right), \qquad \text{...................(20)}$$

where M is a constant multiplier, the Transformation is effected.

But if U and V are both of the nth degree, or if one of the nth and the other of the $(n-1)$th degree, so that either a_n or b_n (not both) is zero, this is necessarily the case; for any square factor in $(U, V)^4$ will appear as a linear factor of

$$\frac{dU}{dx}V - U\frac{dV}{dx},$$

which is also of the $(2n-2)$th degree, and can therefore only differ from T by a constant factor M.

The Transformation is now said to be of the nth order.

By taking X of the sixth, instead of the fourth degree, Mr. W. Burnside has derived hyperelliptic integrals (*Proc. L. M. S.*, XXIII.) from the elliptic element $dy/\surd Y$, similar to the hyperelliptic integrals of §§ 159, 160, by means of substitutions of the second, third, and higher orders.

Now denoting by $\alpha, \beta, \gamma, \delta$ the roots of the quartic $X=0$, and by $\alpha', \beta', \gamma', \delta'$ those of $Y=0$; so that, resolved into factors,

$$X= a(x-\alpha)(x-\beta)(x-\gamma)(x-\delta),$$
$$Y=A(y-\alpha')(y-\beta')(y-\gamma')(y-\delta');$$

then $A(U-\alpha'V)(U-\beta'V)(U-\gamma'V)(U-\delta'V)$

$$=aT^2(x-\alpha)(x-\beta)(x-\gamma)(x-\delta);$$

and now a factor, such as $U-\alpha'V$, must be composed of linear factors, such as $x-\alpha$, and of the squares of factors of T.

In the expression $y=U/V$ there are at most $2n+1$ arbitrary constants; and in determining U and V so as to satisfy relation (19) we determine $2n-2$ of these arbitrary constants; thus there remain at disposal three arbitrary constants, corresponding to the three constants involved in an arbitrary linear transformation, such as that obtained by writing (§ 139)

$$(lx+m)/(l'x+m') \text{ for } x,$$

as exemplified in §§ 153, 160, where the constants l, m, l', m' are chosen so as to make X and Y quadratic functions of x^2 and y^2.

When X and Y reduce to quadratic functions of x and y, the elliptic functions degenerate into circular and hyperbolic functions: and now there is no Theory of Transformation, except for the change from circular to hyperbolic functions, as in § 16.

287. Jacobi, in his *Fundamenta nova*, works throughout with the differential relation for the sn function (§ 35)

$$\frac{Mdy}{\sqrt{(1-y^2 \,.\, 1-\lambda^2 y^2)}} = \frac{dx}{\sqrt{(1-x^2 \,.\, 1-\kappa^2 x^2)}} = du \,; \quad \text{.....(21)}$$

connecting $x=\operatorname{sn}(u, \kappa)$ and $y=\operatorname{sn}(u/M, \lambda)$.

Now, if $$y=U/V,$$

then, since $u=0$ makes $x=0$ and $y=0$, y and therefore U must be an odd function of x, the other, V, being an even function; so that for an odd order of the transformation

$$U=a_1 x + a_3 x^3 + \ldots + a_n x^n, \quad V=b_0 + b_2 x^2 + \ldots + b_{n-1} x^{n-1}.$$

Since $x=1$, $y=1$; $x=1/\kappa$, $y=1/\lambda$; etc., are simultaneous values of x and y, the relation connecting x and y may be written in any one of the following forms,

$$\begin{aligned} 1+y &= (1+x)A^2/V, \text{ or } & V+U &= (1+x)A^2; \\ 1-y &= (1-x)A'^2/V, & V-U &= (1-x)A'^2; \\ 1+\lambda y &= (1+\kappa x)C^2/V, & V+\lambda U &= (1+\kappa x)C^2; \\ 1-\lambda y &= (1-\kappa x)C'^2/V, & V-\lambda U &= (1-\kappa x)C'^2; \quad \text{.....(22)} \end{aligned}$$

where A and C are rational integral functions of x, of the $\frac{1}{2}(n-1)$th degree, which become changed into A' and C' when x is changed into $-x$; so that we may put

$$A=P+Qx, \quad A'=P-Qx,$$
$$C=P'+Q'x, \quad C'=P'-Q'x,$$

where P, Q, P', Q' are even functions of x; and therefore

$$\frac{1-y}{1+y} = \frac{1-x}{1+x}\left(\frac{P-Qx}{P+Qx}\right)^2, \quad \frac{1-\lambda y}{1+\lambda y} = \frac{1-\kappa x}{1+\kappa x}\left(\frac{P'-Q'x}{P'-Q'x}\right)^2;$$

giving $$y = x\frac{P^2+2PQ \quad +Q^2x^2}{P^2+2PQx^2+Q^2x^2} = \frac{x}{\lambda}\,\frac{\kappa P'^2+2P'Q'+\kappa Q'^2x^2}{P'^2+2\kappa P'Q'x^2+Q'^2x^2}. \quad \text{...(23)}$$

When the order n of transformation is even, we put

$$U=a_1 x + a_3 x^3 + \ldots + a_{n-1} x^{n-1}, \quad V=b_0 + b_2 x^2 + \ldots + b_n x^n;$$

and now $$V+U=(1+x)(1+\kappa x)B^2, \quad V+\lambda U=D^2,$$
$$V-U=(1-x)(1-\kappa x)B'^2, \quad V-\lambda U=D'^2; \quad \text{.........(24)}$$

where B, D are rational integral functions of x, of the $(\frac{1}{2}n-1)$th degree, changing into B' and D' when x is changed into $-x$; so that we may put

$$B=R+Sx, \quad B'=R-Sx;$$
$$D=R'+S'x, \quad D'=R'-S'x;$$

where R, S, R', S' are even functions of x.

288. The number of independent constants represented by the a's and b's in U and V can be immediately halved by noticing that a change of u into $u+K'i$ has the effect of changing x into $1/\kappa x$ and y into $1/\lambda y$ (§ 239); and therefore of interchanging U and V.

An algebraical simplification is thus introduced by writing $x/\sqrt{\kappa}$ for x and $y/\sqrt{\lambda}$ for y, as in § 143; the differential relation now becomes of the form (Cayley, *American Journal of Mathematics*, vol. 9)

$$\frac{dy}{\sqrt{(1-2\beta y^2+y^4)}}=\frac{\rho dx}{\sqrt{(1-2\alpha x^2+x^4)}}, \qquad (25)$$

and
$$2\alpha=\kappa+1/\kappa,\quad 2\beta=\lambda+1/\lambda, \qquad (26)$$

connecting
$$x=\frac{\operatorname{sn}(u,\kappa)}{\sqrt{\kappa}},\quad y=\frac{\operatorname{sn}(\rho u,\lambda)}{\sqrt{\lambda}};$$

and now, if
$$y=U/V,$$

$$U=B_{n-1}x+\dots B_2x^{n-2}+B_0x^n,\quad V=B_0+B_2x^2+\dots B_{n-1}x^{n-1};$$

for an odd order n of transformation, involving only n coefficients B_0, B_1, ..., B_{n-1}, and therefore $n-1$ arbitrary constants in y; also $B_{n-1}=\rho B_0$.

It follows then that, in the original relation $y=U/V$, connecting $x=\operatorname{sn}(u,\kappa)$ and $y=\operatorname{sn}(u/M,\lambda)$, if a^2-x^2 is a factor of U, then $1-\kappa^2a^2x^2$ must be a corresponding factor of V; and we thus obtain the expression of y as a function of x given in equation (8), and in addition the relation

$$\lambda=M^2\kappa^n\Pi a^2, \qquad (27)$$

so that we may write

$$y=M\frac{\kappa}{\lambda}x\,\Pi\frac{x^2-a^2}{x^2-1/\kappa^2a^2}. \qquad (28)$$

Professor Cayley writes equation (25) in the form

$$(1+S_1y^2+S_2y^4+\dots)dy=\rho(1+R_1x^2+R_2x^4+\dots)dx,$$
$$y+\tfrac{1}{3}S_1y^3+\tfrac{1}{3}S_2y^5+\dots=\rho(x+\tfrac{1}{3}R_1x^3+\tfrac{1}{5}R_2x^5+\dots),$$

where the R's and S's are the zonal harmonics of α and β.

289. Writing this equation (28) in the form

$$x\,\Pi(x^2-a^2)-\frac{\lambda}{\kappa M}\operatorname{sn}\left(\frac{u}{M},\lambda\right)\Pi\left(x^2-\frac{1}{\kappa^2a^2}\right)=0,$$

which is an equation of the nth degree in x, the roots of which are $x=\operatorname{sn}u$, $\operatorname{sn}(u\pm 2\omega)$, ..., $\operatorname{sn}\{u\pm(n-1)\omega\}$, where $\omega = 2K/n$ or $2K'i/n$ for the two real transformations, we find that the sum of the roots

$$\frac{\lambda}{\kappa M}\mathrm{sn}\left(\frac{u}{M}, \lambda\right) = \sum_{s=-\frac{1}{2}(n-1)}^{s=\frac{1}{2}(n-1)} \mathrm{sn}(u+2s\omega), \quad \ldots\ldots\ldots\ldots\ldots\ldots(29)$$

or combining the equal positive and negative values of s,

$$\frac{\lambda}{\kappa M}\mathrm{sn}\left(\frac{u}{M}, \lambda\right) = \mathrm{sn}\, u + \Sigma \frac{2\,\mathrm{sn}\, u\, \mathrm{cn}\, 2s\omega\, \mathrm{dn}\, 2s\omega}{1-\kappa^2 \mathrm{sn}^2 2s\omega\, \mathrm{sn}^2 u}$$

or

$$\frac{\lambda y}{\kappa M} = x + \Sigma \frac{2x\sqrt{(1-a^2 \,.\, 1-\kappa^2 a^2)}}{1-\kappa^2 a^2 x^2}, \quad \ldots\ldots(30)$$

the expression for y when the product in equation (8) is resolved into its partial fractions; and similar expressions hold for the cn and dn functions (Jacobi, *Werke*, I., p. 429; Cayley, *Elliptic Functions*, p. 256).

290. We need not therefore confine ourselves, with Jacobi, to the Transformations of the sn function; but we may sometimes find it preferable to seek the relations connecting

$$x = \mathrm{cn}(u, \kappa) \text{ and } y = \mathrm{cn}(u/M, \lambda),$$

when (§ 35; Abel, *Œuvres*, I., p. 363)

$$\frac{M dy}{\sqrt{(1-y^2 \,.\, \lambda'^2 + \lambda^2 y^2)}} = \frac{dx}{\sqrt{(1-x^2 \,.\, \kappa'^2 + \kappa^2 x^2)}} = du\,; \quad \ldots(31)$$

or the relations connecting

$$x = \mathrm{dn}(u, \kappa) \text{ and } y = \mathrm{dn}(u/M, \lambda),$$

when

$$\frac{M dy}{\sqrt{(1-y^2 \,.\, y^2 - \lambda'^2)}} = \frac{dx}{\sqrt{(1-x^2 \,.\, x^2 - \kappa'^2)}} = du\,; \quad \ldots\ldots(32)$$

relations already given in (4), (5), (13), (14) of §§ 282, 284.

But Prof. Klein points out (*Math. Ann.*, XIV., p. 116) that it is the differential form of § 38 (really Riemann's form), connecting $z = \mathrm{sn}^2(u, \kappa)$ and $t = \mathrm{sn}^2(u/M, \lambda)$, and leading to the relation, on writing k for κ^2 and l for λ^2,

$$\frac{M dt}{\sqrt{(4t \,.\, 1-t \,.\, 1-lt)}} = \frac{dz}{\sqrt{(4z \,.\, 1-z \,.\, 1-kz)}} = du, \quad \ldots\ldots(33)$$

which is the most fundamental in the theory of the elliptic functions sn, cn, and dn; the periods now being $2K$ and $2K'i$, instead of $4K$ and $2K'i$, etc. (§ 239); the *quadric* transformations (of the second order)

$$z = x^2, \quad 1-x^2, \quad \text{or} \quad 1-\kappa^2 x^2,$$
$$t = y^2, \quad 1-y^2, \quad \text{or} \quad 1-\lambda^2 y^2, \quad \ldots\ldots\ldots\ldots\ldots(34)$$

leading immediately to the preceding transformations of the sn, cn, and dn functions.

291. The Theory of Transformation may be developed entirely from the algebraical point of view; but Abel has shown how the form of the transformation of the nth order may be inferred from the elliptic functions of the nth parts of the periods, called by Klein, *modular functions.*

Thus taking the first real transformation connecting

$$z=\text{sn}^2(u, \kappa) \text{ and } t=\text{sn}^2(u/M, \lambda)$$

in relation (33), then

$$t= \frac{z}{M^2}\Pi\left(1-\frac{z}{a}\right)^2 \div D,$$

$$1-t=(1-z)\,\Pi\left(1-\frac{z}{\beta}\right)^2 \div D,$$

$$1-lt=(1-kz)\Pi\,(1-k\beta z)^2 \div D,$$

$$D=\Pi\,(1-kaz)^2, \qquad\text{.....................(35)}$$

where $\qquad a=\text{sn}^2 2sK/n, \quad \beta=\text{sn}^2(2s-1)K/n,$

and the products extend for all integral values of s from 1 to $\frac{1}{2}(n-1)$.

The form of the factors is inferred by Abel from the consideration that

(i.) when $t=0, \quad u/M=2s\Lambda+2s'\Lambda'i,$

where s and s' are integers; and, from equation (3),

$$u=2sK/n+2s'K'i,$$

$$z=\text{sn}^2 2sK/n=0, \text{ or } a;$$

(ii.) when $t=1, \quad u/M=(2s-1)\Lambda+2s'\Lambda'i,$

$$u=(2s-1)K/n+2s'K'i,$$

$$z=\text{sn}^2(2s-1)K/n=\beta \text{ or } 1;$$

(iii.) when $t=1/l, \; u/M=(2s-1)\Lambda+(2s'-1)\Lambda'i,$

$$u=(2s-1)K/n+(2s'-1)K'i,$$

$$z=\text{sn}^2\{(2s-1)K/n-K'i\}=1/k\beta \text{ or } 1/k.$$

(iv.) when $t=\infty, \; u/M=2s\Lambda+(2s'-1)\Lambda'i,$

$$u=2sK/n+(2s'-1)K'i,$$

$$z=\text{sn}^2(2sK/n-K'i)=1/ka, \text{ or } \infty.$$

Similarly the relations can be inferred connecting

$$z=\text{cn}^2(u, \kappa) \text{ and } t=\text{cn}^2(u/M, \lambda),$$

or $\qquad z=\text{dn}^2(u, \kappa) \text{ and } t=\text{cn}^2(u/M, \lambda),$

not only for the first real transformation, depending on equation (3), but also for the second real transformation, depending on equation (11), and also for any one of the imaginary transformations of the nth order.

292. In Weierstrass's form the relation is

$$\frac{Mdy}{\sqrt{(4y^3-\gamma_2 y-\gamma_3)}}=\frac{dx}{\sqrt{(4x^3-g_2x-g_3)}}=du,$$

connecting $x=\wp(u;\, g_2, g_3)$ and $y=\wp(u/M;\, \gamma_2, \gamma_3)$, by a relation of the form

$$y=U/V;$$

and this must be equivalent to relations of the form

$$y-\epsilon_\alpha=(x-e_\alpha)A^2/V, \text{ or } (x-e_\beta)B^2/V, \text{ or } (x-e_\gamma)C^2/V, \quad (36)$$

for a transformation of odd order; giving

$$4y^3-\gamma_2 y-\gamma_3=(4x^3-g_2x-g_3)(ABC)^2/V^3;\ \ldots\ldots(37)$$

so that V must be a perfect square; thus leading to the requisite number of equations for the determination of the arbitrary coefficients in U and V, and an equation over, which relation may be made to connect the absolute invariants J and J', and corresponds to the *modular* equation.

For a transformation of even order, we shall have

$$y=\frac{U}{(x-e_\alpha)T^2},$$

equivalent to relations of the form

$$y-\epsilon_\alpha=\frac{A^2}{(x-e_\alpha)T^2}, \text{ or } \frac{x-e_\beta}{x-e_\alpha}\frac{B^2}{T^2}, \text{ or } \frac{x-e_\gamma}{x-e_\alpha}\frac{C^2}{T^2}, \ \ldots\ldots(38)$$

and therefore

$$4y^3-\gamma_2 y-\gamma_3=\frac{4x^3-\gamma_2 x-\gamma_3}{(x-e_\alpha)^4}\frac{(ABC)^3}{T^6}\ldots\ldots\ldots(39)$$

293. In the Weierstrassian form we determine the relation connecting $x=\wp(u, J)$ and $y=\wp(u/M, J')$.

But without altering J' we may write (§ 196)

$$\wp(u/M, J')=M^2\wp(u, J');$$

and now, if ω, ω' denote the real and imaginary half periods of $\wp(u, J)$ or $\wp u$, we may take ω/n, ω' as the periods of $\wp(u, J')$ in the first real transformation of the nth order; and ω, ω'/n as the periods in the second real transformation (Felix Muller, *De transformatione functionum ellipticarum*; Berlin, 1867).

The first real transformation, of odd order n, may now be written

$$\wp(u, J')=\wp u+\sum_{s=1}^{s=n-1}{}' \left\{\wp\left(u-\frac{2s\omega}{n}\right)-\wp\frac{2s\omega}{n}\right\}\ldots\ldots\ldots(40)$$

similar to equation (30) for the sn function, and obtained in a similar manner.

By integration of this equation (§ 195)

$$\zeta(u, J') = 2G_1 u + \zeta u + \sum_{s=1}^{s=\frac{1}{2}(n-1)} \zeta(u - 2s\omega/n) + \zeta(u + 2s\omega/n), \quad (41)$$

where $G_1 = \frac{1}{2}\sum_{s=1}^{s=n-1} \wp(2s\omega/n) = \sum_{s=1}^{s=\frac{1}{2}(n-1)} \wp(2s\omega/n)$; ....(42)

and integrating again,

$$\log \sigma(u, J') = G_1 u^2 + \log \sigma u \,\Pi\, \sigma(u - 2s\omega/n)\sigma(u + 2s\omega/n),$$

$$\sigma(u, J') = Ce^{G_1 u^2} \sigma u \,\Pi\, \sigma(u - 2s\omega/n)\sigma(u + 2s\omega/n). \quad \text{..........(43)}$$

The constant C is determined by putting $u=0$, when

$$C = \text{lt}\, e^{-G_1 u^2} \frac{\sigma(u, J')}{\sigma u} \Pi \frac{1}{\sigma(u - 2s\omega/n)\sigma(u + 2s\omega/n)}$$

$$= \Pi \frac{1}{\sigma(-2s\omega/n)\sigma(2s\omega/n)};$$

and now

$$\sigma(u, J') = e^{G_1 u^2} \sigma u \prod_{s=1}^{s=\frac{1}{2}(n-1)} \frac{\sigma(2s\omega/n - u)\sigma(2s\omega/n + u)}{\sigma^2(2s\omega/n)}$$

$$= e^{G_1 u^2} (\sigma u)^n \Pi(\wp u - \wp 2s\omega/n), \quad \text{..................(44)}$$

by formula (K) of § 200.

Thus, for instance, with $n=3$,

$$\sigma(u, J') = e^{G_1 u^2} (\sigma u)^3 (\wp u - G_1), \quad \text{.............(45)}$$

where $\qquad G_1 = \wp\frac{2}{3}\omega = \wp\frac{4}{3}\omega,$

and therefore satisfies the equation of § 149

$$G_1 = \frac{(G_1{}^2 + \frac{1}{4}g_2)^2 + 2g_3 G_1}{4G_1{}^3 - g_2 G_1 - g_3};$$

or

$$G_1{}^4 - \tfrac{1}{2}g_2 G_1{}^2 - g_3 G_1 - \tfrac{1}{48}g_2{}^2 = 0. \quad \text{.............(46)}$$

Denoting by G_2 and G_3 the transformed values of g_2 and g_3, they are found by a comparison of coefficients in the expansion of both sides of equation (44) in ascending powers of u (§ 195).

Thus, if $J=0$, or $g_2=0$, then $G_1=0$ or $\sqrt[3]{g_3}$; and taking the value $G_1=0$, then $J'=0$, $G_2=0$, $G_3=-27g_3$; and

$$\sigma(u;\ 0,\ -27g_3) = (\sigma u)^3 \wp u. \quad \text{..................(47)}$$

Employing the principle of Homogeneity of § 196, this equation may be written

$$\sigma(ui\sqrt{3}) = i\sqrt{3}(\sigma u)^3 \wp u, \quad \text{.............(48)}$$

leading by differentiation to

$$i\sqrt{3}\zeta(ui\sqrt{3}) = 3\zeta u + \wp' u/\wp u, \quad \text{............(49)}$$

and

$$3\wp(ui\sqrt{3}) = -3\wp u + \frac{\wp'' u}{\wp u} - \frac{\wp'^2 u}{\wp^2 u} = -\wp u + \frac{g_3}{\wp^2 u}, \quad \text{.....(50)}$$

since $g_2=0$, as in § 47.

Thus, if g_3 is positive, and ω_2, ω_2' the real and imaginary half periods (§ 62), then $\omega_2'/\omega_2 = i\sqrt{3}$; and if we take $u = \frac{2}{3}\omega_2$, then $\wp^3 u = g_3$ (§§ 166, 233); so that $\wp\frac{2}{3}\omega_2' = 0$.

Again, putting $u = \omega_2$ in equation (49) gives

$$\eta_2' i\sqrt{3} = 3\eta_2. \quad \text{.........................(51)}$$

Making use of the last equation of § 202, we find

$$\eta_2\omega_2 = \tfrac{1}{3}\eta_2'\omega_2' = \tfrac{1}{6}\pi\sqrt{3}.$$

As a numerical exercise the student may construct the following table, and also fill in the values for $u = \omega_2$, ω_2', $\frac{1}{2}\omega_2$, $\frac{1}{2}\omega_2'$, $\frac{4}{3}\omega_2$, $\frac{4}{3}\omega_2'$, ...; taking $g_2 = 0$, $g_3 = 1$; these numerical results are useful in the problem of the Trajectory for the Cubic Law of Resistance, discussed in §§ 227–234.

u	$\wp u$	$\wp' u$	ζu	σu
$\frac{1}{3}\omega_2$	$\frac{1}{3}(\sqrt[3]{2}+1)^3$	$-\frac{1}{3}\sqrt{3}(\sqrt[3]{2}+1)^4$	$\frac{1}{3}\eta_2 + \dfrac{(\sqrt{2}+1)^3}{3\sqrt{3}\sqrt[3]{2}}$	$\frac{1}{3}\sqrt[12]{3}(\sqrt[3]{2}+1)^3 e^{\frac{\pi\sqrt{3}}{108}}$
$\frac{1}{3}\omega_2'$	$-\sqrt[3]{2}$	$-3i$	$\frac{1}{3}\eta_2' - \frac{1}{2}i\sqrt[3]{4}$	$\dfrac{i}{\sqrt[4]{3}} e^{\frac{\pi\sqrt{3}}{36}}$
$\frac{2}{3}\omega_2$	1	$-\sqrt{3}$	$\frac{2}{3}\eta_2 + \frac{1}{3}\sqrt{3}$	$\dfrac{1}{\sqrt[6]{3}} e^{\frac{\pi\sqrt{3}}{27}}$
$\frac{2}{3}\omega_2'$	0	$-i$	$\frac{2}{3}\eta_2'$	$i e^{\frac{\pi\sqrt{3}}{9}}$
ω_2	....	....	....	$\dfrac{\sqrt[3]{2}}{\sqrt[4]{3}} e^{\frac{\pi\sqrt{3}}{12}}$

The Linear Transformation.

294. In Chapter II. the general elliptic differential $dx/\sqrt{X}$ has been reduced to Legendre's standard form

$$(1 - \kappa^2\sin^2\phi)^{-\frac{1}{2}} d\phi$$

and to Jacobi's, or rather Riemann's standard form (11) of § 38,

$$dz/\sqrt{(4z \,.\, 1-z \,.\, 1-kz)}$$

by various substitutions, in §§ 39, 40, 41, 42, 43, etc., which are practical illustrations of the Linear Transformation.

In § 160, the six linear transformations are given which, according to Mr. R. Russell, reduce

$$dx/\sqrt{X} \text{ to the form } dz/\sqrt{(Az^4 + 6Cz^2 + E)}.$$

In determining the linear transformations, of the form

$$y = U/V = (\alpha x + \beta)/(\gamma x + \delta), \quad \text{................(52)}$$

which satisfy Riemann's differential relation

$$\frac{M dy}{\sqrt{(4y \,.\, 1-y \,.\, 1-ly)}} = \frac{dx}{\sqrt{(4x \,.\, 1-x \,.\, 1-kx)}} = du,$$

connecting $x = \text{sn}^2(u, \kappa)$ and $y = \text{sn}^2(u/M, \lambda)$, we notice, by § 139, that the absolute invariant J is unchanged; so that, according to § 68, there are six values of l, given by

$$l=k,\quad \frac{k}{k-1},\quad \frac{1}{k},\quad \frac{1}{1-k},\quad 1-k,\quad 1-\frac{1}{k}; \quad\text{........(53)}$$

and six corresponding linear transformations, in which

$$\frac{\Lambda' i}{\Lambda}=\frac{aK+bK'i}{cK+dK'i},\quad\text{and}\quad bc-ad=1; \quad\text{..........(54)}$$

$$\begin{vmatrix} a, b \\ c, d \end{vmatrix} \equiv \begin{vmatrix} 0 & 1 \\ 1 & 0 \end{vmatrix} \begin{vmatrix} 1 & 1 \\ 1 & 0 \end{vmatrix} \begin{vmatrix} 0 & 1 \\ 1 & 1 \end{vmatrix} \begin{vmatrix} 1 & 0 \\ 1 & 1 \end{vmatrix} \begin{vmatrix} 1 & 0 \\ 0 & 1 \end{vmatrix} \begin{vmatrix} 1 & 1 \\ 0 & 1 \end{vmatrix} \text{ mod. 2.}$$

295. But if we change to Jacobi's form by the *quadric transformation*, which changes x into x^2, and y into y^2, then

$$\frac{Mdy}{\sqrt{(1-y^2 \,.\, 1-\lambda^2y^2)}}=\frac{dx}{\sqrt{(1-x^2 \,.\, 1-\kappa^2x^2)}}=du, \quad\text{......(55)}$$

and now, forming according to § 75 the invariants g_2, g_3, Δ, and J of the quartic $\quad 1-x^2 \,.\, 1-\kappa^2x^2,$

$$g_2=\frac{1+14k+k^2}{12},\quad g_3=\frac{1-33k-33k^2-33k^3}{216},\quad \Delta=\frac{k(1-k)^4}{16};$$

and

$$J=\frac{(1+14k+k^2)^3}{108k(1-k)^4}. \quad\text{........................(56)}$$

Professor Klein writes η^4 for k or κ^2, and calls η the *Octahedron Irrationality;* and now the absolute invariant being unaltered by a linear transformation,

$$J=\frac{(1+14l+l^2)^3}{108l(1-l)^4}=\frac{(1+14\eta^4+\eta^8)^3}{108\eta^4(1-\eta^4)^4}, \quad\text{..............(57)}$$

and the roots of this equation in l are found to be

$$l=\eta^4,\quad \frac{1}{\eta^4},\quad \left(\frac{1\pm\eta}{1\mp\eta}\right)^4,\quad \left(\frac{1\pm i\eta}{1\mp i\eta}\right)^4; \quad\text{..............(58)}$$

giving the six corresponding linear transformations of Abel (*Œuvres*, I., pp. 459, 568).

In the reductions of Chapter II. that linear transformation has been chosen which makes k or l positive and less than unity, and also gives a real value to the multiplier M.

The corresponding values of the multiplier are given by

$$1/M^2=1,\quad \eta^2,\quad -\tfrac{1}{4}(1\pm\eta)^4,\quad -\tfrac{1}{4}(1\pm i\eta)^4,$$

the linear transformations being, as may be verified.

$$y=\pm x,\quad \pm\eta^2x,\quad \pm\frac{1\pm\eta}{1\mp\eta}\,\frac{1\pm\eta x}{1\mp\eta x},\quad \pm\frac{1\pm i\eta}{1\mp i\eta}\,\frac{1\pm i\eta x}{1\mp i\eta x}.$$

Landen's Transformation of the Second Order.

296. The point L (§ 28) in figs. 2 and 3 has been called *Landen's point,* because of the use made of it by Landen (*Phil. Trans.*, 1771, 1775) for his transformation, important historically as the first case investigated of the *Transformation of Elliptic Functions,* being the *Quadric Transformation,* or of the *second degree.*

The ratio AD/AE being $\sin^2\frac{1}{2}\alpha$ or κ^2, while $EL/EA=\cos\alpha$ or κ'; therefore, if C is the middle point of AD,

$$\frac{LC}{CA}=\frac{AL-AC}{AC}=\frac{AE-EL-\frac{1}{2}AD}{\frac{1}{2}AD}$$

$$=\frac{1-\cos\alpha-\frac{1}{2}\sin^2\alpha}{\frac{1}{2}\sin^2\alpha}=\frac{(1-\cos\frac{1}{2}\alpha)^2}{\sin^2\frac{1}{2}\alpha}=\frac{1-\cos\frac{1}{2}\alpha}{1+\cos\frac{1}{2}\alpha}=\tan^2\tfrac{1}{4}\alpha.$$

The ratio LC/CA is denoted by λ; so that

$$\lambda=\frac{1-\kappa'}{1+\kappa'},\quad \kappa'=\frac{1-\lambda}{1+\lambda},\quad \kappa=\frac{2\sqrt{\lambda}}{1+\lambda},\quad \lambda'=\frac{2\sqrt{\kappa'}}{1+\kappa'},\quad (1+\kappa')(1+\lambda)=2,$$

$$\sqrt{\lambda}=(1-\kappa')/\kappa,\quad \sqrt{\kappa'}=(1-\lambda)/\lambda',\ \text{and}\ \kappa\lambda'=2\sqrt{(\kappa'\lambda)},\ \ldots(59)$$

different forms of the *modular equation of the second order.*

Still denoting the angle ADQ in fig. 2 by ϕ, we denote the angle ALQ by ψ; and now (§ 28) since the velocity of Q is $n(1+\kappa')LQ$, perpendicular to CQ, therefore the component velocity of Q, perpendicular to LQ,

$$LQ\,d\psi/dt=n(1+\kappa')LQ\cos LQC,$$

or

$$d\psi/dt=n(1+\kappa')\cos LQC.$$

But since $\dfrac{\sin LQC}{\sin\psi}=\dfrac{LC}{CQ}=\lambda$, therefore

$$\sin LQC=\lambda\sin\psi,\quad \cos LQC=\sqrt{(1-\lambda\sin^2\psi)}=\Delta(\psi,\lambda);$$

and

$$d\psi/dt=n(1+\kappa')\Delta(\psi,\lambda),$$

or

$$\psi=\text{am}\{(1+\kappa')nt,\lambda\}.\ \ldots\ldots\ldots\ldots(60)$$

Now, since the angle $LQC=2\phi-\psi$, therefore

$$\sin(2\phi-\psi)=\lambda\sin\psi;\ \ldots\ldots\ldots\ldots(61)$$

and

$$\kappa'=\frac{1-\lambda}{1+\lambda}=\frac{\sin(2\phi-\psi)-\sin\psi}{\sin(2\phi-\psi)+\sin\psi}=\frac{\tan(\phi-\psi)}{\tan\phi},\ \ldots(62)$$

or

$$\tan\psi=\frac{(1+\kappa')\tan\phi}{1-\kappa'\tan^2\phi},\ \ldots\ldots\ldots\ldots(63)$$

$$\sin\psi=(1+\kappa')\sin\phi\cos\phi/\Delta\phi,$$

as in equation (92), § 67.

Putting $nt=u$, $(1+\kappa')nt=v$, then $\sin\phi=\text{sn}\,u$, $\sin\psi=\text{sn}\,v$; and we obtain the formulas (90) to (98) of § 67.

297. Landen starts with the relation (61); so that, differentiating logarithmically,

$$\cot(2\phi-\psi)(2d\phi-d\psi)=\cot\psi\, d\psi,$$

$$2\cot(2\phi-\psi)d\phi=\{\cot(2\phi-\psi)+\cot\psi\}d\psi$$

$$=\frac{\sin 2\phi\, d\psi}{\sin\psi\sin(2\phi-\psi)}.$$

$$\frac{2d\phi}{\sin 2\phi\operatorname{cosec}\psi}=\frac{d\psi}{\cos(2\phi-\psi)}.$$

Now $\cos(2\phi-\psi)=\sqrt{(1-\lambda^2\sin^2\psi)}=\Delta(\psi,\lambda)$;

while $\sin 2\phi\cot\psi-\cos 2\phi=\lambda$,

$$\cot\psi=\cot 2\phi+\lambda\operatorname{cosec}2\phi,$$

$$\operatorname{cosec}^2\psi=1+(\cot 2\phi+\lambda\operatorname{cosec}2\phi)^2,$$

$$\sin^2 2\phi\operatorname{cosec}^2\psi=\sin^2 2\phi+(\cos 2\phi+\lambda)^2$$

$$=1+2\lambda\cos 2\phi+\lambda^2$$

$$=(1+\lambda)^2-4\lambda\sin^2\phi,$$

or $\sin 2\phi\operatorname{cosec}\psi=(1+\lambda)\sqrt{(1-\kappa^2\sin^2\phi)}=(1+\lambda)\Delta(\phi,\kappa)$,

where $\kappa=2\sqrt{\lambda}/(1+\lambda)$; so that, finally,

$$\frac{d\phi}{\Delta(\phi,\kappa)}=\frac{\frac{1}{2}(1+\lambda)d\psi}{\Delta(\psi,\lambda)},\quad\text{or}\quad\frac{d\psi}{\Delta(\psi,\lambda)}=\frac{(1+\kappa')d\phi}{\Delta(\phi,\kappa)};\ \text{...(64)}$$

so that, if $\phi=\operatorname{am}(nt,\kappa)$, then $\psi=\operatorname{am}\{(1+\kappa')nt,\lambda\}$, and the angle ψ may be made to represent pendulum motion on the circle CRL, on CL as diameter, LQ meeting this circle in R.

The velocity of R will then be due to the level of L', a point on CE produced, such that $CL'=CL/\lambda^2$; and now we find that

$$EL'=CL'-CE=EL,$$

after reduction, so that L and L' are the limiting points of the circle AQD with respect to the horizontal line through E; but now the value of g in the motion of R on the circle CRL must, in accordance with § 20, be reduced to $\frac{1}{2}g(1-\kappa')^4$.

Again, $$\frac{L'Q}{LQ}=\frac{L'D}{LD}=\frac{EL+ED}{EL-ED}=\frac{\kappa'+\kappa'^2}{\kappa'-\kappa'^2}=\frac{1+\kappa'}{1-\kappa'};$$

so that (§ 28) the velocity of Q is

$$n(1+\kappa')LQ,\ \text{or}\ n(1-\kappa')L'Q.\ \text{................(65)}$$

The period of R in the circle CRL is half the period of Q in the circle AQD; so that, if Λ denotes the real quarter period of the elliptic functions of modulus λ,

$$\Lambda=\tfrac{1}{2}(1+\kappa')K,\ \text{or}\ (1+\lambda)\Lambda=K.\ \text{.............(66)}$$

298. Conversely, as in § 123, we can express the elliptic functions of modulus κ and argument $(1+\lambda)v$ in terms of the elliptic functions of modulus λ and argument v; or starting with the motion of R, we can deduce the motion of Q.

But considering the motion of Q as defining in a similar way the motion on a larger circle, to a larger modulus γ, we change λ into κ and κ into γ, where

$$\kappa=\frac{1-\gamma'}{1+\gamma'},\quad \gamma'=\frac{1-\kappa}{1+\kappa},\quad \gamma=\frac{2\sqrt{\kappa}}{1+\kappa},\quad (1+\gamma')(1+\kappa)=2,$$

$$\sqrt{\kappa}=(1-\gamma')/\gamma,\quad \sqrt{\gamma'}=(1-\kappa)/\kappa',\quad \text{and}\quad \kappa'\gamma=2\sqrt{\kappa\gamma'});\quad (67)$$

and now, from § 123,

$$\left.\begin{aligned}\text{dn}(1+\kappa\,.\,u,\gamma)&=\frac{1-\kappa\,\text{sn}^2(u,\kappa)}{1+\kappa\,\text{sn}^2(u,\kappa)},\\ \text{sn}(1+\kappa\,.\,u,\gamma)&=\frac{(1+\kappa)\text{sn}\,(u,\kappa)}{1+\kappa\,\text{sn}^2(u,\kappa)},\\ \text{cn}(1+\kappa\,.\,u,\gamma)&=\frac{\text{cn}(u,\kappa)\text{dn}(u,\kappa)}{1+\kappa\,\text{sn}^2(u,\kappa)},\end{aligned}\right\}\quad\ldots\ldots\ldots\ldots(68)$$

called *Landen's Second Transformation.*

With $x=\text{sn}(u,\kappa)$, $y=\text{sn}(1+\kappa\,.\,u,\gamma)$, where $\gamma=2\sqrt{\kappa}/(1+\kappa)$,

then
$$y=\frac{(1+\kappa)x}{1+\kappa x^2},$$

$$\begin{aligned}1+\ y&=(1+x)(1+\kappa x)\div V,\\ 1-\ y&=(1-x)(1-\kappa x)\div V,\\ 1+\gamma y&=(1+x\sqrt{\kappa})^2\quad\div V,\\ 1-\gamma y&=(1-x\sqrt{\kappa})^2\quad\div V,\\ V&=1+\kappa x^2,\quad\ldots\ldots\ldots\ldots(69)\end{aligned}$$

and
$$\frac{dy}{\sqrt{(1-y^2\,.\,1-\gamma^2y^2)}}=\frac{(1+\kappa)dx}{\sqrt{(1-x^2\,.\,1-\kappa^2x^2)}}.$$

Or, with $x=\text{dn}(u,\kappa)$, $y=\text{dn}(1+\kappa\,.\,u,\kappa)$,

$$y=\frac{-1+\kappa+x^2}{1+\kappa-x^2},$$

$$\begin{aligned}1+y\ &=2\kappa\qquad\qquad\div V,\\ 1-y\ &=2(1-x^2)\ \div V,\\ y+\gamma'&=2\kappa x^2\qquad\ \div V(1+\kappa),\\ y-\gamma'&=2(x^2-\kappa'^2)\div V(1+\kappa),\\ V&=1+\kappa-x^2;\ldots\ldots\ldots\ldots(70)\end{aligned}$$

leading to the differential relation, (3) of § 35,

$$\frac{dy}{\sqrt{(1-y^2\,.\,y^2-\gamma'^2)}}=\frac{(1+\kappa)dx}{\sqrt{(1-x^2\,.\,x^2-\kappa'^2)}}.$$

299. Denoting by Γ the real quarter-period of the elliptic functions to modulus γ, then $x=1$ makes $y=1$, or $u=K$ makes $(1+\kappa).u=\Gamma$; so that

$$(1+\kappa)K=\Gamma,$$

or (66) $$(1+\lambda)\Lambda=K=\tfrac{1}{2}(1+\gamma')\Gamma. \qquad\qquad (71)$$

Also, Λ', K$'$, Γ' denoting the corresponding quarter periods to modulus λ', κ', γ', the imaginary transformations of § 238 show that, with $iu=v$,

$$\operatorname{sn}(1+\kappa'\,.\,v,\lambda')=\frac{(1+\kappa')\operatorname{sn}(v,\kappa')}{1+\kappa'\operatorname{sn}^2(v,\kappa')},$$

$$\operatorname{sn}(1+\kappa\,.\,v,\gamma')=\frac{(1+\kappa)\operatorname{sn}(v,\kappa')\operatorname{cn}(v,\kappa')}{\operatorname{dn}(v,\kappa')},$$

$$\operatorname{cn}(1+\kappa'\,.\,v,\lambda')=\frac{\operatorname{cn}(v,\kappa')\operatorname{dn}(v,\kappa')}{1+\kappa'\operatorname{sn}^2(v,\kappa')},$$

$$\operatorname{cn}(1+\kappa\,.\,v,\gamma')=\frac{1-(1+\kappa)\operatorname{sn}^2(v,\kappa')}{\operatorname{dn}(v,\kappa')},$$

$$\operatorname{dn}(1+\kappa'\,.\,v,\lambda')=\frac{1-\kappa'\operatorname{sn}^2(v,\kappa')}{1+\kappa'\operatorname{sn}^2(v,\kappa')},$$

$$\operatorname{dn}(1+\kappa\,.\,v,\gamma')=\frac{1-(1-\kappa)\operatorname{sn}^2(v,\kappa')}{\operatorname{dn}(v,\kappa')}; \qquad\qquad (72)$$

so that $$\Lambda'=(1+\kappa')\mathrm{K}', \quad \Gamma'=\tfrac{1}{2}(1+\kappa)\mathrm{K}',$$

or $$\tfrac{1}{2}(1+\lambda)\Lambda'=\mathrm{K}'=(1+\gamma')\Gamma'; \qquad\qquad (73)$$

and therefore $$\frac{1}{2}\frac{\Lambda'}{\Lambda}=\frac{\mathrm{K}'}{\mathrm{K}}=2\frac{\Gamma'}{\Gamma}. \qquad\qquad (74)$$

An inspection of Landen's formulas shows that the dn function has always a *rational* Quadric Transformation.

Mr. R. Russell shows (*Proc. L. M. S.*, XVIII.) that the general rational quadric transformations which reduce

$$dx/\sqrt{X} \text{ to the form } dz/\sqrt{(Az^4+6Cz^2+E)}$$

are always of the form

$$z=\frac{m\,P_2+n\,P_3}{m'P_2+n'P_3}, \text{ etc.} \qquad\qquad (75)$$

P_1, P_2, P_3 denoting the quadratic factors of G, the sextic covariant of X (§ 160).

Thus if $$X=1-x^2\,.\,1-\kappa^2x^2,$$

the sextic covariant may be written

$$G=x(1-\kappa x^2)(1+\kappa x^2),$$

leading to Landen's transformations, given above.

300. Landen's Transformation is useful, as employed by Gauss, for the numerical calculation of K; for if we put (fig. 2)

$LA=a$, $LD=b$; and $CA=a_1$, $CL=\sqrt{(a_1^2-b_1^2)}=\frac{1}{2}(a-b)$;

then $a_1=\frac{1}{2}(a+b)$, $b_1=\sqrt{(ab)}$; and $\kappa'=b/a$, $\lambda'=b_1/a_1$. ...(76)

Now, denoting ψ by ϕ_1, and λ by κ_1, equation (64) becomes

$$\frac{2d\phi}{\sqrt{(a^2\cos^2\phi+b^2\sin^2\phi)}}=\frac{d\phi_1}{\sqrt{(a_1^2\cos^2\phi_1+b_1^2\sin^2\phi_1)}};\quad\text{...(77)}$$

while $\phi_1=\pi$, when $\phi=\frac{1}{2}\pi$;

so that

$$\int_0^{\frac{1}{2}\pi}\frac{d\phi}{\sqrt{(a^2\cos^2\phi+b^2\sin^2\phi)}}=\int_0^{\pi}\frac{\frac{1}{2}d\phi_1}{\sqrt{(a_1^2\cos^2\phi_1+b_1^2\sin^2\phi_1)}}$$

$$=\int_0^{\frac{1}{2}\pi}\frac{d\phi_1}{\sqrt{(a_1^2\cos^2\phi_1+b_1^2\sin^2\phi_1)}},$$

or

$$K=K_1a/a_1=K_1(1+\kappa_1).\quad\text{......................(78)}$$

Continuing this process with ϕ_1, a_1, and b_1, so as to obtain a continuous series, given by (§ 296, equation 62).

$$\tan(\phi_n-\phi_{n+1})=\frac{b_n}{a_n}\tan\phi_n,$$

$$a_{n+1}=\tfrac{1}{2}(a_n+b_n),\quad b_{n+1}=\sqrt{(a_nb_n)};\quad\text{............(79)}$$

then a_n and b_n tend to equality; so that, putting

$$a_\infty=b_\infty=\mu,\text{ and }\phi_\infty=\psi,$$

$$\int_0^{\frac{1}{2}\pi}\frac{d\phi}{\sqrt{(a^2\cos^2\phi+b^2\sin^2\phi)}}=\int_0^{\frac{1}{2}\pi}\frac{d\phi_n}{\sqrt{(a_n^2\cos^2\phi_n+b^2_n\sin^2\phi_n)}}$$

$$=\int_0^{\frac{1}{2}\pi}\frac{d\psi}{\sqrt{(\mu^2\cos^2\psi+\mu^2\sin^2\psi)}}=\frac{\frac{1}{2}\pi}{\mu},$$

or

$$\frac{K}{a}=\frac{K_n}{a_n}=\frac{\frac{1}{2}\pi}{\mu},$$

$$K=K_n\prod_{r=1}^{r=n}(1+\kappa_r)=\tfrac{1}{2}\pi\prod_{r=1}^{r=\infty}(1+\kappa_r).\quad\text{................(80)}$$

Denoting the modular angle of κ_n by θ_n, then

$$\kappa_{n+1}=\sin\theta_{n+1}=\tan^2\tfrac{1}{2}\theta_n;$$

$$\cos\theta_{n+1}=\sec^2\tfrac{1}{2}\theta_n\sqrt{(\cos\theta_n)},$$

and

$$1+\kappa_{n+1}=\sec^2\tfrac{1}{2}\theta_n=\frac{\cos\theta_{n+1}}{\sqrt{(\cos\theta)}};$$

so that

$$K=\tfrac{1}{2}\pi\sec\theta\sqrt{(\cos\theta\cos\theta_1\cos\theta_2\cos\theta_3\ldots)},\quad\text{......(84)}$$

a formula suitable for the logarithmic calculation of K.

The Transformation of the Third Order, and of higher Orders.

301. According to Jacobi's method, the transformation may be written

$$\frac{1-y}{1+y}=\frac{1-x}{1+x}\left(\frac{1-ax}{1+ax}\right)^2, \quad\text{..............(82)}$$

connecting $x=\text{sn}(u, \kappa)$ and $y=\text{sn}(u/M, \lambda)$; and then

$$y=x\frac{2a+1+a^2x^2}{1+(a^2+2a)x^2}=\frac{x}{M}\frac{1-x^2/a^2}{1-\kappa^2a^2x^2}, \quad\text{..................(83)}$$

so that

$$1/M=2a+1,$$

and

$$\frac{1-\lambda y}{1+\lambda y}=\frac{1-\kappa x}{1+\kappa x}\left(\frac{a-\kappa x}{a+\kappa x}\right), \quad\text{...............(84)}$$

leading to the differential relation

$$\frac{dy}{\surd(1-y^2\,.\,1-\lambda^2y^2)}=\frac{(2a+1)dx}{\surd(1-x^2\,.\,1-\kappa^2x^2)}. \quad\text{.........(85)}$$

We shall find that, expressed in terms of a,

$$\kappa^2=\frac{a^4+2a^3}{2a+1}, \qquad \lambda^2=a\left(\frac{a+2}{2a+1}\right)^3,$$

and

$$\kappa'^2=\frac{(1-a)(1+a)^3}{1+2a}, \quad \lambda'^2=\frac{(1+a)(1-a)^3}{1+2a},$$

so that

$$\surd(\kappa\lambda)=\frac{2a+a^2}{1+2a}, \qquad \surd(\kappa'\lambda')=\frac{1-a^2}{1+2a},$$

leading to the *Modular Equation of the Third Order.*

$$\surd(\kappa\lambda)+\surd(\kappa'\lambda')=1. \quad\text{......................(86)}$$

We shall also find that this transformation may be written

$$\frac{1-\text{cn}(u/M, \lambda)}{1+\text{cn}(u/M, \lambda)}=\frac{1-\text{cn}\,u}{1+\text{cn}\,u}\left(\frac{a+1+a\,\text{cn}\,u}{a+1-a\,\text{cn}\,u}\right)^2, \quad\text{........(87)}$$

$$\frac{1-\text{dn}(n/M, \lambda)}{1+\text{dn}(u/M, \lambda)}=\frac{1-\text{dn}\,u}{1+\text{dn}\,u}\left(\frac{a+1+\text{dn}\,u}{a+1-\text{dn}\,u}\right)^2. \quad\text{..........(88)}$$

As a numerical exercise the student may work out the case of

$$a=\tfrac{1}{2}(\surd 3-1).$$

In Legendre's notation, with $x=\sin\phi$, $y=\sin\psi$, he finds that these relations are equivalent to

$$\tan\tfrac{1}{2}(\phi+\psi)=(a+1)\tan\phi. \quad\text{.................(89)}$$

The Transformation of the Third Order was the highest to which Legendre attained, until it was pointed out by Jacobi in the *Astronomische Nachrichten*, No. 123, 1827, that Transformations exist of the fourth, fifth, or any other higher order, as already explained.

Thus the transformation of the fifth order may be written

$$\frac{1-y}{1+y}=\frac{1-x}{1+x}\left(\frac{1-ax+\beta x^2}{1+ax+\beta x^2}\right)^2, \quad \text{.....................(90)}$$

and of the seventh order

$$\frac{1-y}{1+y}=\frac{1-x}{1+x}\left(\frac{1-ax+\beta x^2-\gamma x^3}{1+ax+\beta x^2+\gamma x^3}\right)^2, \quad \text{............(91)}$$

and so on.

302. When the transformation of the third order in § 157 is employed for the reduction of the integral in equation (6), § 227, then

$$s^3=-K^3/P^2, \quad \text{..................................(92)}$$

where

$$P=p^3-3p^2\sin^2\alpha+3p, \quad \text{......................(93)}$$

and

$$K=p^2\cos^2\alpha+p\sin\alpha-1, \quad \text{..................(94)}$$

as in equation (27), § 233; so that $K=0$ and $s=0$ at the points of minimum velocity.

Now, with this substitution of § 157,

$$s=\wp(gx/w^2\,;\ 0,\ -\Delta), \quad \text{......................(95)}$$

where

$$\Delta=4-3\sin^2\alpha=27g_3, \quad \text{......................(96)}$$

(§ 228); and denoting

$$\int_{-\sqrt[3]{(\frac{1}{4}\Delta)}}^{\infty} ds/\sqrt{(4s^3+\Delta)} \text{ by } \Omega_2, \quad \zeta\Omega_2 \text{ by } H_2;$$

then $\wp\frac{2}{3}\Omega_2=0$, $\wp'\frac{2}{3}\Omega_2=-\sqrt{\Delta}$, and $H_2\Omega_2=\frac{1}{2}\pi\sqrt{3}$ (§ 293).

Again (§ 157), $\wp'(gx/w^2)=J/P$,

where $J=p^3(3\sin\alpha-2\sin^3\alpha)-3p^2(2-\sin^2\alpha)+3p\sin\alpha-2$,

and

$$J+P\sqrt{\Delta}=2\{\tfrac{1}{2}(\sin\alpha+\sqrt{\Delta})p-1\}^3,$$

$$J-P\sqrt{\Delta}=2\{\tfrac{1}{2}(\sin\alpha-\sqrt{\Delta})p-1\}^3. \quad \text{...........(97)}$$

Now from § 233,

$$\sqrt{\Delta} = \cos\alpha(\tan\beta+\cot\beta),$$

$$\tfrac{1}{2}(\sin\alpha+\sqrt{\Delta})=\tfrac{1}{2}\cos\alpha(\tan\alpha+\tan\beta+\cot\beta)=\cos\alpha\tan\beta,$$

$$\tfrac{1}{2}(\sin\alpha-\sqrt{\Delta})=\tfrac{1}{2}\cos\alpha(\tan\alpha-\tan\beta-\cot\beta)=-\cos\alpha\cot\beta,$$

while

$$p=\frac{\sin\theta}{\cos(\alpha-\theta)}.$$

Therefore

$$\left\{\frac{\wp'(u\,;\ 0,\ -\Delta)-\wp'\frac{2}{3}\Omega_2}{\wp'(u\,;\ 0,\ -\Delta)+\wp'\frac{2}{3}\Omega_2}\right\}^{\frac{1}{3}}=\left(\frac{J+\sqrt{\Delta}P}{J-\sqrt{\Delta}P}\right)^{\frac{1}{3}}=\frac{\frac{1}{2}(\sin\alpha+\sqrt{\Delta})p-1}{\frac{1}{2}(\sin\alpha-\sqrt{\Delta})p-1}$$

$$=\frac{\cos\alpha\tan\beta\sin\theta-\cos(\alpha-\theta)}{-\cos\alpha\cot\beta\sin\theta-\cos(\alpha-\theta)}=\frac{\tan(\beta-\theta)}{\tan\beta}=\frac{\tan\phi}{\tan\beta}$$

$$=\frac{\wp'(u\,;\ 0,\ g_3)-\wp'\frac{2}{3}\omega_2}{\wp'(u\,;\ 0,\ g_3)+\wp'\frac{2}{3}\omega_2}, \text{ or } X \quad \text{........(98)}$$

(§ 234) a curious result of this transformation.

Again, since $\wp'\frac{2}{3}\omega_2 = -\wp'\frac{4}{3}\omega_2$, we may put

$$X = -\frac{\wp'\frac{4}{3}\omega_2 + \wp' u}{\wp'\frac{4}{3}\omega_2 - \wp' u},$$

and then, making use of relation (17) of § 229,

$$X = -\frac{\sigma(\frac{4}{3}\omega_2 + u)\sigma(\frac{4}{3}\omega\omega_2 + u)\sigma(\frac{4}{3}\omega^2\omega_2 + u)}{\sigma(\frac{4}{3}\omega_2 - u)\sigma(\frac{4}{3}\omega\omega_2 - u)\sigma(\frac{4}{3}\omega^2\omega_2 - u)}$$

$$= -\frac{\sigma(\frac{4}{3}\omega_2 + u)\sigma^2(\frac{2}{3}\omega_2 - u)\wp(\frac{2}{3}\omega_2 - u)}{\sigma(\frac{4}{3}\omega_2 - u)\sigma^2(\frac{2}{3}\omega_2 + u)\wp(\frac{2}{3}\omega_2 + u)}$$

by means of (K) § 200, and the relation $\wp\frac{2}{3}\omega_2' = 0$; and this again, by equation (CC) § 279 and by § 293, reduces to

$$X = \frac{\sigma^3(\frac{2}{3}\omega_2 - u)}{\sigma^3(\frac{2}{3}\omega_2 + u)}\frac{\wp(\frac{2}{3}\omega_2 - u)}{\wp(\frac{2}{3}\omega_2 + u)} e^{4\eta_2 u}$$

$$= \frac{\sigma(\frac{2}{3}\omega_2' - ui\surd 3)}{\sigma(\frac{2}{3}\omega_2' + ui\surd 3)} e^{\frac{4}{3}\eta_2' ui\surd 3}$$

$$= \frac{\sigma(\frac{2}{3}\Omega_2 - u;\ 0,\ -\Delta)}{\sigma(\frac{2}{3}\Omega_2 + u;\ 0,\ -\Delta)} e^{\frac{4}{3}\mathrm{H}_2 u}, \quad\text{.....................(99)}$$

The Transformation of the Theta Functions.

303. Taking the θ function, as defined in §§ 263, 265 in the factorial form,

$$\theta(x, q) = \phi(q) \prod_{r=1}^{r=\infty} (1 - 2q^{2r-1}\cos 2x + q^{4r-2}), \quad\text{......(100)}$$

where $\phi(q)$ is a certain function of q which § 264 shows can be written $\qquad \phi(q) = \Pi(1 - q^{2r})$, ...(101)

then changing x into nx, and q into q^n,

$$\theta(nx, q^n) = \phi(q^n)\Pi(1 - 2q^{2nr-n}\cos 2nx + q^{4nr-2n})$$

$$= \phi(q^n) \prod_{r=1}^{r=\infty} \prod_{s=0}^{s=n-1} \{1 - 2q^{2r-1}\cos(2x + 2s\pi/n) + q^{4r-2}\}$$

(by Cotes's Theorem of the Circle of § 270)

$$= \frac{\phi(q^n)}{\{\phi(q)\}^n} \prod_{s=0}^{s=n-1} \theta(x + s\pi/n, q). \quad\text{.....................(102)}$$

Similarly, with $\mu = 1, 2, 3$,

$$\theta_\mu(nx, q^n) = \frac{\phi(q^n)}{\{\phi(q)\}^n} \prod_{s=0}^{s=n-1} \theta_\mu(x + s\pi/n, q). \quad\text{..................(103)}$$

Forming the quotients, and writing x for $\frac{1}{2}\pi u/K$, then (§ 263)

$$\mathrm{sn}\, u = \frac{1}{\surd\kappa}\frac{\theta_1 x}{\theta x}, \quad \mathrm{cn}\, u = \frac{\surd\kappa'}{\surd\kappa}\frac{\theta_2 x}{\theta x}, \quad \mathrm{dn}\, u = \surd\kappa'\frac{\theta_3 x}{\theta x}, \quad (104)$$

and thence we obtain the formulas for the Transformation of the Elliptic Functions of § 283.

Similar considerations will show that, when q is changed to $q^{1/n}$ or $e^{2p\pi i/n}q^{1/n}$,

$$\theta_\mu(x,\, e^{2p\pi i/n}q^{1/n}) = \frac{\phi(e^{2p\pi i/n}q^{1/n})}{\{\phi(q)\}^n} \prod_{s=-\frac{1}{2}(n-1)}^{s=\frac{1}{2}(n-1)} \theta_\mu\left(x + s\frac{2p\pi - i\log q}{n}\right), \quad (105)$$

where $\mu = 0, 1, 2, 3$; this is left as an exercise (Enneper, *Elliptische Functionen*, § 38).

Examples.

1. Prove that a transformation of the fourth order is

$$\frac{1-y}{1+y} = \frac{1-x}{1+x} \cdot \frac{1-\kappa x}{1+\kappa x}\left(\frac{1-x\surd\kappa}{1+x\surd\kappa}\right)^2,$$

and prove that the relation between λ and κ is then

$$1-\lambda^2 = \left(\frac{1-\surd\kappa}{1+\surd\kappa}\right)^4,$$

and

$$M = (1+\surd\kappa)^2.$$

2. Prove that, by means of the substitutions

$$\tan\tfrac{1}{2}\theta = \frac{\cosh\frac{1}{2}u \sinh\phi}{\surd(\cosh u + \sinh u \cosh\phi)},$$

or

$$\sin\tfrac{1}{2}\theta = \frac{\cosh\frac{1}{2}u \sinh\phi}{\sinh\frac{1}{2}u + \cosh\frac{1}{2}u\cosh\phi},$$

$$\int_0^\infty \frac{d\phi}{\surd(\cosh u + \sinh u \cosh\phi)}$$

$$= \frac{1}{\surd 2}\int_0^\pi \frac{d\theta}{\surd(\cosh u + \cos\theta)} = \operatorname{sech}\tfrac{1}{2}u\, F_1(\operatorname{sech}\tfrac{1}{2}u).$$

$$\int_0^\infty \frac{\cosh m\phi\, d\phi}{(\cosh u + \sinh u\cosh\phi)^{n+\frac{1}{2}}}$$

$$= \frac{1\,.\,3\,.\,5\ldots 2m-1}{2n-1\,.\,2n-3\ldots 2n-2m-1}\frac{1}{\surd 2}\int_0^\pi \frac{(\sinh u)^m \cos n\theta\, d\theta}{(\cosh u + \cos\theta)^{n+\frac{1}{2}}}.$$

3. Prove that, with the *homogeneous* variables x_1, x_2 of § 155, and writing X_1 for $\partial X/\partial x_1$, X_2 for $\partial X/\partial x_2$, the general cubic transformation which reduces $dx/\surd X$ to the form

$$dz/\surd(Az^4 + 6Cz^2 + E)$$

is of the form $z = (lX_1 + mX_2)/(l'X_1 + m'X_2)$ (ex. 8, p. 174).

Prove also that the general quartic transformation may be written $\quad z = (lX + mH)/(l'X + m'H)$,

where H denotes the Hessian of the quartic X (§ 75).

(R. Russell, *Proc. L. M. S.*, vol. XVIII.)

4. Prove that (Cayley)

$$y = \frac{\rho x + 7x^3 + 2\rho x^5 + x^7}{1 + 2\rho x^2 + 7x^4 + \rho x^6}$$

satisfies the relation

$$\frac{dy}{\sqrt{(1 - \frac{3}{4}\rho y^2 + y^4)}} = \frac{\rho dx}{\sqrt{(1 + \frac{3}{4}\rho x^2 + x^4)}}.$$

Modular Equations.

304. In the Transformations of the nth order, which connect the Elliptic Functions of modulus λ with those of modulus κ, and make $r = q^n$, or $q^{1/n}$, or $\omega^p q^{1/n}$ (§ 285),

$$\frac{\Lambda' i}{\Lambda} = n\frac{K'i}{K}, \text{ or } \frac{1}{n}\frac{K'i}{K}, \text{ or } \frac{2pK + K'i}{nK}, \text{ or generally } \frac{aK + bK'i}{cK + dK'i}, \quad (106)$$

where

$$bc - ad = n,$$

the Modular Equation, which determines λ in terms of κ, is of the $(n+1)$th order, as already stated, when n is prime, and has two real and $n-1$ imaginary roots.

We shall content ourselves with merely stating the Modular Equations of simple order, connecting κ, λ and κ', λ', adopting the form and classification employed by Mr. R. Russell in the *Proc. London Math. Society*, Vol. XXI.

Class I. $n = 15$, mod. 16;

$P = \sqrt[4]{(\kappa\lambda)} + \sqrt[4]{(\kappa'\lambda')} + 1,$

$Q = \sqrt[4]{(\kappa\lambda\,\kappa'\lambda')} + \sqrt[4]{(\kappa\lambda)} + \sqrt[4]{(\kappa'\lambda')},$

$R = 4\sqrt[4]{(\kappa\lambda\,\kappa'\lambda')}.$

$n = 15, \quad P^3 - 4PQ + R = 0.$

$n = 31, \quad (P^2 - 4Q)^2 - PR = 0.$

$n = 47, \quad P^2 - 4Q - P(R)^{\frac{1}{3}} - 2(R)^{\frac{2}{3}} = 0.$

Class II. $n = 7$, mod. 16;

$P = \sqrt[4]{(\kappa\lambda)} + \sqrt[4]{(\kappa'\lambda')} - 1,$

$Q = \sqrt[4]{(\kappa\lambda\,\kappa'\lambda')} - \sqrt{(\kappa\lambda)} - \sqrt[4]{(\kappa'\lambda')},$

$R = -4\sqrt[4]{(\kappa\lambda\,\kappa'\lambda')}.$

$n = 7, \quad P = 0, \quad$ or $\sqrt[4]{(\kappa\lambda)} + \sqrt[4]{(\kappa'\lambda')} = 1,$ (Guetzlaff).

$n = 23, \quad P - R^{\frac{1}{3}} = 0,$ or $\sqrt[4]{(\kappa\lambda)} + \sqrt[4]{(\kappa'\lambda')} + (256\kappa\lambda\,\kappa'\lambda')^{\frac{1}{12}} = 1.$

$n = 71, \quad P^3 - 4R^{\frac{1}{3}}(P^2 - Q) + 2PR^{\frac{2}{3}} - R = 0.$

$n = 119, \quad P^6 - R^{\frac{1}{3}}(7P^5 - 28P^3Q + 16PQ^2) + R^{\frac{2}{3}}(\ldots)\ldots = 0.$

CLASS III. $n=3$, mod. 8;

$P=\sqrt{(\kappa\lambda)}+\sqrt{(\kappa'\lambda')}-1$,

$Q=\sqrt{(\kappa\lambda\,\kappa'\lambda')}-\sqrt{(\kappa\lambda)}-\sqrt{(\kappa'\lambda')}$,

$R=-16\sqrt{(\kappa\lambda\,\kappa'\lambda')}$.

$n=3$, $P=0$, or $\sqrt{(\kappa\lambda)}+\sqrt{(\kappa'\lambda')}=1$, (Legendre).

$n=11$, $P-R^{\frac{1}{3}}=0$, or $\sqrt{(\kappa\lambda)}+\sqrt{(\kappa'\lambda')}+(256\kappa\lambda\,\kappa'\lambda')^{\frac{1}{6}}=1$.

$n=19$, $P^5-7P^2R+16QR=0$.

$n=35$, $P^4-R^{\frac{1}{3}}(5P^3-16PQ)+2R^{\frac{2}{3}}P^2-RP-R^{\frac{4}{3}}=0$.

$n=43$, $P^{11}+\ldots=0$.

$n=59$, $P^5+\ldots=0$.

$n=83$, $P^7+\ldots=0$.

CLASS IV. $n=1$, mod. 4;

$P=\kappa\lambda+\kappa'\lambda'-1$,

$Q=\kappa\lambda\,\kappa'\lambda'-\kappa\lambda-\kappa'\lambda'$,

$R=-32\,\kappa\lambda\,\kappa'\lambda'$.

$n=1$, $P=0$.

$n=9$, $P^6-14P^3R+64PQR-3R^2=0$.

$n=17$, $P^3-R^{\frac{1}{3}}(10P^2-64Q)+26R^{\frac{2}{3}}P+12R=0$.

$n=41$, ...

$n=5$, $P-R^{\frac{1}{3}}=0$, or $\kappa\lambda+\kappa'\lambda'+(32\kappa\lambda\,\kappa'\lambda')^{\frac{1}{3}}=1$.

$n=13$, $P^{\frac{1}{2}}(P^3+8R)\pm R^{\frac{1}{2}}(11P^2-64Q)=0$.

$n=29$, $P^{\frac{1}{2}}(P^2+17R^{\frac{1}{3}}P-9R^{\frac{2}{3}})$
$\pm R^{\frac{1}{6}}(9P^2-64Q-13R^{\frac{1}{3}}P+15R^{\frac{2}{3}})=0$.

$n=37$, ...

$n=53$, $P^{\frac{1}{2}}\{P^4+R^{\frac{1}{3}}(413P^3-2^{16}PQ)+\ldots\}\pm R^{\frac{1}{6}}\{35P^4\ldots\}=0$.

305. According to Professor Klein (*Proc. L. M. S.*, X.; *Math. Ann.*, XIV.) these Modular Equations are replaced by relations between the absolute invariant J and its transformed value J', by the intermediate of quantities τ and τ', such that J is a certain function of τ, and J' the same function of τ'; and now,

$n=2$; $J:J-1:1=(4\tau-1)^3:(\tau-1)(8\tau+1)^2:27\tau$,
$\tau\tau'=1$ (§ 60).

$n=3$; $J:J-1:1=(\tau-1)(9\tau-1)^3:(27\tau^2-18\tau-1)^2:-64\tau$,
$\tau\tau'=1$.

$n=4$; $J:J-1:1=(\tau^2+14\tau+1)^3:$
$(\tau^3-33\tau^2-33\tau+1)^2:108\tau(1-\tau)^4$,
$\tau+\tau'=1$.

$n=5$; $J:J-1:1=(\tau^2-10\tau+5)^3$
$:(\tau^2-22\tau+125)(\tau^2-4\tau-1)^2:-1728\tau,$
$\tau\tau'=125.$

$n=7$; $J:J-1:1=(\tau^2+13\tau+49)(\tau^2+5\tau+1)^3$
$:(\tau^4+14\tau^3+63\tau^2+70\tau-7)^2:1728\tau,$
$\tau\tau'=49.$

$n=13$; $J:J-1:1=(\tau^2+5\tau+13)(\tau^4+7\tau^3+20\tau^2+19\tau+1)^3$
$:(\tau^2+6\tau+13)(\tau^6+10\tau^5+46\tau^4+108\tau^3+122\tau^2+38\tau-1)^2:1728\tau,$
$\tau\tau'=13.$

The Multiplication of Elliptic Functions.

306. If we perform the second real transformation upon the first real transformation, we obtain a transformation of the order n^2, leading back again to the original modulus κ; because the first real transformation changes q into q^n, and the second real transformation changes q^n back again to q.

We then obtain the elliptic functions of argument

$$u/MM'=nu, \quad \text{since } M=K/n-\Lambda, \quad M'=\Lambda/K,$$

in terms of the elliptic functions of argument u, by a transformation of the order n^2, and thus obtain the formulas for Multiplication of the argument.

Thus multiplication by 2 or 3 can be obtained by two successive transformations of the second or third order; and so on.

Knowing that the order of the transformation is n^2, we infer in Abel's manner the factors of the numerator and denominator of the transformation, involving the *modular functions*, the elliptic functions of the nth part of the periods.

Thus we infer, with the notation of § 258, that, for an odd value of n,

$$\text{sn}\, nu=U/V, \quad \ldots\ldots\ldots\ldots(107)$$

where
$$U=n\,\text{sn}\,u\,\Pi'\Pi'\left(1-\frac{\text{sn}^2 u}{\text{sn}^2\Omega/n}\right),$$

$$V=\Pi'\Pi'\left(1-\frac{\text{sn}^2 u}{\text{sn}^2\Omega_3/n}\right)$$

$$=\Pi'\Pi'(1-\kappa^2\text{sn}^2 u\,\text{sn}^2\Omega/n),$$

where $m, m'=0, \pm 1, \pm 2, \pm 3, \ldots, \pm\frac{1}{2}(n-1)$;

the simultaneous zero values of m and m' being excluded, as denoted by the accents, so that the number of factors is $\frac{1}{2}(n^2-1)$.

Combining the factors by formula (7) of § 137,

$$\text{sn}\, nu = A\, \text{sn}\, u\, \Pi'\Pi' \text{sn}(u+\Omega/n)\text{sn}(u-\Omega/n), \quad (108)$$

where A is a constant factor; and this may be written

$$\text{sn}\, nu = A\, \Pi\Pi\, \text{sn}(u+\Omega/n)\,; \quad \text{.....................}(109)$$

where $m, m' = 0, \pm 1, \pm 2, \ldots, \pm\frac{1}{2}(n-1)$;

the simultaneous zero values of m and m' being now admissible.

Similar considerations will show that

$$\text{cn}\, nu = B\, \Pi\Pi\, \text{cn}(u+\Omega/n), \quad \text{......................}(110)$$

$$\text{dn}\, nu = C\, \Pi\Pi\, \text{dn}(u+\Omega/n). \quad \text{......................}(111)$$

To determine the constant factors, change u into $u+K$ or $u+K'i$, when we shall find (Cayley, *Elliptic Functions*, § 368)

$$A = (-1)^{\frac{1}{2}(n-1)}\kappa^{\frac{1}{2}(n^2-1)}, \quad B = (\kappa/\kappa')^{\frac{1}{2}(n^2-1)}, \quad C = (1/\kappa')^{\frac{1}{2}(n^2-1)}.$$

By taking in § 259 a rectangle $OA_nB_nC_n$, in which $OA_n = na$, $OB_n = nb$, and therefore containing n^2 elementary rectangles, we obtain a physical representation of the formulas (109), (110), (111) for Multiplication of the argument by n.

Writing u/n for u, and making n indefinitely great, we deduce in a rigorous manner the doubly factorial expressions for sn u, cn u, dn u in (1), (2), (3) of § 258.

Again, by putting $\kappa=0$ or $\kappa=1$, the student may deduce as an exercise the trigonometrical formulas for the resolution of the circular and hyperbolic functions into factors.

(Hobson, *Trigonometry*, Chap. XVII.)

The Complex Multiplication of Elliptic Functions.

307. When $K'/K = \surd D$, and D is an integer, we may suppose the multiplier n resolved, by the solution of the *Pellian equation*, into two complementary imaginary factors, so that

$$n = (a+ib\surd D)(a-ib\surd D) = a^2+b^2D\,;$$

and now the multiplication by n can be effected by two successive multiplications by the complex multipliers $a+ib\surd D$ and $a-ib\surd D$, each leading to an imaginary transformation of the nth order, not changing q or the modulus κ.

(Abel, *Œuvres*, I., p. 377; Jacobi, *Werke*, I., p. 489.)

The first requirement then in Complex Multiplication is a knowledge of the value of κ for which $K'/K = \surd D$; and this is found by putting $\kappa = \lambda'$, $\kappa' = \lambda$ in the corresponding Modular Equation of the order D (§ 304).

The equation is now, according to Abel, always solvable algebraically by radicals; so that, returning to the question of the pendulum in § 15, it is possible to determine by a geometrical construction the position of two horizontal BB', bb', as in fig. 1, cutting off arcs below them, such that the period of swing from B to B' is $\sqrt{D}$ times the period from b to b'.

Thus the Modular Equation of the second order being written

$$\lambda=(1-\kappa')/(1+\kappa'),$$

we find, on putting $\kappa'=\lambda$,

$$\lambda^2+2\lambda=1, \text{ or } \lambda=\sqrt{2}-1, \text{ when } \Lambda'/\Lambda=\sqrt{2}.$$

Putting $\kappa=\lambda'$, $\kappa'=\lambda$ in the Modular Equation of the third order (§ 304),

$$2\sqrt{(\kappa\kappa')}=1, \text{ or } 2\kappa\kappa'=\tfrac{1}{2}=\sin\tfrac{1}{6}\pi, \text{ when } K'/K=\sqrt{3};$$

so that the modular angle is $\frac{1}{12}\pi$ or 15°.

When $K'/K=2$, $\kappa=(\sqrt{2}-1)^2$ (§ 71);

obtained by putting $\Gamma'/\Gamma=1$, $\gamma=\gamma'=\frac{1}{2}\sqrt{2}$ in §§ 298, 299.

When $K'/K=\sqrt{5}$, $2\kappa\kappa'=\sqrt{5}-2$, $\sqrt[3]{(2\kappa\kappa')}=\frac{1}{2}(\sqrt{5}-1)$,

or

$$(2\kappa\kappa')^{-\frac{1}{3}}-(2\kappa\kappa')^{\frac{1}{3}}=1.$$

When $K'/K=\sqrt{7}$, $2\sqrt[4]{(\kappa\kappa')}=1$, $2\kappa\kappa'=\frac{1}{8}$, $\sqrt[3]{(2\kappa\kappa')}=\frac{1}{2}$.

Collections of these *singular moduli* required in Complex Multiplication are given by Kronecker in the *Berlin Sitz.*, 1857, 1862, in the *Proc. L. M. S.*, XIX., p. 301; also by Kiepert in the *Math. Ann.*, XXVI., XXXIX., and by H. Weber in his *Elliptische Functionen*, 1891.

308. In the expression of $y=\text{sn}(a+ib\sqrt{D})u$ as a rational function of $x=\text{sn}\,u$, leading to the differential relation

$$\frac{Mdy}{\sqrt{(1-y^2\,.\,1-\kappa^2y^2)}}=\frac{dx}{\sqrt{(1-x^2\,.\,1-\kappa^2x^2)}}, \text{ where } 1/M=a+ib\sqrt{D},$$

Jacobi finds (*Werke*, t. I.; *de multiplicatione functionum ellipticarum per quantitatem imaginariam pro certo quodam modulorum systemate*) that we must restrict a to be an odd integer, and b to be an even integer; but these restrictions disappear if we work with the cn functions; and we can even suppose that $2a$ and $2b$ are odd integers.

Let us determine then the relations connecting

$$x=\text{cn}\,u \quad\text{and}\quad y=\text{cn}\,\tfrac{1}{2}(-1+i\sqrt{D})u,$$

so that

$$1/M=-\tfrac{1}{2}+\tfrac{1}{2}i\sqrt{D},$$

leading to the differential relation

$$\frac{dy}{\sqrt{(1-y^2\,.\,y^2+c^2)}}=\frac{(-\frac{1}{2}+\frac{1}{2}i\sqrt{D})dx}{\sqrt{(1-x^2\,.\,x^2+c^2)}},$$

where $c=\kappa'/\kappa$, the cotangent of the modular angle.

If $D=4n-1$, and we denote $(K+K'i)/n$ by ω, we shall then find that, when n is odd,

$$\frac{1-y}{1-\dfrac{y}{ic}}=-\sqrt{(ic)}\frac{1-x}{1+x}\prod_{r=1}^{r=\frac{1}{2}(n-1)}\left(\frac{x-\text{cn}\,2r\omega}{x+\text{cn}\,2r\omega}\right)^2,$$

$$\frac{1+y}{1+\dfrac{y}{ic}}=\sqrt{(ic)}\frac{1+\dfrac{x}{ic}}{1-\dfrac{x}{ic}}\prod\left\{\frac{x-\text{cn}(2r-1)\omega}{x+\text{cn}(2r-1)\omega}\right\}^2;\ \ldots.(112)$$

but, when n is even,

$$\frac{1-y}{1-\dfrac{y}{ic}}=-\sqrt{(ic)}\frac{1-x}{1+x}\,\frac{1+\dfrac{x}{ic}}{1-\dfrac{x}{ic}}\prod_{r=1}^{r=\frac{1}{2}n-1}\left(\frac{x-\text{cn}\,2r\omega}{x+\text{cn}\,2r\omega}\right)^2$$

$$\frac{1+y}{1+\dfrac{y}{ic}}=\sqrt{(ic)}\prod\left\{\frac{x-\text{cn}(2r-1)\omega}{x+\text{cn}(2r-1)\omega}\right\}^2.\ \ldots\ldots\ldots\ldots\ldots\ldots(113)$$

The arithmetical verification for the simple cases of $D=3$, 7, or 15 is left as an exercise for the student (*Proc. Cam. Phil. Society*, Vol. V.).

Formulas (112) and (113) are inferred by putting

(1) $y=1$,

when $\frac{1}{2}(-1+i\sqrt{D})u=2mK+2m'K'i$ $(m+m'$ even);

and then $u=4m'K-(m+m')\omega,\quad x=\text{cn}\,2r\omega.$

(2) $y=-1$,

$\frac{1}{2}(-1+i\sqrt{D})u=2mK+2m'K'i$ $(m+m'$ odd);

and then $x=\text{cn}(2r-1)\omega.$

(3) $y=ic$,

$\frac{1}{2}(-1+i\sqrt{D})u=(2m+1)K+(2m'+1)K'i$ $(m+m'$ odd);

$u=(4m'+2)K-(m+m'+1)\omega,\ x=-\text{cn}\,2r\omega.$

(4) $y=-ic$,

$\frac{1}{2}(-1+i\sqrt{D})u=(2m+1)K+(2m'+1)K'i$ $(m+m'$ even);

and then $x=-\text{cn}(2r-1)\omega.$

309. When $D=4n+1$ or 1, mod. 4, the relation connecting $x=\text{cn}\,u$ and $y=\text{cn}\,\frac{1}{2}(-1+i\sqrt{D})u$ cannot be rational; but Mr. G. H. Stuart has shown (*Q. J. M.*, Vol. XX.) that it may be written in the irrational form

$$y=\sqrt{(ic)}\sqrt{\left(\frac{ic+x}{ic-x}\right)}\prod_{r=1}^{r=n}\frac{\text{cn}(2r-1)\omega-x}{\text{cn}(2r-1)\omega+x},$$

where $\omega=(K+K'i)/(2n+1)$,

a transformation of the order $n+\frac{1}{2}$; and this is equivalent to

$$1-y^2=(1-ic)(1-x)\Pi\left(1-\frac{x}{\text{cn }2r\omega}\right)^2\div V,$$

$$1+\frac{y^2}{c^2}=\left(1+\frac{i}{c}\right)(1+x)\Pi\left(1+\frac{x}{\text{cn }2r\omega}\right)^2\div V,$$

$$V=\left(1-\frac{x}{ic}\right)\Pi\left\{1+\frac{x}{\text{cn}(2r-1)\omega}\right\}^2; \quad \text{......(114)}$$

this is inferred in the same manner as formulas (111) and (112).

For instance, with $n=0$, $D=1$, and $\kappa=\frac{1}{2}\surd 2$, $c=1$;

$$\text{cn }\tfrac{1}{2}(-1+i)u=\surd(i)\sqrt{\left(\frac{i+\text{cn }u}{i-\text{cn }u}\right)}$$

equivalent to, with $u=(1+i)v$,

$$\text{cn}(1-i)v=i\frac{1-i\text{ cn}^2 v}{1+i\text{ cn}^2 v}.$$

With $n=1$, $D=5$, $2\kappa\kappa'=\surd 5-2$, $c=\surd 5+2+2\surd(\surd 5+2)$,

$$\surd c=\frac{\surd 5+1}{2}+\sqrt{\frac{\surd 5+1}{2}};$$

and $$\text{cn }\tfrac{1}{2}(-1+i\surd 5)u=\surd(ic)\sqrt{\left(\frac{1+\frac{x}{ic}}{1-\frac{x}{ic}}\right)\frac{1-\frac{x}{a}}{1+\frac{x}{a}}},$$

where $a=\text{cn }\frac{1}{3}(K+K'i)$.

310. Generally in the expression of $y=\wp u/M$ as a function of $x=\wp u$, where

$$\omega'/\omega \text{ or } K'i/K=\surd(-D),$$

and the multiplier $1/M$ is complex, of the form

$$1/M=a+b\surd(-D),$$

it is convenient to consider four classes of D.

Class A, $D\equiv 3$, mod. 8;
Class B, $D\equiv 7$, mod. 8;
Class C, $D\equiv 1$, mod. 4;
Class D, $D\equiv 2$, mod. 4;

the class for $D\equiv 0$, mod. 4, not requiring separate consideration.

It is convenient also to consider the discriminant D (§ 53) as *negative*; a change to a positive discriminant being effected by the method of § 59; now $\omega'_2/\omega_2=i\surd D$.

We can also *normalize* the integrals (§§ 196, 252) by taking $g_2{}^3-27g_3{}^2=-1$, so that $g_2=\sqrt[3]{(-J)}$.

CLASS A. $D \equiv 3$, mod. $8 = 8p + 3$ or $4n - 1$, if $n = 2p + 1$.

$$1/M = \tfrac{1}{2}(-1 + i\sqrt{\Delta}).$$

The relation connecting x and y can be written in one of the three equivalent forms

$$y - e_1 = M^2(x - e_2) \prod_{r=1}^{r=p} \{x - \wp(\omega_2 + 2r\omega_3/n)\}^2 \div V,$$
$$y - e_2 = M^2(x - e_3) \Pi \{x - \wp(\omega_3 - 2r\omega_3/n)\}^2 \div V,$$
$$y - e_3 = M^2(x - e_1) \Pi \{x - \wp(\omega_1 + 2r\omega_3/n)\}^2 \div V,$$
$$V = \Pi \{x - \wp(2r\omega_3/n)\};$$

leading to the differential relation

$$\frac{Mdy}{\sqrt{(4y^3 - g_2 y - g_3)}} = \frac{dx}{\sqrt{(4x^3 - g_2 x - g_3)}}.$$

This verifies in the particular case of $p = 0$, when

$$D = 3, \quad J = 0, \quad g_2 = 0, \quad 1/M = \tfrac{1}{2}(-1 + i\sqrt{3}) = m;$$

and then $\qquad e_1 = me_2, \quad e_3 = m^2 e_2.$

This is the simplest case of Complex Multiplication, mentioned in § 196, and employed in § 227 in the determination of the Trajectory for the cubic law of resistance.

The form of the general transformation is inferred from the consideration of the series of values of u which make

$$y \text{ or } \wp(u/M) = e_1, e_2, e_3, \text{ and } \infty.$$

(i.) When $y = e_1$,

$$\begin{aligned}
u/M &= (2q+1)\omega_1 + 2r\omega_3 \\
&= (q + \tfrac{1}{2})(\omega_2 - \omega_2') + r(\omega_2 + \omega_2') \\
&= (q + r + \tfrac{1}{2})\omega_2 - (q - r + \tfrac{1}{2})\omega_2'; \\
u &= \{(q + r + \tfrac{1}{2})\omega_2 - (q - r + \tfrac{1}{2})\omega_2'\}/(-\tfrac{1}{2} + \tfrac{1}{2}i\sqrt{D}) \\
&= \frac{\omega_2}{2n}\{(q + r + \tfrac{1}{2}) - (q - r + \tfrac{1}{2})i\sqrt{D}\}(-1 - i\sqrt{D}) \\
&= \frac{-q - r - \tfrac{1}{2} - (q - r + \tfrac{1}{2})(4n - 1)}{2n}\omega_2 - \frac{q + r + \tfrac{1}{2} - q + r - \tfrac{1}{2}}{2n}\omega_2' \\
&= -2q\omega_2 + 2r\omega_2 - \omega_2 - r(\omega_2 + \omega_2')/n \\
&= -2q\omega_2 + 2r\omega_2 - \omega_2 - 2r\omega_3/n,
\end{aligned}$$

so that x or $\wp u = e_1$ or $\wp(\omega_2 + 2r\omega_3/n)$.

(ii.) When $y = e_2$,

$$\begin{aligned}
u/M &= (2q+1)\omega_1 + (2r+1)\omega_3 \\
&= (q + r + 1)\omega_2 - (q - r)\omega_2', \\
u &= -2q\omega_2 + 2r\omega_2 - (2r+1)\omega_3/n, \\
\wp u &= e_3, \text{ or } \wp(2r+1)\omega_3/n = \wp(\omega_3 - 2r\omega_3/n).
\end{aligned}$$

(iii.) When $y=e_3$,

$$u/M=2q\omega_1+(2r+1)\omega_3$$
$$=(q+r+\tfrac{1}{2})\omega_2-(q-r-\tfrac{1}{2})\omega_2',$$
$$u=-2q\omega_2+2r\omega_2-\omega_2-(2r-1)\omega_3/n,$$
$$\wp u=e_1, \text{ or } \wp\{\omega_2+(2r-1)\omega_3/n\} \text{ or } \wp(\omega_1+2r\omega_3/n).$$

(iv.) When $y=\infty$,

$$u/M=2q\omega_1+2r\omega_3$$
$$=(q+r)\omega_2-(q-r)\omega_2',$$
$$u=-2q\omega_2+2r\omega_2+2r\omega_3/n,$$

and $$\wp u=\wp(2r\omega_3/n).$$

Hence the form of the Transformation is inferred.

By addition, we find

$$y=M^2\frac{x^n-A_1x^{n-1}+A_2x^{n-2}\ldots}{(x^p-G_1x^{p-1}+G_2x^{p-2}\ldots)^2},$$

where $n=2p+1$; and we shall find that $A_1=2G_1$; and the A's and G's are symmetrical functions of e_1, e_2, e_3, and therefore functions of g_2, g_3 or J; while G_1 has the same signification as in § 293.

By employing the Modular Equations given above, or employing Hermite's results (*Theorie des equations modulaires*), we find

$D=3$, $J=0$, $g_2=0$, $\surd(g_2+1)=1$, $g_3=\frac{1}{9}\surd 3$.

$D=11$, $J=-\dfrac{2^9}{3^3}$, $g_2=\dfrac{8}{3}$, $g_3=\dfrac{7\surd 11}{27}$;

$A_1=2G_1=-\frac{1}{3}(\surd 11+i)$, $A_2=\dfrac{11+7i\surd 11}{18}$, $A_3=-\dfrac{\surd 11+14i}{27}$.

$D=19$, $J=-2^9$, $g_2=8$, $\surd(g_2+1)=3$, $g_3=\surd 19$;

$A_1=2G_1=-\surd 19-i$, $A_2=\frac{1}{2}(25+5i\surd 19)$, $A_3=-\frac{1}{2}(\surd 19+6i)$,

$A_4=\frac{1}{2}(21+9i\surd 19)$, $A_5=-\frac{1}{2}(\surd 19+11i)$;

these values of A_2, A_3, A_4, A_5 were calculated by Rev. J. Chevallier, Fellow of New College, Oxford, who has also verified the case of $D=11$.

$D=27$, $J=-2^9\times 5^3\div 3^2$, etc.

$D=35$, $g_2=\frac{8}{3}\surd 5\{\frac{1}{2}(\surd 5+1)\}^4$, $g_2+1=\frac{7}{3}\{\frac{1}{2}(\surd 5+1)\}^6$.

$D=43$, $J=-2^{12}\times 5^3$, $g_2=80$, $\surd(g_2+1)=3^2$,

$g_3=3\times 7\times\surd 43$ (Hermite).

$A_1=2G_1=-6(\surd 43+i)$, $G_2=\frac{1}{2}(279+11i\surd 43)$,

$A_2=1051+73i\surd 43$, etc.

$D=51$, $J=-64(5+\surd 17)^3(\surd 17+4)^2$ (Kiepert).

$D=67,\quad J=-2^9\times 5^3\times 11^3,\quad g_2=440,\quad \surd(g_2+1)=3\times 7,$
$g_3=7\times 31\times\surd 67$ (Hermite).

$D=163,\quad J=-2^{12}\times 5^3\times 23^3\times 29^3,\quad \surd(g_2+1)=3\times 7\times 11,$
$g_3=7\times 11\times 19\times 127\times\surd 163$ (Hermite).

CLASS B. $D\equiv 7$, mod. $8=8p+7=4n-1$, if $n=2p+2$.

The relations connecting $y=\wp(u/M)$ and $x=\wp u$, where

$$1/M=-\tfrac{1}{2}+\tfrac{1}{2}i\surd D,$$

are found, in a manner similar to that employed in Class A;

$$y-e_1=M^2(x-e_1)(x-e_2)\prod_{r=1}^{r=p}\{x-\wp(\omega_2+2r\omega_3/n)\}^2\div V,$$

$$y-e_2=\qquad M^2\prod_{r=0}^{r=p}\{x-\wp(\omega_3-2r\omega_3/n)\}^2\div V,$$

$$y-e_3=\qquad M^2\prod_{r=0}^{r=p}\{x-\wp(\omega_1+2r\omega_3/n)\}^2\div V,$$

$$V=\qquad (x-e_3)\prod_{r=1}^{r=p}\{x-\wp(2r\omega_3/n)\}^2.$$

As simple numerical applications,

$$D=7,\quad 2\kappa\kappa'=\frac{1}{8},\quad J=-\frac{5^3}{2^6},\quad g_2=\frac{5}{4},\quad g_2=\frac{\surd 7}{8},$$

$$e_1=\tfrac{1}{8}(-\surd 7+i),\quad e_2=\tfrac{1}{4}\surd 7,\quad e_3=\tfrac{1}{8}(-\surd 7-i).$$

$D=15,\qquad \surd\kappa\kappa'=\sin 18^\circ$ (Joubert).

In these cases the Jacobian notation is almost more simple, as given in § 308.

CLASS C. $D\equiv 1$, mod. $4=4n+1$.

The relations connecting $x=\wp u$ and $y=\wp(u/M)$, where

$$1/M=-\tfrac{1}{2}+\tfrac{1}{2}i\surd D,$$

cannot now be rational; but, according to Mr. G. H. Stuart, we can express the relations in the irrational form

$$\frac{y-\wp\,\tfrac{1}{2}\omega_2'}{y-\wp\,\tfrac{1}{2}\omega_2}=\left(\frac{x-e_1}{x-e_3}\right)^{\frac{1}{2}}\prod_{r=1}^{r=n}\frac{x-\wp\left(\omega_2-\dfrac{4r+1}{2n+1}\omega_3\right)}{x-\wp\left(\dfrac{4r+1}{2n+1}\omega_3\right)},$$

a relation which may be said to be of the order $n+\tfrac{1}{2}$; and this is equivalent to

$$\frac{(y-e_1)(y-e_3)}{y-e_2}=M^2(x-e_2)\prod_{n=1}^{r=n}\left\{\frac{x-\wp\left(\dfrac{4r}{2n+1}\omega_3\right)}{x-\wp\left(\dfrac{4r+1}{2n+1}\omega_3\right)}\right\}^2.$$

CLASS D. *D* an even number.

In this class the simplest function to employ is the sn function; for instance, with

$$K'/K=\sqrt{2}, \text{ then } \kappa=\sqrt{2}-1;$$

and
$$\operatorname{sn}(1+i\sqrt{2})u=(1+i\sqrt{2})\operatorname{sn} u\frac{1-\dfrac{\operatorname{sn}^2 u}{\operatorname{sn}^2 2\omega}}{1-\kappa^2\operatorname{sn}^2 2\omega \operatorname{sn}^2 u},$$

where
$$\omega=\tfrac{1}{3}(K-iK');$$

leading to the equations

$$\frac{1-y}{1+y}=\frac{1+\kappa x}{1-\kappa x}\left(\frac{1-\dfrac{x}{\operatorname{sn}\omega}}{1+\dfrac{x}{\operatorname{sn}\omega}}\right)^2,$$

$$\frac{1-\kappa y}{1+\kappa y}=\frac{1-x}{1+x}\left(\frac{1+\kappa x\operatorname{sn}\omega}{1-\kappa x\operatorname{sn}\omega}\right)^2,$$

connecting
$$x=\operatorname{sn} u \quad\text{and}\quad y=\operatorname{sn}(1+i\sqrt{2})u.$$

Also
$$\operatorname{sn}\omega=\sqrt{(-i)}, \quad \operatorname{sn}^2 2\omega=\frac{\sqrt{2}-i}{\sqrt{2}-1}.$$

These transformations show that it is not possible to express $\operatorname{cn}(1+i\sqrt{2})u$ in terms of $\operatorname{cn} u$, or $\operatorname{dn}(1+i\sqrt{2})u$ in terms of u, by a rational transformation.

With $K'/K=2$, then $\kappa=(\sqrt{2}-1)^2$ (§ 71),
and the relation connecting $x=\operatorname{sn} u$ and $y=\operatorname{sn}(1+2i)u$ may be written

$$y=(1+2i)x\frac{\left(1-\dfrac{x^2}{\operatorname{sn}^2 2\omega}\right)\left(1-\dfrac{x^2}{\operatorname{sn}^2 4\omega}\right)}{(1-\kappa^2x^2\operatorname{sn}^2 2\omega)(1-\kappa^2x^2\operatorname{sn}^2 4\omega)},$$

where
$$\omega=\tfrac{1}{5}(K-iK');$$

equivalent to the relations

$$\frac{1-y}{1+y}=\frac{1-\kappa x}{1+\kappa x}\left(\frac{1-\dfrac{x}{\operatorname{sn}\omega}}{1+\dfrac{x}{\operatorname{sn}\omega}}\right)^2\left(\frac{1+\dfrac{x}{\operatorname{sn}3\omega}}{1-\dfrac{x}{\operatorname{sn}3\omega}}\right)^2,$$

$$\frac{1-\kappa y}{1+\kappa y}=\frac{1-x}{1+x}\left(\frac{1-\dfrac{x}{\operatorname{sn}2\omega}}{1+\dfrac{x}{\operatorname{sn}2\omega}}\right)^2\left(\frac{1+\dfrac{x}{\operatorname{sn}4\omega}}{1-\dfrac{x}{\operatorname{sn}4\omega}}\right)^2;$$

so that $\operatorname{cn}(1+2i)u$ has a factor $\operatorname{dn} u$, and $\operatorname{dn}(1+2i)u$ has a factor $\operatorname{cn} u$.

When $K'/K=\sqrt{6}$, then $\kappa=(\sqrt{3}-\sqrt{2})(2-\sqrt{3})$;
and the corresponding relation between $\operatorname{sn} u$ and $\operatorname{sn}(1+i\sqrt{6})u$ to be written down is left as an exercise.

(*Proc. Cam. Phil. Soc.*, Vols. IV., V.)

It can also be shown, in the preceding manner, that the relation connecting $x=\wp u$ and $y=\wp(u/M)$ where

$$1/M=-1+i\sqrt{D},$$

and D is an even number $2m$, can be expressed by the relations

$$y-e_1=M^2(x-e_2)\prod_{r=1}^{r=m}\left\{x-\wp\left(\frac{2r-1}{2m+1}\omega_2'\right)\right\}^2 \div V,$$

$$y-e_2=M^2(x-e_1)\,\Pi\left\{x-\wp\left(\omega_1-\frac{2r}{2m+1}\omega_2'\right)\right\}^2\div V,$$

$$y-e_3=M^2(x-e_3)\,\Pi\left\{x-\wp\left(\omega_1-\frac{2r-1}{2m+1}\omega_2'\right)\right\}^2\div V,$$

$$V=\quad\Pi\left\{x-\wp\left(\frac{2r}{2m+1}\omega_2'\right)\right\}^2.$$

As numerical exercises, we may take

(i.) $D=2$, when $g_2=30$, $g_3=28$, $G_1=-1+\frac{1}{2}i\sqrt{2}$;

(ii.) $D=4$, when $g_2=11$, $g_3=7$, $G_1=-2+i$.

311. In conclusion we may quote from Schwarz some general remarks on doubly periodic functions.

Every analytic function ϕu of a single variable u for which an algebraical relation connects $\phi(u+v)$ with ϕu and ϕv is said to have an Algebraical Addition Theorem; and then $\phi' u$ must be an algebraical function of ϕu (Chap. V.).

Every such function is then an algebraical function, or an exponential function (circular or hyperbolic function), or an elliptic function, which can be expressed rationally by $\wp u$ and $\wp' u$ (Chap. VII.).

Elliptic functions are *doubly periodic*. A function of a single variable cannot have more than two distinct periods, one real and one imaginary, or both complex. For if a third period was possible, the three sets of period parallelograms obtained by taking the periods in pairs would reach every point of the plane, so that the function would have the same value at all points of the plane, and would therefore reduce to a *constant* (Bertrand, *Calcul intégral*, p. 602).

Abel, in generalising these theorems, was led to the discovery of the hyperelliptic and Abelian functions.

Thus if X in § 169 is of the fifth or sixth degree, we obtain functions of 2 variables and 4 periods; if of the 7th or 8th degree, of 3 variables and 6 periods; and generally, if X is of the degree $2p+1$ or $2p+2$, there are p variables and $2p$ periods; but this would lead us beyond the scope of the present treatise, and the reader who wishes to follow up this development is recommended to study Professor Klein's articles "*Hyperelliptische Sigmafunctionen*," *Math. Ann.*, XXVII., XXXIII., etc.

APPENDIX.

I. *The Apsidal Angle in the small oscillations of a Top.*

The expression given by Bravais in Note VII. of Lagrange's *Mécanique analytique*, t. II., p. 352, for the apsidal angle in the small oscillations of a Spherical Pendulum about its lowest position is readily extended to the more general case of the Top or Gyrostat, if we employ the expression on p. 261, § 242, as the basis of our approximation.

We divide the apsidal angle Ψ into two parts, Ψ_1 and Ψ_2, such that

$$i\Psi_1 = a\eta_1 - \omega_1 \zeta a,$$
$$i\Psi_2 = b\eta_1 - \omega_1 \zeta b\,;$$

and now put $\quad a = \omega_3 - s\omega_3, \quad b = \omega_1 + q\omega_3,$

where q and s are small numbers; so that, expanding by Taylor's Theorem as far as the first powers of q and s, we may put

$$\zeta a \approx \eta_3 + s\omega_3 \wp\omega_3 = \eta_3 + s\omega_3 e_3,$$
$$\zeta b \approx \eta_1 - q\omega_3 \wp\omega_1 = \eta_1 - q\omega_3 e_1\,;$$

and now, by means of Legendre's relation of p. 209,

$$i\Psi_1 \approx (\omega_3 - s\omega_3)\eta_1 - \omega_1(\eta_3 + s\omega_3 e_3) = \tfrac{1}{2} i\pi - s\omega_3(\eta_1 + e_3\omega_1),$$
$$i\Psi_2 \approx (\omega_1 + q\omega_3)\eta_1 - \omega_1(\eta_1 - q\omega_3 e_1) = \qquad q\omega_3(\eta_1 + e_1\omega_1).$$

But, from equation (B), § 51,

$$\mathrm{dn}^2\sqrt{(e_1 - e_3)}u = \frac{\wp u - e_2}{\wp u - e_3} = 1 - \frac{e_2 - e_3}{\wp u - e_3}$$
$$= 1 - \frac{\wp(u + \omega_3) - e_3}{e_1 - e_3} = \frac{e_1 - \wp(u + \omega_3)}{e_1 - e_3}\,;$$

so that, integrating between the limits 0 and ω_1,

$$e_1\omega_1 + \zeta(\omega_1 + \omega_3) - \zeta\omega_3 = (e_1 - e_3)\int_0^{\omega_1} \mathrm{dn}^2\sqrt{(e_1 - e_3)}u\,du,$$

or $\qquad \eta_1 + e_1\omega_1 = \sqrt{(e_1 - e_3)}E$ (Schwarz, § 29).

Also (§ 51) $\quad (e_1 - e_3)\omega_1 = \sqrt{(e_1 - e_3)}K\,;$

so that $\qquad \eta_1 + e_3\omega_1 = -\sqrt{(e_1 - e_3)}(K - E)\,;$

and therefore

$$i\Psi_1 = \tfrac{1}{2} i\pi + s\omega_3\sqrt{(e_1 - e_3)}(K - E),$$
$$i\Psi_2 = \qquad q\omega_3\sqrt{(e_1 - e_3)}E.$$

But, from § 210, when α and β are very nearly π, their approximate values are given by

$$\cot^2\tfrac{1}{2}\alpha = \frac{e_3 - \wp a}{\wp b - e_3}$$

$$\risingdotseq \frac{e_3 - e_3 - \frac{1}{2}s^2\omega_3{}^2\wp''\omega_3}{e_1 - e_3 + \frac{1}{2}q^2\omega_3{}^2\wp''\omega_1}$$

$$\risingdotseq -s^2\omega_3{}^2(e_2 - e_3) \risingdotseq -\kappa^2 s^2\omega_3{}^2(e_1 - e_3);$$

since $$\wp''\omega_3 = 2(e_1 - e_3)(e_2 - e_3),$$

and $$\kappa^2 = \frac{e_2 - e_3}{e_1 - e_3}, \quad \kappa'^2 = \frac{e_1 - e_2}{e_1 - e_3} \quad (\S\ 52);$$

$$\cot^2\tfrac{1}{2}\beta = \frac{e_2 - \wp a}{\wp b - e_2} \risingdotseq \frac{e_2 - e_3}{e_1 - e_2} = \frac{\kappa^2}{\kappa'^2};$$

and therefore $$(e_1 - e_3)s^2\omega_3{}^2 \risingdotseq -\frac{\kappa'^2}{\kappa^4}\cot^2\tfrac{1}{2}\alpha\cot^2\tfrac{1}{2}\beta.$$

Also (§ 210)

$$\frac{G - Cr}{G + Cr} = \frac{\wp' b}{-\wp' a} = \frac{\wp'(\omega_1 + q\omega_3)}{-\wp'(\omega_3 - s\omega_3)}$$

$$\risingdotseq \frac{q\omega_3\wp''\omega_3}{s\omega_3\wp''\omega_1} = \frac{q}{s}\frac{e_1 - e_2}{e_2 - e_3} = \frac{q}{s}\frac{\kappa'^2}{\kappa^2};$$

so that $$(e_1 - e_3)q^2\omega_3{}^2 \risingdotseq -\left(\frac{G - Cr}{G + Cr}\right)^2 \frac{1}{\kappa'^2}\cot^2\tfrac{1}{2}\alpha\cot^2\tfrac{1}{2}\beta.$$

Therefore $$\Psi_1 \risingdotseq \tfrac{1}{2}\pi + \frac{K - E}{\kappa^2}\kappa'\cot\tfrac{1}{2}\alpha\cot\tfrac{1}{2}\beta,$$

$$\Psi_2 \risingdotseq \left(\frac{G - Cr}{G + Cr}\right)^2 \frac{E}{\kappa'}\cot\tfrac{1}{2}\alpha\cot\tfrac{1}{2}\beta.$$

But, ultimately, when $\kappa = 0$ and $\kappa' = 1$,

then $E = \frac{1}{2}\pi$, and $\operatorname{lt}(K - E)/\kappa^2 = \frac{1}{4}\pi$ (§ 11, 170);

so that $$\Psi_1 \risingdotseq \tfrac{1}{2}\pi + \tfrac{1}{4}\pi\cot\tfrac{1}{2}\alpha\cot\tfrac{1}{2}\beta,$$

$$\Psi_2 \risingdotseq \left(\frac{G - Cr}{G + Cr}\right)\tfrac{1}{2}\pi\cot\tfrac{1}{2}\alpha\cot\tfrac{1}{2}\beta.$$

This reduces for the Spherical Pendulum, in which $Cr = 0$, to

$$\Psi \risingdotseq \tfrac{1}{2}\pi(1 + \tfrac{3}{2}\cot\tfrac{1}{2}\alpha\cot\tfrac{1}{2}\beta) \risingdotseq \tfrac{1}{2}\pi(1 + \tfrac{3}{8}\sin\alpha\sin\beta),$$

when α and β are nearly π, thus agreeing with Bravais's result.

When $\alpha = \pi$ and $G + Cr = 0$, this approximation fails; but the student may now prove that the apsidal angle is

$$\tfrac{1}{2}\pi\left\{1 - \sqrt{\left(\frac{C^2r^2}{4AWgh}\right)}\right\}.$$

This will be the apsidal angle when the Top is spinning in the vertical position with small angular velocity r, and is then struck with a slight horizontal blow.

II. *The Motion of a Solid of Revolution in infinite frictionless liquid.*

The reductions of the Elliptic Integral of the Third Kind in § 282 in consequence of the relation

$$a+b=\omega_a,$$

in connexion with the Top and Spherical Pendulum, are useful also in constructing degenerate cases of the motion of a Solid of Revolution in infinite liquid, as mentioned in § 211.

We refer to Basset's *Hydrodynamics*, Vol. I., Chapters VIII., IX., and Appendix III., also to Halphen's *Fonctions elliptiques*, II., Chap. IV., for an explanation of the notation; and now T the kinetic energy of the system due to the component velocities u, v, w of the centre O of the body along rectangular axes OA, OB, OC, fixed in the body, OC being the axis of figure, and to component angular velocities p, q, r about OA, OB, OC is given by

$$T=\tfrac{1}{2}P(u^2+v^2)+\tfrac{1}{2}Rw^2+\tfrac{1}{2}A(p^2+q^2)+\tfrac{1}{2}Cr^2 \quad \text{......(A)}$$

(to which the terms

$$P'(up+vq)+P'wr$$

may be added in the case of a body like a four-bladed screw propeller, or like a rifled projectile provided with studs or spiral convolutions on the exterior).

Then the Hamiltonian equations of motion are

$$\frac{d}{dt}\frac{\partial T}{\partial u}-r\frac{\partial T}{\partial v}+q\frac{\partial T}{\partial w}=X, \quad \text{.........................(1)}$$

$$\frac{d}{dt}\frac{\partial T}{\partial v}-p\frac{\partial T}{\partial w}+r\frac{\partial T}{\partial u}=Y, \quad \text{.........................(2)}$$

$$\frac{d}{dt}\frac{\partial T}{\partial w}-q\frac{\partial T}{\partial u}+p\frac{\partial T}{\partial v}=Z, \quad \text{.........................(3)}$$

$$\frac{d}{dt}\frac{\partial T}{\partial p}-r\frac{\partial T}{\partial q}+q\frac{\partial T}{\partial r}-w\frac{\partial T}{\partial v}+v\frac{\partial T}{\partial w}=L, \quad \text{.........(4)}$$

$$\frac{d}{dt}\frac{\partial T}{\partial q}-p\frac{\partial T}{\partial r}+r\frac{\partial T}{\partial p}-u\frac{\partial T}{\partial w}+w\frac{\partial T}{\partial u}=M, \quad \text{.........(5)}$$

$$\frac{d}{dt}\frac{\partial T}{\partial r}-q\frac{\partial T}{\partial p}+p\frac{\partial T}{\partial q}-v\frac{\partial T}{\partial u}+u\frac{\partial T}{\partial v}=N. \quad \text{.........(6)}$$

When no forces act, so that X, Y, Z, L, M, N vanish, then equation (6) shows that Cr or r is constant.

Multiplying equations (1) to (6) by u, v, w, p, q, r in order, adding and integrating, shows that T in (A) is constant.

Multiplying (1), (2), (3) by $\frac{\partial T}{\partial u}, \frac{\partial T}{\partial v}, \frac{\partial T}{\partial w}$, adding and integrating, proves that

$$\left(\frac{\partial T}{\partial u}\right)^2+\left(\frac{\partial T}{\partial v}\right)^2+\left(\frac{\partial T}{\partial w}\right)^2 \text{ is constant; or}$$

$$P^2(u^2+v^2)+R^2w^2=F^2, \quad \text{.....................(B)}$$

F being a constant, representing the resultant linear momentum of the system.

Similarly, it is shown that

$$\frac{\partial T}{\partial u}\frac{\partial T}{\partial p}+\frac{\partial T}{\partial v}\frac{\partial T}{\partial q}+\frac{\partial T}{\partial w}\frac{\partial T}{\partial v} \text{ is constant; or}$$

$$AP(up+vq)+CRwr=G, \quad \text{....................(C)}$$

where G is a constant, representing the resultant angular momentum of the system.

From equations (A) and (B),

$$A(p^2+q^2)=2T-Cr^2-Rw^2-P(u^2+v^2)$$

$$=2T-Cr^2-\frac{F^2}{R}+F^2\left(\frac{1}{R}-\frac{1}{P}\right)(F^2-R^2w^2),$$

and, from equation (3),

$$R^2\frac{dw^2}{dt^2}=P^2(uq-vp)^2=P^2(u^2+v^2)(p^2+q^2)-P^2(up+vq)^2$$

$$=\frac{F^2}{A}\left(\frac{1}{R}-\frac{1}{P}\right)(F^2-R^2w^2)^2$$

$$+\left(2T-Cr^2-\frac{F^2}{R}\right)\frac{F^2-R^2w^2}{A}-\left(\frac{G-CRwr}{A}\right)^2; \text{...(D)}$$

so that w or Rw is an elliptic function of t.

Taking the axis Oz in the direction of the resultant impulse F, and denoting by $\gamma_1, \gamma_2, \gamma_3$ the cosines of the angles between Oz and OA, OB, OC, so that

$$Pu=F\gamma_1, \quad Pv=F\gamma_2, \quad Rw=F\gamma_3;$$

then, with Euler's coordinate angles θ, ϕ, ψ,

$$\gamma_1=-\sin\theta\cos\phi, \quad \gamma_2=\sin\theta\sin\phi, \quad \gamma_3=\cos\theta,$$

$$P(up+vq)=F\sin\theta(-p\cos\phi+q\sin\phi)=F\sin^2\theta\frac{d\psi}{dt};$$

so that

$$\frac{d\psi}{dt}=\frac{G-CFr\cos\theta}{AF\sin^2\theta}$$

$$=\frac{G+CFr}{2AF}\frac{1}{1+\cos\theta}+\frac{G-CFr}{2AF}\frac{1}{1-\cos\theta}=\frac{d\psi_1}{dt}+\frac{d\psi_2}{dt},$$

suppose; and then

$$\frac{d\phi}{dt}=r-\cos\theta\frac{d\psi}{dt}=\left(1-\frac{C}{A}\right)r+\frac{CFr-G\cos\theta}{AF\sin^2\theta}$$

$$=\left(1-\frac{C}{A}\right)r+\frac{d\psi_1}{dt}-\frac{d\psi_2}{dt}.$$

The equations given by Kirchhoff (*Vorlesungen über mathematische Physik*, p. 240) for α, β, γ, the coordinates of O with respect to fixed axes $O'\alpha$, $O'\beta$, $O'\gamma$ ($O'\gamma$ parallel to Oz) are

$$F\alpha=\beta_1\frac{\partial T}{\partial p}+\beta_2\frac{\partial T}{\partial q}+\beta_3\frac{\partial T}{\partial r}, \quad\text{.................(7)}$$

$$F\beta=-\alpha_1\frac{\partial T}{\partial p}-\alpha_2\frac{\partial T}{\partial q}-\alpha_3\frac{\partial T}{\partial r}, \quad\text{.................(8)}$$

$$F\frac{d\gamma}{dt}=u\frac{\partial T}{\partial u}+v\frac{\partial T}{\partial v}+w\frac{\partial T}{\partial w}; \quad\text{.................(9)}$$

where α_1, α_2, α_3 denote the cosines of the angles between $O'\alpha$ and OA, OB, OC; and β_1, β_2, β_3, the cosines of the angles between $O'\beta$ and OA, OB, OC.

Expressed by Euler's coordinate angles,

$$\alpha_1=\cos\theta\cos\phi\cos\psi-\sin\phi\sin\psi,$$
$$\alpha_2=-\cos\theta\sin\phi\cos\psi-\cos\phi\sin\psi,$$
$$\alpha_3=\sin\theta\cos\psi;$$
$$\beta_1=\cos\theta\cos\phi\sin\psi+\sin\phi\cos\psi,$$
$$\beta_2=-\cos\theta\sin\phi\sin\psi+\cos\phi\cos\psi,$$
$$\beta_3=\sin\theta\sin\psi;$$

while

$$p=\sin\phi\,\dot\theta-\sin\theta\cos\phi\,\dot\psi,$$
$$q=\cos\phi\,\dot\theta+\sin\theta\sin\phi\,\dot\psi,$$
$$r=\dot\phi+\cos\theta\,\dot\psi;$$

so that, after reduction,

$$F\alpha=A\cos\psi\,\dot\theta+(Cr-A\cos\theta\,\dot\psi)\sin\theta\sin\psi,$$
$$F\beta=A\sin\psi\,\dot\theta-(Cr-A\cos\theta\,\dot\psi)\sin\theta\cos\psi,$$
$$F\frac{d\gamma}{dt}=\frac{F^2}{P}\sin^2\theta+\frac{F^2}{R}\cos^2\theta.$$

Writing Fx for $F\cos\theta$ or Rw, equation (D) becomes

$$\frac{dx^2}{dt^2}=n^2\left\{(x^2-1)^2-\left(2T-Cr^2-\frac{F^2}{R}\right)\frac{x^2-1}{An^2}-\left(\frac{CrFx-G}{AFn}\right)^2\right\}=n^2X,$$

suppose, where $\quad n^2=\dfrac{F^2}{A}\left(\dfrac{1}{R}-\dfrac{1}{P}\right).$

Denoting the roots of the quartic $X=0$ by x_0, x_1, x_2, x_3, we may put, according to §§ 151, 152,

$$x-x_0=\frac{-\wp'c}{\wp u-\wp c},$$

$$x-x_1=\frac{-\wp'c}{\wp u-\wp c}\frac{\wp u-e_1}{\wp c-e_1},$$

$$x-x_2=\frac{-\wp'c}{\wp u-\wp c}\frac{\wp u-e_2}{\wp c-e_2},$$

$$x-x_3=\frac{-\wp'c}{\wp u-\wp c}\frac{\wp u-e_3}{\wp c-e_3};$$

and now, when x oscillates between x_2 and x_3,

$$u=nt+\omega_3.$$

The letter u has been used here in two senses, to agree with the ordinary notation; this need not however lead to confusion.

Differentiating,

$$\sqrt{X}=\frac{dx}{du}=\frac{\wp'u\,\wp'c}{(\wp u-\wp c)^2}=\wp(u-c)-\wp(u+c);$$

$$x=\zeta(u+c)-\zeta(u-c)-\zeta 2c$$

$$=\frac{1}{2}\frac{\wp'(u-c)-\wp'2c}{\wp(u-c)-\wp 2c}$$

$$=\frac{1}{2}\frac{\wp'(u-c)+\wp'(u+c)}{\wp(u-c)-\wp(u+c)},$$

$$x^2=\wp 2c+\wp(u-c)+\wp(u+c);$$

so that we must write v for $2c$ and u for $u-c$, to agree with Halphen's notation.

Now, to determine γ,

$$F\frac{d\gamma}{dt}=\frac{F^2}{P}+\left(\frac{F^2}{R}-\frac{F^2}{P}\right)\cos^2\theta$$

$$=\frac{F^2}{P}+An^2\{\wp 2c+\wp(u-c)+\wp(u+c)\},$$

$$F\gamma=\left(\frac{F^2}{P}+An^2\wp 2c\right)t-An\{\zeta(u-c)+\zeta(u+c)\}$$

$$=\left(\frac{F^2}{nP}+An\wp 2c\right)(u-\omega_3)-2An\zeta u-An\frac{\wp'u}{\wp u-\wp c};$$

so that, in a complete period $2\omega_1$ of the motion, the point O will have advanced parallel to $O'\gamma$ a distance

$$\left(\frac{F^2}{nP}+An\wp 2c\right)2\omega_1-4An\eta_1;$$

also (§ 152) $\qquad 6\wp 2c=$ coefficient of $-x^2$ in X.

We now suppose that $u=a$ makes $x=1$, and $u=b$ makes $x=-1$; then

$$1-x=\frac{-\wp'c(\wp u-\wp a)}{(\wp a-\wp c)(\wp u-\wp c)},\quad 1+x=\frac{-\wp'c(\wp b-\wp u)}{(\wp b-\wp c)(\wp u-\wp c)},$$

$$\frac{\wp'a\,\wp'c}{(\wp a-\wp c)^2}=-i\frac{G-CFr}{AFn},\quad \frac{\wp'b\,\wp'c}{(\wp b-\wp c)^2}=i\frac{G+CFr}{AFn}.$$

Then

$$\begin{aligned}\frac{di\psi_1}{du}&=\frac{-\frac{1}{2}\wp'a(\wp u-\wp c)}{(\wp a-\wp c)(\wp u-\wp a)}\\&=\frac{-\frac{1}{2}\wp'a}{\wp a-\wp c}+\frac{-\frac{1}{2}\wp'a}{\wp u-\wp a}\\&=-\tfrac{1}{2}\zeta(a-c)-\tfrac{1}{2}\zeta(a+c)+\zeta a\\&\quad-\tfrac{1}{2}\zeta(u-a)+\tfrac{1}{2}\zeta(u+a)-\zeta a;\end{aligned}$$

$$i\psi_1=-\tfrac{1}{2}\{\zeta(a-c)+\zeta(a+c)\}u+\tfrac{1}{2}\log\frac{\sigma(u+a)}{\sigma(u-a)};$$

and similarly

$$i\psi_2=-\tfrac{1}{2}\{\zeta(b-c)+\zeta(b+c)\}u+\tfrac{1}{2}\log\frac{\sigma(b+u)}{\sigma(b-u)};$$

and therefore

$$i\psi=-\tfrac{1}{2}Pu+\tfrac{1}{2}\log\frac{\sigma(u+a)\sigma(b+u)}{\sigma(u-a)\sigma(b-u)},$$

where

$$P=\zeta(a-c)+\zeta(a+c)+\zeta(b-c)+\zeta(b+c).$$

Also

$$\begin{aligned}\sin^2\theta&=1-x^2=(1+x)(1-x)\\&=\frac{\wp'^2c(\wp u-\wp a)(\wp b-\wp u)}{(\wp a-\wp c)(\wp b-\wp c)(\wp u-\wp c)^2}\\&=\frac{\sigma^2 2c\sigma(u-a)\sigma(u+a)\sigma(b-u)\sigma(b+u)}{\sigma(a-c)\sigma(a+c)\sigma(b-c)\sigma(b+c)\sigma^2(u-c)\sigma^2(u+c)},\end{aligned}$$

so that

$$\sin\theta e^{i\psi}=Ce^{-\frac{1}{2}Pu}\frac{\sigma(u+a)\sigma(b+u)}{\sigma(u-c)\sigma(u+c)},$$

giving the projection on a plane perpendicular to Oz of the motion of a point on the axis OC, relatively to O; also

$$P(u+vi)=-F\sin\theta e^{-\phi i},$$
$$p+qi=(-\sin\theta\,\dot\psi+i\,\dot\theta)e^{-\phi i}.$$

We find also, as in § 224, that if the values a_1 and b_1 of u correspond to

$$x^2=1+\left(2T-Cr^2-\frac{F^2}{R}\right)\frac{1}{An^2},$$

then $$a_1-b_1=a-b.$$

But now introduce the condition

$$a+b=\omega_a,$$

when, according to § 282, ψ becomes *pseudo-elliptic*.

Putting $\xi=\tan^{-1}\dfrac{i\wp' a}{\wp a-e_a}\dfrac{\wp u-e_a}{\wp' u}$,

or $$\xi=\tan^{-1}\sqrt{\left(-\frac{1+x_\beta\,.\,1+x_\gamma}{1+x_0\,.\,1+x_a}\cdot\frac{x-x_0\,.\,x-x_a}{x-x_\beta\,.\,x-x_\gamma}\right)};$$

and, employing b instead of a, this may also be written

$$\xi=\tan^{-1}\sqrt{\left(-\frac{1-x_\beta\,.\,1-x}{1-x_0\,.\,1-x_a}\cdot\frac{x-x_0\,.\,x-x_a}{x-x_\beta\,.\,x-x_\gamma}\right)};$$

so that $$\frac{1+x_\beta\,.\,1+x_\gamma}{1+x_0\,.\,1+x_a}=\frac{1-x_\beta\,.\,1-x_\gamma}{1-x_0\,.\,1-x_a},$$

and therefore each is equal to -1, and

$$x_0x_a+x_\beta x_\gamma+2=0,$$

since $$x_0+x_a+x_\beta+x_\gamma=0;$$

and, changing to the complementary angle,

$$\xi=\tan^{-1}\sqrt{\frac{x-x_\beta\,.\,x-x_\gamma}{x-x_0\,.\,x-x_a}}$$

$$=\sin^{-1}\sqrt{\frac{x_\beta-x\,.\,x-x_\gamma}{2-2x^2}}=\cos^{-1}\sqrt{\frac{x-x_0\,.\,x_a-x}{2-2x^2}},$$

with $x_a>x_\beta>x>x_\gamma>x_0$.

Differentiating,

$$\frac{d\xi}{dx}=\frac{(x_\beta+x_\gamma)(1+x^2)-2(1+x_\beta x_\gamma)x}{(2-2x^2)\sqrt{X}},\quad\text{while}\quad\frac{dx}{dt}=n\sqrt{X},$$

so that $$\frac{d\xi}{dt}=n\frac{x_\beta+x_\gamma-(1+x_\beta x_\gamma)x}{1-x^2}-\tfrac{1}{2}n(x_\beta+x_\gamma)$$

$$=\tfrac{1}{2}n(x_0+x_a)-n\frac{x_0+x_a-(1+x_0x_a)x}{1-x^2}.$$

Then $$\frac{d\xi}{dt}+\frac{d\psi}{dt}=\tfrac{1}{2}n(x_0+x_a),$$

provided that $n(x_0+x_a)=G/AF$, $n(1+x_0x_a)=Cr/A$.

The quartic X must therefore break up into the two quadratics $x^2-\dfrac{Gx}{AFn}+\dfrac{Cr}{An}-1$ and $x^2+\dfrac{Gx}{AFn}-\dfrac{Cr}{A}-1$; and

$$X=(x^2-1)^2-\left(\frac{Gx-CrF}{AFn}\right)^2;$$

so that the requisite relation when $a+b=\omega_a$, is

$$2T-Cr^2-\frac{F^2}{R}=\frac{G^2-C^2r^2F^2}{AF^2}\qquad\text{(E)}$$

Now

$$\sin\theta\cos\xi = \sqrt{\frac{x-x_0\,.\,x_a-x}{2}} = \sqrt{\left\{\frac{1}{2}\left(1-\frac{Cr}{An}+\frac{G\cos\theta}{AFn}-\cos^2\theta\right)\right\}},$$

$$\sin\theta\sin\xi = \sqrt{\frac{x_\beta-x\,.\,x-x_\gamma}{2}} = \sqrt{\left\{\frac{1}{2}\left(1+\frac{Cr}{An}-\frac{G\cos\theta}{AFn}-\cos^2\theta\right)\right\}},$$

so that $\sin^2\theta\sin 2\xi = \sqrt{X}$, $\sin^2\theta\cos 2\xi = \dfrac{G\cos\theta - CrF}{AFn}$;

and $$\xi = mt - \psi,$$

where $$m = \tfrac{1}{2}n(x_0+x_a) = \tfrac{1}{2}G/AF.$$

Also, from (7) and (8),

$$F(a\cos\psi+\beta\sin\psi) = A\dot\theta$$
$$= An\sqrt{X}/\sin\theta = An\sin\theta\sin 2\xi;$$
$$F(a\sin\psi-\beta\cos\psi) = (Cr - A\cos\theta\dot\psi)\sin\theta$$
$$= \frac{CrF - G\cos\theta}{F\sin\theta} = -An\sin\theta\cos 2\xi.$$

Therefore $$Fa = An\sin\theta(\sin 2\xi\cos\psi - \cos 2\xi\sin\psi)$$
$$= An\sin\theta\sin(2\xi-\psi)$$
$$= An\sin\theta\sin(2mt-3\psi);$$
$$F\beta = An\sin\theta\cos(2\xi-\psi)$$
$$= An\sin\theta\cos(2mt-3\psi).$$

Now in the motion of a point on OC, relative to O,

$$\sin\theta\, e^{i\psi} = \sin\theta\cos(mt-\xi) + i\sin\theta\sin(mt-\xi)$$
$$= e^{imt}\left(\sqrt{\frac{x-x_0\,.\,x_a-x}{2}} - i\sqrt{\frac{x_\beta-x\,.\,x-x_\gamma}{2}}\right),$$

where $$x = \cos\theta.$$

When $b-a=\omega_a$, and $\psi_1-\psi_2$ or ϕ is *pseudo-elliptic*, we shall find that G and Cr are interchanged, and

$$n(x_0+x_a) = Cr/A,$$
$$n(1+x_0x_a) = G/AF;$$

and then $$2T - Cr^2 - \frac{F^2}{R} = 0; \qquad \text{(F)}$$

so that $$P^2(u^2+v^2) = F^2\sin^2\theta,$$
$$p^2+q^2 = n^2\sin^2\theta.$$

As a numerical exercise, we may take, in addition to (F),

$$G = 4AFn,\quad Cr = 2\sqrt{7}An;$$

then $$X = x^4 - 30x^2 + 16\sqrt{7}x - 15$$
$$= (x^2 - 2\sqrt{7}x+3)(x^2+2\sqrt{7}x-5);$$

$x_0 = \sqrt{7}+2$, $x_3 = -\sqrt{7}+2\sqrt{3}$, $x_2 = \sqrt{y}-2$, $x_1 = -\sqrt{7}-2\sqrt{3}$;

$g_2 = 60$, $g_3 = 88$, $e_1 = 1+2\sqrt{3}$, $e_2 = -2$, $e_3 = 1-2\sqrt{3}$;

$\wp a = -3$, $\wp b = 1$; $a = \frac{2}{3}\omega_3$, $b = \omega_1 - \frac{1}{3}\omega_3$ (§ 225);
$\wp c = 2\sqrt{7}+3$, $\wp' c = -8\sqrt{7}-20$, $\wp 2c = 5$, $\wp' 2c = 4\sqrt{7}$.
Now we shall find that

$$nt - \psi = \frac{1}{3}\cos^{-1}\left(\frac{x_0 - x \,.\, x - x_2}{2 - 2x^2}\right)^{\frac{3}{2}} = \text{etc.}$$

$$\sin^3\theta \cos 3(nt-\psi) = (-\tfrac{3}{2} + \sqrt{7}\cos\theta - \tfrac{1}{2}\cos^2\theta)^{\frac{3}{2}},$$

$$\sin^3\theta \sin 3(nt-\psi)$$
$$= (\tfrac{1}{2}\sqrt{7} - 2\cos\theta + \tfrac{1}{2}\sqrt{7}\cos^2\theta)\sqrt{(\tfrac{5}{2} - \sqrt{7}\cos\theta - \tfrac{1}{2}\cos^2\theta)}.$$

MISCELLANEOUS EXAMPLES.

1. Construct a Table exhibiting the connexion between the twelve elliptic functions

$$\begin{array}{llll} \text{sn}\, u, & \text{ns}\, u, & \text{dc}\, u, & \text{cd}\, u\,; \\ \text{cn}\, u, & \text{ds}\, u, & \text{nc}\, u, & \text{sd}\, u\,; \\ \text{dn}\, u, & \text{cs}\, u, & \text{sc}\, u, & \text{nd}\, u. \end{array}$$

2. Construct a Table of the values of the sn, cn, dn of $u + mK + nK'i$ in terms of sn u, cn u, dn u; also of the elliptic functions of $\frac{1}{2}(mK + nK'i)$, for $m, n = 0, 1, 2, \ldots$.

3. Prove that, accents denoting differentiation,

(i.) $$\text{sn}\, u\, \text{dn}'' u - \text{sn}'' u\, \text{dn}\, u = \text{sn}\, u\, \text{dn}\, u, \text{ etc.}$$

(ii.) $$\begin{vmatrix} (\text{sn}\, u)^2, & \text{sn}\, u\, \text{sn}' u, & (\text{sn}' u)^2 \\ (\text{cn}\, u)^2, & \text{cn}\, u\, \text{cn}' u, & (\text{cn}' u)^2 \\ (\text{dn}\, u)^2, & \text{dn}\, u\, \text{dn}' u, & (\text{dn}' u)^2 \end{vmatrix} = \kappa'^2 \text{sn}\, u\, \text{cn}\, u\, \text{dn}\, u.$$

(G. B. Mathews.)

4. Denoting by (m, n) the function

$$\frac{\text{sn}(u_m - u_n)\text{cn}(u_m + u_n)}{\text{cn}(u_m - u_n)\text{sn}(u_m + u_n)},$$

prove that

$$(4, 1)(4, 2)(4, 3)(2, 3)(3, 1)(1, 2) + (4, 1)(2, 3) + (4, 2)(3, 1) + (4, 3)(1, 2) = 0.$$

Denoting by A, B, C the functions

$$\frac{\text{sn}(t-x)\text{sn}(y-z)}{\text{sn}(t+x)\text{sn}(y+z)},\quad \frac{\text{sn}(t-y)\text{sn}(z-x)}{\text{sn}(t+y)\text{sn}(z+x)},\quad \frac{\text{sn}(t-z)\text{sn}(x-y)}{\text{sn}(t+z)\text{sn}(x-y)},$$

prove that $ABC + A + B + C = 0$.

5. Prove that

(i.) $$\int_0^{2u} \kappa \operatorname{sn} v dv = 2 \tanh^{-1}(\kappa \operatorname{sn}^2 u).$$

(ii.) $$\int_0^{u} \kappa \operatorname{sn}(2u+a)du = \tanh^{-1}\{\kappa \operatorname{sn} u \operatorname{sn}(u+a)\}.$$

(iii.) $$\int_0^{K} \log \operatorname{ns} u du = \tfrac{1}{4}\pi K' - \tfrac{1}{2}K \log 1/\kappa.$$

6. Determine the orbit in which

$P = h^2(u^3 + a^2u^5)$, the apsidal distance being a.

7. Rectify $$r^{\frac{2}{3}} = a^{\frac{2}{3}}\cos\tfrac{2}{3}\theta.$$

8. Prove that the perimeter of the Cassinian Oval of § 161

is either $$\frac{4\beta^2 K}{a}, \quad \kappa = \frac{1}{2}\sqrt{\left(1+\frac{\beta^2}{a^2}\right)} - \frac{1}{2}\sqrt{\left(1-\frac{\beta^2}{a^2}\right)},$$

or $$4aK, \quad \kappa = \frac{1}{2}\sqrt{\left(1+\frac{a^2}{\beta^2}\right)} - \frac{1}{2}\sqrt{\left(1-\frac{a^2}{\beta^2}\right)}:$$

and draw the corresponding curves.

9. Prove that the length of the curve of intersection of two circular cylinders, of radius a and b, whose axes intersect at right angles, is $$8a\int_0^{\frac{1}{2}\pi}\left(\frac{1-\kappa^2\sin^4\phi}{1-\kappa^2\sin^2\phi}\right)^{\frac{1}{2}} d\phi, \quad \kappa^2 = a^2/b^2;$$

and verify the result when $a = b$.

10. Prove that K and K' satisfy the differential equation

$$\frac{d}{dk}\left\{k(1-k)\frac{dK}{dk}\right\} - \tfrac{1}{4}K = 0.$$

Deduce the relation

$$\frac{dK}{dk}K' - K\frac{dK'}{dk} = \frac{\pi}{4k(1-k)};$$

and thence deduce Legendre's relation (§ 171).

11. Prove that ϖ_1 and ϖ_2 of § 252 satisfy the differential

equation $$J(J-1)\frac{d^2\varpi}{dJ^2} + \frac{4-7J}{6}\frac{d\varpi}{dJ} - \frac{\varpi}{144} = 0.$$

12. Deduce the Fourier series for $\operatorname{sn} u$, $\operatorname{cn} u$, $\operatorname{dn} u$ of §§ 266, 267 from the series for Zu of § 268, making use of Landen's Transformations and of equations (28), (29), (30) of § 264.

13. Prove that

(i.) $$\frac{\wp'' u}{\wp' u}+\frac{\wp''(u+\omega_1)}{\wp'(u+\omega_1)}+\frac{\wp''(u+\omega_2)}{\wp'(u+\omega_2)}+\frac{\wp''(u+\omega_3)}{\wp'(u+\omega_3)}=0;$$

(ii.) $$\frac{1}{4}\left(\frac{\wp'' u}{\wp' u}\right)^2=\wp 2u+2\wp u;$$

(iii.) $$\frac{\wp u-\wp(u+a+b)}{\wp(u-a)-\wp(u-b)}\,\frac{\wp u-\wp(u-a-b)}{\wp(u+a)-\wp(u+b)}=\left\{\frac{\wp u-\wp(a+b)}{(\wp u-\wp a)(\wp u-\wp b)}\right\}^2.$$

14. Prove that, if a variable straight line meets the curve

$$Ax^3+By^2+Cx+D=0$$

in $(x_1, y_1)(x_2, y_2)(x_3, y_3)$, then (§ 166)

$$\frac{dx_1}{y_1}+\frac{dx_2}{y_2}+\frac{dx_3}{y_3}=0.$$

15. Denoting the integral

$$\int_0^x \frac{xy\,dx}{y^2-ax} \text{ by } \mathrm{f}x,$$

where y is given as a function of x by the equation

$$x^3+y^3-3axy=1,$$

prove that, for three collinear points,

$$\mathrm{f}x_1+\mathrm{f}x_2+\mathrm{f}x_3=3a.$$

16. Prove or verify that, with $g_2=0$, the solution of Lamé's differential equation

(i.) $\dfrac{1}{y}\dfrac{d^2y}{du^2}=2\wp u$ is $y=\{\wp' u\pm\sqrt{(-g_3)}\}^{\frac{1}{3}}$;

(ii.) $\dfrac{1}{y}\dfrac{d^2y}{du^2}=6\wp u+\sqrt{(3g_3)}$ is $y=\wp u-\frac{1}{6}\sqrt{(3g_3)}$;

(iii.) $\dfrac{1}{y}\dfrac{d^2y}{du^2}=\frac{3}{4}\wp u$ is $y=(A+B\wp\tfrac{1}{2}u)(\wp'\tfrac{1}{2}u)^{-\frac{1}{2}}$.

(Halphen, *Mémoire sur la réduction des équations différentielles*, 1884.)

17. Determine, by means of elliptic functions, the motion of liquid filling a rectangular box, due to component angular velocities about axes through the centre parallel to the edges.

(*Q. J. M.*, XV., p. 144; W. M. Hicks, *Velocity and Electric Potentials between parallel planes*, p. 274.)

18. Prove that, with $x=\frac{1}{2}\pi u/\omega$ and $A=\frac{1}{2}\eta/\omega$ (§ 278),

$$\sigma u=\frac{\omega}{\frac{1}{2}\pi}e^{Au^2}\frac{\theta_1 x}{\theta_1'0},\quad \sigma_1 u=e^{Au^2}\frac{\theta_2 x}{\theta_2 0},\quad \sigma_2 u=e^{Au^2}\frac{\theta_3 x}{\theta_3 0},\quad \sigma_3 u=e^{Au^2}\frac{\theta x}{\theta 0};$$

and thence convert the formulas (M) to (T) of § 249 into Jacobi's notation.

19. Prove that (§ 264, 20*)

$$1/Q=\prod_{r=1}^{r=\infty}(1-q^{2r})=\sum_{m=0}^{m=\infty}q^{3m^2+m},$$

$$q^{\frac{1}{12}}/Q=\Sigma q^{\frac{(6m+1)^2}{12}}.$$

20. Prove that

$$\text{(i.)}\quad \kappa=4q^{\frac{1}{2}}\Pi\left(\frac{1+q^{2r}}{1+q^{2r-1}}\right)^4;$$

$$\text{(ii.)}\quad \kappa'=\Pi\left(\frac{1-q^{2r-1}}{1+q^{2r-1}}\right)^4;$$

$$\text{(iii.)}\quad \frac{K}{\frac{1}{2}\pi}=\Pi\frac{\tanh^2 r\pi K'/K}{\tanh^2(r-\frac{1}{2})\pi K'/K}.$$

21. Prove that, in Appendix II., p. 346,

$$\wp 2c-\wp(a+b)=\frac{G^2}{4A^2F^2n^2},$$

$$\wp 2c-\wp(a-b)=\frac{C^2r^2}{4A^2n^2};$$

$$\wp'(a+b)=\frac{iG}{4AFn}\left(\frac{G^2-C^2r^2F^2}{A^2F^2n^2}-\frac{2T-Cr^2-F^2/R}{An^2}\right);$$

$$\wp'(a-b)=\frac{iCr}{4An}\frac{2T-Cr^2-F^2/R}{An^2}.$$

Work out the case of

$$2T-Cr^2-F^2/R=0,$$

$$G=2AFn,\quad Cr=2\sqrt{2}An.$$

INDEX.

GLASGOW: PRINTED AT THE UNIVERSITY PRESS BY ROBERT MACLEHOSE.

Printed in Dunstable, United Kingdom

91220112R00211